Advances and Applications of Carbon Nanotubes

Advances and Applications of Carbon Nanotubes

Editors

Simone Morais
Konstantinos Spyrou

 Basel • Beijing • Wuhan • Barcelona • Belgrade • Novi Sad • Cluj • Manchester

Editors
Simone Morais
REQUIMTE–LAQV, ISEP
Porto, Portugal

Konstantinos Spyrou
University of Ioannina
Ioannina, Greece

Editorial Office
MDPI
St. Alban-Anlage 66
4052 Basel, Switzerland

This is a reprint of articles from the Topic published online in the open access journals *Nanomaterials* (ISSN 2079-4991), *Applied Sciences* (ISSN 2076-3417), and *Materials* (ISSN 1996-1944) (available at: https://www.mdpi.com/topics/Carbon_Nanotubes).

For citation purposes, cite each article independently as indicated on the article page online and as indicated below:

Lastname, A.A.; Lastname, B.B. Article Title. *Journal Name* **Year**, *Volume Number*, Page Range.

ISBN 978-3-0365-9352-4 (Hbk)
ISBN 978-3-0365-9353-1 (PDF)
doi.org/10.3390/books978-3-0365-9353-1

© 2024 by the authors. Articles in this book are Open Access and distributed under the Creative Commons Attribution (CC BY) license. The book as a whole is distributed by MDPI under the terms and conditions of the Creative Commons Attribution-NonCommercial-NoDerivs (CC BY-NC-ND) license.

Contents

About the Editors ix

Simone Morais
Advances and Applications of Carbon Nanotubes
Reprinted from: *Nanomaterials* **2023**, *13*, 2674, doi:10.3390/nano13192674 1

Chil Hyoung Lee, Go Bong Choi, Eun Mi Kim, Jongho Lee, Jaegeun Lee, Hi Gyu Moon, et al.
Gas Barrier Performance of Hexagonal Boron Nitride Monolayers Grown on Copper Foils with
Electrochemical Polishing
Reprinted from: *Appl. Sci.* **2021**, *11*, 4599, doi:10.3390/app11104599 5

**Muhammad Sohail Khan, Sun Mei, Shabnam, Nehad Ali Shah, Jae Dong Chung,
Aamir Khan and Said Anwar Shah**
Steady Squeezing Flow of Magnetohydrodynamics Hybrid Nanofluid Flow Comprising
Carbon Nanotube-Ferrous Oxide/Water with Suction/Injection Effect
Reprinted from: *Nanomaterials* **2022**, *12*, 660, doi:10.3390/nano12040660 13

Yu Liu, Zhicheng Lin, Pengfei Wang, Feng Huang and Jia-Lin Sun
Measurement of the Photothermal Conversion Efficiency of CNT Films Utilizing a Raman
Spectrum
Reprinted from: *Nanomaterials* **2022**, *12*, 1101, doi:10.3390/nano12071101 29

**Echeverry-Cardona Laura, Cabanzo Rafael, Quintero-Orozco Jorge,
Castillo-Cuero Harvi Alirio, Rodríguez-Restrepo Laura Victoria and
Restrepo-Parra Elisabeth**
Effects of Molarity and Storage Time of MWCNTs on the Properties of Cement Paste
Reprinted from: *Materials* **2022**, *15*, 9035, doi:10.3390/ma15249035 37

Wei-Jen Chen and I-Ling Chang
A Thermal Transport Study of Branched Carbon Nanotubes with Cross and T-Junctions
Reprinted from: *Appl. Sci.* **2021**, *11*, 5933, doi:10.3390/app11135933 55

**Ammar I. Alsabery, Tahar Tayebi, Ali S. Abosinnee, Zehba A. S.Raizah, Ali J. Chamkha and
Ishak Hashim**
Impacts of Amplitude and Local Thermal Non-Equilibrium Design on Natural Convection
within NanoflUid Superposed Wavy Porous Layers
Reprinted from: *Nanomaterials* **2021**, *11*, 1277, doi:10.3390/nano11051277 69

Yuliang Mao, Zheng Guo, Jianmei Yuan and Tao Sun
1D/2D van der Waals Heterojunctions Composed of Carbon Nanotubes and a GeSe Monolayer
Reprinted from: *Nanomaterials* **2021**, *11*, 1565, doi:10.3390/nano11061565 93

Fei Teng, Jian Wu, Benlong Su and Youshan Wang
Enhanced Tribological Properties of Vulcanized Natural Rubber Composites by Applications of
Carbon Nanotube: A Molecular Dynamics Study
Reprinted from: *Nanomaterials* **2021**, *11*, 2464, doi:10.3390/nano11092464 105

**Muhammad Sohail Khan, Sun Mei, Shabnam, Unai Fernandez-Gamiz, Samad Noeiaghdam,
Said Anwar Shah and Aamir Khan**
Numerical Analysis of Unsteady Hybrid Nanofluid Flow Comprising
CNTs-Ferrousoxide/Water with Variable Magnetic Field
Reprinted from: *Nanomaterials* **2022**, *12*, 180, doi:10.3390/nano12020180 123

Dongli Wang, Jun Xiao, Xiangwen Ju, Mingyue Dou, Liang Li and Xianfeng Wang
Numerical and Experiment Studies of Different Path Planning Methods on Mechanical Properties of Composite Components
Reprinted from: *Materials* **2021**, *14*, 6100, doi:10.3390/ma14206100 . . . . **143**

Qi Yuan, Heng Chen, Hong Nie, Guang Zheng, Chen Wang and Likai Hao
Soft-Landing Dynamic Analysis of a Manned Lunar Lander Em-Ploying Energy Absorption Materials of Carbon Nanotube Buckypaper
Reprinted from: *Materials* **2021**, *14*, 6202, doi:10.3390/ma14206202 . . . . **161**

Bochra Bejaoui Kefi, Imen Bouchmila, Patrick Martin and Naceur M'Hamdi
Titanium Dioxide Nanotubes as Solid-Phase Extraction Adsorbent for the Determination of Copper in Natural Water Samples
Reprinted from: *Materials* **2022**, *15*, 822, doi:10.3390/ma15030822 . . . . **183**

Fenghui Cao, Jia Xu, Xinci Zhang, Bei Li, Xiao Zhang, Qiuyun Ouyang, et al.
Tuning Dielectric Loss of SiO_2@CNTs for Electromagnetic Wave Absorption
Reprinted from: *Nanomaterials* **2021**, *11*, 2636, doi:10.3390/nano11102636 . . . . **197**

Alvaro R. Adrian, Daniel Cerda, Leunam Fernández-Izquierdo, Rodrigo A. Segura, José Antonio García-Merino and Samuel A. Hevia
Tunable Low Crystallinity Carbon Nanotubes/Silicon Schottky Junction Arrays and Their Potential Application for Gas Sensing
Reprinted from: *Nanomaterials* **2021**, *11*, 3040, doi:10.3390/nano11113040 . . . . **209**

Da Zhang, Yuanzheng Tang, Chuanqi Zhang, Qianpeng Dong, Wenming Song and Yan He
One-Step Synthesis of SnO_2/Carbon Nanotube Nanonests Composites by Direct Current Arc-Discharge Plasma and Its Application in Lithium-Ion Batteries
Reprinted from: *Nanomaterials* **2021**, *11*, 3138, doi:10.3390/nano11113138 . . . . **225**

Dennis Röcker, Tatjana Trunzer, Jasmin Heilingbrunner, Janine Rassloff, Paula Fraga-García and Sonja Berensmeier
Design of 3D Carbon Nanotube Monoliths for Potential-Controlled Adsorption
Reprinted from: *Appl. Sci.* **2021**, *11*, 9390, doi:10.3390/app11209390 . . . . **237**

Takayuki Watanabe, Satoshi Yamazaki, Satoshi Yamashita, Takumi Inaba, Shun Muroga, Takahiro Morimoto, et al.
Comprehensive Characterization of Structural, Electrical, and Mechanical Properties of Carbon Nanotube Yarns Produced by Various Spinning Methods
Reprinted from: *Nanomaterials* **2022**, *12*, 593, doi:10.3390/nano12040593 . . . . **253**

Giuseppe Nocito, Emanuele Luigi Sciuto, Domenico Franco, Francesco Nastasi, Luca Pulvirenti, Salvatore Petralia, et al.
Physicochemical Characterization and Antibacterial Properties of Carbon Dots from Two Mediterranean Olive Solid Waste Cultivars
Reprinted from: *Nanomaterials* **2022**, *12*, 885, doi:10.3390/nano12050885 . . . . **265**

Ysmael Verde-Gómez, Elizabeth Montiel-Macías, Ana María Valenzuela-Muñiz, Ivonne Alonso-Lemus, Mario Miki-Yoshida, Karim Zaghib, et al.
Structural Study of Sulfur-Added Carbon Nanohorns
Reprinted from: *Materials* **2022**, *15*, 3412, doi:10.3390/ma15103412 . . . . **277**

Weifeng Jin, Ying Tao, Xin Wang and Zheng Gao
The Effect of Carbon Nanotubes on the Strength of Sand Seeped by Colloidal Silica in Triaxial Testing
Reprinted from: *Materials* **2021**, *14*, 6119, doi:10.3390/ma14206119 . . . . **289**

Shama Parveen, Bruno Vilela, Olinda Lagido, Sohel Rana and Raul Fangueiro
Development of Multi-Scale Carbon Nanofiber and Nanotube-Based Cementitious Composites for Reliable Sensing of Tensile Stresses
Reprinted from: *Nanomaterials* **2022**, *12*, 74, doi:10.3390/nano12010074 307

Hui Wang, Fan Zhang, Yue Wang, Fangquan Shi, Qingyao Luo, Shanshan Zheng, et al.
DNAzyme-Amplified Electrochemical Biosensor Coupled with pH Meter for Ca^{2+} Determination at Variable pH Environments
Reprinted from: *Nanomaterials* **2022**, *12*, 4, doi:10.3390/nano12010004 329

Guifeng Liu, Huadi Zhang, Jianpeng Liu, Shuqi Xu and Zhengfa Chen
Experimental Study on the Salt Freezing Durability of Multi-Walled Carbon Nanotube Ultra-High-Performance Concrete
Reprinted from: *Materials* **2022**, *15*, 3188, doi:10.3390/ma15093188 349

About the Editors

Simone Morais

Simone Morais has a Ph.D. (1998) in Chemical Engineering from the faculty of Engineering of the University of Porto and the Habilitation in Chemical and Biological Engineering (2022). She is Associate Professor at the Polytechnic Institute of Porto, ISEP, Portugal, where she teaches several courses for the Chemical Engineering degree (1st cycle), Bioresources degree (1st cycle) and the Master's in Bioresources- (Bio)Technologies for Circular Economy.

She develops her research activity at the Associated Laboratory for Green Chemistry of the Network of Chemistry and Technology (REQUIMTE) in the fields of sensors and biosensors, nanotechnology, environmental sciences including human biomonitoring, green technologies, and the circular economy. She has been collaborating with several institutions, both national and international, of Higher Education and Research & Development centers, having (co)supervised several industrial Internships, MSc, PhD and post-doctoral students. She has acted as a reviewer for top-ranked journals and for international funding agencies in her areas of expertise. She has participated in funded projects: as a coordinator of 5 national/international competitively financed projects, and as a research member of 7 international R&D projects, 7 R&TD projects in collaboration with the Portuguese Industry, and 22 national R&D projects.

She (co)authored about 200 articles published in international journals with an impact factor (h-index: 42 - Scopus ID 7007053747), and 40 book chapters in international publishers (Elsevier, Springer, etc.). She has been also serving as an Editor of several peer-reviewed journals (with impact factor) and books in international publishers.

Simone Morais has been included in Stanford University's publicly available databases of the top 2% most-cited scientists (October 2023, September 2022 and August 2021 data-updates).

Konsantinos Spyrou

Konsantinos Spyrou obtained a B.Sc Degree in the Materials Science and Engineering department at the University of Ioannina (UOI). He finished his PhD studies at the University of Groningen, Zernike Institute for Advanced Materials (the Netherlands). He worked for two years as a Post-doc Associate at the University of Cornel, (USA) in the Materials Science and Engineering Department. He was a visiting scientist at King Fahd University of Petroleum and Minerals. He worked as an Experienced Researcher Marie Curie Fellow at the National Center of Scientific Research Demokritos (Athens). His work experience includes participation in 15 research programs, and he is the author of 71 international publications in peer-reviewed journals (*JACS, ACS Nano, Adv. Funct. Mater.*, etc) with more than 2200 citations. He has also participated in numerous national and international conferences. His scientific research focuses on the chemistry and engineering of carbon composite materials and phyllomorphous (2D) materials, such as graphene, germanane, and TMDs. Accordingly, his research activities interconnect organic chemistry, organic and organometallic catalysis, and the science of carbon and polymeric nanocomposites.

Editorial

Advances and Applications of Carbon Nanotubes

Simone Morais

REQUIMTE/LAQV, ISEP, Polytechnic of Porto, rua Dr. António Bernardino de Almeida, 4249-015 Porto, Portugal; sbm@isep.ipp.pt

Carbon nanotubes (CNT) (single-walled CNT, multiwalled CNT, non-covalently functionalized and covalently functionalized CNT, and/or CNT tailored with chemical or biological recognition elements) are by far the most popular nanomaterials thanks to their high electrical and thermal conductivities and mechanical strength, specific optical and sorption properties, low cost, and easy preparation, among other interesting characteristics [1–4]. Current applications comprise the use of CNT-based building blocks in the design of (mechanical, thermal, optical, magnetic, chemical, and biological) sensors, pre-concentration and clean-up schemes in analytical chemistry, (electro)catalysis, environmental protection (e.g., water quality control and treatment, adsorption of contaminants, desalinization, etc.), energy conversion including batteries, electromagnetic absorption and shielding materials, and in other novel tailored (nano)composites or hybrid nanostructures. However, other potential applications are constantly being discovered by ongoing investigations. Thus, the main aim of this Topic is to outline the recent advances in and applications of CNT by gathering a set of multidisciplinary research articles.

In total, twenty-three articles are published on this Topic in the participating journals *Nanomaterials*, *Applied Sciences*, and *Materials*. The presented knowledge embraces numerical simulations [2,5–11] and/or experimental data [1,12–23].

The research community has been particularly active in modelling and simulating different scenarios to better understand the behaviour of natural convection, energy absorption, and thermal and mass transport, as well as the mechanical, electronic, and optical properties.

Khan et al. [2,9] conducted a thorough numeral analysis of the heat transfer and fluid movement between plates under the influence of magnetic fields; for this purpose, a hybrid nanofluid composed of CNT, ferrous oxide, and water was used. Alsabery et al. [6] also studied convective flow and heat transfer, but by using alumina nanoparticles water-based nanofluid inside a cavity with vertical walls. These characterizations [2,6,9] are particularly important for the design of insulation systems, solar energy storage, and cooling systems, among others.

Molecular dynamic simulations have provided deep theoretical insights into the heat transfer mechanisms of branched CNT [5] and improved the tribological features of CNT/vulcanized natural rubber composites for aeronautics [8] and the energy absorption capacity of CNT buckypaper [11]. An additional study [7] used density functional theory to explore the combination of CNT and 2D monolayer germanium selenide (a semiconductor) for potential optoelectronic applications. Numerical and experimental methods were coupled to assess the impact of path planning methods on the mechanical performance of the components [10]. All these theoretical studies pave the way for novel applications of CNT, particularly in the engineering field and in industry.

The other articles on this Topic delve into different experimental approaches for the synthesis of new (nano)materials or the development of new analytical tools. A large set of new (nano)materials is under scrutiny, namely, silica-microsphere-supported N-doped CNT for electromagnetic wave absorption [13], low crystallinity CNT arrays as potential gas sensors [14], synthesis of tin dioxide/CNT nanonests for use in lithium-ion batteries [15],

Citation: Morais, S. Advances and Applications of Carbon Nanotubes. *Nanomaterials* **2023**, *13*, 2674. https://doi.org/10.3390/nano13192674

Received: 31 August 2023
Accepted: 26 September 2023
Published: 29 September 2023

Copyright: © 2023 by the author. Licensee MDPI, Basel, Switzerland. This article is an open access article distributed under the terms and conditions of the Creative Commons Attribution (CC BY) license (https://creativecommons.org/licenses/by/4.0/).

3D CNT monoliths for controlled adsorption [16], and six different CNT yarns prepared by various spinning methods [17], among others [18,19]. Further developments were focused on materials for civil infrastructures such as sand–gel composites [20], cementitious composites [4,21], and ultra-high-performance concrete [23].

It is worth mentioning that analytical strategies were also addressed in this Topic by designing a novel biosensor based on SWNT/DNAzyme for calcium determination in milk [22] and characterization of the photothermal features of CNT films [3].

The remarkable applications of CNT are already uncountable, but their potential is undoubtedly yet to be fully developed. Advances in their synthesis, characterization, and application, as concluded by reading the reported studies on this Topic, involve merging multi- and transdisciplinary knowledge ranging from fundamental science to technological innovations.

Funding: This research was funded by the Portuguese FCT—Foundation for Science and Technology, Ministério da Ciência, Tecnologia e Ensino Superior (MCTES) by national funds, project NATURIST 2022.07089.PTDC.

Data Availability Statement: No new data were created in this study. Data sharing is not applicable to this article.

Acknowledgments: All (co)authors and reviewers, as well as the MDPI assistant, Wing Wang, are deeply acknowledged for their valuable contributions.

Conflicts of Interest: The author declares no conflict of interest.

References

1. Lee, C.; Choi, G.; Kim, E.; Lee, J.; Lee, J.; Moon, H.; Kim, M.; Kim, Y.; Seo, T. Gas Barrier Performance of Hexagonal Boron Nitride Monolayers Grown on Copper Foils with Electrochemical Polishing. *Appl. Sci.* **2021**, *11*, 4599. [CrossRef]
2. Khan, M.; Mei, S.; Shabnam; Ali Shah, N.; Chung, J.; Khan, A.; Shah, S. Steady Squeezing Flow of Magnetohydrodynamics Hybrid Nanofluid Flow Comprising Carbon Nanotube-Ferrous Oxide/Water with Suction/Injection Effect. *Nanomaterials* **2022**, *12*, 660. [CrossRef] [PubMed]
3. Liu, Y.; Lin, Z.; Wang, P.; Huang, F.; Sun, J. Measurement of the Photothermal Conversion Efficiency of CNT Films Utilizing a Raman Spectrum. *Nanomaterials* **2022**, *12*, 1101. [CrossRef] [PubMed]
4. Laura, E.-C.; Rafael, C.; Jorge, Q.-O.; Harvi Alirio, C.-C.; Laura Victoria, R.-R.; Elisabeth, R.-P. Effects of Molarity and Storage Time of MWCNT on the Properties of Cement Paste. *Materials* **2022**, *15*, 9035. [CrossRef] [PubMed]
5. Chen, W.; Chang, I. A Thermal Transport Study of Branched Carbon Nanotubes with Cross and T-Junctions. *Appl. Sci.* **2021**, *11*, 5933. [CrossRef]
6. Alsabery, A.; Tayebi, T.; Abosinnee, A.; Raizah, Z.; Chamkha, A.; Hashim, I. Impacts of Amplitude and Local Thermal Non-Equilibrium Design on Natural Convection within Nanofluid Superposed Wavy Porous Layers. *Nanomaterials* **2021**, *11*, 1277. [CrossRef] [PubMed]
7. Mao, Y.; Guo, Z.; Yuan, J.; Sun, T. 1D/2D van der Waals Heterojunctions Composed of Carbon Nanotubes and a GeSe Monolayer. *Nanomaterials* **2021**, *11*, 1565. [CrossRef] [PubMed]
8. Teng, F.; Wu, J.; Su, B.; Wang, Y. Enhanced Tribological Properties of Vulcanized Natural Rubber Composites by Applications of Carbon Nanotube: A Molecular Dynamics Study. *Nanomaterials* **2021**, *11*, 2464. [CrossRef]
9. Khan, M.; Mei, S.; Shabnam; Fernandez-Gamiz, U.; Noeiaghdam, S.; Shah, S.; Khan, A. Numerical Analysis of Unsteady Hybrid Nanofluid Flow Comprising CNT-Ferrousoxide/Water with Variable Magnetic Field. *Nanomaterials* **2022**, *12*, 180. [CrossRef]
10. Wang, D.; Xiao, J.; Ju, X.; Dou, M.; Li, L.; Wang, X. Numerical and Experiment Studies of Different Path Planning Methods on Mechanical Properties of Composite Components. *Materials* **2021**, *14*, 6100. [CrossRef]
11. Yuan, Q.; Chen, H.; Nie, H.; Zheng, G.; Wang, C.; Hao, L. Soft-Landing Dynamic Analysis of a Manned Lunar Lander Em-Ploying Energy Absorption Materials of Carbon Nanotube Buckypaper. *Materials* **2021**, *14*, 6202. [CrossRef] [PubMed]
12. Bejaoui Kefi, B.; Bouchmila, I.; Martin, P.; M'Hamdi, N. Titanium Dioxide Nanotubes as Solid-Phase Extraction Adsorbent for the Determination of Copper in Natural Water Samples. *Materials* **2022**, *15*, 822. [CrossRef] [PubMed]
13. Cao, F.; Xu, J.; Zhang, X.; Li, B.; Zhang, X.; Ouyang, Q.; Zhang, X.; Chen, Y. Tuning Dielectric Loss of SiO2@CNT for Electromagnetic Wave Absorption. *Nanomaterials* **2021**, *11*, 2636. [CrossRef] [PubMed]
14. Adrian, A.; Cerda, D.; Fernández-Izquierdo, L.; Segura, R.; García-Merino, J.; Hevia, S. Tunable Low Crystallinity Carbon Nanotubes/Silicon Schottky Junction Arrays and Their Potential Application for Gas Sensing. *Nanomaterials* **2021**, *11*, 3040. [CrossRef] [PubMed]

15. Zhang, D.; Tang, Y.; Zhang, C.; Dong, Q.; Song, W.; He, Y. One-Step Synthesis of SnO2/Carbon Nanotube Nanonests Composites by Direct Current Arc-Discharge Plasma and Its Application in Lithium-Ion Batteries. *Nanomaterials* **2021**, *11*, 3138. [CrossRef] [PubMed]
16. Röcker, D.; Trunzer, T.; Heilingbrunner, J.; Rassloff, J.; Fraga-García, P.; Berensmeier, S. Design of 3D Carbon Nanotube Monoliths for Potential-Controlled Adsorption. *Appl. Sci.* **2021**, *11*, 9390. [CrossRef]
17. Watanabe, T.; Yamazaki, S.; Yamashita, S.; Inaba, T.; Muroga, S.; Morimoto, T.; Kobashi, K.; Okazaki, T. Comprehensive Characterization of Structural, Electrical, and Mechanical Properties of Carbon Nanotube Yarns Produced by Various Spinning Methods. *Nanomaterials* **2022**, *12*, 593. [CrossRef]
18. Nocito, G.; Sciuto, E.; Franco, D.; Nastasi, F.; Pulvirenti, L.; Petralia, S.; Spinella, C.; Calabrese, G.; Guglielmino, S.; Conoci, S. Physicochemical Characterization and Antibacterial Properties of Carbon Dots from Two Mediterranean Olive Solid Waste Cultivars. *Nanomaterials* **2022**, *12*, 885. [CrossRef]
19. Verde-Gómez, Y.; Montiel-Macías, E.; Valenzuela-Muñiz, A.; Alonso-Lemus, I.; Miki-Yoshida, M.; Zaghib, K.; Brodusch, N.; Gauvin, R. Structural Study of Sulfur-Added Carbon Nanohorns. *Materials* **2022**, *15*, 3412. [CrossRef]
20. Jin, W.; Tao, Y.; Wang, X.; Gao, Z. The Effect of Carbon Nanotubes on the Strength of Sand Seeped by Colloidal Silica in Triaxial Testing. *Materials* **2021**, *14*, 6119. [CrossRef]
21. Parveen, S.; Vilela, B.; Lagido, O.; Rana, S.; Fangueiro, R. Development of Multi-Scale Carbon Nanofiber and Nanotube-Based Cementitious Composites for Reliable Sensing of Tensile Stresses. *Nanomaterials* **2022**, *12*, 74. [CrossRef]
22. Wang, H.; Zhang, F.; Wang, Y.; Shi, F.; Luo, Q.; Zheng, S.; Chen, J.; Dai, D.; Yang, L.; Tang, X.; et al. DNAzyme-Amplified Electrochemical Biosensor Coupled with pH Meter for Ca2+ Determination at Variable pH Environments. *Nanomaterials* **2022**, *12*, 4. [CrossRef]
23. Liu, G.; Zhang, H.; Liu, J.; Xu, S.; Chen, Z. Experimental Study on the Salt Freezing Durability of Multi-Walled Carbon Nanotube Ultra-High-Performance Concrete. *Materials* **2022**, *15*, 3188. [CrossRef]

Disclaimer/Publisher's Note: The statements, opinions and data contained in all publications are solely those of the individual author(s) and contributor(s) and not of MDPI and/or the editor(s). MDPI and/or the editor(s) disclaim responsibility for any injury to people or property resulting from any ideas, methods, instructions or products referred to in the content.

Article

Gas Barrier Performance of Hexagonal Boron Nitride Monolayers Grown on Copper Foils with Electrochemical Polishing

Chil Hyoung Lee [1,†], Go Bong Choi [2,†], Eun Mi Kim [1,3], Jongho Lee [1], Jaegeun Lee [4], Hi Gyu Moon [5], Myung Jong Kim [6], Yoong Ahm Kim [2,*] and Tae Hoon Seo [1,*]

1. Smart Energy & Nano Convergence Research Group, Korea Institute of Industrial Technology, Gwangju 61012, Korea; chlee0901@kitech.re.kr (C.H.L.); kimeunmi@kitech.re.kr (E.M.K.); jholee@kitech.re.kr (J.H.)
2. Department of Polymer Engineering, Graduate School, Alan G. MacDiarmid Energy Research Institute & School of Polymer Science and Engineering, Chonnam National University, Gwangju 61186, Korea; uppermost.peak@gmail.com
3. School of Materials Science & Engineering, Chonnam National University, 77 Yongbong-ro, Buk-gu, Gwangju 61186, Korea
4. Department of Organic Material Science and Engineering, Pusan National University, 2, Busandaehak-ro 63 beon-gil, Geumjeong-gu, Busan 46241, Korea; jglee@pusan.ac.kr
5. Department of Inhalation Toxicology Research Center, Korea Institute of Toxicology, Jeongeup 56212, Korea; higyu.moon@kitox.re.kr
6. Department of Chemistry, Gachon University, 1342 Seongnam-daero, Sujeong-gu, Seongnam-si 13120, Korea; myungjongkim@gachon.ac.kr
* Correspondence: yak@chonnam.ac.kr (Y.A.K.); thseo@kitech.re.kr (T.H.S.)
† These authors contributed equally to this work.

Abstract: The demand for high-performance two-dimensional gas barrier materials is increasing owing to their potential for application in optoelectronic devices. These materials can help the devices maintain their properties over a long period. Therefore, in this study, we investigated the gas barrier performance of hexagonal boron nitride (h-BN) monolayers grown on copper foils via electrochemical polishing (ECP). The ECP treatment helped reduce the surface roughness of the copper foils. As a result, the nucleation density was reduced and highly crystalline h-BN monolayers were produced. The gas barrier performance of h-BN monolayers on copper foils with ECP was comparable to that of graphene. Our finding demonstrates the potential of monolayer h-BN as a high-performance and economical gas barrier material for organic-based optoelectronic devices.

Keywords: two-dimensional; h-BN; Gas barrier; Electro-chemical polishing

1. Introduction

The development of organic-based optoelectronic devices has revolutionized the electronics industry owing to their superior material properties such as high quantum efficiency, high carrier mobility, good transparency, and high flexibility [1,2]. Despite the advantages of organic materials in optoelectronic device applications, their material properties become seriously degraded over time when they are exposed to water or oxygen [3]. Hence, the use of gas barriers is required to sustain their material properties. Though several materials have been suggested as candidates for gas barriers, such as metal, glass, SiO_x, and Al_2O_3 [4,5], two-dimensional (2D) materials have come into the limelight because they can utilize the properties of organic materials such as flexibility, lightness, and high optical transparency. Among various 2D materials, graphene has evident potential and is widely used as a gas barrier on account of its great barrier properties and its thermal and chemical stability [6,7].

However, graphene has limitations as a gas barrier. The most critical one is that it can facilitate the oxidation of materials [8]. According to recent reports, the high conductivity of graphene can contribute to the supply of electrons for oxidation around defects, instead of serving as a gas barrier [9,10]. In addition, if the number of layers of graphene increases to meet the high criteria of modern packaging applications, an optical loss results because graphene reduces the transmittance by ~2.3% per layer [11]. This reduced transmittance adversely affects the optical properties of organic devices, such as light extraction in light emitting diodes and absorption in solar cells.

Hexagonal boron nitride (h-BN) monolayers have recently emerged as an alternative. An h-BN monolayer has several merits as a gas barrier. First, h-BN monolayers have good dielectric properties which can prevent electron transfer for oxidation and protect from strong electric shock [10,12]. Second, h-BN monolayers have great transparency due to the wide energy bandgap (~6 eV) [13,14]. Furthermore, h-BN monolayers possess the stable material properties and high thermal stability needed for high-power device applications. We selected h-BN monolayers for study because of these advantages. However, one disadvantage is that water or oxygen molecules can penetrate through the defects [15,16], so a technique to minimize defect density is required to enhance the gas barrier performance to the commercial level. Several methods to produce large-grain monolayer h-BN have been reported, such as using an alloy catalyst [17], annealing the catalyst with hydrogen [18], and reducing the roughness of the catalyst by electrochemical polishing (ECP) [19].

In this work, we investigated the gas barrier performance of monolayer h-BN synthesized on electrochemically polished copper foils. Out of various available techniques for reducing the defect density of h-BN monolayers, ECP of copper foil was utilized because it is cheap and simple, and thus suitable for mass production. Scanning electron microscopy (SEM), Raman spectroscopy, and X-ray photoelectron spectroscopy (XPS) measurements were conducted to verify the growth of h-BN monolayers. To study the spatial distribution of the defects and the defect state, atomic force microscopy (AFM) and contact angle analysis of h-BN monolayers were performed. Finally, we measured water vapor transmission rate (WVTR) and water vapor permeability of h-BN monolayers. The performance of an h-BN monolayer as gas barrier on copper foils with ECP was comparable to that of a graphene monolayer [7].

2. Methods

2.1. Electro-Chemical Polishing

To planarize the surface of the copper foils (Alfar Aesar, 046986.RF 0.025 mm thick), ECP was conducted in a solution with phosphoric acid (Sigma-Aldrich, Saint Louis, MO, USA, 85 wt.% in H_2O, 99.99% trace metals basis, 345245-100 ML) and water at 1.8 V for 10 min using Cu plate as a cathode and copper foils (100 mm × 100 mm) as a working electrode, as shown in Figure S1 of the supporting information.

2.2. Synthesis of h-BN Layers

The h-BN layers investigated in this work were synthesized on copper foils (Nippon Mining Co. Ltd., Hitachishi, Japan) via chemical vapor deposition (CVD). Borazine ($B_3N_3H_6$) was used as a precursor for growing the h-BN and was kept in a chiller at $-10\ °C$ in a canister with a bubbler system. Then, the copper foil with ECP treatments was placed at the center position of a quartz tube and heated by a split-tube furnace. The annealing process, under 15 standard cubic centimeters per minute (sccm) of H_2 gas at low pressure (specific value), was performed at 1040 °C for 60 min. 0.3 sccm of borazine and 70 sccm of H_2 were supplied at 1040 °C for 90 s under a growth pressure of 5 mTorr. These are the optimal conditions for obtaining high quality and uniform h-BN, as shown in Figure S2 of the supporting information. As a final step, the samples were rapidly cooled down to room temperature under a hydrogen atmosphere. The as-grown h-BN monolayers were transferred by a method similar to graphene transfer onto various substrates, such as PET film, glass, and SiO_2/Si, to characterize and investigate the gas barrier properties of the h-BN. Details can be found in our previous report [7].

2.3. Structure Characterizations

Field emission scanning electron microscopy (FESEM, Quanta 200 FEG, FEI Company, Hillsboro, OR, USA) was used to observe the domain and surface morphology of the h-BN studied in this work. The quality of graphene was characterized by Raman spectroscopy (RX210 Analyser, Renishaw, Wotton-under-Edge, UK) using the 514 nm line of an Ar ion laser as an excitation source. X-ray photoelectron spectroscopy (XPS, K-Alpha spectrophotometer, Thermo Fisher, Waltham, MA, USA) was conducted with an AXIS Ultra DLD model with a monochromatic Al Kα line at 1486.69 eV. The surface topography of h-BN on copper foils after a film-induced frustrated etching (FIFE) test was investigated with an atomic force microscope (AFM, XE-200 System, PSIA, Suwon, Korea) in tapping mode. The contact angle was measured using a water contact measurement (PHX300, Surface Electro Optics, Suwon, Korea). The water vapor transmission rate (WVTR) was evaluated by a commercial AQUATRAN model 3 WVTR analyzer (MOCON, Minneapolis, MN, USA) at room temperature under 1 atm.

3. Results and Discussion

Figure 1 shows SEM images of h-BN monolayers on copper foils with and without ECP treatment. In both cases, h-BN monolayer fully covered the copper foils after 90 s' growth (Figure 1a,b). In these samples, wrinkles were commonly observed for thermal stress minimization. They could stem from the nucleation of defects on the step edges of Cu terraces during cooling process, and their presence is indirect evidence of the successful continuous growth of h-BN monolayer [20].

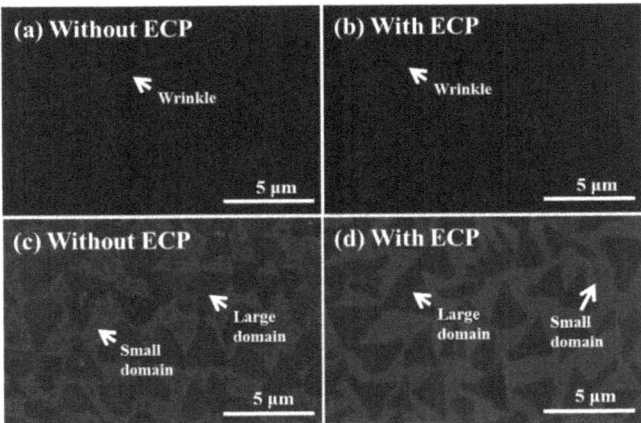

Figure 1. Scanning electron microscopic images of h-BN monolayers synthesized on copper foils (**a**,**b**) for 90 seconds and (**c**,**d**) for 60 seconds; (**a**,**c**) are without ECP treatment and (**b**,**d**) are with ECP treatment.

Once the copper foils were fully covered by h-BN monolayers, it was not possible to analyze the difference in the growth of h-BN monolayers on the two copper foil samples with and without ECP. To investigate how ECP influences the growth of an h-BN monolayer, we grew h-BN monolayers for a shorter time: 60 s (Figure 1c,d). At the initial stage of growth, we could detect small domains as well as large ones in both samples. The shape of the small domains was random, while that of the large domains was triangular. The areal number density of domains was noticeably different in the two cases; it was higher on copper foil without ECP than with ECP. The root-mean-square (RMS) roughness of the copper foils was measured by AFM (Figure S3). The RMS roughness of copper foils with ECP was measured to be 37 nm while that of copper foils without ECP was 78 nm, which is consistent with SEM images. It is well known that surface irregularities such as wrinkles, steps, grain boundaries, and defects have much higher surface energy than a flat surface, and such features could serve as nucleation sites [21–23]. Hence, the higher irregularity of the surface of copper foils without ECP caused a higher areal nucleation density of h-BN monolayer than copper foils with ECP [7]. If the areal number density of initial domains is higher, it will lead to a higher defect density when the domains eventually merge as the growth proceeds [24–26]. As a result, h-BN monolayers grown on copper foils with ECP are expected to show better gas barrier performance owing to a lower defect density.

Figure 2a shows Raman spectra of the h-BN monolayers grown on copper foils without ECP (black line) and with ECP (red line). The peaks observed at 1369.8 cm^{-1} from both the samples are indicative of an h-BN monolayer [27,28]. Typically, to compare the crystallinity of h-BN monolayers, the intensities of the spectra are compared. However, it would be hard to compare the crystallinity directly using this method because of the scattering of the laser spot by the surface roughness of the copper foils. Instead, we compared the crystalline quality of h-BN monolayers on copper foils based on the full width at half maximum (FWHM). The FWHM values of h-BN monolayer grown on copper foils without ECP and with ECP were ~25.02 cm^{-1} and ~17.41 cm^{-1} respectively, indicating that the crystallinity of h-BN monolayers on copper foils with ECP was higher than that of h-BN monolayers on copper foils without ECP. To further confirm the growth of the h-BN monolayers, we measured the 1s core level XPS spectra of boron and nitrogen in the h-BN monolayers on copper foils with and without ECP (Figure 2b,c). From both samples, peaks in the XPS spectra were observed at ~190.8 eV, which corresponds to the binding energy of boron atoms [29,30]. Peaks were also observed at ~398.2 eV in both samples. This energy corresponds to the binding energy of nitride atoms [29,30].

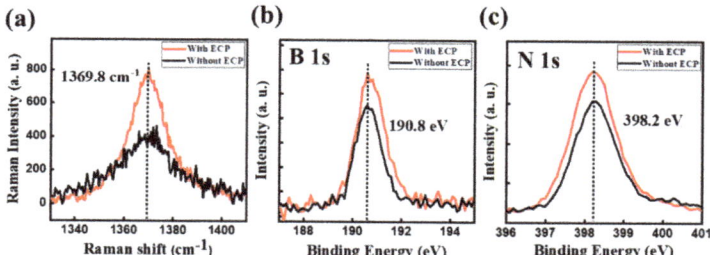

Figure 2. (a) Raman spectra of h-BN monolayers on copper foils without ECP (black line) and with ECP (red line). (b,c) show the 1s core level XPS spectra of boron and nitrogen of h-BN monolayers on copper foils without ECP and with ECP, respectively.

To investigate the spatial distribution of the defects, the surface morphology of h-BN monolayers on copper foils without ECP and with ECP was examined by AFM (Figure 3a,b). Before the measurements, a film-induced frustrated etching (FIFE) was completed to reveal the spatial distribution of the defects. The etched fit densities for h-BN monolayers on copper foils without ECP and with ECP were estimated to be $1.3 \times 10^8/cm^2$ and $3 \times 10^7/cm^2$, respectively (Figure 3c). In addition, etched to total area ratios were obtained by an image processing program (Image J 1.52p, National Institute of Health, Bethesda, MD, USA) based on a randomly selected area (indicated by a white rectangle in Figure 3a,b) of the AFM images of both samples [31]. The etched to total area ratios of h-BN monolayers on copper foils without ECP and with ECP were 23% and 3%, respectively. It is obvious that the h-BN monolayer on copper foils with ECP had a lower etched pit density with a smaller etched area, indicating that the ECP methods significantly reduced the defect density.

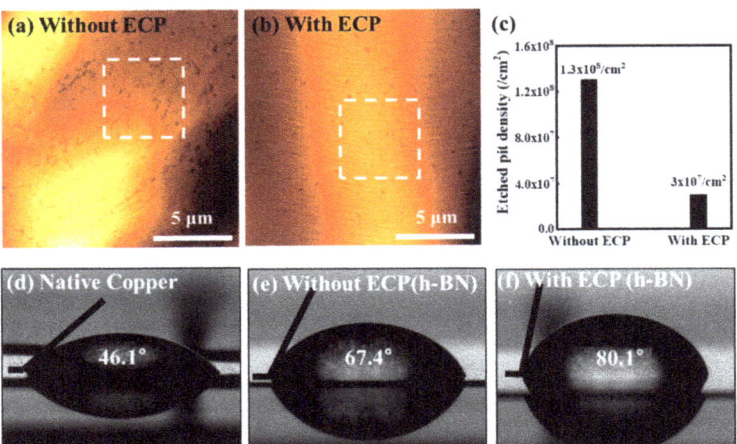

Figure 3. AFM images of the surface of h-BN monolayers on copper foils (**a**) without ECP and (**b**) with ECP after a film-induced frustrated etching. (**c**) Etched pit density observed from AFM images. (**d**–**f**) Contact angles of water on a (**d**) native copper foil, (**e**) an h-BN monolayer on copper foils without ECP, and (**f**) that with ECP, respectively.

We also evaluated contact angles of water on native copper foils, an h-BN monolayer on copper foil without ECP, and one with ECP, respectively (Figure 3d–f). Contact angle analysis is a fast and straightforward technique for estimating the defect density of 2D materials such as graphene and h-BN monolayers [32–34]. The measured contact angle of water on a native copper foil was 46.1°, which was low due to the high surface energy between water and copper foil. The contact angles were higher in the other samples as the copper substrates were passivated by the h-BN monolayers; they were found to be 67.4° without ECP and 80.1° with ECP. It is well known that h-BN is totally charge neutral and nonpolar, satisfying the octet rule. However, when the defects are created, the resulting dangling bonds break the charge neutrality. Because of this, defects in h-BN monolayer contribute to a change in the contact angle by increasing the surface energy and surface polarities [35]. The h-BN monolayer on the copper foil without ECP had higher surface energy than that with ECP, implying that the defect density of h-BN monolayer was higher on the copper foil without ECP than that with ECP. This result is consistent with our discussion of defect density based on the results of SEM, Raman spectroscopy, and AFM analyses.

WVTR was measured with varying time for bare polyethylene terephthalate (PET), an h-BN monolayer without ECP on PET, an h-BN monolayer with ECP on PET, and graphene on PET by a commercial MOCON's proprietary AQUATRAN model 3 WVTR analyzer (Figure 4a). Recently, we studied the gas barrier properties of graphene grown on copper foils with ECP treatment [7]. Here, we compare the gas barrier performance of h-BN monolayers and graphene. All four samples arrived at a steady state of WVTR within 1 day. The WVTR values of h-BN monolayers without and with ECP were measured to be, respectively, 0.798 and 0.774 g m^{-2} day^{-1}. These values were slightly higher than that of graphene on PET (0.728 g m^{-2} day^{-1}), but significantly lower than that of bare PET (1.101 g m^{-2} day^{-1}). The gas barrier properties of h-BN are slightly lower than that of graphene, but after a long term, we believe that gas barrier properties of h-BN will be improved than that of graphene. This is because h-BN is not easily oxidized by the absence of itinerant electron. Between h-BN monolayers, the one with ECP showed a lower WVTR value than that without ECP. We expect that the lower WVTR value of the h-BN monolayer with ECP is associated with its reduced defects and grain boundaries. The results indicate that the h-BN monolayer with ECP successfully acted as a protective layer for PET against water. Recently, the gas-barrier performance of a wafer-scale single-crystal h-BN monolayer on PET was reported, and was 0.60 g m^{-2} day^{-1} [36]. A single-crystal h-BN monolayer is ideal for the gas barrier application since it can minimize the diffusion of gas molecules through the grain boundary. However, the synthesis of a single-crystalline h-BN monolayer is very expensive and not generic yet. Thus, our approach to use poly-crystalline h-BN monolayers composed of large grains as gas barriers is more realistic since they are more economical and can be readily mass produced.

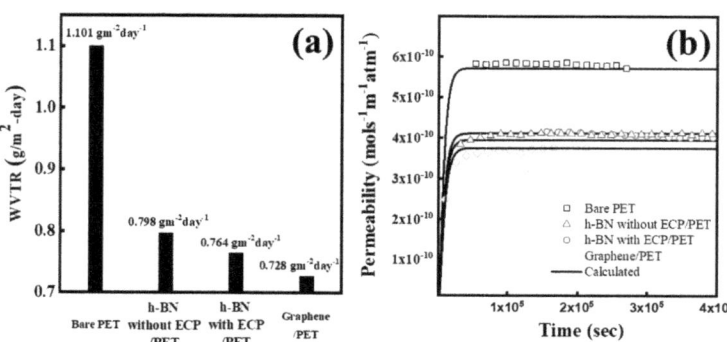

Figure 4. (a) WVTR and (b) water vapor permeability of pure PET, an h-BN monolayer without ECP on PET, an h-BN monolayer with ECP on PET, graphene on PET.

The water permeability of samples can be obtained from Equation (1) [37]

$$P° = P \left(\frac{4d^2}{\pi D t} \right)^{0.5} \sum_{n=0}^{\infty} exp \left[\frac{-d^2}{4Dt} (2n+1)^2 \right] \quad (1)$$

Here, J and J$_s$ are the water vapor molar flux at time t and at steady state, respectively, d is the sample thickness, P is the permeability, D is the diffusivity, S is the solubility, and P^0 = Jd, P = J$_s$d = SD. The permeability, diffusivity, and solubility of all samples were determined from the best fitting of Fick's second law of diffusion in Figure 4b. Table 1 shows the estimated permeability, diffusivity and solubility of bare PET, a h-BN monolayer without ECP on PET, a h-BN monolayer with ECP on PET, and graphene on PET, respectively. The permeabilities were 7.430×10^{-14}, 5.36×10^{-14}, 5.14×10^{-14}, and 4.88×10^{-14} mols^{-1} m^{-1} atm^{-1}, respectively. Interestingly, the decreased permeability was not correlated to the diffusivity; the diffusivity of all samples was nearly identical. Meanwhile, the solubilities of the h-BN monolayer without ECP on PET, the h-BN monolayer with ECP on PET, and graphene on PET were decreased by 25.4%, 26.4%, and 27.0%, respectively, compared to the solubility of bare PET. These results indicate that when adopting 2D material as a gas barrier, the reduction of permeability is caused by reduced solubility rather than the diffusion path blocking effect.

Table 1. Estimated permeability, diffusivity and solubility of bare PET, h-BN monolayer without ECP on PET, h-BN monolayer with ECP on PET, and graphene on PET.

	Bare PET	h-BN Without ECP/PET	h-BN with ECP/PET	Graphene/PET
Permeability [$\times 10^{-14}$ mol s^{-1} m^{-1} atm^{-1}]	7.43	5.36	5.14	4.88
Diffusivity [$\times 10^{-14}$ m^2 s^{-1}]	0.30	0.28	0.28	0.27
Solubility [mol m^{-3} atm^{-1}]	24.8	18.5	18.2	18.1

4. Conclusions

In summary, we investigated the gas barrier performance of h-BN monolayers on copper foils with ECP treatment. The ECP treatment of copper foils significantly reduced the defect density of h-BN monolayers. WVTR and permeability of h-BN monolayers on PET were significantly lower than those of bare PET and were comparable to those of graphene on PET. In addition, the ECP treatment of copper foils resulted in improved gas barrier performance of h-BN monolayers. The results demonstrate the potential of h-BN monolayers as gas barriers in organic-based optoelectronic devices.

Supplementary Materials: The following are available online at https://www.mdpi.com/article/10.3390/app11104599/s1, Figure S1: Schematics and experimental conditions of electrochemical polishing for Cu foils. Figure S2: The optimization process of synthesis conditions for high quality and uniform h-BN. Figure S3: AFM images and RMS roughness of Cu foils (a) without ECP and (b) with ECP, respectively.

Author Contributions: Conceptualization, C.H.L., G.B.C., and J.L. (Jaegeun Lee); Investigation, C.H.L. and J.L. (Jongho Lee); Data curation, C.H.L. and T.H.S.; Writing-original draft, C.H.L.; Writing, G.B.C.; Methodology, G.B.C., Validation, E.M.K., H.G.M., and Y.A.K.; Formal analysis, E.M.K., J.L. (Jongho Lee), M.J.K., and Y.A.K.; Writing-review & editing, J.L. (Jaegeun Lee) and T.H.S.; Resources, H.G.M. and M.J.K.; Supervision, T.H.S.; Project administration, T.H.S. All authors have read and agreed to the published version of the manuscript.

Funding: This research was supported by the Korea Institute of Industrial Technology (KITECH) and by the National Research Foundation of Korea (NRF) funded by the Ministry of Education, Science and Technology (2019RM3F5A1A02092650). And Y.A.K acknowledges support from the Nano·Material Technology Development Program through the National Research Foundation of Korea (NRF), funded by Ministry of Science and ICT (2017M3A7B4014045).

Institutional Review Board Statement: Not applicable.

Informed Consent Statement: Not applicable.

Data Availability Statement: Not applicable.

Conflicts of Interest: The authors declare no conflict of interest.

References

1. Wang, X.; Zhou, D.; Huang, J.; Yu, J. High performance organic ultraviolet photodetector with efficient electroluminescence realized by a thermally activated delayed fluorescence emitter. *Appl. Phys. Lett.* **2015**, *107*, 043303. [CrossRef]
2. Pradana, A.; Gerken, M. Photonic crystal slabs in flexible organic light-emitting diodes. *Photon. Res.* **2015**, *3*, 32–37. [CrossRef]
3. Yu, D.; Yang, Y.-Q.; Chen, Z.; Tao, Y.; Liu, Y.-F. Recent progress on thin-film encapsulation technologies for organic electronic devices. *Opt. Commun.* **2016**, *362*, 43–49. [CrossRef]
4. Xiao, W.; Hui, D.Y.; Zheng, C.; Yu, D.; Qiang, Y.Y.; Ping, C.; Xiang, C.L.; Yi, Z. A flexible transparent gas barrier film employing the method of mixing ALD/MLD-grown Al2O3 and alucone layers. *Nanoscale Res. Lett.* **2015**, *10*, 130. [CrossRef] [PubMed]
5. Xie, X.; Rieth, L.; Williams, L.; Negi, S.; Bhandari, R.; Caldwell, R.; Sharma, R.; Tathireddy, P.; Solzbacher, F. Long-term relia-bility of Al2O3 and Parylene C bilayer encapsulated Utah electrode array based neural interfaces for chronic implantation. *J. Neural Eng.* **2014**, *11*, 26016. [CrossRef] [PubMed]
6. Ding, D.; Hibino, H.; Ago, H. Grain Boundaries and Gas Barrier Property of Graphene Revealed by Dark-Field Optical Microscopy. *J. Phys. Chem. C* **2018**, *122*, 902–910. [CrossRef]
7. Seo, T.H.; Lee, S.; Cho, H.; Chandramohan, S.; Suh, E.-K.; Lee, H.S.; Bae, S.K.; Kim, S.M.; Park, M.; Lee, J.K.; et al. Tailored CVD graphene coating as a transparent and flexible gas barrier. *Sci. Rep.* **2016**, *6*, 24143. [CrossRef]
8. Cui, C.; Lim, A.T.O.; Huang, J. A cautionary note on graphene anti-corrosion coatings. *Nat. Nanotechnol.* **2017**, *12*, 834–835. [CrossRef]
9. Schriver, M.; Regan, W.; Gannett, W.J.; Zaniewski, A.M.; Crommie, M.F.; Zettl, A. Graphene as a Long-Term Metal Oxidation Barrier: Worse Than Nothing. *ACS Nano* **2013**, *7*, 5763–5768. [CrossRef]
10. Tanjil, R.-E.; Jeong, Y.; Yin, Z.; Panaccione, W.; Wang, M.C. Ångstrom-Scale, Atomically Thin 2D Materials for Corrosion Mitigation and Passivation. *Coatings* **2019**, *9*, 133. [CrossRef]
11. Luo, S.; Wang, Y.; Tong, X.; Wang, Z. Graphene-based optical modulators. *Nanoscale Res. Lett.* **2015**, *10*, 199. [CrossRef] [PubMed]
12. Laturia, A.; Van De Put, M.L.; Vandenberghe, W.G. Dielectric properties of hexagonal boron nitride and transition metal dichalcogenides: From monolayer to bulk. *2D Mater. Appl.* **2018**, *2*, 6. [CrossRef]

13. Wang, J.; Ma, F.; Sun, M. Graphene, hexagonal boron nitride, and their heterostructures: Properties and applications. *RSC Adv.* **2017**, *7*, 16801–16822. [CrossRef]
14. Sun, J.; Lu, C.; Song, Y.; Ji, Q.; Song, X.; Li, Q.; Zhang, Y.; Zhang, L.; Kong, J.; Liu, Z. Recent progress in the tailored growth of two-dimensional hexagonal boron nitride via chemical vapour deposition. *Chem. Soc. Rev.* **2018**, *47*, 4242–4257. [CrossRef] [PubMed]
15. Boutilier, M.S.H.; Sun, C.; O'Hern, S.C.; Au, H.; Hadjiconstantinou, N.G.; Karnik, R. Implications of Permeation through Intrinsic Defects in Graphene on the Design of Defect-Tolerant Membranes for Gas Separation. *ACS Nano* **2014**, *8*, 841–849. [CrossRef]
16. Giesbers, A.; Bouten, P.; Cillessen, J.; Van Der Tempel, L.; Klootwijk, J.; Pesquera, A.; Centeno, A.; Zurutuza, A.; Balkenende, A. Defects, a challenge for graphene in flexible electronics. *Solid State Commun.* **2016**, *229*, 49–52. [CrossRef]
17. Lu, G.; Wu, T.; Yuan, Q.; Wang, H.; Wang, H.; Ding, F.; Xie, X.; Jiang, M. Synthesis of large single-crystal hexagonal boron nitride grains on Cu–Ni alloy. *Nat. Commun.* **2015**, *6*, 6160. [CrossRef] [PubMed]
18. Cho, H.; Park, S.; Won, D.I.; Kang, S.O.; Pyo, S.S.; Kim, D.I.; Kim, S.M.; Kim, H.C.; Kim, M.J. Growth kinetics of white gra-phene (h-BN) on a planarised Ni foil surface. *Sci. Rep.* **2015**, *5*, 11985. [CrossRef]
19. Tay, R.Y.; Griep, M.H.; Mallick, G.; Tsang, S.H.; Singh, R.S.; Tumlin, T.; Teo, E.H.T.; Karna, S.P. Growth of large single-crystalline two-dimensional boron nitride hexagons on electropolished copper. *Nano Lett.* **2014**, *14*, 839–846. [CrossRef]
20. Li, X.; Cai, W.; An, J.; Kim, S.; Nah, J.; Yang, D.; Piner, R.; Velamakanni, A.; Jung, I.; Tutuc, E.; et al. Large-Area Synthesis of High-Quality and Uniform Graphene Films on Copper Foils. *Science* **2009**, *324*, 1312–1314. [CrossRef]
21. Braeuninger-Weimer, P.; Brennan, B.; Pollard, A.J.; Hofmann, S. Understanding and Controlling Cu-Catalyzed Graphene Nucleation: The Role of Impurities, Roughness, and Oxygen Scavenging. *Chem. Mater.* **2016**, *28*, 8905–8915. [CrossRef] [PubMed]
22. Han, G.H.; Güneş, F.; Bae, J.J.; Kim, E.S.; Chae, S.J.; Shin, H.-J.; Choi, J.-Y.; Pribat, D.; Lee, Y.H. Influence of Copper Morphology in Forming Nucleation Seeds for Graphene Growth. *Nano Lett.* **2011**, *11*, 4144–4148. [CrossRef] [PubMed]
23. Griep, M.H.; Sandoz-Rosado, E.; Tumlin, T.M.; Wetzel, E. Enhanced Graphene Mechanical Properties through Ultrasmooth Copper Growth Substrates. *Nano Lett.* **2016**, *16*, 1657–1662. [CrossRef] [PubMed]
24. Tian, W.; Li, W.; Yu, W.; Liu, X. A Review on Lattice Defects in Graphene: Types, Generation, Effects and Regulation. *Micromachines* **2017**, *8*, 163. [CrossRef]
25. Sun, X.-Y.; Hu, H.; Cao, C.; Xu, Y.-J. Anisotropic vacancy-defect-induced fracture strength loss of graphene. *RSC Adv.* **2015**, *5*, 13623–13627. [CrossRef]
26. Yang, B.; Wang, S.; Guo, Y.; Yuan, J.; Si, Y.; Zhang, S.; Chen, H. Strength and failure behavior of a graphene sheet containing bi-grain-boundaries. *RSC Adv.* **2014**, *4*, 54677–54683. [CrossRef]
27. Stehle, Y.; Meyer, I.H.M.; Unocic, R.R.; Kidder, M.; Polizos, G.; Datskos, P.G.; Jackson, R.; Smirnov, S.N.; Vlassiouk, I.V. Synthesis of Hexagonal Boron Nitride Monolayer: Control of Nucleation and Crystal Morphology. *Chem. Mater.* **2015**, *27*, 8041–8047. [CrossRef]
28. Mahvash, F.; Eissa, S.; Bordjiba, T.; Tavares, A.C.; Szkopek, T.; Siaj, M. Corrosion resistance of monolayer hexagonal boron nitride on copper. *Sci. Rep.* **2017**, *7*, 42139. [CrossRef]
29. Lee, S.H.; Jeong, H.; Okello, O.F.N.; Xiao, S.; Moon, S.; Kim, D.Y.; Kim, G.Y.; Lo, J.I.; Peng, Y.C.; Cheng, B.M.; et al. Improvements in structural and optical properties of wafer-scale hexagonal boron nitride film by post-growth annealing. *Sci. Rep.* **2019**, *9*, 10590. [CrossRef]
30. Haider, A.; Ozgit-Akgun, C.; Goldenberg, E.; Okyay, A.K.; Biyikli, N. Low-Temperature Deposition of Hexagonal Boron Nitride via Sequential Injection of Triethylboron and N2/H2 Plasma. *J. Am. Ceram. Soc.* **2014**, *97*, 4052–4059. [CrossRef]
31. Abràmoff, M.D.; Magalhães, P.J.; Ram, S.J. Image Processing with ImageJ. *Biophotonics Int.* **2004**, *11*, 36–42. [CrossRef]
32. Belyaeva, L.A.; van Deursen, P.M.G.; Barbetsea, K.I.; Schneider, G.F. Hydrophilicity of Graphene in Water through Trans-parency to Polar and Dispersive Interactions. *Adv. Mater.* **2018**, *30*, 1703274. [CrossRef]
33. Prydatko, A.V.; Belyaeva, L.A.; Jiang, L.; Lima, L.M.C.; Schneider, G.F. Contact angle measurement of free-standing square-millimeter single-layer graphene. *Nat. Commun.* **2018**, *9*, 4185. [CrossRef] [PubMed]
34. Parobek, D.; Liu, H. Wettability of graphene. *2D Mater.* **2015**, *2*. [CrossRef]
35. Annamalai, M.; Gopinadhan, K.; Han, S.A.; Saha, S.; Park, H.J.; Cho, E.B.; Kumar, B.; Patra, A.; Kim, S.-W.; Venkatesan, T. Surface energy and wettability of van der Waals structures. *Nanoscale* **2016**, *8*, 5764–5770. [CrossRef] [PubMed]
36. Lee, J.S.; Choi, S.H.; Yun, S.J.; Kim, Y.I.; Boandoh, S.; Park, J.-H.; Shin, B.G.; Ko, H.; Lee, S.H.; Lee, Y.H.; et al. Wafer-scale single-crystal hexagonal boron nitride film via self-collimated grain formation. *Science* **2018**, *362*, 817–821. [CrossRef]
37. Kim, H.M.; Lee, J.K.; Lee, H.S. Transparent and high gas barrier films based on poly(vinyl alcohol)/graphene oxide composites. *Thin Solid Films* **2011**, *519*, 7766–7771. [CrossRef]

Article

Steady Squeezing Flow of Magnetohydrodynamics Hybrid Nanofluid Flow Comprising Carbon Nanotube-Ferrous Oxide/Water with Suction/Injection Effect

Muhammad Sohail Khan [1], Sun Mei [1,*], Shabnam [1], Nehad Ali Shah [2], Jae Dong Chung [2], Aamir Khan [3,*] and Said Anwar Shah [4]

1. School of Mathematical Sciences, Jiangsu University, Zhenjiang 212013, China; sohailkhan8688@gmail.com (M.S.K.); shabnam8688@gmail.com (S.)
2. Department of Mechanical Engineering, Sejong University, Seoul 05006, Korea; nehadali199@yahoo.com (N.A.S.); jdchung@sejong.ac.kr (J.D.C.)
3. Department of Mathematics and Statistics, University of Haripur, Haripur 22620, KPK, Pakistan
4. Department of Basic Sciences and Islamiat, University of Engineering and Technology Peshawar, Peshawar 25120, KPK, Pakistan; anwarshah@uetpeshawar.edu.pk
* Correspondence: sunm@ujs.edu.cn (S.M.); aamir.khan@uoh.edu.pk (A.K.)

Abstract: The main purpose of the current article is to scrutinize the flow of hybrid nanoliquid (ferrous oxide water and carbon nanotubes) (CNTs + Fe_3O_4/H_2O) in two parallel plates under variable magnetic fields with wall suction/injection. The flow is assumed to be laminar and steady. Under a changeable magnetic field, the flow of a hybrid nanofluid containing nanoparticles Fe_3O_4 and carbon nanotubes are investigated for mass and heat transmission enhancements. The governing equations of the proposed hybrid nanoliquid model are formulated through highly nonlinear partial differential equations (PDEs) including momentum equation, energy equation, and the magnetic field equation. The proposed model was further reduced to nonlinear ordinary differential equations (ODEs) through similarity transformation. A rigorous numerical scheme in MATLAB known as the parametric continuation method (PCM) has been used for the solution of the reduced form of the proposed method. The numerical outcomes obtained from the solution of the model such as velocity profile, temperature profile, and variable magnetic field are displayed quantitatively by various graphs and tables. In addition, the impact of various emerging parameters of the hybrid nanofluid flow is analyzed regarding flow properties such as variable magnetic field, velocity profile, temperature profile, and nanomaterials volume fraction. The influence of skin friction and Nusselt number are also observed for the flow properties. These types of hybrid nanofluids (CNTs + Fe_3O_4/H_2O) are frequently used in various medical applications. For the validity of the numerical scheme, the proposed model has been solved by another numerical scheme (BVP4C) in MATLAB.

Keywords: steady; hybrid nanofluid flow; variable magnetic field; parametric continuation method (PCM); BV4C Schemes

1. Introduction

Heat transfer through the flow of fluid on the plate surface or on the surface of a revolving disk is gaining incredible interest from researchers due to its many uses in the aeronautical sciences and engineerings including chemical processes, thermal-energy-producing systems, geothermal industry, gas turbine rotators, medical equipment, rotating machinery, and computer storage. The squeezing flow produces by the motion of the boundaries play a significant role in polymer processing, hydrodynamical machines, lubrication equipment, etc. Due to its wide range of applications in many modern technologies, it can be considered a good source of heat transmission. Researchers have also updated the squeezing flow through the introduction of new ideas known as nanofluids. Nanofluids are widely used in the fields of micromanufacturing, cancer treatment, power generation, microelectronics, microchannels, thermal therapy, drug delivery, and metallurgical sectors. Choi [1] was probably the first person in technology to work on nanofluid for cooling purposes. From his study, he concluded that putting some nanoparticles into the base fluids (oil, water, and blood) make the fluid more efficient for transferring thermal conductivity. Many researchers in this field have used Choi's idea while working on nanofluids. Shahid et al. [2] numerically analyzed nanofluid by gyrostatic microorganisms in the porous medium on a stretched surface. Turkyilmazoglu [3] analyzed the slip flow pattern theoretically between concentric circular pipes. Xu et al. [4] performed a numerical study of the pulse flow of GOP water nanofluids in a microchannel for heat transmission. Ganji and Dogonchi [5] scrutinized the behavior of time-dependent MHD squeezing flow of nanoliquids between two plates regarding heat transfer using the Cattaneo–Christov heat flux model and thermal radiation. Reddy and Sreedevi [6] analyzed the impact of the chemical reaction and double classification at porous stretching sheets through nanofluid with thermal radiation regarding mass and heat transport. Arrigo et al. [7] studied the behavior of multipurpose hybrid nanofluid using carbon nanotubes (CNTs). Khan et al. [8] observed the behavior of heat transmission on the plate surface using nanofluids formulated by (CNT) nanomaterial. Water and engine oil have been used as the base fluids by Rehman et al. [9] to analyze the behavior of single and multiwall carbon nanotube nanofluids. Flow characteristics with convective heat transfer via Cu-Ag/water hybrid-nanofluid are investigated by Hassan et al. [10]. The numerical and analytical solution of electro-MHD hybrid nanofluid flow in the porous medium with entropy generation has been investigated by Ellahi et al. [11]. Using CNT, Raza et al. [12] studied the development of heat conduction via peristaltic flow in a porous channel in the presence of a magnetic field. Majeed et al. [13] analyzed the heat transfer behavior with dipole effect for magnetite (Fe_3O_4) nanomaterials that are injected into the following basic fluids, namely refrigerated, water, and kerosene. Hafeez et al. [14] recently studied the flow characteristics at a revolving disk. Using Fortran Code 21, a numerical simulation of non-Newtonian liquid/Al_2O_3 nanofluids for (0–4) nanosize particles in a two-dimensional square gap with hot and cold lid-driven movement is performed by the Richardson method [15]. Magnetohydrodynamics (MHD) plays a significant role in a fluid motion and has many applications in different technology based on fluid flow. This is the reason most researchers are interested to study the effects of magnetic fields on the motion of the fluid. Regulating the movement of fluid is a basic principle of MHD, such as for the purpose of proper cooling. This approach can also be used in many technologies that involve the improvement of the thermal conductivity and heat transfer rate. One of the leading applications of this type of study is the merging process of the metal in the furnace under the magnetic field. Brain therapy, malignant tumors, blood pressure, and arthritis are well-known uses of magnetic fields. Hsu [16] analyzed the transient Couette flow between two parallel plates regarding heat transfer in the presence of a magnetic field. Siddiqui et al. [17] studied the movement of the MHD fluid flow in order to monitor various diseases through the respiratory tract. A numerical scheme called the Keller box algorithm has been used to solve MHD flow problems at the porous media [18]. Subhani and Nadeem in [19] considered the fluid theory to study the flow of MHD time-dependent hybrid nanofluids at a porous rotating surface. The numerical study of the 3D flow of Casson nanoliquid with chemical reactions on the stretching surface has been performed by Lokesh et al. [20]. The flow properties of Cu-Al_2O_3/water hybrid nanofluid with Joule heating and MHD on the shrinking/expanding surface has been studied in [21]. The properties of hybrid nanoliquid flow in the presence of a magnetic field on the stretching surface with slip condition has been studied by Tlili et al. [22]. In the current era of technology, the use of hybrid nanoliquids has gained much attention among researchers due to their excellent thermal properties. Hybrid nanofluids increase the rate of heat transfer compared to simple nanofluids and provide better results. Experimental studies of dissolving nanomaterials with volume fraction (1–100) nm have been conducted to improve heat transfer rate and thermal conductivity [23]. An experimental study has been conducted to analyze the temperature and concentration profile of engine oil using small particles of ZnO-MWCNTs [24]. Bovand et al. [25] investigated the properties of aluminium oxide–water nanoliquid flow at a constant temperature in the gap of two long parallel plates. In his proposed model, he also analyzed the thermophoresis power produced by the temperature variation between walls and liquid. The blending of small components into conventional fluids is a popular technique used to improve the thematic properties, which lead to enhancing the drag force [26]. In addition, a recent study of some other types of nanomaterials used in hybrid nanofluids was conducted by Waini et al. [27,28]. It has also been found that 5 percent of the volume of nanomaterials in the principle fluid is more efficient for maximum heat transfer rate. It is concluded from several studies that 5 percent of the volume fraction of nanomaterials in the principle fluid is more efficient for maximum heat transfer rate. In water-based Fe_3O_4 nanoliquids, 12–15 percent volume fraction has shown a significant influence on the Nusselt number [29–33]. Yahaya et al. [34] examined heat transfer via $Cu-Al_2O_3/H_2O$ hybrid nanofluid on the stretching plate. The ongoing work is filling the research gap in the current literature by studying the incompressible steady flow of hybrid nanoliquid between two parallel porous plates with the variable magnetic field. The problem under consideration has not been researched and is being researched for the first time. According to the author's knowledge, the current study is a novelty in the field. The authors have also considered several flow properties in the proposed flow model, such as suction/injection, stretching, and nanomaterial volume fractions, so that the effect of different emerging parameters regarding velocity profile, temperature profile, Nusselt number, and skin friction are studied.

2. Mathematical Formulation of the Problem

We consider hybrid nanofluids flow in the gap of two horizontal parallel plates separated by a distance of $H(t) = H\sqrt{1-\alpha t}$, where H represent the gap between plates at $t = 0$. For $\alpha > 0$, the two plates are squeezed until they touch at $t = \frac{1}{\alpha}$ and for $\alpha < 0$, the two plates are separated, as depicted in Figure 1. Furthermore, the top plate is moving toward the bottom plate with velocity $v = \frac{dh}{dt}$. Khan et al. [35] in their work used both discs are perfectly conducted. In the present problem, electrical forces are ignored as they are much smaller than magnetic forces. An applied magnetic field is used to generate the induced magnetic field $(b1_x, 0, b2_z)$. Single and multiwalled carbon nanotubes are blended in base fluid water to form a hybrid nanofluid ($CNT-Fe_3O_4/H_2O$). Cartesian coordinates system is taken at the center of the bottom plate, where the x-axis lies along the horizontal axis and the z-axis is orthogonal to the plate. The bottom and top plates are kept at a fixed temperature T_0 and T_H, respectively. The flow properties of the hybrid nanofluid $CNT-Fe_3O_4/H_2O$ under consideration are not time-dependent, resulting in more influence of the variable magnetic field. This sort of influence is considered to be due to magnetic characteristics. We have formulated hybrid nanofluid by dissolving the volume fraction of CNT ($\Phi_2 = 0.5$) into the originally formulated ferrofluid (Fe_3O_4/H_2O). The mathematical formulation of the aforementioned hybrid nanofluids by continuity, momentum, magnetic field, and energy conservation equations are as follows [35].

Continuity equation:

$$\frac{\partial u}{\partial x} + \frac{\partial v}{\partial x} = 0, \tag{1}$$

The momentum equations with magnetic effect [35–37]:

$$u\frac{\partial u}{\partial x} + v\frac{\partial u}{\partial y} = -\frac{1}{\rho_{hnf}}\frac{\partial P}{\partial x} + \frac{\mu_{hnf}}{\rho_{hnf}}\left(\frac{\partial^2 u}{\partial x^2} + \frac{\partial^2 u}{\partial y^2}\right) - \frac{b_2 \sigma_{hnf}}{\rho_{hnf}}\left(\frac{\partial b_1}{\partial y} - \frac{\partial b_2}{\partial x}\right) \tag{2}$$

$$u\frac{\partial v}{\partial x} + v\frac{\partial v}{\partial y} = -\frac{1}{\rho_{hnf}}\frac{\partial P}{\partial y} + \frac{\mu_{hnf}}{\rho_{hnf}}\left(\frac{\partial^2 v}{\partial x^2} + \frac{\partial^2 v}{\partial y^2}\right) - \frac{b_1 \sigma_{hnf}}{\rho_{hnf}}\left(\frac{\partial b_1}{\partial y} - \frac{\partial b_2}{\partial x}\right) \tag{3}$$

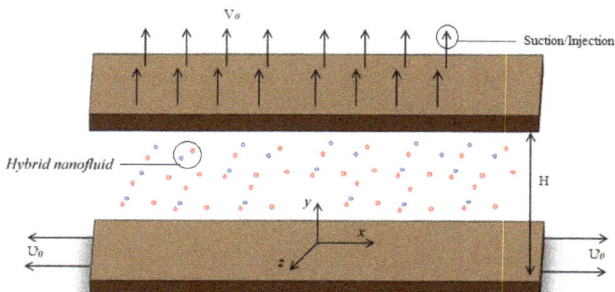

Figure 1. Geometry.

Maxwell Equations [37]:

$$u\frac{\partial b_2}{\partial y} + b_2\frac{\partial u}{\partial y} - v\frac{\partial b_1}{\partial y} - b_1\frac{\partial v}{\partial y} + \frac{1}{\sigma_{hnf}\mu_e}(\frac{\partial^2 b_1}{\partial x^2} + \frac{\partial^2 b_1}{\partial y^2}) = 0 \quad (4)$$

$$v\frac{\partial b_1}{\partial x} + b_1\frac{\partial v}{\partial x} - u\frac{\partial b_2}{\partial x} - b_2\frac{\partial u}{\partial x} + \frac{1}{\sigma_{hnf}\mu_e}(\frac{\partial^2 b_2}{\partial x^2} + \frac{\partial^2 b_2}{\partial y^2}) = 0 \quad (5)$$

The energy Equation [36]:

$$u\frac{\partial T}{\partial x} + v\frac{\partial T}{\partial y} = \frac{\kappa_{hnf}}{(\rho C_p)_{hnf}}(\frac{\partial^2 T}{\partial x^2} + \frac{\partial^2 T}{\partial y^2}) \\ + \frac{\mu_{hnf}}{(\rho C_p)_{hnf}}(4(\frac{\partial v}{\partial y})^2 + (\frac{\partial u}{\partial y})^2 + (\frac{\partial v}{\partial x})^2 + 2\frac{\partial u}{\partial y}\frac{\partial v}{\partial x}) \quad (6)$$

where b_1, b_2 are the components of the magnetic field, $(\rho C_p)_{hnf}$ is the heat capacity of the hybrid nanofluid, P is fluid pressure, T is the temperature, ρ_{hnf} is fluid density of hybrid nanofluid, σ_{hnf} is electrical conductivity of hybrid nanofluid, and μ_{hnf} is kinematic viscosity of hybrid nanofluid.

Nanofluid are defined as [31,36]:

$$\begin{aligned} \nu_{hnf} &= \frac{\mu_h nf}{\rho_h nf}, \quad \frac{\rho_{hnf}}{\rho_f} = (1-\Phi_2)((1-\Phi_1 + \frac{\Phi_1 \rho_{MS}}{\rho_f}) + \frac{\Phi_2 \rho_{CNT}}{\rho_f}), \\ \frac{\kappa_{hnf}}{\kappa_{bf}} &= \frac{\kappa_{CNT} + (m-1)\kappa_{bf} + \Phi_2(\kappa_{bf} - \kappa_{CNT})}{\kappa_{CNT} + (m-1)\kappa_{bf} + \Phi_2(m-1)(\kappa_{bf} - \kappa_{CNT})} \\ \frac{\kappa_{bf}}{\kappa_f} &= (\frac{\kappa_{MS} + (m-1)\kappa_f + \Phi_1(\kappa_f - \kappa_{MS})}{\kappa_{MS} + (m-1)\kappa_f + \Phi_1(m-1)(\kappa_f - \kappa_{MS})}) \\ \frac{(\rho C_p)_{hnf}}{(\rho C_p)_f} &= (1-\Phi_2)((1-\Phi_1 + \frac{\Phi_1 (\rho C_p)_{MS}}{(\rho C_p)_f}) + \frac{\Phi_2 (\rho C_p)_{CNT}}{(\rho C_p)_f}), \\ \frac{\sigma_{hnf}}{\sigma_f} &= (\frac{\sigma_{CNT} + 2\sigma_f\sigma_{bf} - 2\Phi_2(\sigma_{bf} - \sigma_{CNT})}{\sigma_{CNT} + 2\sigma_f\sigma_{bf} + \Phi_2(\sigma_{bf} - \sigma_{CNT})}), \quad \frac{\mu_{hnf}}{\mu_f} = \frac{1}{(1-\Phi_1)^{2.5}(1-\Phi_2)^{2.5}}, \\ \frac{\sigma_{bf}}{\sigma_f} &= (\frac{\sigma_{MS} + 2\sigma_f - 2\Phi_1(\sigma_f - \sigma_{MS})}{\sigma_{MS} + 2\sigma_f + \Phi_2(\sigma_f - \sigma_{MS})}), \end{aligned} \quad (7)$$

with κ_{hnf} is the thermal conductivity of hybrid nanofluid, κ_{bf} is the thermal conductivity of the Fe_3O_4-nanofuid, and Φ_1, Φ_2 are the volume fraction of CNTs.

3. Boundary Conditions

The boundary conditions are chosen as [36]:

$$\begin{aligned} u = 0, \ v = V_0, \ b_1 = ax, \ b_2 = -aH, \ T = T_H \quad \text{at} \quad y = h(t) \\ u = U_0 = ax, \ v = 0, \ T = T_0, \ b_1 = b_2 = 0, \quad \text{at} \quad y = 0 \end{aligned} \tag{8}$$

The following similarity transformations [36] are considered for the reducing PDEs (1)–(6) to ODEs system,

$$\begin{aligned} u = axf'(\eta), \ v = -aHf(\eta), \ b_1 = axK'(\eta), \\ b_2 = -aHK(\eta), \ \eta = \frac{y}{H}, \ \theta(\eta) = \frac{T - T_H}{T_0 - T_H}, \end{aligned} \tag{9}$$

Therefore, Equation (1) is satisfied and the remaining Equations (2)–(6) transform into the following form

$$f'''' = \frac{\rho_{hnf}}{\rho_f} \frac{\mu_f}{\mu_{hnf}} (S(f'f'' - ff''')) + M^2 SRe_m (\frac{\sigma_{hnf}}{\sigma_f})^2 \frac{\mu_f}{\mu_{hnf}} (fK^2) \\ - M^2 S^2 Re_m^2 (\frac{\sigma_{hnf}}{\sigma_f})^3 \frac{\mu_f}{\mu_{hnf}} (ff'K^2 - f^2 KK'), \tag{10}$$

$$K'' = Re_m S \frac{\sigma_{hnf}}{\sigma_f} (f'K - fK'), \tag{11}$$

$$\theta'' = -SPr \frac{(\rho Cp)_{hnf}}{(\rho C_p)_f} \frac{\kappa_f}{\kappa_{hnf}} (\theta' f) - \frac{\mu_{hnf}}{\mu_f} \frac{\kappa_f}{\kappa_{hnf}} PrEc(4\delta f'^2 + f''^2), \tag{12}$$

and the transform boundary conditions are

$$\begin{aligned} f(0) = 0, \ f'(0) = 1, \ K(0) = 0, \ \theta(0) = 1, \\ f(1) = A, \ f'(1) = 0, \ K(1) = 1, \ \theta(1) = 0, \end{aligned} \tag{13}$$

where A is the suction/injection parameter, $M^2 = \frac{H^2 a \sigma_f}{\rho_f \nu_f}$ magnetic parameter, $S = \frac{aH^2}{2\nu_f}$ squeeze number, $Re_m = \sigma_f \nu_f \mu_e$ Rynold's Magnetic parameter, $Pr = \frac{\nu_f (\rho Cp)_f}{\kappa_f}$ Prandtl number, $Ec = \frac{1}{(Cp)_f T_H} (ax)^2$ Eckert number, and $\delta = \frac{H^2}{x^2}$.

Required physical parameters are the Nusselt number and skin friction coefficient, which can be defined as,

$$C_f = \frac{\mu_{nf}}{\rho_{nf}(\frac{-al}{2\sqrt{1-at}})^2} (\frac{\partial u}{\partial z} + \frac{\partial u}{\partial r})_{z=h(t)}, \quad N_u = -\frac{\kappa_{nf}(\frac{\partial T}{\partial z})_{z=h(t)}}{k_f (T_0 - T_u)}, \tag{14}$$

In case of Equation (16), we get

$$\frac{H^2 (1-\Phi_1)^{2.5} (1-\Phi_2)^{2.5} ((1-\Phi_2)((1-\Phi_1 + \frac{\Phi_1 \rho_{MS}}{\rho_f}) + \frac{\Phi_2 \rho_{CNT}}{\rho_f}))}{r^2} ReC_f = f''(1),$$

$$-\theta'(0) = \frac{H\sqrt{1-\alpha t}}{\frac{\kappa_{CNT} + (m-1)\kappa_{bf} + \Phi_2 (\kappa_{bf} - \kappa_{CNT})}{\kappa_{CNT} + (m-1)\kappa_{bf} + \Phi_2 (m-1)(\kappa_{bf} - \kappa_{CNT})}} N_u. \tag{15}$$

4. Numerical Solution by PCM

In this section, optimal choices of continuation parameters are made through the algorithm of PCM for the solution of nonlinear Equations (10)–(12) with boundary conditions in Equation (13):

- First order of ODE
  To transform the Equations (10)–(12) into first order of ODEs, consider the following

$$f = P_1, \quad f' = P_2, \quad f'' = P_3, \quad f''' = P_4$$
$$K = P_5, \quad K' = P_6, \quad \theta = P_7, \quad \theta' = P_8 \tag{16}$$

putting these transformations in Equations (10)–(12), which becomes

$$P_4' = \frac{\rho_{hnf}}{\rho_f}\frac{\mu_f}{\mu_{hnf}}(S(P_2P_3 - P_1P_4)) + M^2SRe_m(\frac{\sigma_{hnf}}{\sigma_f})^2\frac{\mu_f}{\mu_{hnf}}(P_1P_5^2)$$
$$- M^2S^2Re_m^2(\frac{\sigma_{hnf}}{\sigma_f})^3\frac{\mu_f}{\mu_{hnf}}(P_1P_2P_5^2 - P_1^2P_5P_6'), \tag{17}$$

$$P_6' = Re_mS\frac{\sigma_{hnf}}{\sigma_f}(P_2P_5 - P_1P_6), \tag{18}$$

$$P_8' = -SPr\frac{(\rho C p)_{hnf}}{(\rho C_p)_f}\frac{\kappa_f}{\kappa_{hnf}}(P_1P_8) - \frac{\mu_{hnf}}{\mu_f}\frac{\kappa_f}{\kappa_{hnf}}PrEc(4\delta P_2^2 + P_3^2), \tag{19}$$

and the boundary conditions become

$$P_1(0) = 0, \quad P_2(0) = 1, \quad P_1(1) = A, \quad P_2(1) = 0,$$
$$P_5(0) = 0, \quad P_5(1) = 1, \quad P_7(0) = 1, \quad P_7(1) = 0, \tag{20}$$

- Introducing parameter q, we obtained ODEs in a q-parameter group,
  To get ODEs in a q-parameter group, we use q-parameter in Equations (17)–(19), and, therefore,

$$P_4' = \frac{\rho_{hnf}}{\rho_f}\frac{\mu_f}{\mu_{hnf}}(S(P_2P_3 - P_1(P_4 - 1)q)) + M^2SRe_m(\frac{\sigma_{hnf}}{\sigma_f})^2\frac{\mu_f}{\mu_{hnf}}(P_1P_5^2)$$
$$- M^2S^2Re_m^2(\frac{\sigma_{hnf}}{\sigma_f})^3\frac{\mu_f}{\mu_{hnf}}(P_1P_2P_5^2 - P_1^2P_5P_6'), \tag{21}$$

$$P_6' = Re_mS\frac{\sigma_{hnf}}{\sigma_f}(P_2P_5 - P_1(P_6 - 1)q)), \tag{22}$$

$$P_8' = -SPr\frac{(\rho C p)_{hnf}}{(\rho C_p)_f}\frac{\kappa_f}{\kappa_{hnf}}(P_1(P_8 - 1)q) - \frac{\mu_{hnf}}{\mu_f}\frac{\kappa_f}{\kappa_{hnf}}PrEc(4\delta P_2^2 + P_3^2), \tag{23}$$

- Differentiation by q reaches at the following system w.r.t. the sensitivities to the parameter-q
  Differentiating the Equations (21)–(23) w.r.t. by q

$$d_1' = h_1 d_1 + e_1 \tag{24}$$

where h_1 is the coefficient matrix, e_1 is the remainder, and $d_1 = \frac{dP_i}{d\tau}, 1 \leq i \leq 8$.
- Cauchy Problem

$$d_1 = y_1 + a1v_1, \tag{25}$$

where y_1, v_1 are vector functions. By resolving the two Cauchy problems for every component, we then automatically satisfy the ODEs

$$e_1 + h_1(a1v_1 + y_1) = (a1v_1 + y_1)' \tag{26}$$

and leave the boundary conditions.
- Using by Numerical Solution

An absolute scheme has been used for the resolution of the problem

$$\frac{v_1^{i+1} - v_1^i}{\triangle \eta} = h_1 v_1^{i+1} \tag{27}$$

$$\frac{y^{i+1} - y^i}{\triangle \eta} = h_1 y^{i+1} + e_1 \tag{28}$$

- Taking of the corresponding coefficients
  As given boundaries are usually applied for P_i, where $1 \leq i \leq 8$, for the solution of ODEs, we required to apply $d_2 = 0$, which seems to be in matrix form as

$$l_1.d_1 = 0 \quad \text{or} \quad l_1.(a1v_1 + y_1) = 0 \tag{29}$$

where $a1 = \frac{-l_1.y_1}{l_1.v_1}$.

5. Error Analysis

Error analysis is performed to study the reliability of the proposed model solution. The solution of the proposed model by PCM method is validated by BVP4c in MATLAB. In addition, to further support the validity of the solution of the proposed model, the numerical results of several parameters are tabulated (2) and (3) and displayed graphically. Table 1 illustrates the thermophysical properties of water and nanomaterial. Table 2 shows the comparison of the proposed model solution with the solution in [35]. A closed agreement has been found in the two results for different values of S and Φ. Further support for validating the model solution is provided in Table 3 for the numerical values of $f''(1)$ and $-\theta'(1)$. It is evident that the model solution by two methods PCM and BVP4C are correct up to two decimal places.

Table 1. The thermophysical properties of water base fluid and hybrid nanoparticles.

	ρ	C_p	κ	σ
H_2O	997.1	4179	0.613	5.5×10^{-6}
Fe_3O_4	5200	670	6	9.74×10^6
SWCNT	2600	425	6600	10^6
MWCNT	1600	796	3000	10^7

Table 2. Comparison of the numerical results for Nusselt number when $Ec = 0$.

	S	Present	Ali et al. [35]
Φ = 0%	0.1	1.075221	1.078381
	0.5	1.401148	1.403658
	1.0	1.810361	1.813100
	1.5	2.206271	2.201327
Φ = 5%	0.1	1.292331	1.298621
	0.5	1.613601	1.619052
	1.0	2.021148	2.026519
	1.5	2.423006	2.422539
Φ = 10%	0.1	1.572331	1.573849
	0.5	1.883601	1.889474
	1.0	2.291148	2.292857
	1.5	2.693006	2.691972

Table 3. Comparison of the numerical results by two methods PCM and BVP4C for skin friction and Nusselt number, with various physical parameters.

Φ_1	M	S	PCM $f''(1)$	BVP4C $f''(1)$	PCM $-\theta'(1)$	BVP4C $-\theta'(1)$
0.0	0.4	0.2	−3.9734	−3.9704	5.7209	5.7252
0.3			−3.9967	−3.9909	5.7042	5.7038
0.5			−3.9956	−3.9905	0.7386	0.7314
	0.6		−3.9388	−3.9317	0.7410	0.7452
	0.8		−3.8789	−3.8752	0.7434	0.7481
		0.6	−3.6286	−3.6232	0.7397	0.7372
		1.2	−3.8337	−3.8323	0.7082	0.7075

6. Results and Discussions

In this section, we discuss the impact of various emerging parameters of the proposed model quantitatively via different graphs and tables at the velocity and magnetic profile. Hybrid nanoflow flow is observed between two long parallel plates with variable magnetic fields and phenomena of heat transfer. The impact of different key involved parameters, including Prandtl number (Pr), squeezing parameter (S), Eckert number (Ec), magnetic Reynolds number (Re_m), magnetic parameter (M), and nanomaterials volume fraction (Φ_1, Φ_2) are studied to analyze the behavior of mass and heat transfer. Table 1 presents the complete sketch of the thermophysical properties of various nanomaterials. The numerical outcomes of two key flow parameters such as skin friction and Nusselt number are shown in Table 2. For the validity of the numerical solution, the proposed model has been solved through two different numerical schemes (BVP4C, PCM) in MATLAB, and their numerical outcomes are displayed in Table 3. The results obtained from the solution of the model reveals that the thermal flow rate of (Fe_3O_4-SWCNTs-water) and (Fe_3O_4-MWCNTs-water) rises from 0.8206 percent to 2.5233 percent and 0.9526 percent to 2.8758 percent, respectively, when the volume fraction of nanomaterial increases from 0.01 to 0.03 as depicted in Table 4.

Table 4. The heat transfer has been calculated percent wise as for the various nanoparticles Pr = 6.2, S = 1.8, Ec = 0.8, using the percentage formula %increase = $\frac{With\ Nanoparticle}{Without\ Nanoparticle} \times 100$ = Result, Result-100 = %enhancment.

Φ_1, Φ_2	$-\theta'(1)$ for Fe_3O_4, SWCNT	$-\theta'(1)$ for Fe_3O_4, MWCNT
0.0	5.5324	5.5324
0.01	5.5778	5.5851
	(0.8206% increase)	(0.9526% increase)
0.02	5.6242	5.6381
	(1.6593% increase)	(1.9106% increase)
0.03	5.6720	5.6915
	(2.5233% increase)	(2.8758% increase)

In fact, as the surfaces move, fluids are squeezed in the channels, resulting in an increase in the velocity of the boundary region. Figure 2b demonstrates the impact of magnetic parameter M on the transverse velocity. The magnetic parameter is defined as the ratio of fluid flux to the magnetic diffusivity and it plays an important role in determining the diffusion in the magnetic field along streamlines [36]. From the figure, it has been noticed that the transverse velocity is declining due to a rise in the magnetic parameter M. In the same way, it has been noticed that the creation of Lorentz force due to magnetic field surging the resistance in the flow region. As the plates move forward, the flow of fluid in the channel wall is suppressed, resulting in increased velocity in the boundary area. Figure 3a illustrates the variational impact of the squeezing parameter S on the axial velocity. The squeezing Reynolds number, $S = \frac{\beta l^2}{2v_f}$, is the ratio of the normal velocity on the upper plate to the kinematic viscosity of the fluid. It is important to note that the large or small values of S indicate slow or rapid movement of the lower plate toward the upper plate. The positive values of S indicate that the bottom plate moves away from the top plate, or the distance between the plates is increased, while the negative values of S indicate that the upper plate moves away from the lower plate, or the distance between two plates is decreased. The figure illustrates that the area that lies in the domain $0 \leq \eta < 0.55$ represents the area near to the lower wall and the area that lies in the domain $0.55 \leq \eta \leq 1$ represents the area adjacent to the upper wall. It has been noticed that the increase in flow velocity is improving the velocity profile in the domain $\eta < 0.55$ It has been observed that the fluids pass rapidly through the narrow channel when the surfaces are compressed, in this way, the velocity profile decline in the domain $\eta > 0.55$, and the fluid encounter additional resistance in the wide channel. The crossflow exists in the centre of the fluid domain. It has been demonstrated in Figure 3b that the velocity profile does not vary at the critical point $\eta = 0.55$ with the fluctuation in the squeezing parameter and diminish for the magnetic parameter M. In addition, for the injection processes, the rising values of the magnetic parameter M reduce the $f'(\eta)$.

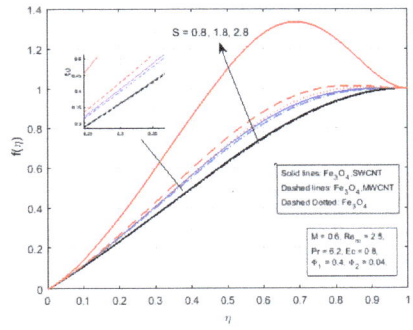

(a)

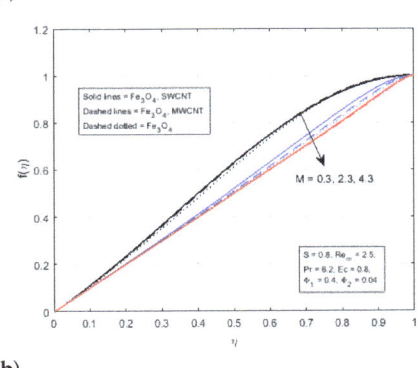

(b)

Figure 2. Effect of $f(\eta)$ for (**a**) S and (**b**) M.

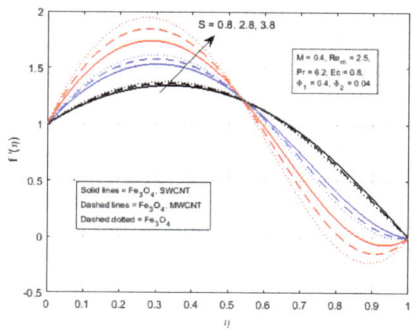

(a)

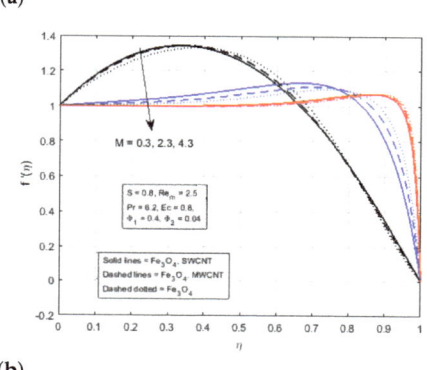

(b)

Figure 3. Effect of $f'(\eta)$ for (**a**) S and (**b**) M.

The impact of two nanomaterials (CNTs and Fe_3O_4) with volume fraction (Φ_1 and Φ_2) have been demonstrated in Figure 4. It reveals that the fluid flow is positively influenced by the increase in volume fraction parameters. The augmenting values of nanomaterial volume fraction Φ_1 and Φ_2 reduce the boundary layer thickness of the flow, thus increasing the axial velocity $f'(\eta)$ of the fluid flow in the interval $\eta < 0.56$, and decreasing in the interval $\eta > 0.56$. It has been observed that the axial velocity profile is cross-flow in the centre of the boundary layer. Figure 5a demonstrates the impact of the squeezing parameter of the magnetic field profile $K(\eta)$, it reflects that the magnetic profile $K(\eta)$ is almost linear at $S = 0.1$, and it becomes parabolic at a larger value of S, and approach to the maximum value at the middle of the channel. Figure 5b illustrates the impact of Re_m on the magnetic field profile $K(\eta)$, showing that the rising value of the magnetic Reynolds number increments the kinematic viscosity of the fluid flow. A rise in the kinematic viscosity means a fall in fluid density, which increases the magnetic field from the bottom to the upper plate, as displayed in Figure 5b.

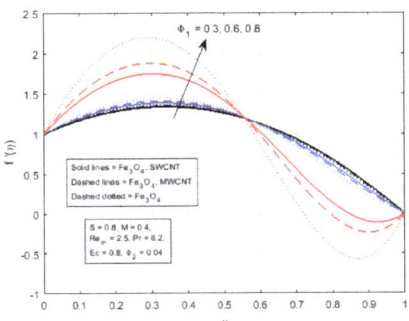

Figure 4. Effect of $f'(\eta)$ for Φ_1.

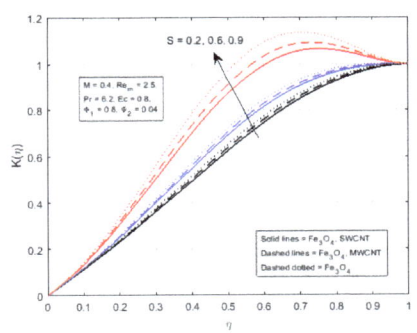

(a)

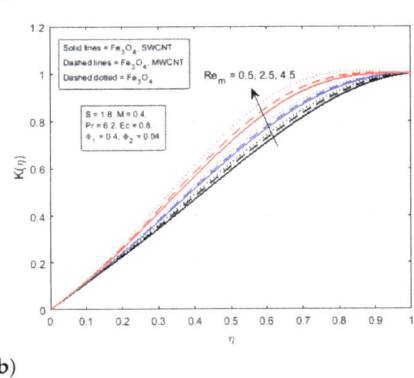

(b)

Figure 5. Effect of $K(\eta)$ for (**a**) S and (**b**) Re_m.

Figure 6a is plotted to notice the impact of squeezing parameter S on the temperature profile. The incrementing value of the squeezing parameter S for the fluid flow produces friction, which generates a certain amount of heat, as well as increases the kinematics energy of the fluid, which also produces heat. As a result, the mean temperature of the hybrid nanofluid flow in the middle of the channel increases. A rise in Eckert number Ec lead to an increase in the thermal flow rate, as displayed in Figure 6b. In fact, due to the increasing values of Eckert number, fluid friction in nanomaterial is produced with greater intensity. In this physical process, kinetic energy is converted into thermal energy, which ultimately assists the increase in temperature profile, and for high values of Eckert number, the profile of $\theta(\eta)$ becomes parabolic and attains the maximum value at the middle of the channel. In addition, increasing the volume fraction of nanomaterials leads to an increment in the density of fluid particles. In this physical process, fluid flow is appreciated and the thermal properties of the fluid flow particles reduce. Therefore, the increasing value of Φ_1 lead to diminishing the temperature profile, as portrayed in Figure 7, and at higher values of Φ_1, the $\theta(\eta)$ profile becomes parabolic, and approaches to the bottom at the middle of the channel.

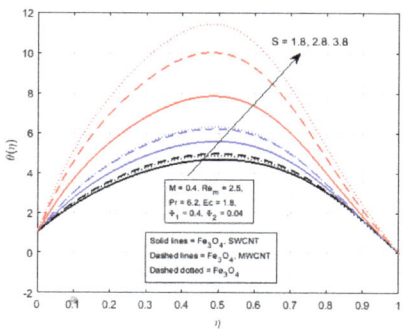

(a)

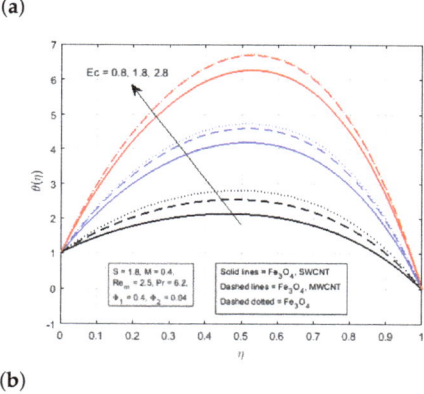

(b)

Figure 6. Effect of $\theta(\eta)$ for (**a**) S and (**b**) Ec.

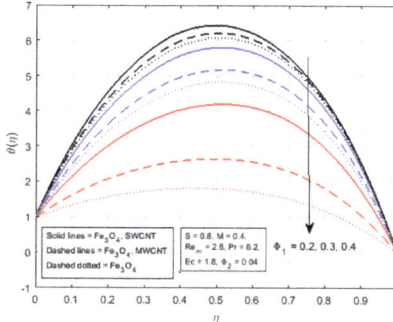

Figure 7. Effect of $\theta(\eta)$ for Φ_1.

7. Concluding Remarks

Hybrid fluid flow between two plates in the presence of variable magnetic field is considered using continuity, Navier–Stokes, Maxwell's, energy, and transport equations. The governing equations of the proposed flow model are converted into highly nonlinear systems of ODEs using similarity transformation. The numerical scheme PCM has been used for the solution of the flow model. The impacts of many involved parameters in the flow model are discussed via different graphs and tables on the velocity profile, temperature profile, and magnetic profile. The main findings of this research work are:

- As the magnetic field parameter M is increasing, the transverse velocity of the hybrid nanofluid is decreasing because of the increase of electromagnetic force.
- The Lorentz force, which is produced by a magnetic field, has been observed to increase flow resistance.
- The velocity profile is reduced in the region $\eta > 0.55$ because the fluid faces more resistance in the wide channel.
- The increase in the volume fraction of nanoparticles has a favorable impact on fluid flow. If $\Phi 1$ and $\Phi 2$ are increased, the thickness of the boundary layer is reduced, resulting in an increase in axial velocity $f'(\eta)$ for $\eta < 0.56$, otherwise it decreases.
- The fluid friction increases as the Eckert number rises, converting kinetic energy into thermal energy and raising the temperature profile. The Eckert number $\theta(\eta)$ becomes parabolic as the value of $\theta(\eta)$ increases, approaching the maximum value at the center.

Author Contributions: Formal analysis, J.D.C.; Funding acquisition, N.A.S.; Investigation, S.A.S.; Methodology, S.; Supervision, S.M.; Writing—original draft, M.S.K.; Writing—review & editing, A.K. All authors have read and agreed to the published version of the manuscript.

Funding: We acknowledge the insightful comments of editorial board to enhance this work. We also acknowledge the financial support provided by the Postdoctoral research support fund of School of Mathematical Sciences, Jiangsu University, Zhenjiang, 212013, China.

Institutional Review Board Statement: Not applicable.

Informed Consent Statement: Not applicable.

Data Availability Statement: Not applicable.

Acknowledgments: We acknowledge the insightful comments of editorial board to make this work more beautiful. We also acknowledge the financial support provided by the Postdoctoral research support fund of School of Mathematical Sciences, Jiangsu University, Zhenjiang, 212013, China. This work was supported by Korea Institute of Energy Technology Evaluation and Planning (KETEP) grant funded by the Korea government (MOTIE) (No. 20192010107020, Development of hybrid adsorption chiller using unutilized heat source of low temperature).

Conflicts of Interest: The authors declare no conflict of interest.

References

1. Choi, S.U.; Eastman, J.A. *Enhancing Thermal Conductivity of Fluids With Nanoparticles*; CONF-951135-29 ON: DE96004174; TRN: 96:001707; Argonne National Lab.: Lemont, IL, USA, 1995. Available online: https://ANL/MSD/CP-84938 (accessed on 8 May 2021).
2. Shahid, A.; Huang, H.; Bhatti, M.M.; Zhang, L.; Ellahi, R. Numerical investigation on the swimming of gyrotactic microorganisms in nanofluids through porous medium over a stretched surface. *Mathematics* **2020**, *8*, 380. [CrossRef]
3. Turkyilmazoglu, M. Anomalous heat transfer enhancement by slip due to nanofluids in circular concentric pipes. *Int. J. Heat Mass Transfer* **2015**, *85*, 609–614. [CrossRef]
4. Xu, C.; Xu, S.; Wei, S.; Chen, P. Experimental investigation of heat transfer for pulsating flow of GOPs-water nanofluid in a microchannel. *Int. Commun. Heat Mass Transfer* **2015**, *110*, 104403. [CrossRef]
5. Dogonchi, A.S.; Ganji, D.D. Investigation of MHD nanofluid flow and heat transfer in a stretching/shrinking convergent/divergent channel considering thermal radiation. *J. Mol. Liq.* **2016**, *220*, 592–603. [CrossRef]
6. Reddy, P.S.; Sreedevi, P. Impact of chemical reaction and double stratification on heat and mass transfer characteristics of nanofluid flow over porous stretching sheet with thermal radiation. *Int. J. Ambient Energy* **2020**, *126*, 1–11. [CrossRef]

7. Arrigo, R.; Bellavia, S.; Gambarotti, C.; Dintcheva, N.T.; Carroccio, S. Carbon nanotubes-based nanohybrids for multifunctional nanocomposites. *J. King Saud Univ. Sci.* **2017**, *29*, 502–509. [CrossRef]
8. Khan, U.; Ahmed, N.; Asadullah, M.; Mohyud-din, S.T. Effects of viscous dissipation and slip velocity on two-dimensional and axisymmetric squeezing flow of Cu-water and Cu-kerosene nanofluids. *Propul. Power Res.* **2015**, *4*, 40–49. [CrossRef]
9. Rehman, A.U.; Mehmood, R.; Nadeem, S.; Akbar, N.S.; Motsa, S.S. Effects of single and multi-walled carbon nano tubes on water and engine oil based rotating fluids with internal heating. *Adv. Powder Technol.* **2017**, *28*, 1991–2002. [CrossRef]
10. Hassan, M.; Marin, M.; Ellahi, R.; Alamri, S.Z. Exploration of convective heat transfer and flow characteristics synthesis by CuAg/water hybridnanofluids. *Heat Transfer Res.* **2018**, *49*, 1837–1848. [CrossRef]
11. Ellahi, R.; Sait, S.M.; Shehzad, N.; Ayaz, Z. A hybrid investigation on numerical and analytical solutions of electro-magnetohydrodynamics flow of nanofluid through porous media with entropy generation. *Int. J. Numer. Methods Heat Fluid Flow* **2019**, *30*, 834–854. [CrossRef]
12. Raza, M.; Ellahi, R.; Sait, S.M.; Sarafraz, M.M.; Shadloo, M.S.; Waheed, I. Enhancement of heat transfer in peristaltic flow in a permeable channel under induced magnetic field using different CNTs. *J. Therm. Anal. Calorim.* **2020**, *140*, 1277–1291. [CrossRef]
13. Majeed, A.; Zeeshan, A.; Bhatti, M.M.; Ellahi, R. Heat transfer in magnetite (Fe3O4) nanoparticles suspended in conventional fluids: Refrigerant-134A (C2H2F4), kerosene (C10H22), and water (H2O) under the impact of dipole. *Heat Transfer Res.* **2020**, *51*, 217–232. [CrossRef]
14. Hafeez, A.; Khan, M.; Ahmed, J. Flow of Oldroyd-B fuid over a rotating disk with Cattaneo-Christov theory for heat and mass fuxes. *Comput. Methods Programs Biomed.* **2020**, *191*, 1053–1074. [CrossRef] [PubMed]
15. Nazari, S.; Ellahi, R.; Sarafraz, M.M.; Safaei, M.R.; Asgari, A.; Akbari, O.A. Numerical study on mixed convection of a non-Newtonian nanofuid with porous media in a two lid-driven square cavity. *J. Term. Anal. Calorim.* **2020**, *140*, 1121–1145. [CrossRef]
16. Hsu, J.-P.; Kao, C.-Y.; Tseng, S.; Chen, C.-J. Electrokinetic flow through an elliptical microchannel: Effects of aspect ratio and electrical boundary conditions. *J. Colloid Interface Sci.* **2002**, *248*, 176–184. [CrossRef]
17. Siddiqui, A.M.; Manzoor, N.; Maqbool, K.; Mann, A.B.; Shaheen, S. Magnetohydrodynamic fow induced by ciliary movement: An application to lower respiratory track diseases. *J. Magn. Magn. Mater.* **2019**, *480*, 164–170. [CrossRef]
18. Hafidzuddin, M.E.; Nazar, R.; Arifin, N.M. Application of the Keller-box method to magnetohydrodynamic rotating flow over a permeable shrinking surface. *Embrac. Math. Divers.* **2019**, *130*, 36–41. [CrossRef]
19. Subhani, M.; Nadeem, S. Numerical investigation into unsteady magnetohydrodynamics flow of micropolar hybrid nanofluid in porous medium. *Phys. Scr.* **2019**, *94*, 105–120. [CrossRef]
20. Lokesh, H.J.; Gireesha, B.J.; Kumar, K.G. Characterization of chemical reaction on magnetohydrodynamics flow and nonlinear radiative heat transfer of Casson nanoparticles over an exponentially sheet. *J. Nanofluids* **2019**, *8*, 1260–1266. [CrossRef]
21. Khashi'ie, N.S.; Arifin, N.M.; Nazar, R.; Hafidzuddin, E.H.; Wahi, N.; Pop, I. Magnetohydrodynamics (MHD) axisymmetric flow and heat transfer of a hybrid nanofluid past a radially permeable stretching/shrinking sheet with Joule heating. *Chin. J. Phys.* **2020**, *64*, 251–263. [CrossRef]
22. Tlili, I.; Nabwey, H.A.; Ashwinkumar, G.P.; Sandeep, N. 3-D magnetohydrodynamic AA7072-AA7075/methanol hybrid nanofluid flow above an uneven thickness surface with slip effect. *Sci. Rep.* **2020**, *10*, 13. [CrossRef] [PubMed]
23. Khan, U.; Zaib, A.; Khan, I.; Nisar, K.S. Activation energy on MHD flow of titanium alloy (Ti6Al4V) nanoparticle along with a cross flow and streamwise direction with binary chemical reaction and non-linear radiation: Dual solutions. *J. Mater. Res. Technol.* **2020**, *9*, 188–199. [CrossRef]
24. Goodarzi, M.; Toghraie, D.; Reiszadeh, M.; Afrand, M. Experimental evaluation of dynamic viscosity of ZnOMWCNTs/engine oil hybrid nanolubricant based on changes in temperature and concentration. *J. Therm. Anal. Calorim.* **2019**, *30*, 513–525. [CrossRef]
25. Bovand, M.; Rashidi, S.; Ahmadi, G.; Esfahani, J.A. Effects of trap and reflect particle boundary conditions on particle transport and convective heat transfer for duct flow-A two-way coupling of Eulerian-Lagrangian model. *Appl. Therm. Eng.* **2016**, *108*, 368–377. [CrossRef]
26. Bovand, M.; Rashidi, S.; Esfahani, J.A. Optimum interaction between magnetohydrodynamics and nanofluid for thermal and drag management. *J. Thermophys. Heat Transfer* **2017**, *31*, 218–229. [CrossRef]
27. Waini, I.; Ishak, A.; Pop, I. Hybrid nanofluid flow towards a stagnation point on a stretching/shrinking cylinder. *Sci. Rep.* **2020**, *10*, 92–96. https://doi.org/10.1038/s41598-020-66126-2. [CrossRef]
28. Waini, I.; Ishak, A.; Pop, I. Hybrid nanofluid flow and heat transfer over a permeable biaxial stretching/shrinking sheet. *Int. J. Numer. Meth. Heat Fluid Flow* **2019**, *30*, 3497–3513. [CrossRef]
29. Waini, I.; Ishak, A.; Pop, I. Squeezed hybrid nanofluid flow over a permeable sensor surface. *Mathematics* **2020**, *8*, 898–905. [CrossRef]
30. Khan, M.S.; Mei, S.; Fernandez-Gamiz, U.; Noeiaghdam, S.; Shah, S.A.; Khan, A. Numerical Analysis of Unsteady Hybrid Nanofluid Flow Comprising CNTs-Ferrousoxide/Water with Variable Magnetic Field. *Nanomaterials* **2022**, *12*, 180. [CrossRef]
31. Shah, R.A.; Ullah, H.; Khan, M.S.; Khan, A. Parametric analysis of the heat transfer behavior of the nano-particle ionic-liquid flow between concentric cylinders. *Adv. Mech. Eng.* **2021**, *13*, 16878140211024009. [CrossRef]
32. Shah, R.A.; Anjum, M.N.; Khan, M.S. Analysis of unsteady squeezing flow between two porous plates with variable magnetic field. *Int. J. Adv. Eng. Manag. Sci.* **2017**, *3*, 239–756.

33. Khan, A.; Shah, R.A.; Alam, M.K.; Rehman, S.; Shahzad, M.; Almad, S.; Khan, M.S. Flow dynamics of a time-dependent non-Newtonian and non-isothermal fluid between coaxial squeezing disks. *Adv. Mech. Eng.* **2021**, *13*, 16878140211033370. [CrossRef]
34. Yahaya, R.I.; Arifin, N.M.; Nazar, R.; Pop, I. Flow and heat transfer past a permeable stretching/shrinking sheet in Cu-Al2O3/water hybrid nanofluid. *Int. J. Numer. Meth. Heat Fluid Flow* **2019**, *130*, 2345–2353. [CrossRef]
35. Khan, M.S.; Rehan, A.S.; Amjad, A.; Aamir, K. Parametric investigation of the NernstPlanck model and Maxwells equations for a viscous fluid between squeezing plates. *Bound. Value Probl.* **2019**, *2019*, 107. [CrossRef]
36. Chamkha, A.J.; Dogonchi, A.S.; Ganji, D.D. Magneto-hydrodynamic flow and heat transfer of a hybrid nanofluid in a rotating system among two surfaces in the presence of thermal radiation and Joule heating. *AIP Adv.* **2019**, *9*, 025103. [CrossRef]
37. Khan, M.S.; Rehan, A.S.; Aamir, K. Effect of variable magnetic field on the flow between two squeezing plates. *Eur. Phys. J. Plus* **2019**, *134*, 219. [CrossRef]

Article

Measurement of the Photothermal Conversion Efficiency of CNT Films Utilizing a Raman Spectrum

Yu Liu [1], Zhicheng Lin [1], Pengfei Wang [1], Feng Huang [1] and Jia-Lin Sun [2,*]

[1] College of Mechanical Engineering and Automation, Fuzhou University, Fuzhou 350108, China; liuyu19@fzu.edu.cn (Y.L.); 15160454010@163.com (Z.L.); wangpf@fzu.edu.cn (P.W.); huangf@fzu.edu.cn (F.H.)
[2] State Key Laboratory of Low-Dimensional Quantum Physics, Department of Physics, Tsinghua University, Beijing 100084, China
* Correspondence: jlsun@tsinghua.edu.cn

Abstract: Because carbon nanotube (CNT) films have high photothermal conversion efficiency (PTCE), they have been widely used in bolometric and photothermoelectric photodetectors, seawater desalination, and cancer therapy. Here, we present a simple, quick, and non-destructive method to measure the PTCE of CNT films. According to the linear relationship between the Raman shift of the G^+ peak and the temperature of a CNT, the offset of the G^+ peak under varying excitation light power can characterize the changed temperature. Combining the simulation of the temperature distribution, the final value of the PTCE can be obtained. Finally, a CNT film with a high PTCE was chosen to be fabricated as a bolometric photodetector; a quite high responsivity (2 A W^{-1} at 532 nm) of this device demonstrated the effectiveness of our method.

Keywords: CNT film; Raman shift; photothermal conversion efficiency

1. Introduction

Carbon nanotubes (CNTs) are widely used in seawater desalination [1–6], environmental protection [7], thermal energy storage [8–11], photothermal dynamic therapy [12–14], and photodetectors [15–18] because of their efficient photothermal conversion ability. However, the synthesis process of CNTs is difficult to control strictly, which often induces complex morphologies and chirality distribution; simultaneously, some defects and impurities are also introduced. These factors will lead to significant differences between the values of the photothermal conversion efficiency (PTCE) of CNTs. Therefore, the measurement of the PTCE before practical applications is a critical step that can effectively help to identify the CNT materials with a high PTCE.

Previously, the general methods for measuring PTCE used infrared cameras or infrared photodetectors to observe the temperature changes. These methods always exhibited some drawbacks of low accuracy, small dynamic range, poor resolution, and great damage [19–21]. Here, we proposed a simple, quick, and non-destructive method to measure the PTCE of CNT films based on a Raman spectrum. Due to the linear relationship between the Raman shift of the G^+ peak and the temperature of a CNT [22–24], the changed temperature of a CNT film under laser excitation can be equivalently expressed as the offset of the G^+ peak. The value of the PTCE was calculated through integrating the simulated temperature distribution. The results indicated that the impurities and defects can significantly reduce the PTCE of CNT films and fibers. The PTCE of a pure CNT film was 4.8 times larger than that of a dirty one. In another sample, the PTCE for a position with few defects was 2.5 times larger than that for a position with many defects. This method was able to very quickly compare the PCTE values among different CNT materials, and to identify ones with high PTCE. Finally, a CNT film with the highest PTCE was used to fabricate a bolometric photodetector. The responsivity of the device under illumination of 532 nm laser reached 2 A W^{-1}, which was the highest among the bolometric photodetectors based on CNTs [15,16,25].

2. Materials and Methods

The CNT films used in this paper were synthesized following the CVD method in a previous report, and the process of fabricating a suspended structure also followed that work [26]. The CNT fibers were drawn from an as-prepared CNT film. Then, the CNT film or fibers were stretched between two substrates. Alcohol was used to wet the CNT materials to promote intimate contact between the CNT materials and the substrates (deposited Au/Ti on SiO$_2$) [15].

The relationship between the Raman shift of the G$^+$ peak and the temperature of a CNT film was calibrated by a high-resolution Raman spectrometer (HORIBA T64000, Kyoto, Japan) with a tunable thermostatic sample cavity. Then, the Raman shift of the G$^+$ peaks under different excitation powers were measured by a Laser Confocal Raman Microscope (WITec Alpha 300RAS, Stuttgart, Germany). A laser with a wavelength of 532 nm was used as an excitation light, which heated and also probed the samples. An objective (Zeiss EC Epiplan- Neofluar 100×/0.9, Oberkochen, Germany) was used to focus the laser beam to a spot with a diameter of 0.5 μm. The integration time was 0.7 s and the number of accumulations was 10. A SourceMeter (Keithley 2400, Beaverton, OR, USA) unit was used to characterize the photoresponse of the bolometric photodetector under the illumination of a 532 nm laser.

3. Results

Several previous articles demonstrated that the temperature of a single CNT is linearly related to the Raman shift of the G$^+$ peak [22–24]. Further, the temperature of a CNT film remained linearly related to the Raman shift of the G$^+$ peak [27]. Here, we also measured the Raman shift of the G$^+$ peak under different temperatures, as Figure 1 shows [28]. The Raman shift of the G$^+$ peak exhibited a linear relationship with the temperature, with a slope of −0.031 cm^{-1} K^{-1}, which was a little larger than the previously reported value [27]. This difference in slopes is possibly due to the suspended structure, because the stress within suspended CNT film is larger than that in a supported structure, and the stress can also change the Raman shift of the G$^+$ peak [29].

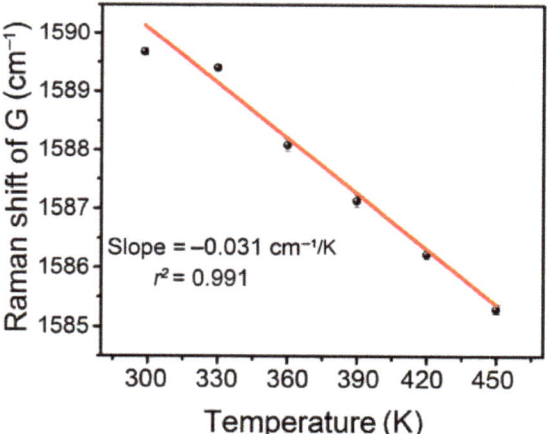

Figure 1. Raman shift of G$^+$ peak as a function of the temperature for CNT films.

Figure 2 shows the optical images and the Raman spectral under different excitation light powers for Sample 1 and Sample 2. Figure 2b,d shows that both of the CNT films have the same experimental phenomenon, i.e., that a higher excitation light power induces a red shift of the G$^+$ peak. If we define a proportionality coefficient $R = \Delta\omega/\Delta p$, where $\Delta\omega$ and Δp represent the offset value of the G$^+$ peak and the increased light power respectively, the R of Sample 1 (8.47 cm^{-1} mW^{-1}) is almost five times larger than that of Sample 2

(1.77 cm^{-1} mW^{-1}). Therefore, together with the result in Figure 1, the ratio of the increased temperature to the light power at the tested point in Sample 1 was obtained as 273 K mW^{-1} and the value in Sample 2 was 57 K mW^{-1}. However, when the changed temperature was high enough, the relationship would be non-linear; this condition was not considered in this work.

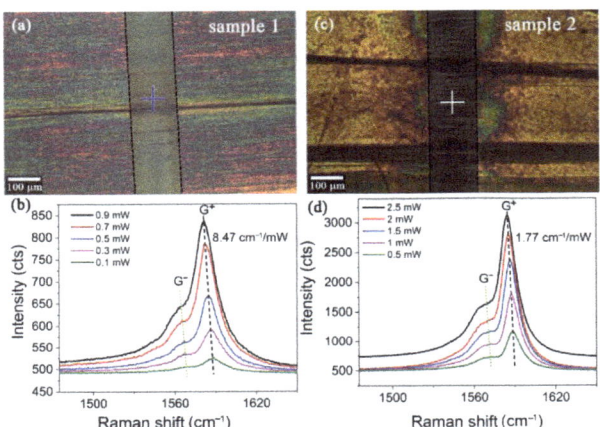

Figure 2. (a) Optical image of Sample 1; (b) Raman spectra of Sample 1 under different excitation powers; (c) Optical image of Sample 2; (d) Raman spectra of Sample 2 under different excitation powers. The middle part between the two black dash lines is the suspended CNT film. The widths of the trenches for Sample 1 and Sample 2 are both 120 μm. The crosses mark the probe points.

The experimental results indicated that Sample 1 could better convert the incident photon energy into heat. Comparing the optical images of Sample 1 and Sample 2, it is obvious that the surface of Sample 1 is cleaner, but a large number of impurities (metal catalysts and amorphous carbon) can be observed on the surface of Sample 2 (see Figure S1a,c). In addition, Figure 3 exhibits that under the same excitation light power, the intensity of the G$^+$ peak in Sample 2 is much larger than that in Sample 1, indicating that these impurities on the surface can greatly enhance the Raman scattering intensity. During the synthesis process of the CNT films, iron and copper nanoparticles were introduced as catalysts (see Figure S1b,d). Hence, the surface-enhanced Raman scattering from iron nanoparticles could dominate the enhanced intensity of the G$^+$ peak for Sample 2 [30–33]. According to previous reports, the ratio of ω_G^- (the Raman shift of the G$^-$ peak) to the temperature is the same as with ω_G^+, which was also demonstrated in our work, as the green dash lines show in Figure 2b,d [22–24]. Together, these results indicate that impurities could dramatically weaken the ability of photothermal conversion for a CNT film.

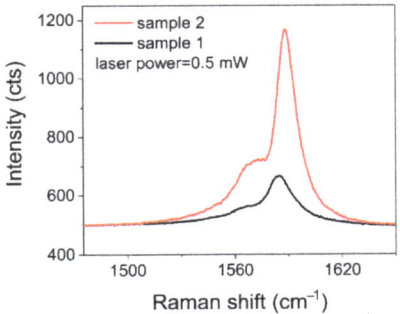

Figure 3. Raman spectra of Sample 1 and Sample 2.

The same experiments were also conducted on a CNT fiber noted as Sample 3, and the experimental results are shown in Figure 4. The R of point 1 in Sample 3 is greater than that of point 2. The Raman spectra of point 2 display some obvious shoulder peaks that show a linear relationship with the excitation light power, but the slope is different from the G^+ peak. Hence, these shoulder peaks were possibly due to defects. Some reported works demonstrated that the existence of defects and impurities in carbon nanotubes had a greatly negative impact on photothermal conversion [34], which was very consistent with our experimental results. According to the optical image of Sample 3, it is obvious that point 2 is closer to the edge of the suspended film. Therefore, we assumed that the defects may be introduced by the strong tensile effect at the edge, or by damage during the transfer processes [35].

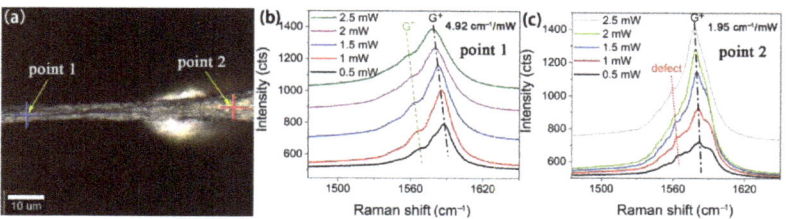

Figure 4. (a) Optical image of Sample 3, in which the position of blue cross marks point 1 and the position of red cross marks point 2; (b) Raman spectra of point 1 at different excitation light powers; (c) Raman spectra of point 2 at different excitation light powers.

4. Discussion

The photothermal conversion efficiency η of CNT film can be expressed as:

$$\eta = Q/P \tag{1}$$

where Q is the heat generating power of the CNT film (generated heat per unit time) and P is the incident light power [36].

Under the illumination of a continuing laser, a CNT film absorbs the light energy and converts it into heat. Some of the heat can transfer to the surrounding air, and the remaining heat will increase the temperature of the CNT film. Therefore, the stored heat within the CNT film per unit time can be expressed as:

$$Q - Q_{surr} = c \times \int_{\Sigma} \rho \cdot h \cdot (dT(x,y)/dt) \cdot ds \tag{2}$$

where Q_{surr} is the heat dissipation power from the CNT film to air, c is the specific heat capacity of the CNT film, ρ is the density of the CNT film, h is the thickness of the CNT film, $T(x, y)$ is the temperature distribution of the CNT film, and t is time.

After a fast process of increasing temperature, the heat generated by the CNT film and the heat dissipating to air will become equal. Therefore, the temperature of the CNT film will remain stable. The heat-generating power at the thermal equilibrium state can be expressed as:

$$Q = Q_{surr} = 2G_0 \times \int_{\Sigma} (T - T_0) \cdot ds \tag{3}$$

where G_0 is the heat transfer coefficient of air and T_0 is the room temperature, 300 K. Therefore, the PTCE can be expressed as the ratio of Q_{surr} to incident light power P.

However, the calculation of an accurate solution to the temperature distribution $T(x, y)$ at a steady state is very complicated. Here, we obtained the temperature distribution through numerical simulations with the following conditions: the length of the CNT film along the x axis was 120 μm; the width of the CNT film along the y axis was infinite; the thickness of the CNT film was 0.15 μm; the thermal conductivity of the CNT film was 6 W m^{-1} K^{-1} in the xy-plane and 0.1 W m^{-1} K^{-1} on the z axis; the density of the CNT film was 850 kg cm^{-3}; the specific heat capacity of the CNT film was 1200 J kg^{-1} K^{-1}; the heat transfer coefficient of air was 140 W m^{-2} K^{-1}; the excitation laser power was 0.1 mW, the laser spot diameter was 0.5 µm; and the temperature of the air was maintained at a constant 300 K [37]. Because the length of the trench was much larger than the diameter of the laser spot, an assumption was introduced that the generated heat would completely transfer to the air before arriving at the substrate, and the influence of the substrate on the final temperature distribution could be neglected. This assumption proved to be true, as Figure S2 shows. The heat power transferred from the CNT film to the substrate was only 3% of the total heat dissipation, which demonstrated that the influence of the substrate was weak enough to be ignored. Figure 5a,c displays the simulated results for the temperature distribution. Figure 5b,d shows the temperature distributions of Sample 1 and Sample 2 along the x axis at the center. Considering that the temperature distribution under the illumination of the point light source indicates central symmetry, Equation (3) can be rewritten as:

$$Q = Q_{surr} = 2G_0 \times \int_0^{l/2} (T - T_0) 2\pi r \cdot dr \qquad (4)$$

where l is the length of the film along x axis. The PTCEs of Sample 1 (η_1 = 0.26) and Sample 2 (η_2 = 0.05) can be calculated by Equation (4). The average temperature within the laser light spot were consistent with the results calculated based on the value of R. Because the impurities in Sample 2 led to a reduced in-plane thermal conductivity, the real PTEC of Sample 2 should be smaller.

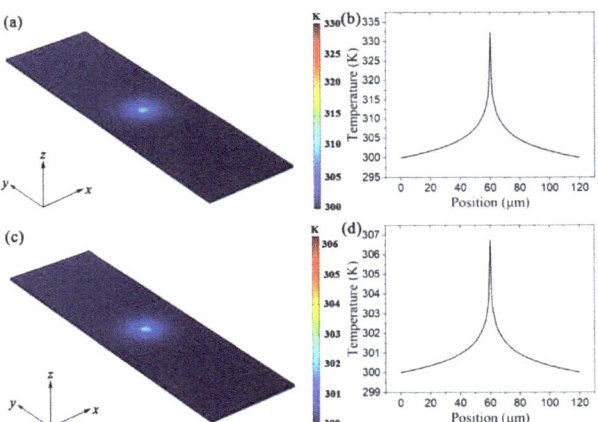

Figure 5. Simulated in-plane temperature distribution for (**a**) Sample 1, and (**c**) Sample 2. The temperature distributions of (**b**) Sample 1 and (**d**) Sample 2 along x axis at the center.

According to the above theoretical analyses and discussion, we tried to fabricate a bolometric photodetector based on a suspended CNT film with extremely high PTCE. We chose a thicker CNT film to reduce the defects caused by transfer, and then soaked the CNT film in a hydrogen peroxide solution and ethanol successively to remove the impurity (metal catalysts and amorphous carbon) in the films. Because the thickness of the CNT film was increased, the volt-ampere characteristic curve had a large linear range, as shown in Figure 6a. The photoresponse curve of the device under the illumination of a 532 nm laser is shown in Figure 6b. The device exhibited a responsivity of up to 2 A W^{-1}, and the resistance change rate of the device was 4% mW^{-1}, while the response time was only 0.2 ms. The performance of the device was much better than that of many photodetectors that only contained carbon nanotubes [15,16,25,37,38].

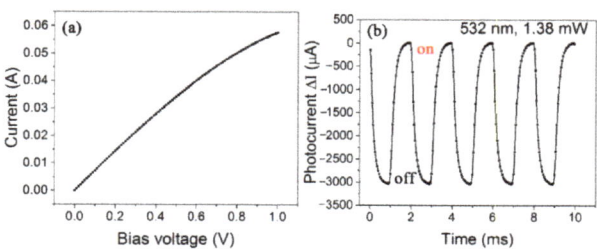

Figure 6. (**a**) *I–V* characteristic curve, whose linear range was extended to 0.8 V; (**b**) Photoresponse curve under 532 nm laser illumination in air; bias voltage was 0.8 V. The dark current was subtracted.

5. Conclusions

In conclusion, we proposed a simple, quick, and non-destructive method to measure the PTCE of different CNT films. Combining the increased temperature within the light spot and the simulated temperature distribution of the CNT films, the final values of the PTCE were calculated accurately. In addition, the impurities from metal catalysis and defects on the CNT film were demonstrated to have a great negative influence on the PTCE of the CNT film. Finally, a CNT film with the highest PTCE was identified by this method to be fabricated as a bolometric photodetector; the high performance (2 A W^{-1} at 532 nm) confirmed the effectiveness of the method. With the expanding applications of CNTs, this measurement method has huge potential in the future.

Supplementary Materials: The following supporting information can be downloaded at: https://www.mdpi.com/article/10.3390/nano12071101/s1, Figure S1: Characterization results of SEM for Sample 1 and Sample 2. Figure S2: Influence of the substrate on the temperature distribution.

Author Contributions: Conceptualization, Y.L. and Z.L.; methodology, Y.L.; software, Z.L.; validation, Y.L., Z.L. and P.W.; investigation, P.W.; resources, F.H.; data curation, Y.L.; writing—original draft preparation, Y.L. and Z.L.; writing—review and editing, Y.L. and Z.L.; supervision, J.-L.S. and F.H.; All authors have read and agreed to the published version of the manuscript.

Funding: This work was supported by the Joint Fund of National Natural Science Foundation of China and China Academy of Engineering Physics (NSAF No. U1730246).

Institutional Review Board Statement: Not applicable.

Informed Consent Statement: Not applicable.

Data Availability Statement: The data presented in this study are available on request from the corresponding author.

Conflicts of Interest: The authors declare no conflict of interest.

References

1. Hu, T.; Chen, K.; Li, L.; Zhang, J. Carbon nanotubes@silicone solar evaporators with controllable salt-tolerance for efficient water evaporation in a closed system. *J. Mater. Chem. A* **2021**, *9*, 17502–17511. [CrossRef]
2. Jian, H.; Qi, Q.; Wang, W.; Yu, D. A Janus porous carbon nanotubes/poly (vinyl alcohol) composite evaporator for efficient solar-driven interfacial water evaporation. *Sep. Purif. Technol.* **2021**, *264*, 118459. [CrossRef]
3. Qin, D.-D.; Zhu, Y.-J.; Chen, F.-F.; Yang, R.-L.; Xiong, Z.-C. Self-floating aerogel composed of carbon nanotubes and ultralong hydroxyapatite nanowires for highly efficient solar energy-assisted water purification. *Carbon* **2019**, *150*, 233–243. [CrossRef]
4. Zhang, Q.; Xu, W.; Wang, X. Carbon nanocomposites with high photothermal conversion efficiency. *Sci. China Mater.* **2018**, *61*, 905–914. [CrossRef]
5. Zhao, Y.; Yuan, H.; Zhang, X.; Xue, G.; Tang, J.; Chen, Y.; Zhang, X.; Zhou, W.; Liu, H. Laser-assisted synthesis of cobalt@N-doped carbon nanotubes decorated channels and pillars of wafer-sized silicon as highly efficient three-dimensional solar evaporator. *Chin. Chem. Lett.* **2021**, *32*, 3090–3094. [CrossRef]
6. Zhu, B.; Kou, H.; Liu, Z.; Wang, Z.; Macharia, D.K.; Zhu, M.; Wu, B.; Liu, X.; Chen, Z. Flexible and Washable CNT-Embedded PAN Nonwoven Fabrics for Solar-Enabled Evaporation and Desalination of Seawater. *ACS Appl. Mater. Interfaces* **2019**, *11*, 35005–35014. [CrossRef] [PubMed]

7. Liang Hu, S.G. Photothermal-Responsive Single-Walled Carbon Nanotube-Based Ultrathin Membranes for OnOff Switchable Separation of Oil-in-Water Nanoemulsions. *ACS Nano* **2015**, *9*, 4835–4842.
8. Chen, Y.; Zhang, Q.; Wen, X.; Yin, H.; Liu, J. A novel CNT encapsulated phase change material with enhanced thermal conductivity and photo-thermal conversion performance. *Sol. Energy Mater. Sol. Cells* **2018**, *184*, 82–90. [CrossRef]
9. Gao, G.; Zhang, T.; Guo, C.; Jiao, S.; Rao, Z. Photo-thermal conversion and heat storage characteristics of multi-walled carbon nanotubes dispersed magnetic phase change microcapsules slurry. *Int. J. Energy Res.* **2020**, *44*, 6873–6884. [CrossRef]
10. Gimeno-Furió, A.; Martínez-Cuenca, R.; Mondragón, R.; Gasulla, A.F.V.; Doñate-Buendía, C.; Mínguez-Vega, G.; Hernández, L. Optical characterisation and photothermal conversion efficiency of a water-based carbon nanofluid for direct solar absorption applications. *Energy* **2020**, *212*, 118763. [CrossRef]
11. Li, B.; Nie, S.; Hao, Y.; Liu, T.; Zhu, J.; Yan, S. Stearic-acid/carbon-nanotube composites with tailored shape-stabilized phase transitions and light–heat conversion for thermal energy storage. *Energy Convers. Manag.* **2015**, *98*, 314–321. [CrossRef]
12. Behnam, M.A.; Emami, F.; Sobhani, Z.; Koohi-Hosseinabadi, O.; Dehghanian, A.R.; Zebarjad, S.M.; Moghim, M.H.; Oryan, A. Novel Combination of Silver Nanoparticles and Carbon Nanotubes for Plasmonic Photo Thermal Therapy in Melanoma Cancer Model. *Adv. Pharm Bull.* **2018**, *8*, 49–55. [CrossRef]
13. Hashida, Y.; Tanaka, H.; Zhou, S.; Kawakami, S.; Yamashita, F.; Murakami, T.; Umeyama, T.; Imahori, H.; Hashida, M. Photothermal ablation of tumor cells using a single-walled carbon nanotube-peptide composite. *J. Control. Release* **2014**, *173*, 59–66. [CrossRef]
14. Murakami, T.; Nakatsuji, H.; Inada, M.; Matoba, Y.; Umeyama, T.; Tsujimoto, M.; Isoda, S.; Hashida, M.; Imahori, H. Photodynamic and photothermal effects of semiconducting and metallic-enriched single-walled carbon nanotubes. *J. Am. Chem. Soc.* **2012**, *134*, 17862–17865. [CrossRef]
15. Liu, Y.; Hu, Q.; Yin, J.; Wang, P.; Wang, Y.; Wen, J.; Dong, Z.; Zhu, J.-L.; Wei, J.; Ma, W.; et al. Bolometric terahertz detection based on suspended carbon nanotube fibers. *Appl. Phys. Express* **2019**, *12*, 096505. [CrossRef]
16. Liu, Y.; Yin, J.; Wang, P.; Hu, Q.; Wang, Y.; Xie, Y.; Zhao, Z.; Dong, Z.; Zhu, J.L.; Chu, W.; et al. High-Performance, Ultra-Broadband, Ultraviolet to Terahertz Photodetectors Based on Suspended Carbon Nanotube Films. *ACS Appl. Mater. Interfaces* **2018**, *10*, 36304–36311. [CrossRef] [PubMed]
17. He, X.; Fujimura, N.; Lloyd, J.M.; Erickson, K.J.; Talin, A.A.; Zhang, Q.; Gao, W.; Jiang, Q.; Kawano, Y.; Hauge, R.H.; et al. Carbon nanotube terahertz detector. *Nano. Lett.* **2014**, *14*, 3953–3958. [CrossRef] [PubMed]
18. Zhang, Y.; Deng, T.; Li, S.; Sun, J.; Yin, W.; Fang, Y.; Liu, Z. Highly sensitive ultraviolet photodetectors based on single wall carbon nanotube-graphene hybrid films. *Appl. Surf. Sci.* **2020**, *512*, 145651. [CrossRef]
19. Harata, A.; Aono, M.; Kitazawa, N.; Watanabe, Y. Correlation of photothermal conversion on the photo-induced deformation of amorphous carbon nitride films prepared by reactive sputtering. *Appl. Phys. Lett.* **2014**, *105*, 051905. [CrossRef]
20. Tam, N.T.; van Trinh, P.; Anh, N.N.; Hong, N.T.; Hong, P.N.; Minh, P.N.; Thang, B.H. Thermal Conductivity and Photothermal Conversion Performance of Ethylene Glycol-Based Nanofluids Containing Multiwalled Carbon Nanotubes. *J. Nanomater.* **2018**, *2018*, 2750168. [CrossRef]
21. Zhou, L.; Wang, X.; Zhang, J.; Yang, S.; Hao, K.; Gao, Y.; Li, D.; Li, Z. Self-suspended carbon nanotube/polyimide composite film with improved photothermal properties. *J. Appl. Phys.* **2020**, *127*, 205103. [CrossRef]
22. Li, Q.; Liu, C.; Wang, X.; Fan, S. Measuring the thermal conductivity of individual carbon nanotubes by the Raman shift method. *Nanotechnology* **2009**, *20*, 145702. [CrossRef] [PubMed]
23. Wu, Q.; Wen, Z.; Zhang, X.; Tian, L.; He, M. Temperature Dependence of G− Mode in Raman Spectra of Metallic Single-Walled Carbon Nanotubes. *J. Nanomater.* **2018**, *2018*, 1–6. [CrossRef]
24. Zhang, Y.; Xie, L.; Zhang, J.; Wu, Z.; Liu, Z. Temperature Coefficients of Raman Frequency of Individual. *J. Phys. Chem. C* **2007**, *111*, 14031–14034. [CrossRef]
25. Itkis, M.E.; Borondics, F.; Yu, A.; Haddon, R.C. Bolometric infrared photoresponse of suspended single-walled carbon nanotube films. *Science* **2006**, *312*, 413–416. [CrossRef]
26. Li, Z.; Jia, Y.; Wei, J.; Wang, K.; Shu, Q.; Gui, X.; Zhu, H.; Cao, A.; Wu, D. Large area, highly transparent carbon nanotube spiderwebs for energy harvesting. *J. Mater. Chem.* **2010**, *20*, 7236–7240. [CrossRef]
27. Duzynska, A.; Taube, A.; Korona, K.P.; Judek, J.; Zdrojek, M. Temperature-dependent thermal properties of single-walled carbon nanotube thin films. *Appl. Phys. Lett.* **2015**, *106*, 183108. [CrossRef]
28. Liu, Y.; Hu, Q.; Wang, P.; Wei, J.; Huang, F.; Sun, J.L. Electrically driven transport of photoinduced hot carriers in carbon nanotube fibers. *Opt. Lett.* **2021**, *46*, 5228–5231. [CrossRef]
29. Gao, B.; Jiang, L.; Ling, X.; Zhang, J.; Liu, Z. Chirality-Dependent Raman Frequency Variation of Single-Walled Carbon Nanotubes under Uniaxial Strain. *J. Phys. Chem. C* **2008**, *112*, 20123–20125. [CrossRef]
30. Liu, M.; Xiang, R.; Cao, W.; Zeng, H.; Su, Y.; Gui, X.; Wu, T.; Maruyama, S.; Tang, Z. Is it possible to enhance Raman scattering of single-walled carbon nanotubes by metal particles during chemical vapor deposition? *Carbon* **2014**, *80*, 311–317. [CrossRef]
31. Shao, Q.; Que, R.; Shao, M.; Cheng, L.; Lee, S.-T. Copper Nanoparticles Grafted on a Silicon Wafer and Their Excellent Surface-Enhanced Raman Scattering. *Adv. Funct. Mater.* **2012**, *22*, 2067–2070. [CrossRef]
32. Cao, P.G.; Yao, J.L.; Ren, B.; Mao, B.W.; Gu, R.A.; Tian, Z.Q. Surface-enhanced Raman scattering from bare Fe electrode surfaces. *Chem. Phys. Lett.* **2000**, *316*, 1–5. [CrossRef]

33. Aghajani, S.; Accardo, A.; Tichem, M. Aerosol Direct Writing and Thermal Tuning of Copper Nanoparticle Patterns as Surface-Enhanced Raman Scattering Sensors. *ACS Appl. Nano Mater.* **2020**, *3*, 5665–5675. [CrossRef]
34. García-Merino, J.A.; Martínez-González, C.L.; Miguel, C.R.T.-S.; Trejo-Valdez, M.; Martínez-Gutiérrez, H.; Torres-Torres, C. Photothermal, photoconductive and nonlinear optical effects induced by nanosecond pulse irradiation in multi-wall carbon nanotubes. *Mater. Sci. Eng. B* **2015**, *194*, 27–33. [CrossRef]
35. Zhang, S.; Mielke, S.L.; Khare, R.; Troya, D.; Ruoff, R.S.; Schatz, G.C.; Belytschko, T. Mechanics of defects in carbon nanotubes: Atomistic and multiscale simulations. *Phys. Rev. B* **2005**, *71*, 115403. [CrossRef]
36. Li, Z.; Johnson, O.; Huang, J.; Feng, T.; Yang, C.; Liu, Z.; Chen, W. Enhancing the photothermal conversion efficiency of graphene oxide by doping with NaYF4: Yb, Er upconverting luminescent nanocomposites. *Mater. Res. Bull.* **2018**, *106*, 365–370. [CrossRef]
37. Suzuki, D.; Kawano, Y. Flexible terahertz imaging systems with single-walled carbon nanotube films. *Carbon* **2020**, *162*, 13–24. [CrossRef]
38. St-Antoine, B.C.; Menard, D.; Martel, R. Position Sensitive Photothermoelectric Effect in Suspended Single-Walled Carbon Nanotube Films. *Nano Lett.* **2009**, *9*, 3053–3508. [CrossRef]

Article

Effects of Molarity and Storage Time of MWCNTs on the Properties of Cement Paste

Echeverry-Cardona Laura [1], Cabanzo Rafael [2], Quintero-Orozco Jorge [3], Castillo-Cuero Harvi Alirio [4], Rodríguez-Restrepo Laura Victoria [1] and Restrepo-Parra Elisabeth [1,*]

[1] Laboratorio de Física del Plasma, Universidad Nacional de Colombia, Sede Manizales, Manizales 170001, Colombia
[2] Laboratorio de Espectroscopía Atómica y Molecular (LEAM), Centro de Materiales y Nanociencias (CMN), Parque Tecnológico Guatiguará, Universidad Industrial de Santander, Bucaramanga 681012, Colombia
[3] Ciencia de Materiales Biológicos y Semiconductores (CIMBIOS), Centro de Materiales y Nanociencias (CMN), Parque Tecnológico Guatiguará, Universidad Industrial de Santander, Bucaramanga 681012, Colombia
[4] Centro de Nanociencias y Nanotecnología, Universidad Nacional Autónoma de México, Km 107 Carretera Tijuana Ensenada, Ensenada 22860, Mexico
* Correspondence: erestrepopa@unal.edu.co; Tel.: +57-315-805-0903

Abstract: Nowadays, nanomaterials in cement pastes are among the most important topics in the cement industry because they can be used for several applications. For this reason, this work presents a study about the influence of changing the molarity of dispersed multiple wall carbon nanotubes (MWCNTs) and varying the number of storage days on the mechanical properties of the cement paste. To achieve this objective, dispersions of 0.35% MWCNTs, varying the molarity of the surfactant as 10 mM, 20 mM, 40 mM, 60 mM, 80 mM, and 100 mM, were performed. The mixture of materials was developed using the sonication process; furthermore, materials were analyzed using UV-Vis, Z-potential, and Raman spectroscopy techniques. Materials with a molarity of 10 mM exhibited the best results, allowing them to also be stored for four weeks. Regarding the mechanical properties, an increase in the elastic modulus was observed when MWCNTs were included in the cement paste for all storage times. The elastic modulus and the maximum stress increased as the storage time increased.

Keywords: MWCNT; dispersion; sonication energy; stability; cement paste

1. Introduction

Carbon nanotubes (CNTs) have been used in a variety of applications due to their versatility. Some of their applications are as additives in polymers, catalysts, autoelectron emission for cathodic rays in illumination components, absorption and filters of electromagnetic waves, energy conversion, anodes of lithium batteries, hydrogen storage, and sensors, among others [1]. CNTs are a carbon allotrope phase that possesses intermediate properties between graphite and fullerenes [1,2]. These materials are composed of sp^2 hybridization carbon bonds and can be produced as structures with a simple wall or multiple walls separated by around 0.35 nm [3]. In this sense, multiple wall carbon nanotubes (MWCNTs) have gained more attention due to their high performance and low cost of production per unit. In addition, their thermodynamic stability and capacity to sustain and improve the electrical properties make them excellent candidates for applications that require these special properties [3].

For instance, the cement industry has used MWCNTs as an additive in cement matrices to improve the electrical and mechanical properties [4]. MWCNTs are added in proportions of 0.2% of the cement weight to enhance the flexural strength; this is an important aspect that may have to be taken into account, because when the mixture is performed using traditional mechanical methods, if they are included in higher ratios, the MWCNT dispersion can present strong drawbacks such as exhibiting possible agglomerations and clusters [5]. To avoid these problems, the ultrasound technique is mostly used because it fragments the MWCNT agglomerations [6–9]. The dispersion is perhaps the most critical factor that influences the mechanical properties of cement pastes [10]. Some experiments have confirmed that MWCNTs can be effectively dispersed in water using ultrasound energy and commercial surfactants [11–13]. In the literature, it is possible to observe works focused on applying surfactants to sustain the dispersion [14]. For example, Mendoza [15] studied several dispersion concentrations of surfactants such as sodium lauryl sulfate, cetylpyridinium chloride, and Triton X-100, finding that the reinforcing effect of MWCNTs is masked by the negative effect of the surfactants.

The main limitations of using MWCNTs are (i) achieving a total dispersion in water and (ii) reaching the stability for a long storage time [16]. Because the van der Waals forces are responsible for these phenomena, the size of the agglomerates can reach the micrometer scale [17–19]. Furthermore, these agglomerations can cause a stress concentration because they behave as weak spots in the cement paste, reducing the fluidity of the material because they absorb the free water [20]. Because of this, different investigations and several approaches have been proposed for their dispersion [21–23]. Physical methods such as sonication and adsorption of the surfactant are the most currently used. Regarding the use of surfactants, they are considered a surface-active agent that has an amphipathic structure, containing a lyophobic (solvent repulsive) and a lyophilic (solvent attractive) group. It has been found that using low surfactant concentrations allows the molecules to be absorbed on the surface or interface, decreasing the interfacial tension and improving the dispersion [24]. This is due to the variation in the dielectric constant of the water depending on the surfactant type used; for example, if it is an ionic surfactant, the particles are stabilized via a repulsive electrostatic force, whereas if it is a nonionic surfactant, an interparticle repulsion via steric-hydration forces is produced [25].

Researchers have evaluated nonionic surfactants for dispersing MWCNTs, graphene, and graphene oxide, among others [26–32]; nevertheless, these studies have focused on using different surfactants without considering changes in the concentration of the surfactant. For example, Blanch et al. [25] reported than the increase in the surfactant concentrations above a certain value led to the flocculation of the CNTs, possibly due to the attractive depletion interactions. This effect generated a poor dispersion. The surfactant concentration influences the dispersion, generating an encapsulation of MWCNTs in cylindrical micelles, adsorption of hemimicelles, or random adsorption; then, each surfactant must be working below the critical micelle concentration. Although it has been demonstrated that optimal concentrations exist for nanotube dispersion [33], few studies have presented a more in-depth analysis of the influence of the surfactant concentration on nanotube dispersion.

In this work, we have produced dispersions of MWCNTs with Triton TX-100 surfactant for introduction into the cement matrix. A detailed analysis of the Triton X-100 concentration for values of 10, 20, 40, 60, 80, and 100 mM was carried out to find the optimal concentration. Firstly, UV-Vis, Z-potential, and Raman spectroscopy analyses were performed to determine the dispersion at different storage times (1, 2, and 4 weeks), and secondly, three types of cylinders of cement paste were produced: (i) without MWCNTs, (ii) with MWCNTs + TX-100 (one week of storage), and (iii) with MWCNTs + TX-100 (four weeks of storage).

2. Materials and Methods

2.1. Materials

For the samples built, a mixture of water type 1, MWCNTs, and Triton TX-100 was carried out.

Water type I was used because it is required to avoid elements that alter the electrical properties of the surfactant and MWCNTs.

Triton TX-100 was used as a surfactant because it exhibits a nonionic character and contributes negative charges; thus, it does not affect the electrostatic repulsion or attraction of the nanotubes.

Finally, industrial grade MWCNTs NC7000 produced by Belgium Nanocyl SA were used.

2.2. Dispersion Procedure

In order to identify the influence of the percentage of surfactant on the behavior of the mixture, the Triton TX-100 surfactant was used with molarities of 10 mM, 20 mM, 40 mM, 60 mM, 80 mM, and 100 mM. The percentage of the MWCNTs was chosen as 0.35%, according to the literature [15]. The steps performed during the materials production were:

- The TX-100 at different molarities (10 mM, 20 mM, 40 mM, 60 mM, 80 mM, and 100 mM) and water type 1 were mixed for 5 min using a magnetic stirrer at room temperature;
- After that, MWCNTs were added;
- The mixture was placed in the ultrasonic cube with a power of 500 W and 40% amplitude, applying an energy of 390 J/g;
- Sonication was performed with 20-s on/off cycles;
- The room temperature was kept constant using a cold bath that consists of immersing the beaker containing the mixture in a larger beaker containing a mixture of ice and water. The temperature was constantly measured and maintained at room temperature;
- Materials were stored for 1, 2, 4, 10, and 13 weeks;
- After that, the materials were characterized in order to determine the stability of the samples;
- The test cylinders were made from the mixtures stored for 1 and 4 weeks because for 2 weeks, the results were very similar to those of 1 week, and for 10 and 13 weeks, the material had already become unstable, according to the UV-Vis spectroscopy analysis.

The sonication time (t_{son}) was calculated with the relationship obtained by Mendoza-Reales [11]. According to this report, the time for dispersing 155 g is 60 min, with an energy ratio of 390 J/g. With these values, the calculated energy is 390 J/g × 155 g = 60,450 J. From this reference, the proposed time of sonication is:

$$t_{son} = \frac{((m_{water} + m_{dis} + m_{MWCNT}) * E_{dis}) * 60}{60,450 \, J} \quad (1)$$

where m_{water}, m_{dis}, and m_{MWCNT} are the masses of the water, dispersant, and MWCNTs, respectively, and E_{dis} is the dispersion energy. Table 1 shows the values of the total mass, dispersion energy, and sonication time for the experiments. A diagram of the experimental setup is shown in Figure 1.

Table 1. Mass of TXT-100, total mass of the mixture, and sonication time for each molarity of TXT-100, following the parameters: energy—390 J/g and 0.35% mass MWCNT of the sample.

Molarity of TXT-100 (mM)	Mass of TXT-100 (g)	Total Mass (g)	Sonication Time (min:s)
10	0.647 ± 0.001	100.997 ± 0.001	39:10 ± 0:0.01
20	1.29 ± 0.001	101.64 ± 0.001	39:35 ± 0:0.01
40	2.59 ± 0.001	102.94 ± 0.001	39:85 ± 0:0.01
60	3.88 ± 0.001	104.23 ± 0.001	40:35 ± 0:0.01
80	5.18 ± 0.001	105.53 ± 0.001	41:25 ± 0:0.01
100	6.47 ± 0.001	106.82 ± 0.001	41:35 ± 0:0.01

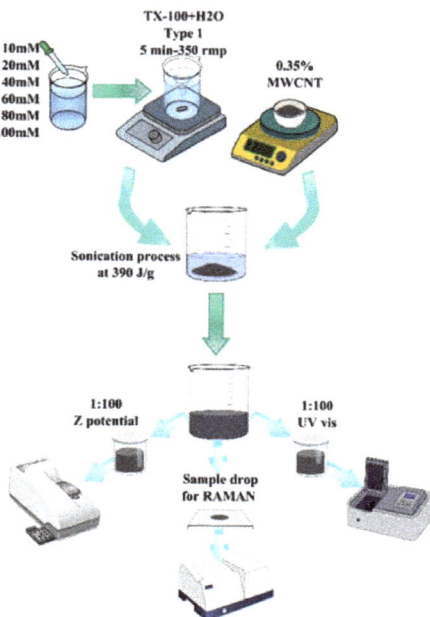

Figure 1. Scheme of the procedure to produce and analyze the dispersion of MWCNTs dissolved in type 1 water, varying the molarity of the surfactant between 10 mM and 100 mM.

2.3. Materials Characterization

A UV-Vis UV2600 (Shimadzu—Chicago, IL, USA) with a 200 to 850 nm spectral range was used to obtain the UV-Vis spectra. Z sizer nano Ze3690 de Malvern was used to obtain the Z-potentials, with water as the solvent and 1.33 refraction indices. For both UV-Vis and Z sizer characterizations, the samples had to be diluted in a ratio of 1 to 100 in type 1 water. Moreover, the measurements were obtained for the six samples by varying the molarity of the surfactant. Furthermore, the measurements on all samples were carried out by varying the weeks of storage (1, 2, 4, 10, and 13 weeks). Each measurement was performed five times for statistical purposes, and the average value and standard deviation were determined.

A Raman Confocal LabRam HR Evolution, Horiba Scientific (YOBIN IVON), was used to obtain the Raman spectra with the following conditions: 532 nm laser, optical microscopy with 10X magnification, and a 1250–1690 cm^{-1} spectral range. Raman spectra were taken by varying the molarity of the TXT-surfactant. Data were acquired for samples stored for one week. For one week, the mixtures exhibited the highest stability; in addition, in the case of using the materials in a particular application, costs must be reduced—for example, those related to storage. This means that the mixture with greater stability was selected, which implies less economic and time costs.

2.4. Construction of the Test Cylinders

Cylinders of the cement paste with a H_2O/cement ratio of 0.4 were built according to the following equation:

$$\frac{MWCNTs/TXT - 100/H_2O}{Cement} = 0.4 \qquad (2)$$

The cylinders were built with a 1-inch diameter and a 2-inch length (C109/C109M ASTM norm) [34]. The samples were made with the NTC 550 norm. Firstly, the cement was mixed with the MWCNTs/TX-100/H_2O solution; secondly, the mixture was introduced into the cylinders through three equal layers using the compaction method. This was carried out using 50 beats for each layer to decrease the porosity. Finally, the cylinders were brought to room temperature and, after 24 h, were introduced to a calcium oxide-cured process (see Figure 2), according to ASTM C192 norm [35]. Figure 3a shows the specimens during the drying process, while Figure 3b presents a photograph of the specimens in the curing and storage processes.

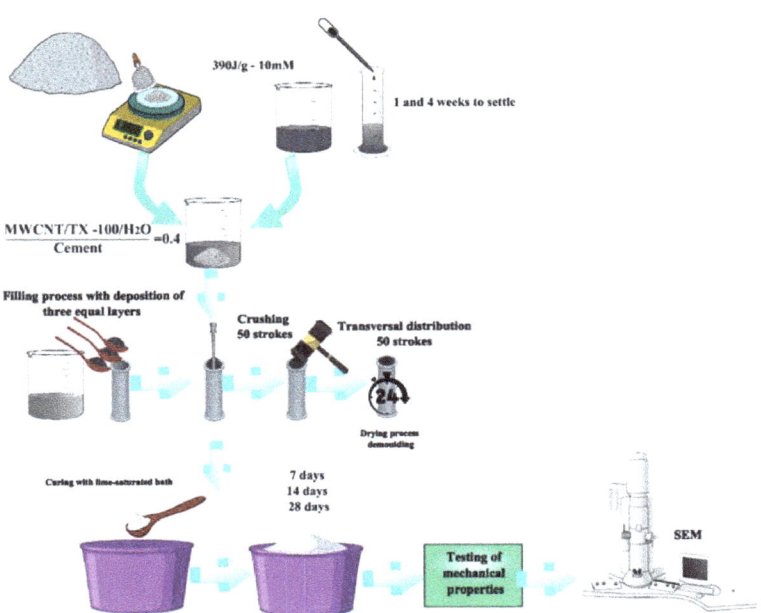

Figure 2. Scheme of the procedure to produce and analyze cement specimens with MWCNT solution dispersed at 390 J/g and with a molarity of surfactant of 10 Mm.

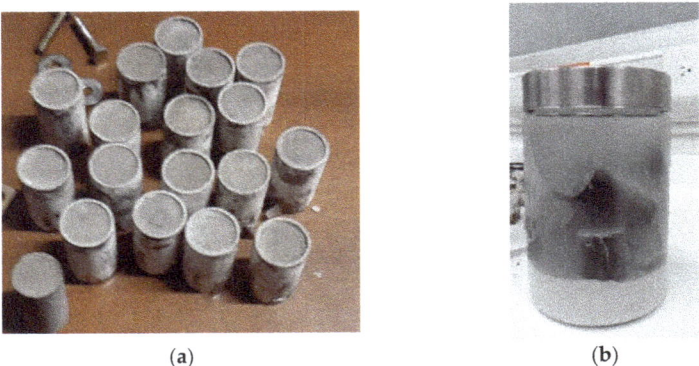

Figure 3. Image of the test cylinder (**a**) during the drying process and (**b**) during the curing and storage process.

2.5. Properties of the Cylinders

Finally, a Humbolt HM 5030 Master Loader (Manizales, Colombia) with a 50 kN capacity load cell was used in the test of the specimens. The established parameters were a speed of 0.25 mm/min, taking the strain data every 0.010 mm, and measuring the load in kN for each strain reading [36,37]. The elasticity modulus and maximum strength were obtained from the stress—strain curve. The parameters used were 0.25 mm/min velocity and data taken each 0.010 mm. A Carl Zeiss EVO MA 10 scanning electron microscope (Oxford model Xact) equipped with a silicon detector of 10 mm was used for morphological examination. Images were taken with a resolution of 5 nm.

3. Results

The first result analyzed was the degree of dispersion of the MWCNTs into the cement using UV-Vis spectroscopy. This analysis was carried out on the samples of MWCNTs mixed with type 1 water and with the TXT-100 dispersant, varying the molarity and storage time. The degree of dispersion of the nanotubes within the TXT-100 dispersant is directly related to the presence of peaks in the spectra, which are an indication of the generation of certain bonds, as will be explained later.

The maximum absorbance in the UV-Vis spectra was identified at 300 nm. It is well known that the agglomerated CNTs absorb in the ultraviolet region at around 300 nm; meanwhile, the individual CNT is active in the Vis region. Hence, it is possible to establish a relationship between the absorbance intensity and the degree of dispersion [36,37]. Moreover, the behavior of each sample was evaluated by varying the number of weeks of storage and determining the stability as a function of time. Figure 4 shows the intensity of the absorbance peak at 300 nm as a function of the molarity and the weeks of storage using UV-Vis. According to this figure, when the molarity is increased, the intensity of the maximum absorbance (at 300 nm) increases. This behavior is due to the presence of a great quantity of benzene rings and alkyne chains that causes many interactions between the surfactant and the MWCNTs (π-π stacking and van der Waals forces). This effect would entail a higher π plasmon resonance [38,39].

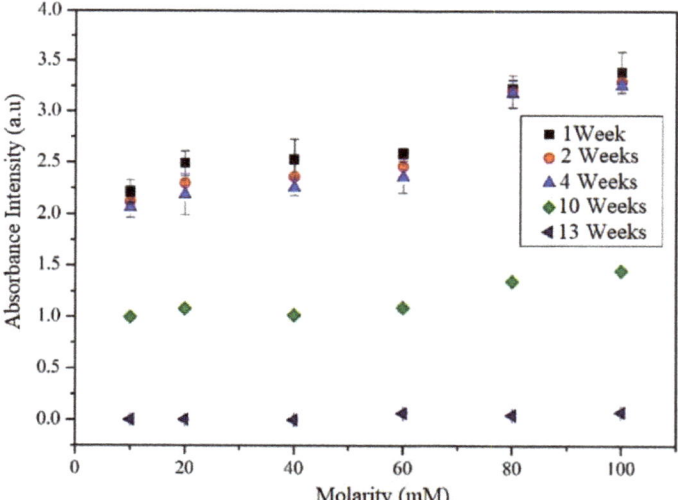

Figure 4. Evolution of the 300 nm absorbance peak depending on the number of weeks of MWCNT dispersion in water with different TX-100 molarities.

The presence of the peak in the UV-Vis spectra is an indication that there is good dispersion and integration between the MWCNTs and the TXT-100 dispersant. By way of their hydrophobic group, the surfactants get adsorbed onto the exterior surface of the MWCNT via noncovalent attraction forces [40], including hydrophobic interaction, hydrogen bonding, π–π stacking, and electrostatic interaction [41], which improve the dispersion of CNTs through steric or/and electrostatic repulsion [42]. It should be noted that the solutions (H$_2$O + TX-100 + MWNTC), varying the TXT-100 molarity, exhibit a good dispersion (stability) for 1, 2, and 4 weeks, showing a high intensity of the absorbance peak at 300 nm. Nevertheless, at the 10th and 13th weeks, the intensity of the absorbance peak abruptly decreases to zero, indicating that the MWCNTs were agglomerated. This is a promising result because as far as we know, there have been no studies about the time for which the solution (H$_2$O + TX-100 + MWNT) remains active. The fact that the absorbance intensity decreases as a function of the weeks of storage indicates that the MWCNTs remained dispersed for a few weeks of storage. For many weeks of storage, the nanotubes tend to agglomerate. MWCNTs produce small clusters/agglomerates due to their high affinity. As the storage time increases, the nanotubes tend to agglomerate, taking into account that just as other nanostructures, they have a large number of free bonds on the surface that are highly reactive. This high reactivity generates a strong attraction between them, causing them to get closer until they agglomerate [18].

On the other hand, the Z-potential spectra were obtained for 1, 2, 4, 10, and 13 weeks for samples with varying TXT-100 molarity. The Z-potential is related to the surface of hydrodynamic shear. When the MWCNTs are immersed in the surfactant, the surfactant layer surrounding the nanotubes can be divided into two parts: an inner region (Stern layer) where the ions are strongly bonded and an outer (diffuse) region where they are less bonded. There exists a notional boundary within the diffuse region where the ions and the nanotubes form a stable interaction. When a nanotube moves, for instance, due to the gravitational force, the boundary is shifted due to the ion movement. Those over-the-limit ions remain with the TXT-100 bulk dispersant. The potential formed at this boundary region is named the Z-potential. A schematic representation for the nanotubes is presented in Figure 5 [43].

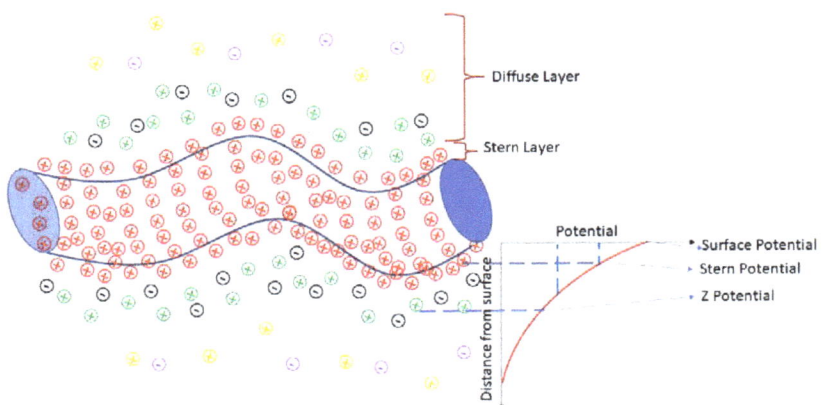

Figure 5. Schematic representation of the zeta potential for nanotubes. The plus and minus signs represent the positive and negative charges, respectively. Colors of charges represent the potentials (charges from surface potential are red; charges from stern potential are green and black; charges from Z-potential are yellow and purple).

The value of the Z-potential gives an indication of the potential stability of the colloidal system. If the nanomaterials in suspension exhibit a large negative or positive zeta potential, they tend to repel each other, avoiding agglomeration. Nevertheless, if nanomaterials present a low zeta potential value, no forces exist to prevent the nanomaterial agglomeration and flocculation. Nanomaterials with Z-potentials more positive than +30 mV or more negative than −30 mV are stable; this depends on the type of dispersant. Considering these aspects, the Z-potential analyses were carried out.

Figure 6 shows an increasing tendency of the Z-potential when the molarity and storage period are increased. In the case of the increase in molarity, the amount of moles present in solution of course increases. This makes that the tense-active micelles produce a decrease in the structural damage and the electrostatic charges present in the MWCNTs surface [38]. It is known that when a minor electrostatic charge is present in the MWCNT surface, the repulsive force and electrostatic attractions also decrease; furthermore, as the molarity of the surfactant is increased, a great quantity of mass is found around the MWCNTs, avoiding their agglomeration. On the other hand, as the period of storage is increased, the Z-potential decreases. It can be explained through the minimum energy principle: a stable system can experience instability when it is subjected to external energy; nevertheless, when this energy is suppressed, the system comes back to its initial state. This instability is generated by electrostatic charges present in the MWCNT surface. This electrostatic charge is produced by the rupture of energetically weak bonds [44,45]. Furthermore, it is well known that colloids tend to precipitate because of the gravitational force; then, as the number of weeks increases, the MWCNTs tend to agglomerate due to the precipitation, and the Z-potential decreases drastically. This means that the samples with better conditions for building the test cylinder in order to determine the mechanical properties are those cured for 1, 2, and 4 weeks, taking into account that the sample cured for one week exhibited the greater Z-potential values. On the other hand, for the case of four weeks, an intermediate behavior was observed. Then, these two samples were chosen for the next stage of the experiment; that is, the mechanical properties evaluation.

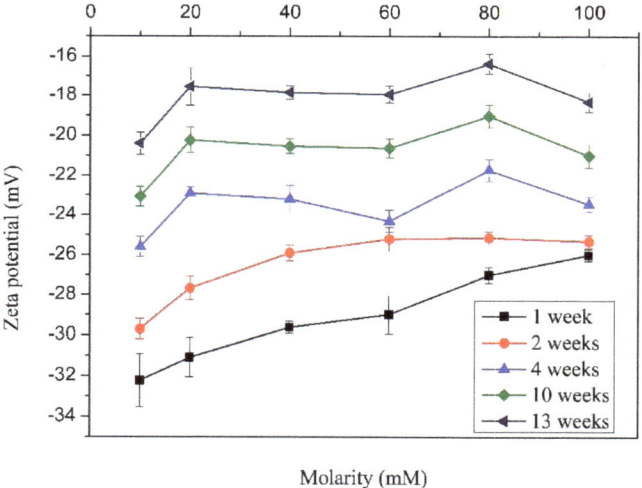

Figure 6. Z-potential of MWCNT dispersions in water with different TX-100 molarities and weeks of storage.

Before the mechanical properties evaluation was carried out, a Raman study was performed to identify the evolution of the samples in which the surfactant molarity was increased. Figure 7 shows a superposition of the Raman spectra belonging to each sample for the case of one week of storage. It is possible to observe three characteristic peaks: (i) The D band at 1344 cm^{-1} is due to the phonon induced by defects associated with the breakdown of kinematic restriction disorders (breathing mode A1) [46]. This band is attributed to the disorder of the solution because of the presence of vacancies and due to lattice defects caused by the mixture of Sp^2 and Sp^3 bonds. (ii) The G band at 1591 cm^{-1} is due to the phonon mode allowed by Raman, which shows the Sp^2 vibration bonds (carbon–carbon), due to the graphene-type bonds [47]. Finally (iii) the G' band at 1622 cm^{-1} is related to the second-order dispersion process that can involve two phonons of the same mode (overtone) or phonons of different modes (combinations) having a similar origin of the D band [48]. Using the I_D/I_G relationship, the structural order of the MWCNTs was estimated; the D band intensity decreases with the decrease in the defect density. To calculate the I_D/I_G ratio, the spectra were deconvoluted with Lorentzian functions.

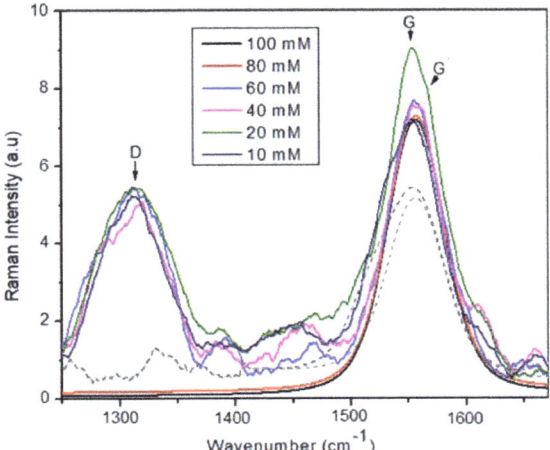

Figure 7. Raman spectra of MWCNT dispersions in water with different TX-100 molarities.

Figure 8 shows the evolution of the I_D/I_G relationship as the molarity increases, showing a tendency to grow. That means that the disorder may increase in the system. According to the literature, spectral properties vary depending on the mechanical conditions of tension by stretching or compression of the MWCNTs and the temperature to which they are subjected. This phenomenon is especially relevant in the case of CNTs mixed with other substances. In this case of the D and G bands in the Raman spectrum of the mixed material, these bands become a sensor of high sensitivity at the microscopic level, giving information about the stress conditions due to the stretching or compression to which they are subjected once they are dispersed [49,50]. Then, as the molarity of the dispersant increases, the stress can increase because of the greater quantity of mass producing greater friction between the nanotubes and dispersant, increasing the disorder. This behavior can be explained by the effect of the dispersant that consists in retaining the separation between MWCNTs when they exhibit a Brownian movement, due to the influence of the ultrasonic tip. For this reason, local cuts are produced in the unraveled MWCNTs, increasing the disorder and then the I_D/I_G relationship; on the other hand, an interaction between surfactant molecules and MWCNTs is produced that increases the friction as the molarity increases [51,52].

This increase in disorder indicates a transformation from Sp^2 and Sp^3 bonds in the solution. Then, it can be concluded that the mixtures of MWCNTs, water, and TXT-100 exhibit a lower disorder for lower values of the surfactant molarities.

In previous works, it has been reported that 10 mM is the most suitable molarity [11,53]. It was possible to see that the tense-active molecules act as (i) an exfoliant in the MWCNT agglomerations and (ii) a delay factor in the reagglomeration of MWCNTs in the period of storage. Based on previous works [52] and the stability observed using Z-potential and Raman analyses, three samples of paste cement were prepared, including MWCNTs, 10 mM molarity (TX-100), and 390 J/g sonication energy. Table 2 includes the nomenclature of the test cylinder (samples). This procedure was carried out to evaluate if the mechanical properties are maintained over time. For this study, the samples were cured for 7, 14, and 21 days according to the ASTM-C39 standard.

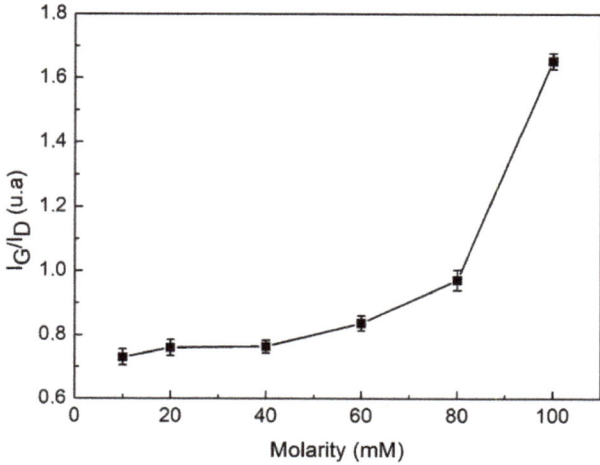

Figure 8. I_D/I_G ratio of MWCNTs dispersed in water with different TX-100 molarities.

Table 2. Description and notation of the samples used for the mechanical properties characterization.

Description of the Mixtures Used for Building the Test Cylinders	Nomenclature of the Samples
cement + H$_2$O + TX-100	S1
cement + H$_2$O + TX-100 + MWCNT (one week of storage)	S2
cement + H$_2$O + TX-100 + MWCNT (four weeks of storage)	S3

Mechanical Test

Stress vs strain curves for each sample were obtained after both one and four weeks of storage. Figure 9 presents a compressive test for S2 and S3, from which the elastic modulus and maximum strength are obtained, according to the procedure described in Figure 10. These results are presented in Figure 11. It can be observed that the Young's modulus (Figure 11a) and the maximum stress (Figure 11b) increase for the case of samples S2 and S3 compared to the S1 sample.

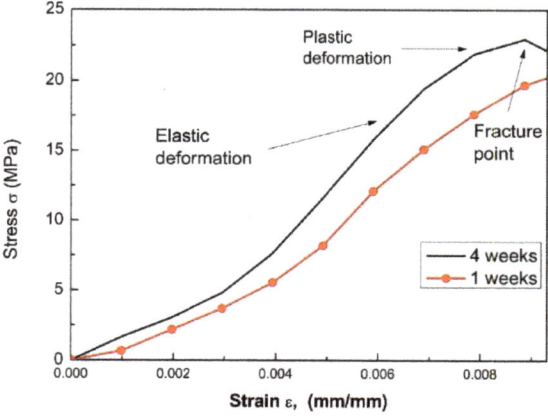

Figure 9. Compressive tests for cement cylinders with the dispersion for S2 and S3.

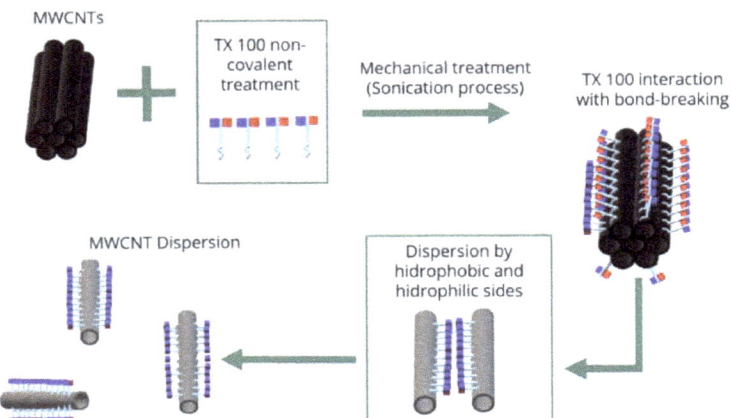

Figure 10. Scheme of the gradual exfoliation processes of the MWCNTs.

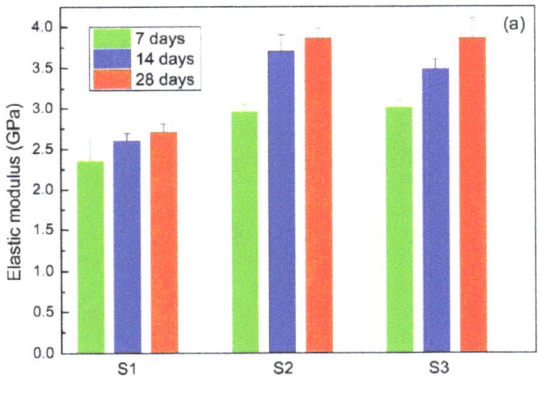

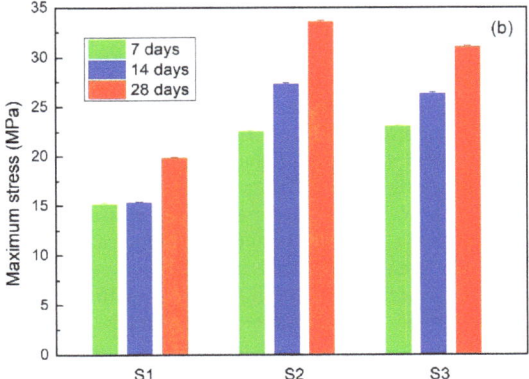

Figure 11. (**a**) Elastic modulus and (**b**) maximum stress to S1, S2, and S3 samples.

An experiment was carried out that consists of measuring the elastic modulus and the maximum stress of the material with three different metal components. For this, measurements were made on three different days, i.e., on day 1, a week later, and 15 days later. Therefore, it is valid to use the design of experiments with the factor (S1, S2, S3) and the factor (day 7, 14, 28).

The factorial design of the experiments was used to check the influence of the material and the day of measurement on the properties of the elastic modulus and maximum strength in two separate experiments. For this design of experiments, it is necessary to check the assumptions of normality, homoscedasticity, and randomness. For this, graphical and statistical diagnostic tests were used, which are not presented in this article because they do not expand on the results. Using a significance of $\sigma = 0.05$ and, therefore, a confidence of 95, the assumptions of normality and equality of variances were confirmed, as shown in Table 3.

Table 3. ANOVA for the results obtained for the Young's modulus and maximum strength.

Factor	p-Value	
	Young's Modulus	Maximum Strength
Sample	0.00215	0.01721
Time of curing	0.01019	0.00414

Given that the assumptions for both experiments were met, two ANOVA tables were made to verify whether the day of measurement or the material significantly influence the elasticity or the maximum force. Using a significance of 0.05, that is, a confidence of 95%, the p-values of the ANOVA tests carried out to validate the influence of the day and the material on both properties are summarized. It is observed that all the p-values are less than 0.05; therefore, it is concluded that both the material and the day of measurement significantly influence the elasticity and maximum strength of the metal.

Using Dunnett's method with a significance of 0.05, it was observed that in the first measurement, on day 1, significantly lower average values of elasticity and maximum strength were obtained. In addition, it was also verified that the maximum force with material S3 is considerably higher than with materials S1 and S2; however, these two are comparably similar. However, for elasticity, the value for material S1 is less than those for S2 and S3, but these two are considered equivalent.

These results can be explained due to the fact that: (i) The MWCNTs unravel during the sonication process (Figure 10). The MWCNTs (reinforcement phase) exhibit a good interaction with the matrix phase (cement paste) through the radicals available on the surface, establishing good binding within the sample. (ii) There is a shortage of secondary bonds and van der Waals forces (as observed in the UV-Vis analysis); that avoids the presence of energetically weak bonds. By including a surfactant, a more significant amount of carboxylic residues are formed, transforming sp^2 bonds into sp^3 (as observed in the Raman analysis) [54]. These bonds interact within the cement matrix with C-S-H phases, generating different bridges (bridge effects) to link capillary pores and inhibit the crack propagation [44,45,55,56]. On the other hand, as the curing time increases, the mechanical properties also increase. This is due to the fact that the hydration is accelerated by the addition of MWCNTs because these act as crystallization centers for the hydrated cement. Moreover, it fills the holes between cement grains, giving as a result an immobilization of water and generating a decrease in the porosity of the samples. Figure 9b shows the behavior of the maximum stress vs. the curing time of the specimens, where the increase in the S2 sample is evident. It manages to reach the maximum effort, so it will have a greater resistance before breaking or reforming. This maximum value of this sample is possible due to it exhibiting a greater stability according to the Z-potential analysis.

In the literature, there are several works that report the enhancement of the mechanical properties of Portland cement by adding MWCNTs; for instance, Yousefi et al. [57] present results showing that the addition of a surfactant employing the mild ultrasonication technique facilitates the homogeneous dispersion of MWCNTs in the cement matrix and enhances the mechanical properties of the hardened concrete. A more recent work by Shahzad et al. [58] reported a study focused on different techniques for dispersing MWCNTs in cementitious materials and the impact on the mechanical properties. As MWCNTs are better dispersed, they tend to fill the micropores, thus increasing the density of the matrix and improving the mechanical properties. Then, it is concluded that there is an enhancement of mechanical properties of MWCNTs' cement pastes for low sonication energies (less than 1000 J/mL). For higher sonication energies, the mechanical properties decrease, especially because of the higher cohesion of the pastes and the consequent higher difficulty of molding, the incorporation of empty spaces, and/or the higher damages suffered by the MWCNTs [59].

Figure 12 shows an SEM image for sample S3 after mechanical failure. In this image, three forms of arrangement of the MWCNTs can be observed, i.e., bridge effect, spiderweb, and cement fragment without MWCNT anchorage. This image was taken with a scanning electron microscope at resolutions of 2 and 10 μm. Based on the results obtained in the mechanical evaluation, it can be deduced that the bridging effect occurs because a large number of MWCNTs have sufficient length to join capillary pores and act as crack bridging as a result of the covalent bonds, inhibiting the propagation of cracks. This behavior generates a better load capacity, ductility, and fracture energy of the pastes. On the other hand, the spiderweb effect is due to the fact that there was not a total dispersion of nanotubes or, at the time of constructing the test cylinders, a great homogeneity was not reached.

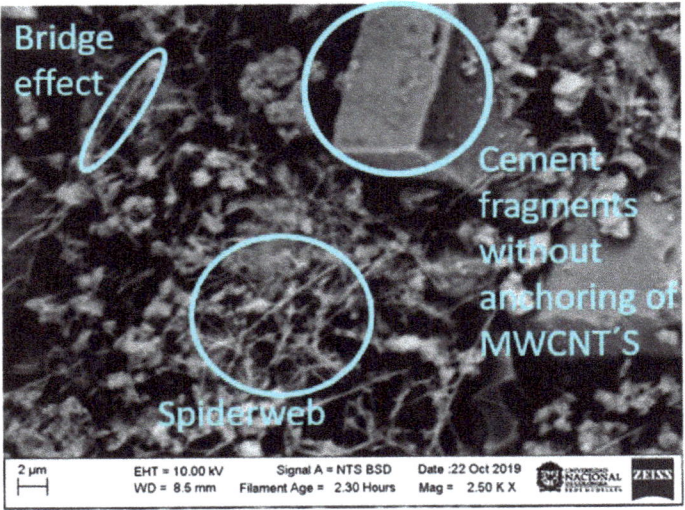

Figure 12. SEM image obtained for a cement paste after the compressive test. This image was taken of sample S3.

As a prospective work, a study of the effect of other different dispersants, possibly more affordable, on the physicochemical, electrical, and mechanical properties of the systems including MWCNTs should be carried out.

Regarding future work, considering that corrosion is a process that affects constructions and buildings, it is necessary to carry out an investigation on the corrosion resistance of steel embedded in cement with the addition of MWCNTs.

A great challenge and limitation for these investigations are the MWCNTs, as stated by another author [60]. Despite being used in various areas, carbon nanotubes still have a high cost, which can be an obstacle to the use of this material in cementitious compounds. It is believed that with the increase in demand and with the possibility of synthesizing CNTs for the manufacture of various applications, the material will become more accessible.

Thus, even though the cost of the material is currently a negative aspect, the tendency is for this drawback to be overcome over time.

Moreover, according to the literature, the addition of MWCNTs improves several properties of the concrete (comprised of water, aggregates, and cement), including the mechanical properties. The results reported by Mohsen et al. [61] indicated that a high CNT content (greater than 15%) would increase the flexural strength of concrete by more than 100%. Furthermore, the results also showed that CNTs would increase the ductility of concrete by about 150%. Adhikary et al. [62] also reported that the utilization of CNTs significantly improves the mechanical performance of lightweight concrete. An almost 41% improvement in the compression of lightweight concrete was observed at 0.6 wt% CNT loading.

4. Conclusions

Multiple wall carbon nanotubes (MWCNTs) were mixed with cement paste and Triton TX-100 surfactant while varying the molarities (10 mM, 20 mM, 40 mM, 60 mM, 80 mM, and 100 mM) and the time of storage (1, 2, 4, 10, and 13 weeks). This procedure was implemented to study the influence of the molarity and time of storage on the mechanical properties of the samples.

The UV-Vis results showed that as the molarity is increased, the intensity of the maximum absorbance (at 300 nm) increases. This peak is due to the presence of the interaction between the surfactant and MWCNTs (π-π stacking and van der Waals forces).

On the other hand, an increasing tendency of the Z-potential as the molarity and storage period were increased was observed. The increase in the molarity generates an increase in the tense-active micelles that decreases both the structural damage and the electrostatic charges present in the MWCNT surface.

According to the Raman results, three characteristic peaks were observed: D band at 1344 cm^{-1} attributed to the disorder of the solution, G band at 1591 cm^{-1} due to the phonon mode of the graphene type bonds, and G' band at 1622 cm^{-1} related to the second-order dispersion process.

The evolution of the I_D/I_G relationship as the molarity increases is an indication of a growth tendency. This indicates that the disorder increases in the system.

Regarding the mechanical properties, it can be observed that the Young's modulus and the maximum stress increased for samples of cement + H_2O + TX-100 with MWCNTs. Furthermore, as the curing time increases, the mechanical properties also increase due to the hydration being accelerated by the MWCNTs.

The SEM image for sample S3 after mechanical failure shows the formation of three arrangements of the MWCNTs: bridge effect, spiderweb, and cement fragment without MWCNT anchorage. The bridging effect occurs because a large number of MWCNTs have sufficient length, acting as a crack bridging as a result of the covalent bonds, inhibiting the propagation of cracks. The spiderweb effect is due to the fact that there was not a total dispersion of nanotubes or, at the time of constructing the test cylinders, a great homogeneity was not reached.

For future work, studies with other surfactants and with concrete containing different aggregates are proposed. In addition, we propose to carry out research with carbon nanotubes produced in our research laboratory. We also propose to make a variation of the sonication energy to reach values close to 1000 J/g.

Author Contributions: Conceptualization, E.-C.L.; Methodology, E.-C.L., C.R., Q.-O.J. and C.-C.H.A.; Data curation, E.-C.L., C.R. and R.-P.E.; Formal analysis, E.-C.L., C.-C.H.A., R.-R.L.V. and R.-P.E.; Resources, C.R., C.-C.H.A. and R.-P.E.; Writing—original draft, E.-C.L. and R.-P.E.; Writing—review & editing, R.-P.E., Q.-O.J. and C.-C.H.A.; Visualization, E.-C.L.; Funding acquisition, R.-P.E.; Supervision, Q.-O.J. and R.-P.E.; Project administration, Q.-O.J. and R.-P.E. All authors have read and agreed to the published version of the manuscript.

Funding: La Facultad de Ciencias Exactas y Naturales at the Universidad Nacional de Colombia—Manizales for the financial support through the postgraduate student's scholarship number 812 of 2018 and the project "Desarrollo de un sistema para la remoción de contaminantes en agua usando nanopartículas obtenidas con reactores de membrana." Supported by the call "CONVOCATORIA DE PROGRAMAS CONECTANDO CONOCIMIENTO 2019".

Institutional Review Board Statement: Not applicable, since no work with animals or humans was included in this study.

Informed Consent Statement: Not applicable.

Data Availability Statement: Not applicable.

Acknowledgments: The authors would like to thank La Facultad de Ciencias Exactas y Naturales at the Universidad Nacional de Colombia—Manizales for the financial support through the postgraduate student's scholarship number 812 of 2018. The authors also want to thank the Universidad Industrial de Santander for the technical support provided in the Guatiguará Technology Park.

Conflicts of Interest: The authors declare no conflict of interest.

References

1. Liu, J.; Liu, L.; Lu, J.; Zhu, H. The formation mechanism of chiral carbon nanotubes. *Phys. B Condens. Matter* **2018**, *530*, 277–282. [CrossRef]
2. Oliveira, T.M.B.F.; Morais, S. New Generation of Electrochemical Sensors Based on Multi-Walled Carbon Nanotubes. *Appl. Sci.* **2018**, *8*, 1925. [CrossRef]
3. Zaporotskova, I.V.; Boroznina, N.P.; Parkhomenko, Y.N.; Kozhitov, L.V. Carbon nanotubes: Sensor properties. A review. *Mod. Electron. Mater.* **2016**, *2*, 95–105. [CrossRef]
4. Bortz, D.R.; Merino, C.; Martin-Gullon, I. Carbon nanofibers enhance the fracture toughness and fatigue performance of a structural epoxy system. *Compos. Sci. Technol.* **2011**, *71*, 31–38. [CrossRef]
5. Senff, L.; Modolo, R.; Tobaldi, D.; Ascenção, G.; Hotza, D.; Ferreira, V.; Labrincha, J. The influence of TiO2 nanoparticles and poliacrilonitrile fibers on the rheological behavior and hardened properties of mortars. *Constr. Build. Mater.* **2015**, *75*, 315–330. [CrossRef]
6. Alafogianni, P.; Dassios, K.; Farmaki, S.; Antiohos, S.; Matikas, T.; Barkoula, N.-M. On the efficiency of UV–vis spectroscopy in assessing the dispersion quality in sonicated aqueous suspensions of carbon nanotubes. *Colloids Surf. A Physicochem. Eng. Asp.* **2016**, *495*, 118–124. [CrossRef]
7. Konsta-Gdoutos, M.S.; Danoglidis, P.A.; Falara, M.G.; Nitodas, S.F. Fresh and mechanical properties, and strain sensing of nanomodified cement mortars: The effects of MWCNT aspect ratio, density and functionalization. *Cem. Concr. Compos.* **2017**, *82*, 137–151. [CrossRef]
8. Chen, S.J.; Qiu, C.Y.; Korayem, A.H.; Barati, M.R.; Duan, W.H. Agglomeration process of surfactant-dispersed carbon nanotubes in unstable dispersion: A two-stage agglomeration model and experimental evidence. *Powder Technol.* **2016**, *301*, 412–420. [CrossRef]
9. Belli, A.; Mobili, A.; Bellezze, T.; Tittarelli, F.; Cachim, P. Evaluating the Self-Sensing Ability of Cement Mortars Manufactured with Graphene Nanoplatelets, Virgin or Recycled Carbon Fibers through Piezoresistivity Tests. *Sustainability* **2018**, *10*, 4013. [CrossRef]
10. Rodriguez, C.; Vélez, E.; Restrepo, J.; Quintero, J.H.; Acuña, R. Testing industrial laboratory dispersion method of Multi-Walled Carbon Nanotubes (MWCNTs) in aqueous medium. *J. Phys. Conf. Ser.* **2019**, *1247*, 012011. [CrossRef]
11. Rodríguez, B.; Quintero, J.H.; Arias, Y.P.; Mendoza-Reales, O.A.; Ochoa-Botero, J.C.; Toledo-Filho, R.D. Influence of MWCNT/surfactant dispersions on the mechanical properties of Portland cement pastes. *J. Phys. Conf. Ser.* **2017**, *935*, 012014. [CrossRef]
12. Rodriguez, B.; Correa, E.; Arias, Y.P.; Quintero, J.; Calderón, J.; Mendoza, O.A. Carbonation study in a cement matrix with carbon nanotubes. *J. Phys. Conf. Ser.* **2019**, *1247*, 012024. [CrossRef]
13. Xie, N. *Mechanical and Environmental Resistance of Nanoparticle-Reinforced Pavement Materials*; Elsevier: Amsterdam, The Netherlands, 2016. [CrossRef]
14. Echeverry-Cardona, L.M.; Álzate, N.; Restrepo-Parra, E.; Ospina, R.; Quintero-Orozco, J.H. Time-Stability Dispersion of MWCNTs for the Improvement of Mechanical Properties of Portland Cement Specimens. *Materials* **2020**, *13*, 4149. [CrossRef]
15. Reales, O.A.M.; Ocampo, C.; Jaramillo, Y.P.A.; Botero, J.C.O.; Quintero, J.H.; Silva, E.C.C.M.; Filho, R.D.T. Reinforcing Effect of Carbon Nanotubes/Surfactant Dispersions in Portland Cement Pastes. *Adv. Civ. Eng.* **2018**, *2018*, 1–9. [CrossRef]
16. Isfahani, F.T.; Li, W.; Redaelli, E. Dispersion of multi-walled carbon nanotubes and its effects on the properties of cement composites. *Cem. Concr. Compos.* **2016**, *74*, 154–163. [CrossRef]
17. Sldozian, R.J.; Tkachev, A.; Burakova, I.; Mikhaleva, Z. Improve the mechanical properties of lightweight foamed concrete by using nanomodified sand. *J. Build. Eng.* **2021**, *34*, 101923. [CrossRef]

18. Rubel, R.I.; Ali, H.; Jafor, A.; Alam, M. Carbon nanotubes agglomeration in reinforced composites: A review. *AIMS Mater. Sci.* **2019**, *6*, 756–780. [CrossRef]
19. Wang, T.; Song, B.; Qiao, K.; Huang, Y.; Wang, L. Effect of Dimensions and Agglomerations of Carbon Nanotubes on Synchronous Enhancement of Mechanical and Damping Properties of Epoxy Nanocomposites. *Nanomaterials* **2018**, *8*, 996. [CrossRef]
20. Wang, J.; Dat Nguyen, T.; Cao, Q.; Wang, Y.; Tan, M.Y.C.; Chan-Park, M.B. Selective Surface Charge Sign Reversal on Metallic Carbon Nanotubes for Facile Ultrahigh Purity Nanotube Sorting. *ACS Nano* **2016**, *10*, 3222–3232. [CrossRef]
21. Ke, F.; Qiu, X. Nanoscale Structure and Interaction of Condensed Phases of DNA–Carbon Nanotube Hybrids. *J. Phys. Chem. C* **2015**, *119*, 15763–15769. [CrossRef]
22. Dassios, K.G.; Alafogianni, P.; Antiohos, S.K.; Leptokaridis, C.; Barkoula, N.-M.; Matikas, T.E. Optimization of Sonication Parameters for Homogeneous Surfactant-Assisted Dispersion of Multiwalled Carbon Nanotubes in Aqueous Solutions. *J. Phys. Chem. C* **2015**, *119*, 7506–7516. [CrossRef]
23. Li, Z.; Kameda, T.; Isoshima, T.; Kobatake, E.; Tanaka, T.; Ito, Y.; Kawamoto, M. Solubilization of Single-Walled Carbon Nanotubes Using a Peptide Aptamer in Water below the Critical Micelle Concentration. *Langmuir* **2015**, *31*, 3482–3488. [CrossRef] [PubMed]
24. Määttä, J.; Vierros, S.; Sammalkorpi, M. Controlling Carbon-Nanotube—Phospholipid Solubility by Curvature-Dependent Self-Assembly. *J. Phys. Chem. B* **2015**, *119*, 4020–4032. [CrossRef] [PubMed]
25. Blanch, A.J.; Lenehan, C.E.; Quinton, J.S. Optimizing Surfactant Concentrations for Dispersion of Single-Walled Carbon Nanotubes in Aqueous Solution. *J. Phys. Chem. B* **2010**, *114*, 9805–9811. [CrossRef]
26. Shih, C.-J.; Lin, S.; Strano, M.S.; Blankschtein, D. Understanding the Stabilization of Single-Walled Carbon Nanotubes and Graphene in Ionic Surfactant Aqueous Solutions: Large-Scale Coarse-Grained Molecular Dynamics Simulation-Assisted DLVO Theory. *J. Phys. Chem. C* **2015**, *119*, 1047–1060. [CrossRef]
27. Bertels, E.; Bruyninckx, K.; Kurttepeli, M.; Smet, M.; Bals, S.; Goderis, B. Highly Efficient Hyperbranched CNT Surfactants: Influence of Molar Mass and Functionalization. *Langmuir* **2014**, *30*, 12200–12209. [CrossRef]
28. Määttä, J.; Vierros, S.; Van Tassel, P.R.; Sammalkorpi, M. Size-Selective, Noncovalent Dispersion of Carbon Nanotubes by PEGylated Lipids: A Coarse-Grained Molecular Dynamics Study. *J. Chem. Eng. Data* **2014**, *59*, 3080–3089. [CrossRef]
29. Sarukhanyan, E.; Milano, G.; Roccatano, D. Coating Mechanisms of Single-Walled Carbon Nanotube by Linear Polyether Surfactants: Insights from Computer Simulations. *J. Phys. Chem. C* **2014**, *118*, 18069–18078. [CrossRef]
30. Homenick, C.M.; Rousina-Webb, A.; Cheng, F.; Jakubinek, M.B.; Malenfant, P.R.L.; Simard, B. High-Yield, Single-Step Separation of Metallic and Semiconducting SWCNTs Using Block Copolymers at Low Temperatures. *J. Phys. Chem. C* **2014**, *118*, 16156–16164. [CrossRef]
31. Blanch, A.J.; Shapter, J.G. Surfactant Concentration Dependent Spectral Effects of Oxygen and Depletion Interactions in Sodium Dodecyl Sulfate Dispersions of Carbon Nanotubes. *J. Phys. Chem. B* **2014**, *118*, 6288–6296. [CrossRef]
32. Sohrabi, B.; Poorgholami-Bejarpasi, N.; Nayeri, N. Dispersion of Carbon Nanotubes Using Mixed Surfactants: Experimental and Molecular Dynamics Simulation Studies. *J. Phys. Chem. B* **2014**, *118*, 3094–3103. [CrossRef] [PubMed]
33. Moniruzzaman, M.; Winey, K.I. Polymer Nanocomposites Containing Carbon Nanotubes. *Macromolecules* **2006**, *39*, 5194–5205. [CrossRef]
34. *C109/C109M–11b*; Standard Test Method for Compressive Strength of Hydraulic Cement Mortars (Using 2-in. or [50-mm] Cube Specimens). ASTM International: West Conshohocken, PA, USA, 2010; pp. 1–9.
35. *C192/C192M–16a*; ASTM International, Standard Practice for Making and Curing Concrete Test Specimens in the Laboratory. ASTM International: West Conshohocken, PA, USA, 2007; pp. 1–8.
36. Ryabenko, A.; Dorofeeva, T.; Zvereva, G. UV-VIS–NIR spectroscopy study of sensitivity of single-wall carbon nanotubes to chemical processing and Van-der-Waals SWNT/SWNT interaction. Verification of the SWNT content measurements by absorption spectroscopy. *Carbon* **2004**, *42*, 1523–1535. [CrossRef]
37. Grossiord, N.; van der Schoot, P.; Meuldijk, J.; Koning, C.E. Determination of the Surface Coverage of Exfoliated Carbon Nanotubes by Surfactant Molecules in Aqueous Solution. *Langmuir* **2007**, *23*, 3646–3653. [CrossRef] [PubMed]
38. Li, H.; Qiu, Y. Dispersion, sedimentation and aggregation of multi-walled carbon nanotubes as affected by single and binary mixed surfactants. *R. Soc. Open Sci.* **2019**, *6*, 190241. [CrossRef]
39. Di Crescenzo, A.; Di Profio, P.; Siani, G.; Zappacosta, R.; Fontana, A. Optimizing the Interactions of Surfactants with Graphitic Surfaces and Clathrate Hydrates. *Langmuir* **2016**, *32*, 6559–6570. [CrossRef]
40. Vaisman, L.; Wagner, H.D.; Marom, G. The role of surfactants in dispersion of carbon nanotubes. *Adv. Colloid Interface Sci.* **2006**, *128–130*, 37–46. [CrossRef]
41. Lin, D.; Liu, N.; Yang, K.; Xing, B.; Wu, F. Different stabilities of multiwalled carbon nanotubes in fresh surface water samples. *Environ. Pollut.* **2010**, *158*, 1270–1274. [CrossRef]
42. Strano, M.S.; Moore, V.C.; Miller, M.K.; Allen, M.J.; Haroz, E.H.; Kittrell, C.; Hauge, R.H.; Smalley, R.E. The Role of Surfactant Adsorption during Ultrasonication in the Dispersion of Single-Walled Carbon Nanotubes. *J. Nanosci. Nanotechnol.* **2003**, *3*, 81–86. [CrossRef]
43. Vold, M.J. Zeta potential in colloid science. In *Principles and Applications*; Hunter, R.J., Ed.; Academic Press: New York, NY, USA; London, UK, 1982; Volume 88, pp. 386–608. [CrossRef]
44. Sanchez, F.; Sobolev, K. Nanotechnology in concrete—A review. *Constr. Build. Mater.* **2010**, *24*, 2060–2071. [CrossRef]
45. Gowripalan, N. Autogenous shrinkage of concrete at early ages. *Lect. Notes Civ. Eng.* **2020**, *37*, 269–276. [CrossRef]

46. Prasankumar, R.P.; Taylor, A.J. *Optical Techniques for Solid-State Materials Characterization*, 1st ed.; CRC Press: Boca Raton, FL, USA, 2011.
47. Hodkiewicz, J. *Characterizing Carbon Materials with Raman Spectroscopy—Application Note*; Thermo Fisher Scientific: Waltham, MA, USA, 2010; pp. 1–5.
48. Ordoñez-Casanova, E.G. Estudio Experimental y Teórico de Nanotubos de Carbono de Pocas Paredes. Ph.D. Thesis, Centro de Investigación en Materiales Avanzados CIMAV, Chihuahua, Mexico, 2013. Available online: https://cimav.repositorioinstitucional.mx/jspui/bitstream/1004/579/1/Tesis%20Elsa%20Gabriela%20Ord%C3%B3%C3%B1ez%20Casanova.pdf (accessed on 1 June 2022).
49. Zein, H. Studying the Influence of Various Geometrical Parameters of Single-Walled Carbon Nano-Tubes of Armchair Chirality Type on Its Mechanical Behavior. *World J. Appl. Chem.* **2018**, *3*, 17. [CrossRef]
50. Cooper, C.; Young, R.; Halsall, M. Investigation into the deformation of carbon nanotubes and their composites through the use of Raman spectroscopy. *Compos. Part A Appl. Sci. Manuf.* **2001**, *32*, 401–411. [CrossRef]
51. Frømyr, T.; Hansen, F.K.; Olsen, T. The Optimum Dispersion of Carbon Nanotubes for Epoxy Nanocomposites: Evolution of the Particle Size Distribution by Ultrasonic Treatment. *J. Nanotechnol.* **2012**, *2012*, 1–14. [CrossRef]
52. Duan, W.H.; Wang, Q.; Collins, F. Dispersion of carbon nanotubes with SDS surfactants: A study from a binding energy perspective. *Chem. Sci.* **2011**, *2*, 1407–1413. [CrossRef]
53. Reales, O.A.M.; Jaramillo, Y.P.A.; Ochoa-Botero, J.C.; Delgado, C.A.; Quintero, J.H.; Filho, R.D.T. Influence of MWCNT/surfactant dispersions on the rheology of Portland cement pastes. *Cem. Concr. Res.* **2018**, *107*, 101–109. [CrossRef]
54. Hang, Z. Preparation and Characterization of Carbon Micro/Nano Hybrids and Their Functional Composites. Ph.D. Thesis, Université Paris-Saclay, Paris, France, 2015.
55. Gao, Y.; Jing, H.; Du, M.; Chen, W. Dispersion of Multi-Walled Carbon Nanotubes Stabilized by Humic Acid in Sustainable Cement Composites. *Nanomaterials* **2018**, *8*, 858. [CrossRef]
56. Han, B.; Yu, X.; Ou, J. Multifunctional and Smart Carbon Nanotube Reinforced Cement-Based Materials. In *Nanotechnology in Civil Infrastructure*; Gopalakrishnan, K., Birgisson, B., Taylor, P., Attoh-Okine, N.O., Eds.; Springer: Berlin/Heidelberg, Germany, 2011; pp. 1–47. [CrossRef]
57. Yousefi, A.; Bunnori, N.M.; Khavarian, M.; Majid, T.A. Dispersion of Multi-Walled Carbon Nanotubes in Portland Cement Concrete Using Ultra-Sonication and Polycarboxylic Based Superplasticizer. *Appl. Mech. Mater.* **2015**, *802*, 112–117. [CrossRef]
58. Shahzad, S.; Toumi, A.; Balayssac, J.-P.; Turatsinze, A.; Mazars, V. Cementitious composites incorporating Multi-Walled Carbon Nanotubes (MWCNTs): Effects of annealing and other dispersion methods on the electrical and mechanical properties. *Matériaux Tech.* **2022**, *110*, 104. [CrossRef]
59. Santiago, E.Q.R.; Lima, P.R.L.; Leite, M.B.; Filho, R.D.T. Mechanical behavior of recycled lightweight concrete using EVA waste and CDW under moderate temperature. *Rev. IBRACON de Estruturas e Mater.* **2009**, *2*, 211–221. [CrossRef]
60. Marcondes, C.G.N.; Medeiros, M.H.F.; Filho, J.M.; Helene, P. Carbon nanotubes in Portland cement concrete: Influence of dispersion on mechanical properties and water absorption. *ALCONPAT J.* **2015**, *5*, 97–113. Available online: https://www.scielo.org.mx/scielo.php?script=sci_issuetoc&pid=2007-683520150002&lng=es&nrm=iso (accessed on 1 June 2022).
61. Mohsen, M.O.; Al Ansari, M.S.; Taha, R.; Al Nuaimi, N.; Abu Taqa, A. Carbon Nanotube Effect on the Ductility, Flexural Strength, and Permeability of Concrete. *J. Nanomater.* **2019**, *2019*, 6490984. [CrossRef]
62. Adhikary, S.K.; Rudžionis, Ž.; Tučkutė, S.; Ashish, D.K. Effects of carbon nanotubes on expanded glass and silica aerogel based lightweight concrete. *Sci. Rep.* **2021**, *11*, 2104. [CrossRef] [PubMed]

Article

A Thermal Transport Study of Branched Carbon Nanotubes with Cross and T-Junctions

Wei-Jen Chen and I-Ling Chang *

Department of Mechanical Engineering, National Cheng Kung University, Tainan 70101, Taiwan; n18011041@mail.ncku.edu.tw
* Correspondence: ilchang@mail.ncku.edu.tw; Tel.: +886-6-2757575 (ext. 62124)

Featured Application: Our findings provide deep insights into the thermal transport mechanisms of branched carbon nanotubes, which may be useful in thermal management applications.

Abstract: This study investigated the thermal transport behaviors of branched carbon nanotubes (CNTs) with cross and T-junctions through non-equilibrium molecular dynamics (NEMD) simulations. A hot region was created at the end of one branch, whereas cold regions were created at the ends of all other branches. The effects on thermal flow due to branch length, topological defects at junctions, and temperature were studied. The NEMD simulations at room temperature indicated that heat transfer tended to move sideways rather than straight in branched CNTs with cross-junctions, despite all branches being identical in chirality and length. However, straight heat transfer was preferred in branched CNTs with T-junctions, irrespective of the atomic configuration of the junction. As branches became longer, the heat current inside approached the values obtained through conventional prediction based on diffusive thermal transport. Moreover, directional thermal transport behaviors became prominent at a low temperature (50 K), which implied that ballistic phonon transport contributed greatly to directional thermal transport. Finally, the collective atomic velocity cross-correlation spectra between branches were used to analyze phonon transport mechanisms for different junctions. Our findings deeply elucidate the thermal transport mechanisms of branched CNTs, which aid in thermal management applications.

Keywords: ballistic; phonon; branched CNT

Citation: Chen, W.-J.; Chang, I.-L. A Thermal Transport Study of Branched Carbon Nanotubes with Cross and T-Junctions. *Appl. Sci.* **2021**, *11*, 5933. https://doi.org/10.3390/app11135933

Academic Editor: Simone Morais

Received: 28 April 2021
Accepted: 20 June 2021
Published: 25 June 2021

Publisher's Note: MDPI stays neutral with regard to jurisdictional claims in published maps and institutional affiliations.

Copyright: © 2021 by the authors. Licensee MDPI, Basel, Switzerland. This article is an open access article distributed under the terms and conditions of the Creative Commons Attribution (CC BY) license (https://creativecommons.org/licenses/by/4.0/).

1. Introduction

Since the discovery of carbon nanotubes (CNTs) in 1991, considerable research has been conducted on their mechanical, optical, and electronic properties. CNTs have been proposed as the most promising building blocks for nanoelectronic devices and thermal management applications due to their high thermal conductivity and mechanical stability [1,2]. However, the quasi-one-dimensional thermal transport of CNTs has restricted their applications. Therefore, branched CNTs [3,4] have been proposed as alternatives to CNTs. Due to their strong sp2 covalent bonds, branched CNTs exhibit lower thermal resistance than those structures with van der Waals interactions (e.g., CNT bundles [5,6], intermolecular junctions [7–9], and CNT networks [10,11]). In addition, high-quality and high-purity branched CNTs have been synthesized using various approaches [12–18]. Moreover, branched CNTs demonstrate potential in rectifying electrical currents [12–14] and heat [3,19]. Therefore, the thermal transport behavior inside branched CNTs should be investigated to identify their possible applications.

Molecular dynamic (MD) simulation is an appropriate tool for characterizing the collective and individual behavior of atomics or molecules at the nanoscale level. Compared with other atomistic methods, such as the use of Boltzmann transport equations, lattice dynamics, and non-equilibrium Green's function, MD simulation can more accurately capture the dynamics of lattice vibration, including harmonic and anharmonic interactions [20]. MD simulations have been widely adopted to investigate the thermal transport mechanisms of pristine CNTs. A frequently researched aspect of pristine CNTs is their length-dependent thermal conductivity, which involves ballistic or diffusive phonon transport behavior at varying lengths [21–23]. However, similar examinations on the phonon transport mechanism or thermal energy transfer inside branched CNTs have rarely been conducted. Branched CNTs possess topological defects and the possibility of multi-directional thermal transfer in junction regions, whereas pristine CNTs do not. Some studies have examined the influences of the aforementioned phenomena on the thermal transport behavior [3,4,24–26]. The topological defects at the junction area of branched CNTs must satisfy Euler's polyhedral formula [27,28]. Topological defects often cause considerable phonon scattering and thus an increase in thermal resistance, which subsequently influences heat transfer [24,26]. Moreover, the connections of different carbon structures, such as CNTs and graphene, must facilitate multi-directional thermal transfer; however, dimensional mismatch occurs at the junction, which causes an increase in resistance, a decrease in transmission, and changes in the incident phonon mode [24,29,30].

Some researchers have investigated thermal ratification behaviors for branched CNTs by conducting experiments [19], heat pulse simulations [3], and wave packet simulations [4]. Chen et al. [31] investigated the thermal transfer inside a branched CNT with T-junctions and observed that the heat preferred to flow straight rather than sideways, especially for short branch lengths. They also noticed that symmetric temperature setups exhibited lower transmission than asymmetric temperature setups because the phonons were forced to change their propagation directions in the symmetric temperature setups. The aforementioned phenomenon has been attributed to the occurrence of ballistic phonon transport and topological defects (i.e., heptagons and octagons). However, the underlying dominant mechanisms of the directional heat transfer inside branched CTNs have yet to be determined.

In the present study, the heat flow inside branched CNTs with cross and T-junctions was investigated using non-equilibrium molecular dynamics (NEMD) simulations. The junctions in the CNTs were constructed through thermal welding. All of the CNT branches had the same chirality and length and were attached to the cross and T-junctions to form branched CNTs. The effects on thermal transfer due to branch length, topological defects at the junction, and temperature were studied. Moreover, the collective atomic velocity cross-correlation spectra between branches were used to analyze the phonon transport mechanisms in the branched CNTs.

2. Methodology

All NEMD simulations were conducted using the Large-scale Atomic/Molecular Massively Parallel Simulator (LAMMPS) package [32]. The adaptive intermolecular reactive bond order (AIREBO) potential [33] was employed in this study to describe the interatomic interaction between carbon atoms. T-junctions and cross-junctions were created using the thermal welding method proposed in [17]. Some atoms were removed intentionally from one pristine 4 nm (6, 6) CNT to create a crossing region. The atoms were removed such that the topological requirement described by the Euler polyhedral formula was satisfied [28]. This formula regulates the numbers of vertices, faces, and edges in a convex three-dimensional polyhedron. The number of pentagons (n_5), heptagons (n_7), and octagons (n_8) at a junction must satisfy the following equality: $2n_8 + n_7 - n_5 = 6$. A 1.5 nm (6, 6) CNT was placed vertically with respect to the crossing region of the defect-containing 4 nm (6, 6) CNT. Subsequently, the two aforementioned CNTs were moved closer and simultaneously heated up to 4000 K to perform thermal welding. A high temperature was adopted to accelerate the creation of interlinks and surface reconstruction. Finally, the T-junction model was cooled to 300 K.

Through the sequential removal of carbon atoms, topological defects in the junction area could be quantitatively controlled to satisfy the Euler polyhedral formula. Thus, T-junctions with two atomic configurations containing different topological defects were constructed, as illustrated in Figure 1a,b. One atomic configuration, which is hereafter referred to as T-junction 1, contained two heptagon rings (one in the back) and two octagons in the corner region, as depicted in Figure 1a. The second atomic configuration, which is hereafter referred to as T-junction 2, contained four heptagon rings (two in the back), two pentagon rings (one in the back), and two octagons in the corner region, as depicted in Figure 1b.

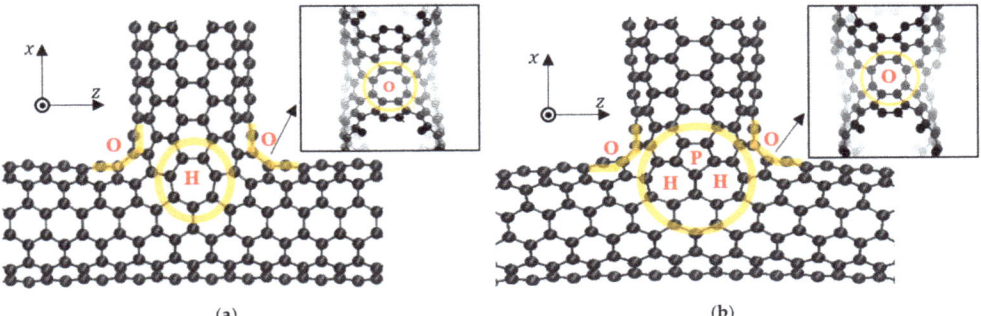

Figure 1. Atomic configurations of the two T-junctions: (**a**) T-junction 1 and (**b**) T-junction 2 (P, H, and O represent pentagons, heptagons, and octagons, respectively).

Cross-junctions were constructed by further removing atoms from the bottom region of the T-junction model, which is hereafter referred to as the defect-containing T-junction, as illustrated in Figure 2. For symmetry, the atoms removed from the bottom of the aforementioned model were the same as those in the top crossing region of this model. A 1.5 nm (6, 6) CNT was assembled with the defect-containing T-junction model by using the thermal welding process to form cross-junctions, as depicted in Figure 3a,b. The cross-junction models depicted in Figure 3a,b were called cross-junctions 1 and 2, respectively. The cross-junction models had twice the number of topological defects that their T-junction counterparts did.

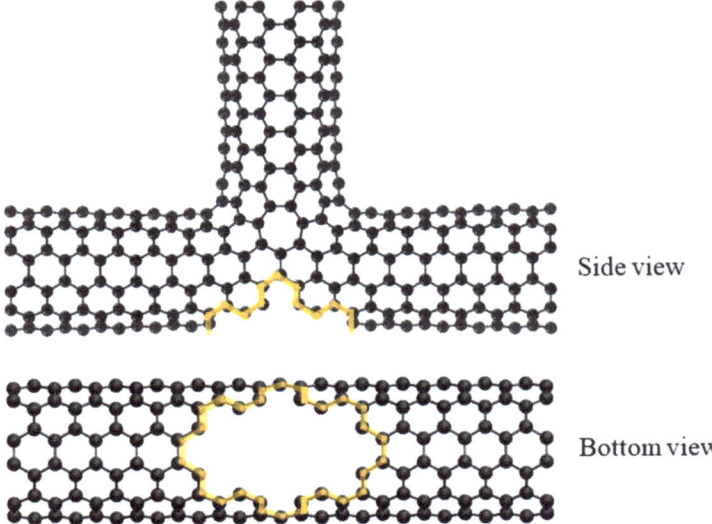

Figure 2. Defect-containing T-junction model.

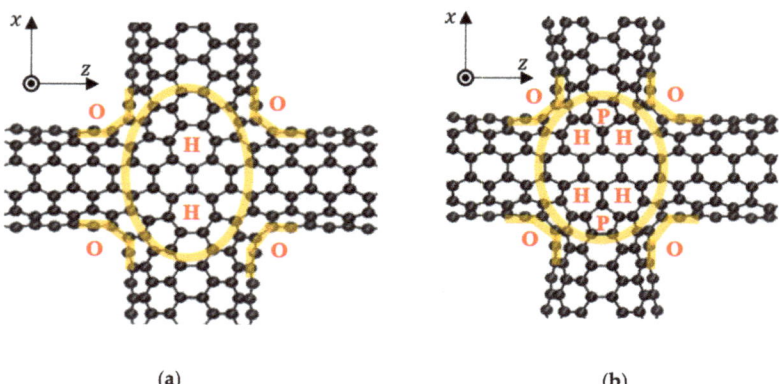

Figure 3. Atomic configurations of two cross-junctions: (**a**) cross-junction 1 and (**b**) cross-junction 2 (P, H, and O represent pentagons, heptagons, and octagons, respectively).

Equal-length (6, 6) CNTs were connected to T-junctions and cross-junctions to form branched CNTs. To investigate the effect of branch length on the heat flow inside the branched CNTs, various branch lengths between 25 and 200 nm were considered for the simulations. The entire system was relaxed using a canonical (NVT) ensemble at the junction region under a temperature of 300 K. The time step used for all simulations was 1 fs. To release the stress inside branched CNTs, the atoms on each individual branch were allowed to move only along the axial direction of the CTNs under the microcanonical (NVE) ensemble and the equilibrating process was run for 1 ns. Through the aforementioned relaxation procedure, both longitudinal and transverse stresses inside the branches were eliminated. Relaxing the structure was crucial because the residual stress from the initial model can influence the material properties and can thus affect the heat flow [30].

The temperature setups of the cross and T-junctions are displayed in Figure 4a,b, respectively. To avoid rigid body motion (i.e., translational movement and rotation) of the branched CNTs, all of the branches were fixed at their ends for a length of 0.5 nm adjacent to a 5 nm temperature-controlled region. The temperatures in the hot and cold regions were controlled at 300 ± 30 K by using the Langevin thermostat setting. To observe the thermal flow, a hot region was created at the end of branch 1 of the branched CNTs whereas cold regions were created at all of the other branch ends. The NVE ensemble was implemented in the free area of branched CNTs for 1 ns to enable the heat current to reach a steady state. All of the results were collected in the following 1 ns. Each model was simulated and averaged six times with different initial velocities.

In the branched CNT with cross-junctions, the heat current flowed from branch 1 into the other three branches (branches 2, 3, and 4; Figure 4a). To investigate the heat flow behavior inside this branched CNT, we defined the heat current ratio beyond the junction as follows:

$$R_i = \frac{I_i}{\sum I_i}, \, i = 2 \sim 4 \tag{1}$$

where I_i is the power removed from the temperature-controlled region of branch i. The heat current ratio for the T-junctions was similarly defined as that in Equation (1).

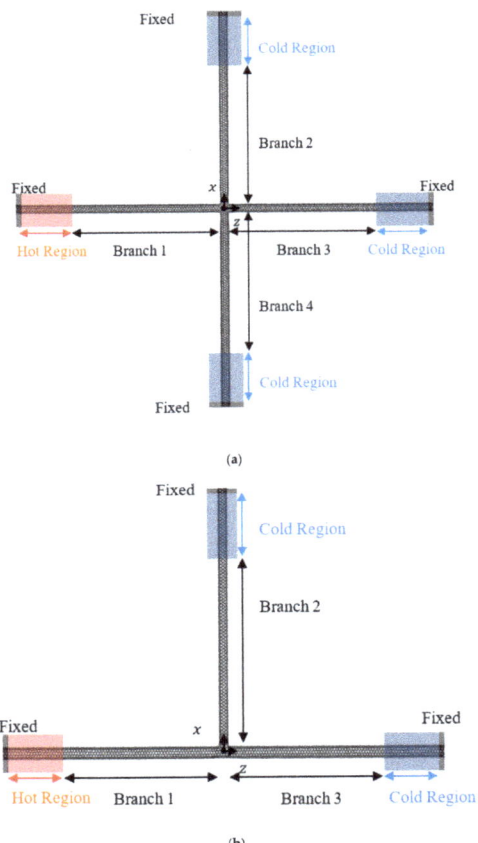

Figure 4. Illustration of the temperature-controlled setup for non-equilibrium molecular dynamics simulations. A hot region was created at branch 1, whereas cold regions were created in the remaining branches of the branched CNTs with (**a**) cross-junction and (**b**) T-junction.

3. Results

The heat current ratios occurring after the cross and T-junctions were examined through NEMD simulations. The effects of branch length, atomic configurations at the junctions, and temperature on the heat flow inside the branched CNTs were examined. The length dependence of the heat current ratios for the two cross-junctions is illustrated in Figure 5a,b. Regardless of the atomic configuration at the junctions, the heat current ratios R_2 and R_4 (for branches 2 and 4, respectively) were approximately equal and larger than R_3. This result indicates that the heat currents did not flow equally into the three branches after the junction even though all of the branches had identical length and chirality. The heat flow into the top (branch 2) and bottom (branch 4) branches was higher than that into the straight branch (branch 3), especially for short branch lengths. However, as the branch length increased, the heat current ratio tended to approach 1/3. This value was equal to the prediction result obtained using conventional thermal circuit theory under the assumptions that the thermal resistances along all the branch directions at the junction were identical and that the thermal properties of all of the branches were the same.

Moreover, the thermal transport behavior was investigated by creating a hot region placed at the bottom branch (branch 4) and cold regions at the ends of the other branches for the branched CNT with cross-junctions. Different temperature settings produced different heat current directions, along which different topological defects were encountered. For instance, the heat current from branch 1 encountered two heptagons in parallel at cross-junction 1; however, the heat flow from branch 4 encountered two heptagons in series. The heat current results for two temperature settings were within the standard deviation error, as shown in Figures 5a and 6a. For both settings, the heat current preferentially flowed sideways rather than straight.

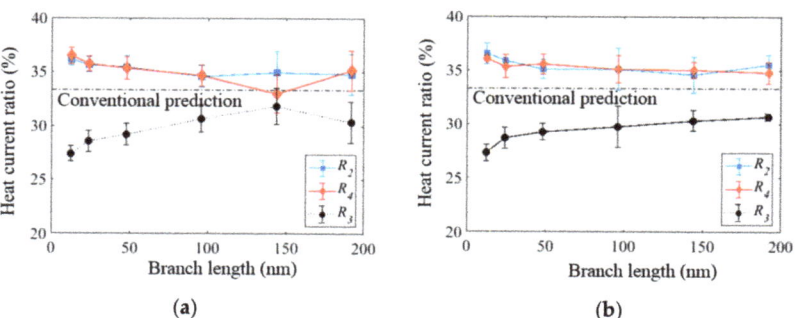

Figure 5. Branch length dependence of the heat current ratio for branched CNTs with two atomic configurations at the cross-junctions: (**a**) cross-junction 1 and (**b**) cross-junction 2.

We attempted to explain the aforementioned observations by using thermal circuit theory and by considering the anisotropic thermal conductance at the junction. As the topological defect locations were horizontally (along branches 1 and 3) and vertically (along branches 2 and 4) asymmetric, the thermal resistances at the junctions along the branches may have been different. A previous study [29] indicated that defects in series have lower thermal resistance than those in parallel. When the asymmetric junction resistance is considered (Figure 6b), the temperature difference between the hot and cold regions (ΔT) (hot region placing at branch 1) can be expressed as follows:

$$\Delta T = T_{hot} - T_{cold} = I_1(\Omega_1 + \Omega_{CNT}) + I_2(\Omega_2 + \Omega_{CNT})$$
$$= I_1(\Omega_1 + \Omega_{CNT}) + I_3(\Omega_3 + \Omega_{CNT}) = I_1(\Omega_1 + \Omega_{CNT}) + I_4(\Omega_4 + \Omega_{CNT}), \quad (2)$$

where Ω_{CNT} is the thermal resistance of the pristine CNT branch and I_i is the heat current at branch i. The following expression is obtained:

$$I_2(\Omega_2 + \Omega_{CNT}) = I_3(\Omega_3 + \Omega_{CNT}) = I_4(\Omega_4 + \Omega_{CNT}) \quad (3)$$

As the heat currents of branches 2 and 4 were larger than that of branch 3 ($I_{2,4} > I_3$), we deduced that the thermal resistance at the junction satisfied the following expression: $\Omega_{2,4} < \Omega_3$. However, a contradictory result (i.e., $\Omega_{1,3} < \Omega_2$) was obtained when we considered the heat current situation in which a hot region was created at branch 4. Our simulation results cannot be explained using an account of conventional thermal circuits with or without the consideration of thermally anisotropic junctions. This finding implies that the obtained results cannot be explained simply by diffusive-based thermal circuit theory.

The thermal transport behavior inside the branched CNT with T-junctions was also simulated. The dependences of the heat current ratios on the branch lengths are displayed in Figure 7a,b, respectively. The heat current preferentially flowed straight into the straight branch rather than sideways (i.e., $R_3 > 50\%$ or $R_2 < 50\%$), especially for the short branch, irrespective of the atomic configurations at the T-junctions. However, as the branch length increased to 200 nm, the heat current ratios of branches 2 and 3 (i.e., R_2 and R_3, respectively) approached 50%, which was predicted by a diffusive-based conventional thermal circuit calculation. Similar to the branched CNT with cross-junctions, the atomic defects of pentagonal, heptagonal, and octagonal rings did not influence the directional heat transfer behaviors. Moreover, the heat current preferentially flowed sideways rather than straight, with one additional bottom branch attached to a T-junction.

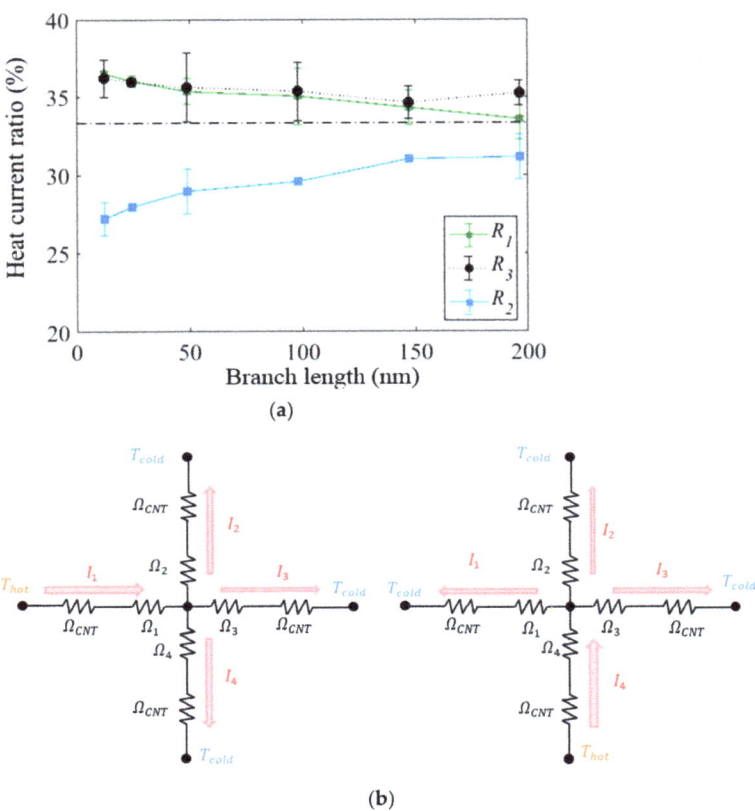

Figure 6. (a) Length dependence of the heat current ratio for branched CNTs with cross-junction 1. A hot region was created at the bottom branch (branch 4), and cold regions were created at the other branches. (b) Conventional thermal circuit when assuming different thermal resistances at the junctions.

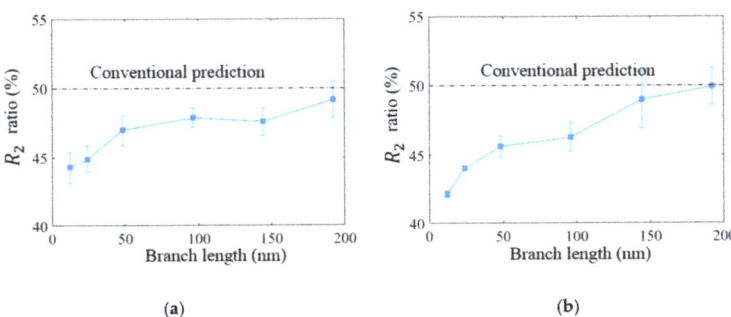

Figure 7. Branch length dependence of the heat current ratios inside branched CNTs with two T-junctions: (a) T-junction 1 and (b) T-junction 2.

4. Discussion

Through NEMD simulation of branched CNTs, directional heat transfer on the branches could be observed. The directional heat transfer behavior in branched CNTs can be attributed to two possible causes: asymmetric polygonal rings in the junction regions and ballistic phonon transport [31]. First, it was reported that phonons exhibit different interactions with asymmetric polygon rings, which resulted in directional flow due to the specific location of the polygon rings [29]. This effect became weaker as the branch length increased because the thermal resistance was influenced more by the branches than the junctions. Therefore, heat currents tended to flow equally into each branch at long branch lengths. However, similar heat current results were obtained regardless of the atomic configuration at the junction (Figure 5 (cross-junctions) and Figure 7 (T-junctions)). Even if the configuration of the polygon rings changed (e.g., from two heptagons in parallel to two heptagons in series) by shifting the hot region to the bottom side, the heat still tended to flow sideways rather than straight, as displayed in Figure 6a. Thus, the results indicated that the directional heat transfer in branched CNTs was not dominated by local asymmetric polygonal rings.

Second, heat may exhibit ballistic transport rather than diffusive scattering inside branched CNTs, especially for short branches. When a branch was longer than the phonon mean free path, ballistic to diffusive crossover of heat flow occurred and the heat was transported predominantly through diffusive scattering; thus, an equal heat current occurred in each branch. As ballistic phonon transport was more pronounced at low temperatures [1,21,34,35], the temperature effect was studied to further investigate the thermal flow mechanism inside branched CNTs with cross and T-junctions. The hot and cold regions were controlled at 50 ± 30 K to observe how the heat currents between branches changed. As illustrated in Figure 8a,b, the heat current ratios for branch 2 exhibited relatively high deviation from the conventional predictions for both junctions, which indicates the thermal transport mechanisms in these branches were less diffusive and more ballistic at lower temperatures. Moreover, the NEMD simulation was conducted on the branched CNT with a defect-containing T-junction (Figure 2) to examine the different directional thermal flow behaviors between cross and T-junctions. Compared with the T-junction, the defect-containing T-junction, which was the intermediate construction stage of a cross-junction, had an additional opening in its bottom region. As displayed in Figure 9, the size dependence of the heat current ratio in branch 2 of the defect-containing T-junction was similar to that on the size of a cross-junction and completely different from that on the size of the T-junction. It is likely that phonons would scatter by the additional opening at the bottom of defect-containing T-junction. Thus, the presence of an additional opening at the bottom caused the heat current to tend to flow sideways into branch 2 rather than straight into branch 3, which was unlike the phenomenon observed for the T-junction.

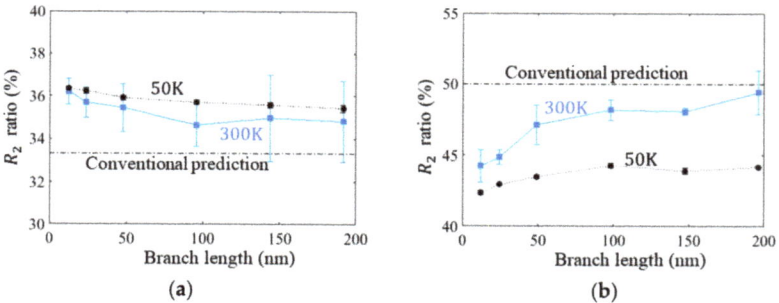

Figure 8. Effect of the temperature on heat flow in branched CNTs with (**a**) cross-junction 1 and (**b**) T-junction 1.

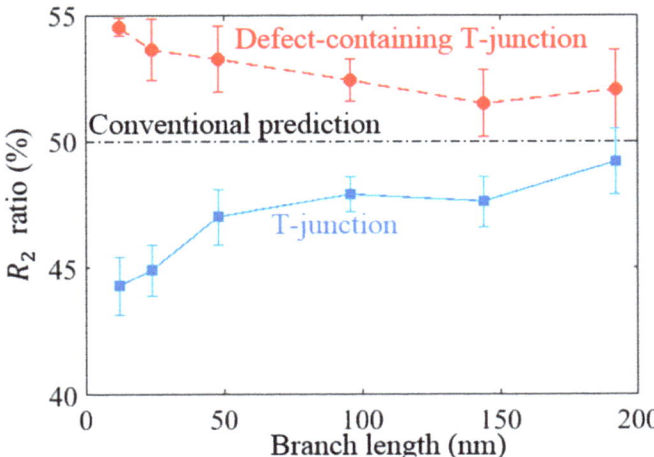

Figure 9. Branch length dependence of the heat current ratio in branch 2 of the branched CNT with the T-junction and defect-containing T-junction.

Moreover, to analyze the thermal transport mechanism inside the branched CNTs with different junctions, the collective atomic velocity cross-correlation spectra between branches were examined to determine whether the phonons still maintained their original vibrational characteristics after passing the junction structure. First, the power spectrum of velocity cross-correlation between a pair of atoms located in two branches was defined as follows:

$$\Gamma_{b,i}^{m\alpha,n\beta}(\omega) = \int dt\, e^{-i\omega t} \left\langle v_{b,i}^{m\alpha}(t) v_{b,i}^{n\beta}(0) \right\rangle, \qquad (4)$$

where $v_{b,i}^{m\alpha}$ was the atomic velocity along the α direction of the bth atom at the ith unit cell located on branch m. The angle in brackets indicates the time moving average. The power spectrum of the time correlation function can provide information on how one atomic vibration was transmitted to another atom in different frequency ranges. As phonons are a collective motion of the periodic elastic arrangement of atoms inside condensed matter, the power spectrum of velocity cross-correlation between a pair of atoms located in two branches was insufficiently informative. Therefore, a group of atoms within n_c unit cells in each branch was selected to calculate the atomic velocity cross-correlation as follows:

$$C^{m\alpha,n\beta}(\omega) = \frac{1}{n_c N_b} \sum_{i}^{n_c} \sum_{j}^{N_b} \frac{\left|\Gamma_{j,i}^{m\alpha,n\beta}(\omega)\right|^2}{\left|\Gamma_{j,i}^{m\alpha,m\alpha}(\omega)\right|\left|\Gamma_{j,i}^{n\beta,n\beta}(\omega)\right|} \qquad (5)$$

where N_b was the number of atoms in one unit cell. In the conducted analysis, n_c was selected as 5. Moreover, the number of atoms in one unit cell, N_b, for a (6, 6) CNT was 24. Our methodology of velocity cross-correlation was similar to the one adopted to calculate coherence length by Latour et al. [36]. Special care was taken to ensure that the distance between each pair of atoms used to calculate the velocity cross-correlation was the same (Figure 10), which was unlike the method used by Latour et al. [36], to ensure that the time for traveling between the calculated atoms would not be different. Different atomic velocity directions were considered to characterize different phonon modes (i.e., the longitudinal and transverse phonon modes). The longitudinal (transverse) phonon mode was identified when the atoms vibrated parallel (perpendicular) to the branch axial direction. The cross-correlation results were averaged over six different runs.

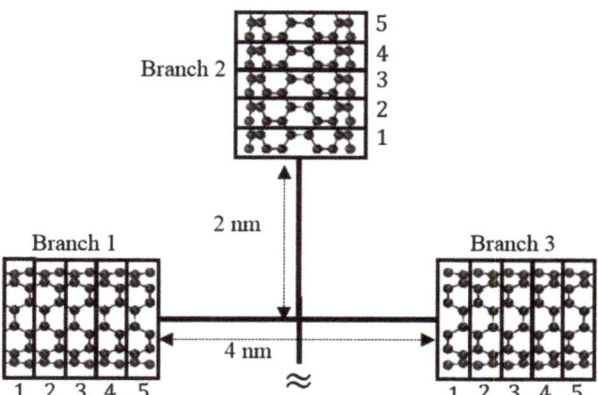

Figure 10. Illustration of atoms at different branches in cross-correlation calculation.

The collective power spectra for the longitudinal and transverse vibrations, which were termed the phonon correlation function, were analyzed for cross-junction 1 (Figure 11) and T-junction 1 (Figure 12). As demonstrated in Figures 11a and 12a, the longitudinal vibrations of branch 1 were more easily transmitted to branch 3 than to branch 2 in both the longitudinal and transverse modes, which indicates that the vibrations in branch 2 originated from scattering and their correlation subsequently reduced. At low frequencies (<1 THz), all of the vibrations between branches exhibited higher correlations, which is consistent with the fact that low-frequency phonons have a long mean free path [36,37]. In addition, the longitudinal vibrations of branch 1 had a higher correlation with the transverse vibration of branch 2 than with the longitudinal vibration of branch 2. This result indicated that the longitudinal phonon may change to a transverse phonon when transported from branch 1 to branch 2. This behavior is similar to the phonon polarization conversion behavior observed by Shi et al. [24,29,30].

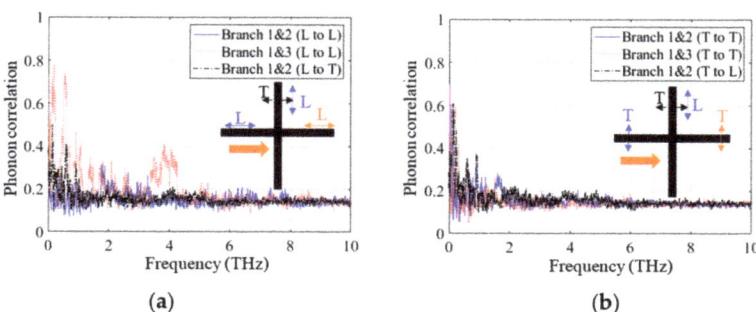

Figure 11. Collective power spectra of velocity cross-correlations of (**a**) longitudinal and (**b**) transverse vibrations of branch 1 with those of branches 2 and 3 for cross-junction 1.

As depicted in Figures 11b and 12b, the transverse vibrations of branch 1 exhibited higher transmittance strength in branch 3 for a T-junction than for a cross-junction, especially at a frequency of approximately 2 THz. However, the transverse vibrations of branch 1 had lower overall correlations with the vibrations of branches 2 and 3 than the longitudinal vibrations of branch 1. This result implies that the transverse phonons scatter more at a junction relative to the longitudinal phonon. Both the longitudinal and transverse phonons could move more and without scattering from branch 1 into branch 3 when encountering T-junctions than when encountering cross-junctions. This result was obtained because the atomic vibration could easily propagate straight into branch 3 when encountering the T-junction without an additional opening and discontinuity at the bottom. It should be pointed out that the phonon correlation function, which describes the resemblance in atomic vibration between branches, only reflects the phonon scattering behavior during thermal transport. This function could not be directly related to the amount of transferred heat because the heat current can be transported by both diffusive and ballistic phonons, and the heat carried by diffusive phonons cannot be reflected by the atomic velocity cross-correlation spectra.

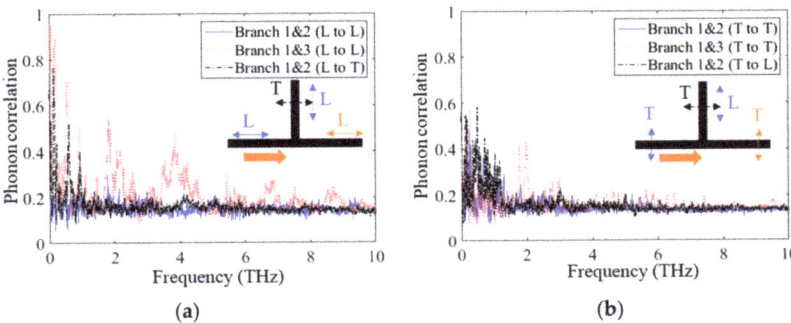

Figure 12. Collective power spectra of the velocity cross-correlation of (**a**) longitudinal and (**b**) transverse vibrations of branch 1 with those of branches 2 and 3 for T-junction 1.

5. Conclusions

In this study, NEMD simulations were conducted to investigate the thermal transport behavior inside branched CNTs with cross and T-junction. Junctions were prepared through the thermal welding of (6, 6) CNTs by removing atoms such that Euler formula-based topological requirements were satisfied. All of the connecting branches had the same length and chirality. The effects of branch length, topological defects at junctions, and temperature on the thermal flow were studied. The simulation results indicated that the heat current preferentially flowed into the side branch rather than the straight one for the cross-junction structures; however, the heat inside the T-junction behaved differently. This directional heat transfer was obvious especially in short branches and low temperatures. Moreover, the atomic configuration at junctions with different topological defects only had a marginal influence on the heat transfer. As the branch length increased, the heat tended to flow equally into each branch. The directional heat transfer cannot be explained using diffusive-based thermal circuit theory even when the anisotropic thermal conductance at the junction was considered. Finally, the collective atomic velocity cross-correlation spectra between branches were used to examine phonon transport mechanisms for the branched CNTs. Furthermore, we found that the phonon correlation function, which describes the similarity of atomic vibration between branches, could only provide information regarding the phonon scattering behavior during thermal transport. Moreover, this function could not be directly related to the amount of transferred heat because the heat current can be transported by both diffusive and ballistic phonons, and the heat carried by diffusive phonons cannot be reflected in the atomic velocity cross-correlation spectra. This finding provides deep insight into the thermal transport mechanisms of branched CNTs, which can be useful when employed in thermal management applications.

Author Contributions: Conceptualization, W.-J.C. and I.-L.C.; methodology, W.-J.C.; software, W.-J.C.; validation, W.-J.C.; formal analysis, W.-J.C. and I.-L.C.; investigation, W.-J.C.; resources, I.-L.C.; data curation, W.-J.C. and I.-L.C.; writing—original draft preparation, W.-J.C. and I.-L.C.; writing—review and editing, W.-J.C. and I.-L.C.; visualization, W.-J.C.; supervision, I.-L.C.; project administration, I.-L.C.; funding acquisition, I.-L.C. Both authors have read and agreed to the published version of the manuscript.

Funding: This research and the APC were funded by the Taiwan Ministry of Science and Technology, MOST 106-2923-E-006-004-MY3 and MOST 105-2628-E-006-003-MY3.

Institutional Review Board Statement: Not applicable.

Acknowledgments: This research was financially supported by the Taiwan Ministry of Science and Technology under the grants MOST 106-2923-E-006-004-MY3 and MOST 105-2628-E-006-003-MY3. The authors want to thank Guan-Shiung Wang for preparing the cross-junction model. We also thank the National Center for High-performance Computing (NCHC) for providing the computational and storage resources.

Conflicts of Interest: The authors declare no conflict of interest. The funders had no role in the design of the study; in the collection, analyses, or interpretation of data; in the writing of the manuscript; or in the decision to publish the results.

References

1. Berber, S.; Kwon, Y.K.; Tománek, D. Unusually high thermal conductivity of carbon nanotubes. *Phys. Rev. Lett.* **2000**, *84*, 4613–4616. [CrossRef] [PubMed]
2. Ruoff, R.S.; Lorents, D.C. Mechanical and thermal properties of carbon nanotubes. *Carbon* **1995**, *33*, 925–930. [CrossRef]
3. Gonzalez, E.; Srivastava, D.; Menon, M. Heat-pulse rectification in carbon nanotube Y junctions. *Phys. Rev. B* **2009**, *79*, 115432. [CrossRef]
4. Park, J.; Lee, J.; Prakash, V. Phonon scattering at SWCNT–SWCNT junctions in branched carbon nanotube networks. *J. Nanoparticle Res.* **2015**, *17*, 59. [CrossRef]
5. Vollebregt, S.; Banerjee, S.; Chiaramonti, A.N.; Tichelaar, F.D.; Beenakker, K.; Ishihara, R. Dominant thermal boundary resistance in multi-walled carbon nanotube bundles fabricated at low temperature. *J. Appl. Phys.* **2014**, *116*, 023514. [CrossRef]
6. Gharib-Zahedi, M.R.; Tafazzoli, M.; Bohm, M.C.; Alaghemandi, M. Transversal thermal transport in single-walled carbon nanotube bundles: Influence of axial stretching and intertube bonding. *J. Chem. Phys.* **2013**, *139*, 184704. [CrossRef]
7. Wang, J.; Chen, D.; Wallace, J.; Gigax, J.; Wang, X.; Shao, L. Introducing thermally stable inter-tube defects to assist off-axial phonon transport in carbon nanotube films. *Appl. Phys. Lett.* **2014**, *104*, 191902. [CrossRef]
8. Liao, D.; Chen, W.; Zhang, J.; Yue, Y. Tuning thermal conductance of CNT interface junction via stretching and atomic bonding. *J. Phys. D Appl. Phys.* **2017**, *50*, 475302. [CrossRef]
9. Aitkaliyeva, A.; Chen, D.; Shao, L. Phonon transport assisted by inter-tube carbon displacements in carbon nanotube mats. *Sci. Rep.* **2013**, *3*, 1–5. [CrossRef]
10. Lian, F.; Llinas, J.P.; Li, Z.; Estrada, D.; Pop, E. Thermal conductivity of chirality-sorted carbon nanotube networks. *Appl. Phys. Lett.* **2016**, *108*, 103101. [CrossRef]
11. Chen, J.; Gui, X.; Wang, Z.; Li, Z.; Xiang, R.; Wang, K.; Wu, D.; Xia, X.; Zhou, Y.; Wang, Q.; et al. Superlow thermal conductivity 3D carbon nanotube network for thermoelectric applications. *ACS Appl. Mater. Interfaces* **2012**, *4*, 81–86. [CrossRef] [PubMed]
12. Kaskela, A.; Nasibulin, A.G.; Timmermans, M.Y.; Aitchison, B.; Papadimitratos, A.; Tian, Y.; Zhu, Z.; Jiang, H.; Brown, D.P.; Zakhidov, A.; et al. Aerosol-synthesized SWCNT networks with tunable conductivity and transparency by a dry transfer technique. *Nano Lett.* **2010**, *10*, 4349–4355. [CrossRef] [PubMed]
13. Sun, D.M.; Timmermans, M.Y.; Tian, Y.; Nasibulin, A.G.; Kauppinen, E.I.; Kishimoto, S.; Mizutani, T.; Ohno, Y. Flexible high-performance carbon nanotube integrated circuits. *Nat. Nanotechnol.* **2011**, *6*, 156–161. [CrossRef] [PubMed]
14. Hu, J.; Ouyang, M.; Yang, P.; Lieber, C.M. Controlled growth and electrical properties of heterojunctions of carbon nanotubes and silicon nanowires. *Nature* **1999**, *399*, 48–51. [CrossRef]
15. Ma, X.; Wang, E.G. CNx/carbon nanotube junctions synthesized by microwave chemical vapor deposition. *Appl. Phys. Lett.* **2001**, *78*, 978–980. [CrossRef]
16. Li, Y.-L.; Kinloch, I.A.; Windle, A.H. Direct spinning of carbon nanotube fibers from chemical vapor deposition synthesis. *Sci. China Technol. Sci.* **2004**, *304*, 276–278. [CrossRef]
17. Terrones, M.; Banhart, F.; Grobert, N.; Charlier, J.C.; Terrones, H.; Ajayan, P. Molecular junctions by joining single-walled carbon nanotubes. *Phys. Rev. Lett.* **2002**, *89*, 075505. [CrossRef] [PubMed]
18. Srivastava, D.; Menon, M.; Ajayan, P.M. Branched carbon nanotube junctions predicted by computational nanotechnology and fabricated through nanowelding. *J. Nanopart. Res.* **2003**, *5*, 395–400. [CrossRef]
19. Aiyiti, A.; Zhang, Z.; Chen, B.; Hu, S.; Chen, J.; Xu, X.; Li, B. Thermal rectification in Y-junction carbon nanotube bundle. *Carbon* **2018**, *140*, 673–679. [CrossRef]
20. Bao, H.; Chen, J.; Gu, X.; Gao, B. A review of simulation methods in micro/nanoscale heat conduction. *Energy Environ. Sci.* **2018**, *1*, 16–55. [CrossRef]
21. Marconnet, A.M.; Panzer, M.A.; Goodson, K.E. Thermal conduction phenomena in carbon nanotubes and related nanostructured materials. *Rev. Mod. Phys.* **2013**, *85*, 1295–1326. [CrossRef]

22. Che, J.; Cagin, T.; Goddard, W.A. Thermal conductivity of carbon nanotubes. *Nanotechnology* **2000**, *11*, 65–69. [CrossRef]
23. Yamamoto, T.; Konabe, S.; Shiomi, J.; Maruyama, S. Crossover from ballistic to diffusive thermal transport in carbon nanotubes. *Appl. Phys. Express* **2009**, *2*, 095003. [CrossRef]
24. Shi, J.; Dong, Y.; Fisher, T.; Ruan, X. Thermal transport across carbon nanotube-graphene covalent and van der Waals junctions. *J. Appl. Phys.* **2015**, *118*, 044302. [CrossRef]
25. Han, Y.; Yang, J.Y.; Hu, M. Unusual strain response of thermal transport in dimerized three-dimensional graphene. *Nanoscale* **2018**, *10*, 5229–5238. [CrossRef] [PubMed]
26. Yang, X.; Chen, D.; Du, Y.; To, A.C. Heat conduction in extended X-junctions of single-walled carbon nanotubes. *J. Phys. Chem. Solids* **2014**, *75*, 123–129. [CrossRef]
27. Varshney, V.; Unnikrishnan, V.; Lee, J.; Roy, A.K. Developing nanotube junctions with arbitrary specifications. *Nanoscale* **2017**, *10*, 403–415. [CrossRef]
28. Zhang, Z.; Kutana, A.; Roy, A.; Yakobson, B.I. Nanochimneys: Topology and thermal conductance of 3D nanotube-graphene cone junctions. *J. Phys. Chem. C* **2017**, *121*, 1257–1262. [CrossRef]
29. Shi, J.; Zhong, Y.; Fisher, T.S.; Ruan, X. Decomposition of the thermal boundary resistance across carbon nanotube–graphene junctions to different mechanisms. *ACS Appl. Mater. Interfaces* **2018**, *10*, 15226–15231. [CrossRef]
30. Shi, J.; Lee, J.; Dong, Y.; Roy, A.; Fisher, T.S.; Ruan, X. Dominant phonon polarization conversion across dimensionally mismatched interfaces: Carbon-nanotube–graphene junction. *Phys. Rev. B Condens. Matter* **2018**, *97*, 134309. [CrossRef]
31. Chen, W.-J.; Chang, I.-L. The atomistic study on thermal transport of the branched CNT. *J. Mech.* **2020**, *36*, 721–727. [CrossRef]
32. Plimpton, S. Fast parallel algorithms for short-range molecular dynamics. *J. Comp. Phys.* **1995**, *117*, 1–19. [CrossRef]
33. Stuart, S.J.; Tutein, A.B.; Harrison, J.A. A reactive potential for hydrocarbons with intermolecular interactions. *J. Chem. Phys.* **2000**, *112*, 6472–6486. [CrossRef]
34. Hone, J.; Whitney, M.; Piskoti, C.; Zettl, A. Thermal conductivity of single-walled carbon nanotubes. *Phys. Rev. B Condens. Matter* **1999**, *59*, R2514. [CrossRef]
35. Dove, M.T. *Introduction to Lattice Dynamics*; Cambridge University Press: Cambridge, UK, 1993.
36. Latour, B.; Volz, S.; Chalopin, Y. Microscopic description of thermal-phonon coherence: From coherent transport to diffuse interface scattering in superlattices. *Phys. Rev. B Condens. Matter* **2014**, *90*, 014307. [CrossRef]
37. Wang, Z.; Ruan, X. On the domain size effect of thermal conductivities from equilibrium and nonequilibrium molecular dynamics simulations. *J. Appl. Phys.* **2017**, *121*, 044301. [CrossRef]

Article

Impacts of Amplitude and Local Thermal Non-Equilibrium Design on Natural Convection within NanoflUid Superposed Wavy Porous Layers

Ammar I. Alsabery [1], Tahar Tayebi [2,3], Ali S. Abosinnee [4], Zehba A. S. Raizah [5] and Ali J. Chamkha [6,7] and Ishak Hashim [8,*]

1. Refrigeration & Air-conditioning Technical Engineering Department, College of Technical Engineering, The Islamic University, Najaf 54001, Iraq; alsabery_a@iunajaf.edu.iq
2. Faculty of Sciences and Technology, Mohamed El Bachir El Ibrahimi University, Bordj Bou Arreridj, El-Anasser 19098, Algeria; tahartayebi@gmail.com
3. Energy Physics Laboratory, Department of Physics, Faculty of Science, Mentouri Brothers Constantine1 University, Constantine 25017, Algeria
4. Computer Technical Engineering Department, College of Technical Engineering, The Islamic University, Najaf 54001, Iraq; abosinnee.ali@gmail.com
5. Department of Mathematics, College of Science, King Khalid University, Abha 61421, Saudi Arabia; zhbarizah@hotmail.com
6. Faculty of Engineering, Kuwait College of Science and Technology, Doha District 35001, Kuwait; achamkha@gmail.com
7. Center of Excellence in Desalination Technology, King Abdulaziz University, P.O. Box 80200, Jeddah 21589, Saudi Arabia
8. Department of Mathematical Sciences, Faculty of Science & Technology, Universiti Kebangsaan Malaysia, UKM Bangi 43600, Selangor, Malaysia
* Correspondence: ishak_h@ukm.edu.my

Citation: Alsabery, A.I.; Tayebi, T.; Abosinnee, A.S.; Raizah, Z.A.S.; Chamkha, A.J.; Hashim, I. Impacts of Amplitude and Local Thermal Non-Equilibrium Design on Natural Convection within NanoflUid Superposed Wavy Porous Layers. *Nanomaterials* 2021, 11, 1277. https://doi.org/10.3390/nano11051277

Academic Editor: Simone Morais

Received: 13 April 2021
Accepted: 10 May 2021
Published: 13 May 2021

Publisher's Note: MDPI stays neutral with regard to jurisdictional claims in published maps and institutional affiliations.

Copyright: © 2021 by the authors. Licensee MDPI, Basel, Switzerland. This article is an open access article distributed under the terms and conditions of the Creative Commons Attribution (CC BY) license (https://creativecommons.org/licenses/by/4.0/).

Abstract: A numerical study is presented for the thermo-free convection inside a cavity with vertical corrugated walls consisting of a solid part of fixed thickness, a part of porous media filled with a nanofluid, and a third part filled with a nanofluid. Alumina nanoparticle water-based nanofluid is used as a working fluid. The cavity's wavy vertical surfaces are subjected to various temperature values, hot to the left and cold to the right. In order to generate a free-convective flow, the horizontal walls are kept adiabatic. For the porous medium, the Local Thermal Non-Equilibrium (LTNE) model is used. The method of solving the problem's governing equations is the Galerkin weighted residual finite elements method. The results report the impact of the active parameters on the thermo-free convective flow and heat transfer features. The obtained results show that the high Darcy number and the porous media's low modified thermal conductivity ratio have important roles for the local thermal non-equilibrium effects. The heat transfer rates through the nanofluid and solid phases are found to be better for high values of the undulation amplitude, the Darcy number, and the volume fraction of the nanofluid, while a limit in the increase of heat transfer rate through the solid phase with the modified thermal ratio is found, particularly for high values of porosity. Furthermore, as the porosity rises, the nanofluid and solid phases' heat transfer rates decline for low Darcy numbers and increase for high Darcy numbers.

Keywords: natural convection; nanofluid-porous cavity; wavy solid wall; darcy-forchheimer model; local thermal non-equilibrium (LTNE)

1. Introduction

Natural convection in a composite cavity, where part of it is porous and the other is fluid (natural convection in many layers of superimposed porous fluids), represents one of the most important topics that has received wide attention from researchers due to its multiple applications in engineering, such as insulation systems for fibrous and granular, packed bed solar energy storage, water conservers, and reactors cooling after the accident, and in geophysics, such as thermal circulation in lakes and contaminant transport in groundwater. Several researchers have dealt with this topic, as it was started by the study of Beavers and Joseph [1], where they presented the boundary conditions between the homogeneous fluid and the porous media in the simple situation. Convective heat transfer in porous beds saturated with a fluid was investigated for various thicknesses and permeabilities of bed [2]. Natural convection heat transfer in an enclosure, which is divided into two regions, one filled with a porous medium and the other with a fluid, was analyzed by Tong and Subramanian [3], for the aim of developing the characteristics of heat transfer for enclosures containing different quantities of porous material. Both studies of Poulikakos et al. [4] and Poulikakos [5] relied on measuring the flow of fluid floating on a porous bed heated from below at Rayleigh numbers above a critical value. The authors in Beckermann et al. [6] presented a two-dimensional study on natural convection, a rectangular fluid enclosure partially filled with different layers of porous material, vertical or horizontal. Numerical investigations were conducted for various enclosure aspect ratios, Rayleigh and Darcy numbers, and ratios and thicknesses of thermal conductivity of the porous region [7]. The authors in Hirata et al. [8] discussed the thermosolutal natural convection onset in horizontal superimposed fluid-porous layers. A numerical and analytical investigation was done for combined thermal and moisture conventions in an enclosure filled with a partially porous medium to enhance the moisture transport in the thermal energy storage unit [9]. The authors in Mikhailenko et al. [10] discussed the mechanism to address the effects of a uniform rotation and a porous layer in a local heat source electronic cabinet, where it studied the impacts of the Rayleigh, Taylor, and Darcy numbers and the porous layer thickness on hydro-thermodynamics. The authors in Saleh et al. [11] investigated the unstable convective flow in a vertical porous layer inside an enclosure due to a flexible fin.

To enhance the fluids' thermal properties, researchers and engineers have used new kinds of particles with a nanometer size, which are named nanoparticles, in traditional fluids, which generated the term "nanofluid". The applications of the heat transfer of nanofluids have been widely used for, e.g., cooling electronics, heating exchangers, car radiators, and machining [12]. The authors in Alsabery et al. [13] provided an explanation for the influence of the Darcy number, Rayleigh number, nanoparticle volume fraction, and power-law index on streamlines, isotherms, and the total heat transfer and on the thermal conductivity of the nanofluid and the porous medium. The numerical analysis of Al-Zamily [14] was implemented to investigate the fluid flow, entropy generation, and heat transfer inside an enclosure with an internal heat generation. The authors in Armaghani et al. [15] presented numerically the natural convection and generation of thermodynamic irreversibility in a cavity containing a partial porous layer filled with a Cu-water nanofluid. The authors in Miroshnichenko et al. [16] utilized a numerical simulation of porous layers' effect on natural convection in an open cavity with a vertical hot wall and filled with a nanofluid.

Four main classifications in the modelling of transportation methods for porous materials include the Local Thermal Non-Equilibrium (LTNE), thermal dispersion, constant porosity, and variant porosity. LTNE assumptions can be used in modelling the heat exchange of convection in porous materials due to the different thermal conductivities in the fluid and porous material [12]. By applying an exact Chebyshev spectral element method, the natural convection in a porous cavity was improved using an LTNE model [17]. By considering LTNE effects, the authors in Ghalambaz et al. [18] addressed the natural convection in a cavity filled with a porous medium with the consideration of the thickness of the solid walls of the cavity. Taking into account the local thermal non-equilibrium model, natural convective circulation in a rotating porous cavity was investigated with a variable volumetric heat generation by [19].

The authors in Sivasankaran et al. [20] analyzed the convective heat and fluid flow of a nanofluid in an inclined cavity saturated with a heat-generating porous medium based on the LTNE model. The authors in Tahmasebi et al. [21] investigated the heat transfer of the natural convection in an enclosure filled with a nanofluid in three different layers of the fluid, the porous medium, and the solid, where the local thermal non-equilibrium model was used to model the porous layer. The studies of [22–24] investigated the natural convection of different nanofluids in each article within a porous cavity depending on a Local Thermal Non-Equilibrium Model (LTNEM). Natural, forced, and Marangoni convective flows in an open cavity partially saturated with a porous medium under the impacts of an inclined magnetic field were studied, where the LTNEM was used to represent the thermal field in the porous layer [25].

The study of natural convection in a wavy porous cavity is an interesting topic due to its wide range of usage in engineering, e.g., for the management of nuclear waste, the cooling of transpiration, building thermal insulators, geothermal power plants, and grain storage, and in geophysics, e.g., for modelling pollutant spreading (radionuclides), the movement of water in geothermal reservoirs, and petroleum reservoirs' enhanced recovery [26]. Free convection in a cubical porous enclosure has been controlled by the wavy shape of the bottom wall and by inserting a conductive square cylinder inside the considered cavity [27]. The natural convection heat transfer inside a square wavy-walled enclosure filled with nanofluid and containing a hot inner corrugated cylinder was simulated by [28]. The enclosure was divided into two layers, one filled with Ag nanofluid and the other with porous media. The authors in Kadhim et al. [29] presented a parametric numerical analysis of the free convection in a porous enclosure with wavy walls filled by a hybrid nanofluid, at several inclination angles. The authors in Alsabery et al. [30] simulated the free convection heat transfer inside a porous cavity filled with water-based nanofluid with the consideration of the LTNE model. They assumed that there was an inner solid cylinder centered in the enclosure and that the bottom wall of the cavity was heated and wavy.

As acknowledged in an earlier literature survey, and to the best of the authors' knowledge, and based on the need to consider the LTNE condition, there is no study dealing with the natural convection flow within nanofluid-superposed wavy porous layers with the local thermal non-equilibrium model. Therefore, this work proposes an understanding of the amplitude's impacts and the local thermal non-equilibrium of a nanofluid-superposed wavy porous layers via the fluid flow and heat transfer features.

2. Mathematical Formulation

The two-dimensional natural convection state within the wavy-walled cavity with length L is explained in Figure 1. The analysed composite cavity is divided into three layers (portions). The first layer (left wavy portion) is solid as brickwork ($k_w = 0.76$ tW/m.°C), the second layer (middle portion) is loaded with a porous medium that is saturated with nanofluid, and the third layer (right wavy portion) is filled with a nanofluid. The wavy (vertical left) solid surface has a fixed hot temperature of T_h, while the vertical right wavy surface is fixed with a cold temperature of T_c. On the other hand, the horizontal top and bottom surfaces are preserved as adiabatic. The edges of the domain (except for the interface surface between the porous-nanofluid layer) are supposed to remain impermeable. The mixed liquid inside the composite cavity performs as a water-based nanofluid holding Al_2O_3 nanoparticles. The Forchheimer-Brinkman-extended Darcy approach and the Boussinesq approximation remain appropriate. In contrast, the nanofluid phase's convection and the solid matrix are not in a local thermodynamic equilibrium condition. The set of porous media applied in the following output is glass balls ($k_m = 1.05$ W/m.°C). Considering the earlier specified hypotheses, the continuity, momentum, and energy equations concerning the Newtonian fluid, laminar, and steady-state flow are formulated as follows:

For the nanofluid layer,

$$\frac{\partial u_{nf}}{\partial x} + \frac{\partial v_{nf}}{\partial y} = 0, \tag{1}$$

$$u_{nf}\frac{\partial u_{nf}}{\partial x} + v_{nf}\frac{\partial u_{nf}}{\partial y} = -\frac{1}{\rho_{nf}}\frac{\partial p}{\partial x} + \frac{\mu_{nf}}{\rho_{nf}}\left(\frac{\partial^2 u_{nf}}{\partial x^2} + \frac{\partial^2 u_{nf}}{\partial y^2}\right), \tag{2}$$

$$u_{nf}\frac{\partial v_{nf}}{\partial x} + v_{nf}\frac{\partial v_{nf}}{\partial y} = -\frac{1}{\rho_{nf}}\frac{\partial p}{\partial y} + \frac{\mu_{nf}}{\rho_{nf}}\left(\frac{\partial^2 v_{nf}}{\partial x^2} + \frac{\partial^2 v_{nf}}{\partial y^2}\right) + \beta_{nf}g(T_h - T_c), \tag{3}$$

$$u_{nf}\frac{\partial T_{nf}}{\partial x} + v_{nf}\frac{\partial T_{nf}}{\partial y} = \frac{k_{nf}}{(\rho C_p)_{nf}}\left(\frac{\partial^2 T_{nf}}{\partial x^2} + \frac{\partial^2 T_{nf}}{\partial y^2}\right). \tag{4}$$

For the porous layer,

$$\frac{\partial u_m}{\partial x} + \frac{\partial v_m}{\partial y} = 0, \tag{5}$$

$$\frac{\rho_{nf}}{\varepsilon^2}\left(u_m\frac{\partial u_m}{\partial x} + v_m\frac{\partial u_m}{\partial y}\right) = -\frac{\partial p}{\partial x} + \frac{\mu_{nf}}{\varepsilon}\left(\frac{\partial^2 u_m}{\partial x^2} + \frac{\partial^2 u_m}{\partial y^2}\right)$$
$$-\left(\frac{\mu_{nf}}{K}u_m - \frac{1.75}{\sqrt{150}\varepsilon^{3/2}}\frac{\rho_{nf}u_m|\mathbf{u}|}{\sqrt{K}}\right), \tag{6}$$

$$\frac{\rho_{nf}}{\varepsilon^2}\left(u_m\frac{\partial v_m}{\partial x} + v_m\frac{\partial v_m}{\partial y}\right) = -\frac{\partial p}{\partial y} + \frac{\mu_{nf}}{\varepsilon}\left(\frac{\partial^2 v_m}{\partial x^2} + \frac{\partial^2 v_m}{\partial y^2}\right)$$
$$-\left(\frac{\mu_{nf}}{K}v_m - \frac{1.75}{\sqrt{150}\varepsilon^{3/2}}\frac{\rho_{nf}v_m|\mathbf{u}|}{\sqrt{K}}\right) + (\rho\beta)_{nf}g(T_h - T_c), \tag{7}$$

$$u_m\frac{\partial T_m}{\partial x} + v_m\frac{\partial T_m}{\partial y} = \frac{\varepsilon k_{nf}}{(\rho C_p)_{nf}}\left(\frac{\partial^2 T_m}{\partial x^2} + \frac{\partial^2 T_m}{\partial y^2}\right) + \frac{h(T_s - T_m)}{(\rho C_p)_{nf}}, \tag{8}$$

$$0 = (1-\varepsilon)k_s\left(\frac{\partial^2 T_s}{\partial x^2} + \frac{\partial^2 T_s}{\partial y^2}\right) + h(T_m - T_s). \tag{9}$$

The energy equation of the wavy left solid surface is

$$\frac{\partial^2 T_w}{\partial x^2} + \frac{\partial^2 T_w}{\partial y^2} = 0. \tag{10}$$

The subscripts nf, m, s, and w correspond to the nanofluid layer, porous layer (nanofluid phase), porous layer (solid phase), and solid wavy surface, respectively. x and y are the fluid velocity elements, $|\mathbf{u}| = \sqrt{u^2 + v^2}$ denotes the Darcy velocity, g displays the acceleration due to gravity, ε signifies the porosity of the medium, and K is the permeability of the porous medium which is determined as

$$K = \frac{\varepsilon^3 d_m^2}{150(1-\varepsilon)^2}. \tag{11}$$

Here, d_m represents the average particle size of the porous bed.

The thermophysical characteristics regarding the adopted nanofluid for the 33 nm particle-size are given by [31]:

$$(\rho C_p)_{nf} = (1-\phi)(\rho C_p)_f + \phi(\rho C_p)_p, \tag{12}$$

$$\rho_{nf} = (1-\phi)\rho_f + \phi\rho_p, \tag{13}$$

$$(\rho\beta)_{nf} = (1-\phi)(\rho\beta)_f + \phi(\rho\beta)_p, \tag{14}$$

$$\frac{\mu_{nf}}{\mu_f} = \frac{1}{1 - 34.87\left(\frac{d_p}{d_f}\right)^{-0.3}\phi^{1.03}}, \tag{15}$$

$$\frac{k_{nf}}{k_f} = 1 + 4.4 \text{Re}_B^{0.4} \text{Pr}^{0.66} \left(\frac{T}{T_{fr}}\right)^{10} \left(\frac{k_p}{k_f}\right)^{0.03} \phi^{0.66}. \tag{16}$$

where Re_B is defined as

$$\text{Re}_B = \frac{\rho_f u_B d_p}{\mu_f}, \quad u_B = \frac{2k_b T}{\pi \mu_f d_p^2}. \tag{17}$$

The molecular diameter of water (d_f) is given as [31]

$$d_f = 0.1 \left[\frac{6M}{N^* \pi \rho_f}\right]^{\frac{1}{3}}. \tag{18}$$

Now, we present the employed non-dimensional variables:

$$(X,Y) = \frac{(x,y)}{L}, \ U_{nf,m} = \frac{u_{nf,m}L}{\alpha_f}, \ V_{nf,m} = \frac{v_{nf,m}L}{\alpha_f}, \ \theta_{nf} = \frac{T_{nf} - T_c}{T_h - T_c},$$

$$\theta_m = \frac{T_m - T_c}{T_h - T_c}, \ P = \frac{pL^2}{\rho_f \alpha_f^2}, \ k_{eff} = \varepsilon k_{nf} + (1-\varepsilon)k_m, \ C_F = \frac{1.75}{\sqrt{150}}. \tag{19}$$

The set scheme leads to the following dimensionless governing equations:
In the nanofluid layer,

$$\frac{\partial U_{nf}}{\partial X} + \frac{\partial V_{nf}}{\partial Y} = 0, \tag{20}$$

$$U_{nf}\frac{\partial U_{nf}}{\partial X} + V_{nf}\frac{\partial U_{nf}}{\partial Y} = -\frac{\partial P}{\partial X} + \text{Pr}\frac{\rho_f}{\rho_{nf}}\frac{\mu_{nf}}{\mu_f}\left(\frac{\partial^2 U_{nf}}{\partial x^2} + \frac{\partial^2 U_{nf}}{\partial Y^2}\right), \tag{21}$$

$$U_{nf}\frac{\partial V_{nf}}{\partial X} + V_{nf}\frac{\partial V_{nf}}{\partial Y} = -\frac{\partial P}{\partial Y} + \text{Pr}\frac{\rho_f}{\rho_{nf}}\frac{\mu_{nf}}{\mu_f}\left(\frac{\partial^2 V_{nf}}{\partial X^2} + \frac{\partial^2 V_{nf}}{\partial Y^2}\right)$$

$$+ \frac{(\rho\beta)_{nf}}{\rho_{nf}\beta_f} Ra \, \text{Pr} \, \theta_{nf}, \tag{22}$$

$$U_{nf}\frac{\partial \theta_{nf}}{\partial X} + V_{nf}\frac{\partial \theta_{nf}}{\partial Y} = \frac{(\rho C_p)_f}{(\rho C_p)_{nf}}\frac{k_{nf}}{k_f}\left(\frac{\partial^2 \theta_{nf}}{\partial X^2} + \frac{\partial^2 \theta_{nf}}{\partial Y^2}\right). \tag{23}$$

In the porous layer,

$$\frac{\partial U_m}{\partial X} + \frac{\partial V_m}{\partial Y} = 0, \tag{24}$$

$$\frac{1}{\varepsilon^2}\left(U_m\frac{\partial U_m}{\partial X} + V_m\frac{\partial U_m}{\partial Y}\right) = -\frac{\partial P}{\partial X} + \frac{\rho_f}{\rho_{nf}}\frac{\mu_{nf}}{\mu_f}\frac{\Pr}{\varepsilon}\left(\frac{\partial^2 U_m}{\partial X^2} + \frac{\partial^2 U_m}{\partial Y^2}\right)$$
$$-\frac{\rho_f}{\rho_{nf}}\frac{\mu_{nf}}{\mu_f}\frac{\Pr}{Da}U_m - \frac{C_F\sqrt{U_m^2 + V_m^2}}{\sqrt{Da}}\frac{U_m}{\varepsilon^{3/2}}, \tag{25}$$

$$\frac{1}{\varepsilon^2}\left(U_m\frac{\partial V_m}{\partial X} + V_m\frac{\partial V_m}{\partial Y}\right) = -\frac{\partial P}{\partial Y} + \frac{\rho_f}{\rho_{nf}}\frac{\mu_{nf}}{\mu_f}\frac{\Pr}{\varepsilon}\left(\frac{\partial^2 V_m}{\partial X^2} + \frac{\partial^2 V_m}{\partial Y^2}\right)$$
$$-\frac{\rho_f}{\rho_{nf}}\frac{\mu_{nf}}{\mu_f}\frac{\Pr}{Da}V_m - \frac{C_F\sqrt{U_m^2 + V_m^2}}{\sqrt{Da}}\frac{V_m}{\varepsilon^{3/2}} + \frac{(\rho\beta)_{nf}}{\rho_{nf}\beta_f}Ra\Pr\theta_m, \tag{26}$$

$$\frac{1}{\varepsilon}\left(U_m\frac{\partial \theta_m}{\partial X} + V_m\frac{\partial \theta_m}{\partial Y}\right) = \frac{k_{eff}}{k_f}\frac{(\rho C_p)_f}{(\rho C_p)_{nf}}\left(\frac{\partial^2 \theta_m}{\partial X^2} + \frac{\partial^2 \theta_m}{\partial Y^2}\right)$$
$$+\frac{(\rho C_p)_f}{(\rho C_p)_{nf}}H(\theta_s - \theta_m), \tag{27}$$

$$0 = \frac{\partial^2 \theta_s}{\partial X^2} + \frac{\partial^2 \theta_s}{\partial Y^2} + \gamma H(\theta_m - \theta_s). \tag{28}$$

In the wavy solid wall,

$$\frac{\partial^2 \theta_w}{\partial X^2} + \frac{\partial^2 \theta_w}{\partial Y^2} = 0. \tag{29}$$

The dimensionless boundary conditions of Equations (20)–(28) are

On the left solid hot wavy surface,
$$U = V = 0, \ \theta_w = 1, \ A(1 - \cos(2N\pi Y)), \ 0 \leq Y \leq 1, \tag{30}$$
On the bottom adiabatic horizontal surface,
$$U = V = 0, \ \frac{\partial \theta_{(nf,m,s,w)}}{\partial Y} = 0, \ 0 \leq X \leq 1, \ Y = 0, \tag{31}$$
On the right cold wavy surface,
$$U = V = 0, \ \theta_{nf} = 0, \ 1 - A(1 - \cos(2N\pi Y)), \ 0 \leq Y \leq 1, \tag{32}$$
On the top adiabatic horizontal surface,
$$U = V = 0, \ \frac{\partial \theta_{(nf,m,s,w)}}{\partial Y} = 0, \ 0 \leq X \leq 1, \ Y = 1, \tag{33}$$

The dimensionless boundary forms toward the interface between the nanofluid and the porous layers will be obtained from (1) the continuity of tangential and normal velocities, (2) shear and normal stresses, and (3) the temperature and the heat flux crossing the central interface and allowing an identical dynamic viscosity ($\mu_{nf} = \mu_m$) into both layers. Therefore, the interface dimensionless boundary conditions can be addressed as the following:

$$\theta_{nf}|_{Y=D^+} = \theta_m|_{Y=D^-}, \tag{34}$$
$$\left.\frac{\partial \theta_{nf}}{\partial Y}\right|_{Y=D^+} = \frac{k_{eff}}{k_{nf}}\left.\frac{\partial \theta_m}{\partial Y}\right|_{Y=D^-}, \tag{35}$$
$$U_{nf}|_{Y=D^+} = U_m|_{Y=D^-}, \tag{36}$$
$$V_{nf}|_{Y=D^+} = V_m|_{Y=D^-}, \tag{37}$$

Here, D denotes the nanofluid layer's thickness, and the subscripts + and − indicate that the corresponding measures are estimated while addressing the interface of the nanofluid and the porous layers, respectively. $Ra = \frac{g\beta_f(T_h-T_c)L^3}{\nu_f \alpha_f}$ and $Pr = \frac{\nu_f}{\alpha_f}$ signify the Rayleigh number and the Prandtl number related to the used base liquid.

The local Nusselt numbers (Nu_{nf} and Nu_s) at the wavy vertical (left) surface for the nanofluid and the solid phases, respectively, are written as follows:

$$Nu_{nf} = \frac{k_{eff}}{k_f}\left(\frac{\partial \theta_{nf}}{\partial n}\right)_n, \tag{38}$$

$$Nu_s = \frac{k_s}{k_f}\left(\frac{\partial \theta_s}{\partial n}\right)_n. \tag{39}$$

Here, n denotes the entire length of the curved heat source.

Lastly, the average Nusselt numbers at the wavy vertical surface within the nanofluid and solid phases are addressed by the following:

$$\overline{Nu}_{nf} = \int_0^n Nu_{nf}\,dn, \tag{40}$$

$$\overline{Nu}_s = \int_0^n Nu_s\,dn. \tag{41}$$

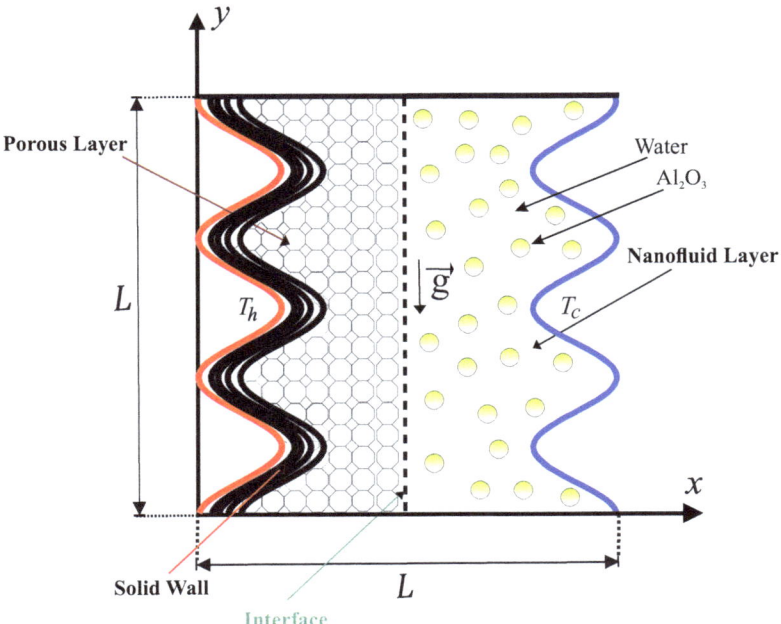

Figure 1. Schematic representation concerning the convection flow in the wavy-walled composite.

3. Numerical Method and Validation

The governing dimensionless equations Equations (20)–(28) ruled with the boundary conditions Equations (30)–(37) are solved by the Galerkin weighted residual finite element technique. The computational region is discretised into small triangular portions as shown in Figure 2.

These small triangular Lagrange components with various forms are applied to each flow variable within the computational region. Residuals for each conservation equation is accomplished through substituting the approximations within the governing equations. The Newton-Raphson iteration algorithm is adopted for clarifying the nonlinear expressions into the momentum equations. The convergence from the current numerical solution is considered, while the corresponding error of each of the variables satisfies the following convergence criteria:

$$\left| \frac{\Gamma^{i+1} - \Gamma^i}{\Gamma^{i+1}} \right| \leq 10^{-6}.$$

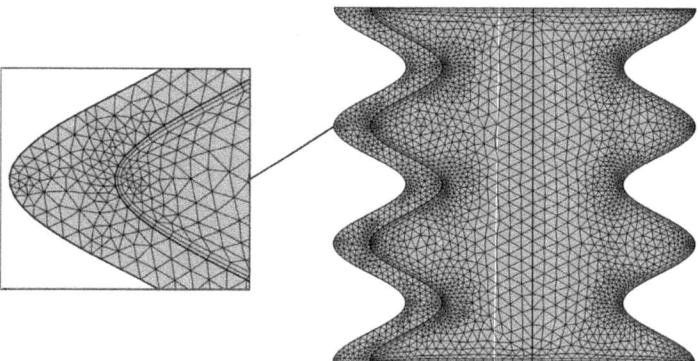

Figure 2. Framework configuration of the FEM for the grid dimension of 5464 components.

To assure the confidence of the existing numerical solution at the grid size of the numerical region, we have adopted various grid dimensions for calculating the minimum strength of the flow circulation (Ψ_{min}), the average Nusselt number of the nanofluid phase ($\overline{Nu}_{nf}$), and the average Nusselt number of the solid phase ($\overline{Nu}_s$) for the case of $Ra = 10^6$, $Da = 10^{-3}$, $\phi = 0.02$, $N = 3$, $\gamma = 10$, $H = 10$, $\varepsilon = 0.5$, and $A = 0.1$. The outcomes are displayed in Table 1, which designates insignificant variations for the G6 grids and higher. Hence, concerning all calculations into this numerical work, the G6 uniform grid is applied.

Table 1. Grid testing for Ψ_{min}, $\overline{Nu}_{nf}$, and $\overline{Nu}_s$ at different grid sizes for $Ra = 10^6$, $Da = 10^{-3}$, $\phi = 0.02$, $N = 3$, $\gamma = 10$, $H = 10$, $\varepsilon = 0.5$, and $A = 0.1$.

Grid Size	Number of Elements	Ψ_{min}	$\overline{Nu}_{nf}$	$\overline{Nu}_s$
G1	3187	−11.812	6.4267	4.677
G2	3686	−11.691	6.4309	4.6961
G3	4096	−11.676	6.4453	4.7096
G4	4576	−11.603	6.4607	4.7197
G5	5464	−11.591	6.4633	4.7257
G6	**10,180**	**−11.528**	**6.4654**	**4.7431**
G7	21,830	−11.503	6.4655	4.7434

Concerning the validation for the current numerical data, the outcomes are examined with earlier published experimental results reported by Beckermann et al. [6] for natural convection within a square cavity including fluid and porous layers, as performed in Figure 3. Besides that, a comparison is obtained for the resulting patterns and the one implemented by Khanafer et al. [32] for the case of natural convection heat transfer in a wavy non-Darcian porous cavity, as displayed in Figure 4. According to the above-achieved comparisons, the numerical outcomes of the existing numerical code are significant to a great degree of reliability.

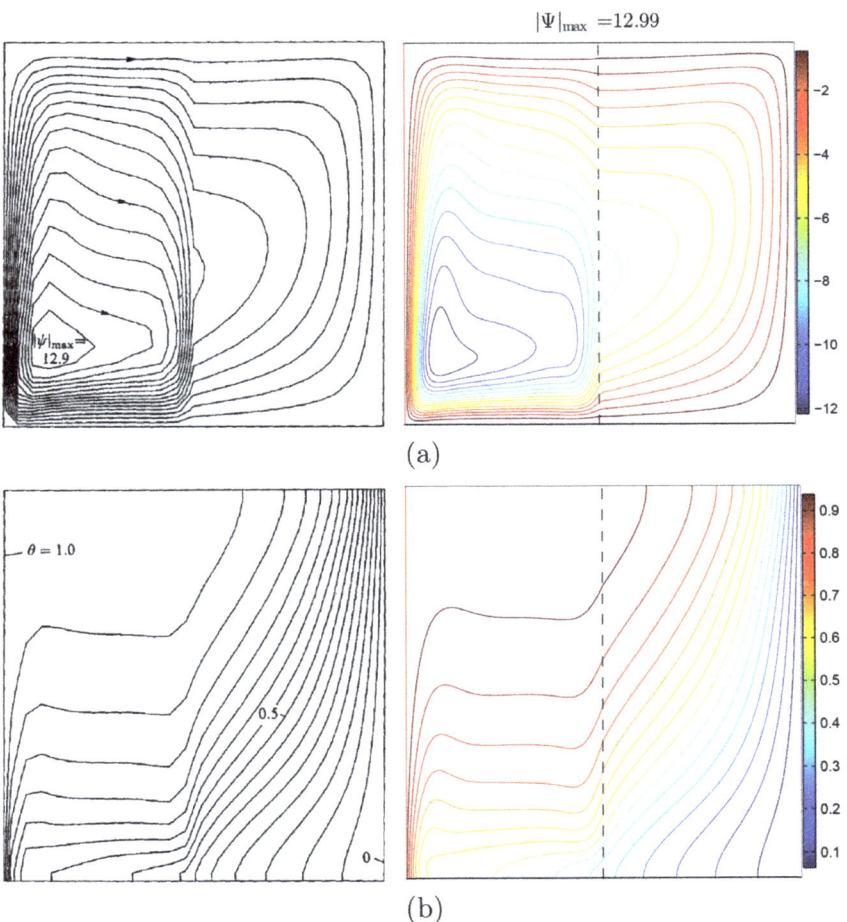

Figure 3. (a) Streamlines of Beckermann et al. [6] (**left**) and the present study (**right**); (b) isotherms for $Ra = 3.70 \times 10^6$, $Da = 1.370 \times 10^{-5}$, $\varepsilon = 0.9$, $D = 0.5$ $N = 0$, $\frac{k_{eff}}{k_f} = 1.362$, and $Pr = 6.44$.

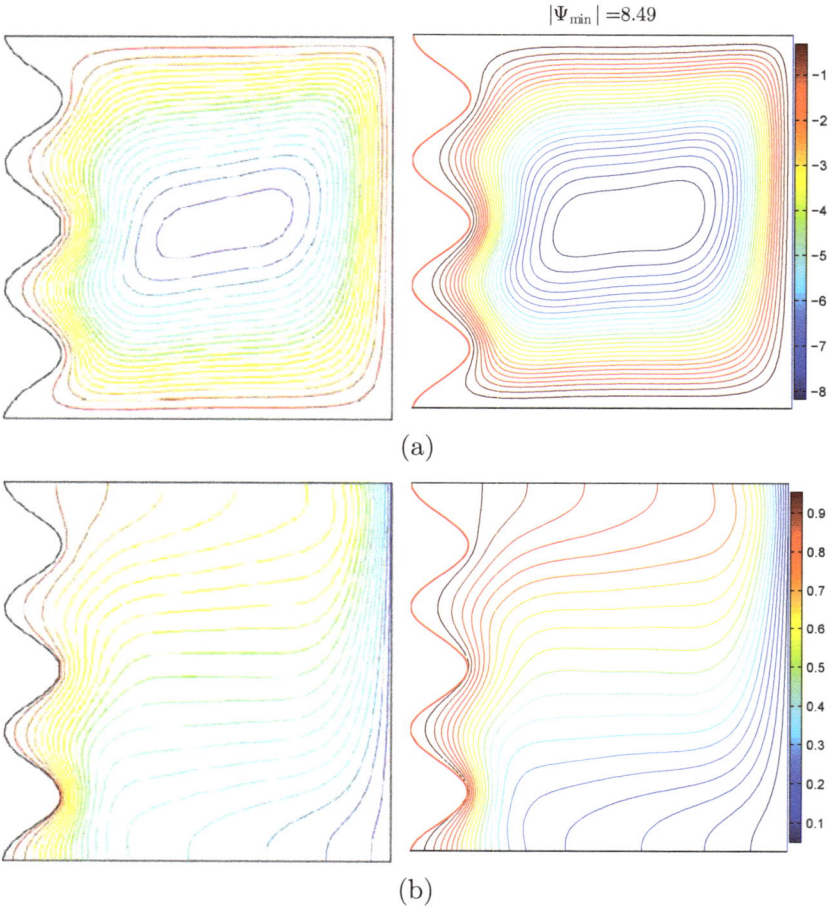

Figure 4. (**a**) Streamlines of (**left**) Khanafer et al. [32] and (**right**) the present study; (**b**) isotherms of (**left**) Khanafer et al. [32] and (**right**) the present study for $Ra = 10^5$, $Da = 10^{-2}$, $\varepsilon = 0.9$, $N = 3$, $D = 0$, and $\Pr = 1$.

4. Results and Discussion

The outcomes described by streamlines, isotherms, and isentropic distributions are addressed within this section. We have modified the following four parameters: the Darcy number ($10^{-6} \leq Da \leq 10^{-2}$), the nanoparticle volume fraction ($0 \leq \phi \leq 0.04$), the modified conductivity ratio ($0.1 \leq \gamma \leq 1000$), the amplitude ($0 \leq A \leq 0.2$), and the porosity of the medium ($0.2 \leq \varepsilon \leq 0.8$). The values of the Rayleigh number, the number of undulations, the coefficient of inter-phase heat transfer, the thickness of the wavy solid wall, and the Prandtl number are fixed at $Ra = 10^6$, $N = 3$, $H = 10$, $W = 0.2$, and $\Pr = 4.623$, respectively. Table 2 displays the thermos-physical properties of the base fluid (water) and the solid Al_2O_3 phases at $T = 310$ K.

Figure 5 displays from left to right, respectively, streamlines, isotherms of the nanofluid phase, and isotherms of the solid phase for various Darcy numbers (Da) when $\phi = 0.02$, $\gamma = 10$, $A = 0.1$, and $\varepsilon = 0.5$. For a low Darcy number, the isotherms are virtually vertical in the porous layer and correspond to the solid phase isotherms, indicating that the heat transfer occurs essentially by the conduction mode because the porous medium becomes less permeable. The porous matrix causes the flow to cease in the porous layer; as a result, the flow is closed or entirely limited to the nanofluid layer and is not able to permeate into the porous medium. The heat transfer in the nanofluid layer is mainly convective, as is shown from the isotherms and the streamlines. When the Darcy number increases, the porous layer provides less resistance to the nanofluid flow, the natural convection increases, and the mechanism of heat transfer shifts from the conduction mechanism at a small Darcy number into the convection mechanism at a high Darcy number in the porous layer as well as in the entire enclosure. A high thermal boundary layer is present at the porous-layer/conducting solid-wall interface. Moreover, by comparing the isothermal lines in the solid part and the nanofluid phase in the porous layer, it is clear that, when the heat transport mode is mainly governed by conduction at low Darcy number values, the Local Thermal-Equilibrium (LTE) state is feasible because the isotherms are conformal (identical and similar). Meanwhile, the effect of the Local Thermal Non-Equilibrium (LTNE) is important at high Darcy numbers since there is a marked difference in the temperature distribution between the solid phase and the nanofluid phase in the porous layer.

To frame the effect of the porous medium permeability, we present in Figure 6 the numerical results given by the profiles of the local velocity with the vertical line at $X = 0.5$ (a), the local Nusselt number of the nanofluid phase (b), and the local Nusselt number of the solid phase (c) for different Da for the case of $\phi = 0.02$, $\gamma = 10$, $A = 0.1$, and $\varepsilon = 0.5$. The velocity profiles indicate the rotational aspect of the nanofluid in the cavity. The nanofluids' circulation rate increases with increasing values of the Darcy number. It is also evident from this figure that the local Nusselt numbers, for both the nanofluid phase and the solid phase, form peaks, and each peak corresponds to a convex boundary of the undulating hot wall. In comparison, the values of the Nusselt number improve by incrementing the Darcy number, and this improvement is more progressive for medium Darcy number levels. It is also worth noting that the heat transfer rate on the lower portion of the wavy wall is bigger than on the top. This is because these regions represent the contact areas for the cold nanofluid returning from the opposite cold wall, so a high temperature difference exists there, which causes a high heat transfer rate.

Table 2. Thermo-physical characteristics concerning pure liquid (water) and Al_2O_3 nanoparticles at $T = 310$ K [33].

Physical Properties	Fluid Phase (Water)	Al_2O_3
C_p (J/kgK)	4178	765
ρ (kg/m^3)	993	3970
k (Wm^{-1}K^{-1})	0.628	40
$\beta \times 10^5$ (1/K)	36.2	0.85
$\mu \times 10^6$ (kg/ms)	695	–
d_p (nm)	0.385	33

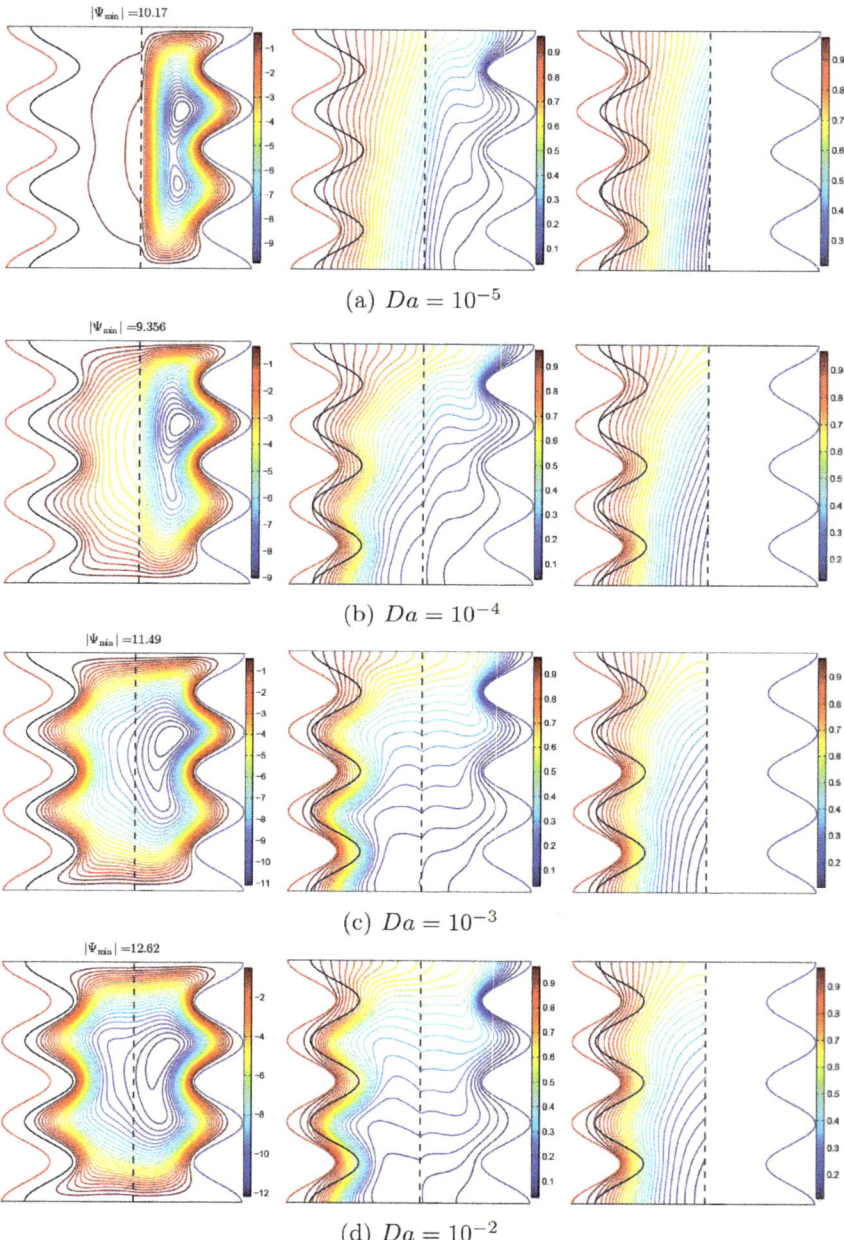

Figure 5. Streamlines (**left**), isotherms of the nanofluid phase (**middle**), and isotherms of the solid phase (**right**) with various Darcy numbers (Da); $\phi = 0.02$, $\gamma = 10$, $A = 0.1$, and $\varepsilon = 0.5$.

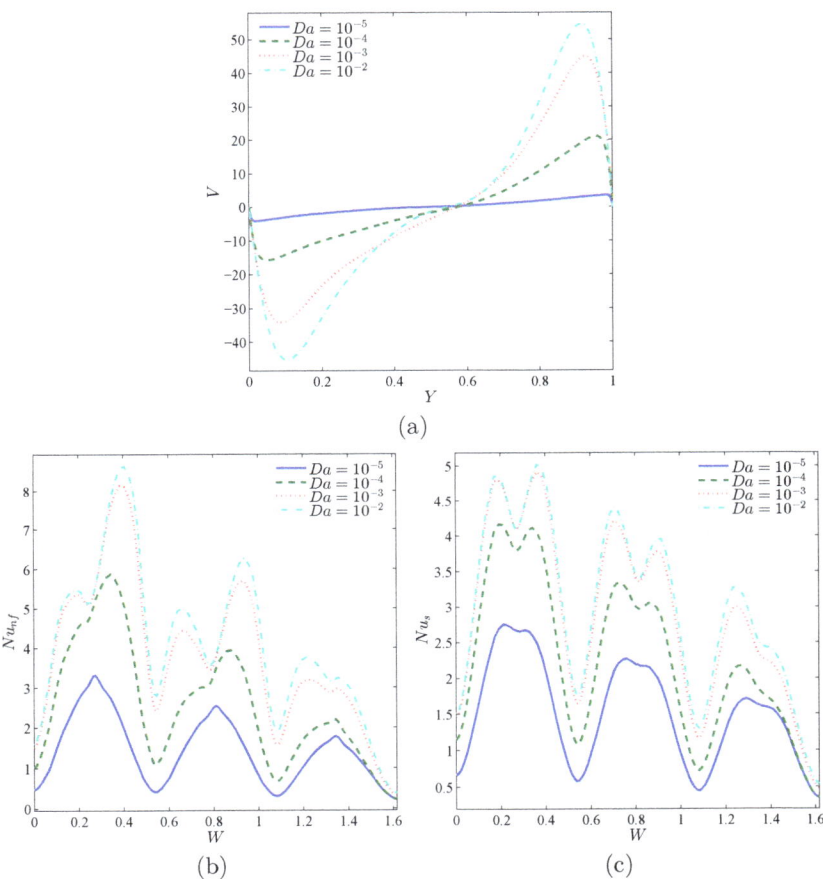

Figure 6. Local velocity (**a**), the local Nusselt number of the nanofluid phase (**b**), and the local Nusselt number of the solid phase (**c**) with the vertical line (Y) for $X = 0.5$ for different Da values; $\phi = 0.02$, $\gamma = 10$, $A = 0.1$, and $\varepsilon = 0.5$.

The differences in the streamlines patterns and isotherms of the nanofluid and solid phases in the different regions with respect to the thermal conductivity ratio in the porous layer (γ) when $Da = 10^{-3}$, $\phi = 0.02$, $A = 0.1$, and $\varepsilon = 0.5$ are depicted in Figure 7. For low values of γ, the core of the flow vortex is found in the nanofluid layer, indicating the nanofluid circulation strength in this region and its weakness in the porous layer region. This is because most of the heat is transmitted through the solid matrix instead of the nanofluid in the porous layer due to the high value of the solid thermal conductivity. The nanofluid-phase and solid-phase isothermal lines are not similar, indicating the existence of the local thermal non-equilibrium case. When raising the value of γ from 0.1 to 1000, the flow vortex expands to cover the porous layer region, and the speed of the circulation strength increases in the porous layer and decreases in the nanofluid layer. It also seems that, as γ increases, the system tends to realize the local thermal equilibrium situation in the porous layer, which is observed by the identicalness between the nanofluid-phase isotherms and those of the solid phase.

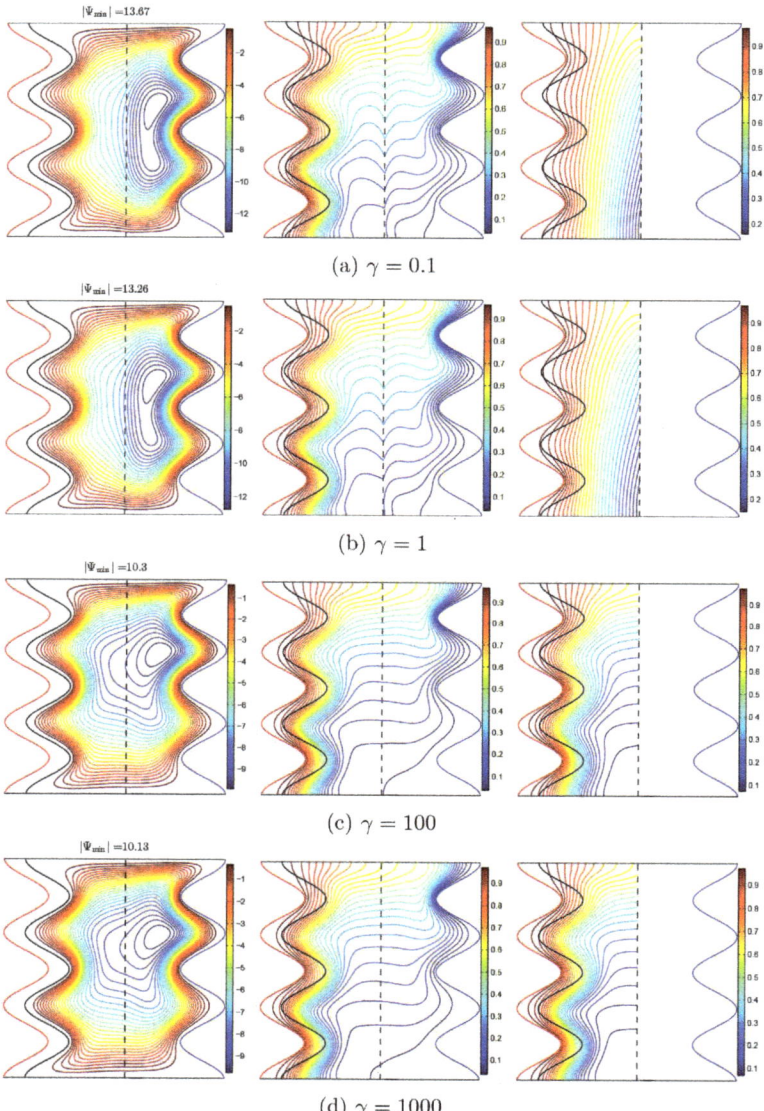

Figure 7. Streamlines (**left**), the isotherms of the nanofluid phase (**middle**), and the isotherms of the solid phase (**right**) with various modified conductivity ratios (γ); $Da = 10^{-3}$, $\phi = 0.02$, $A = 0.1$, and $\varepsilon = 0.5$.

The local velocity profiles in Figure 8a show that, for low values of γ, the velocity profiles are in conformity because the effect of the solid part is dominant compared to the pore space, which obstructs the buoyancy effects and the flow vortex remains at the center of the cavity. The disparity in the profiles at high γ can be explained by the fact that the flow vortex rises up, which results in a low nanofluid velocity at the lower part of the cavity, near the bottom wall. Figure 8b,c illustrates the impact of the modified thermal conductivity ratio, γ, on both the nanofluid-phase and solid-phase local heat transfer rates. Boosting γ results in an improvement in the Nusselt numbers. For a weak γ, the distribution of Nu_s

and Nu_{nf} differs, which means that the non-thermal equilibrium effects are significant. In addition, it is to note that the impacts of γ on the Nusselt numbers are more important for the solid phase than for the nanofluid phase. In fact, the attainment of thermal equilibrium between the nanofluid and the solid matrix leads to a greater heat transfer through the entire porous layer.

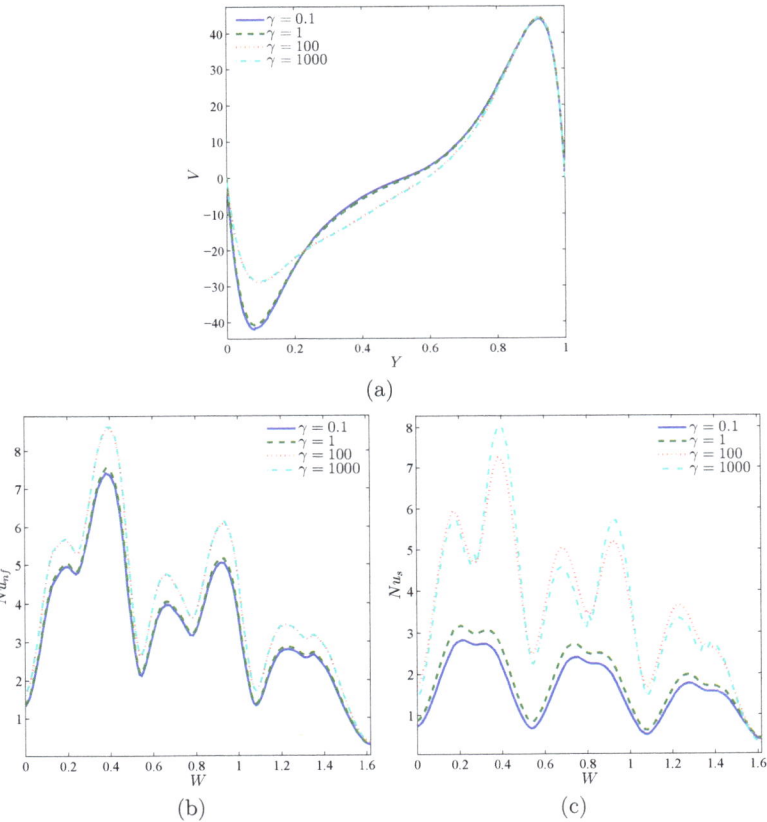

Figure 8. Local velocity (**a**), the local Nusselt number of the nanofluid phase (**b**), and the local Nusselt number of the solid phase (**c**) with the vertical line (Y) for $X = 0.5$ for $Y = 0.5$ for different γ values; $Da = 10^{-3}$, $\phi = 0.02$, $A = 0.1$, and $\varepsilon = 0.5$.

Figure 9 analyzes the effect of the magnitude of the undulations of the solid corrugated wall (A) on the system's thermal and dynamic features. Obviously, the larger magnitude undulations, the more conductive the heat transfer tends to be, since the undulations act to impede the nanofluid circulation inside the cavity. The flow configuration switches from one central flow vortex to a multi-core vortex by raising A, which influences the distribution of the velocity within the cavity (Figure 10a). Considering the great difference in size and form of the heat exchange surface, the amplitude of the undulations significantly impacts the distribution of the local heat transfer over the hot surface, as seen in the profiles of the Nusselt numbers in Figure 10c,d.

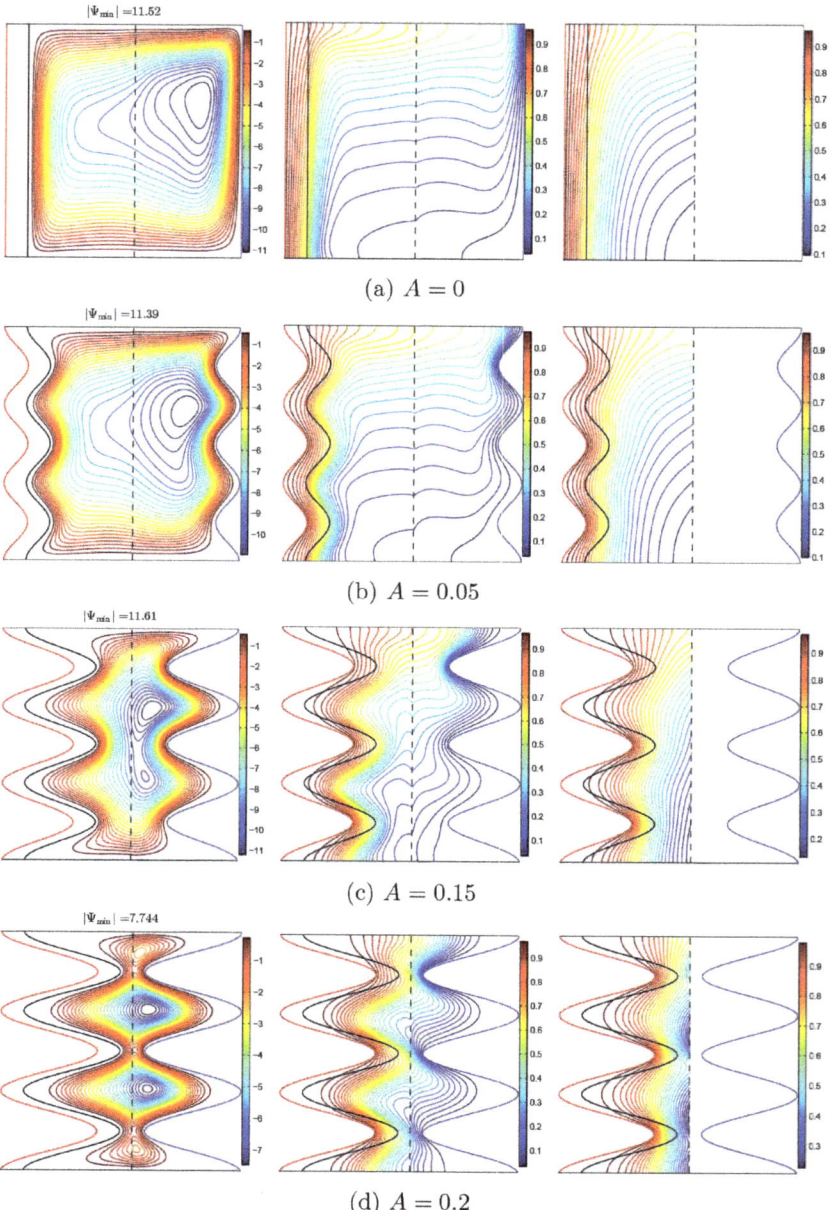

Figure 9. Streamlines (**left**), the isotherms of the nanofluid phase (**middle**), and the isotherms of the solid phase (**right**) with various amplitudes (A); $Da = 10^{-3}$, $\phi = 0.02$, $\gamma = 10$, and $\varepsilon = 0.5$.

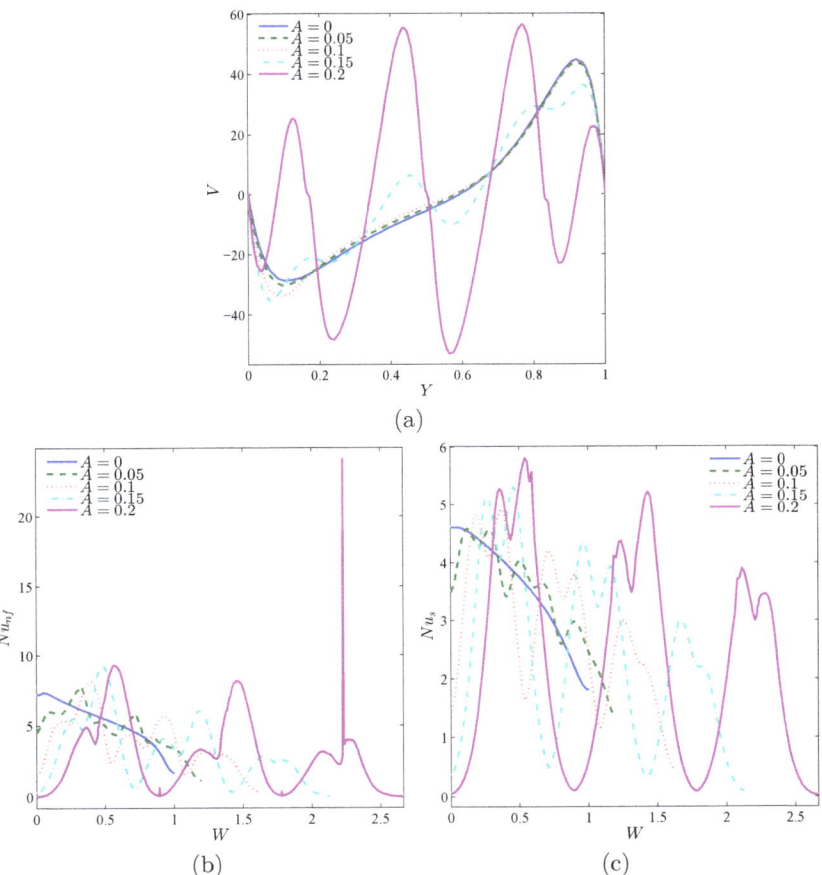

Figure 10. Local velocity (**a**), the local Nusselt number of the nanofluid phase (**b**), and the local Nusselt number of the solid phase (**c**) with the vertical line (Y) for $X = 0.5$ for different N values; $Da = 10^{-3}$, $\phi = 0.02$, $\gamma = 10$, and $\varepsilon = 0.5$.

The varying porosity effects of the porous layer (ε) on the nanofluid and solid phase isotherms and flow patterns in the different regions inside the cavity are demonstrated in Figure 11. As can be seen, an increase in the porosity of the porous layer leads to an increase in the nanofluid circulation within the entire cavity. Indeed, as the porosity increases, the nanofluid movement becomes freer in the cavity, which contributes to a greater heat transfer to the nanofluid layer through the porous layer.

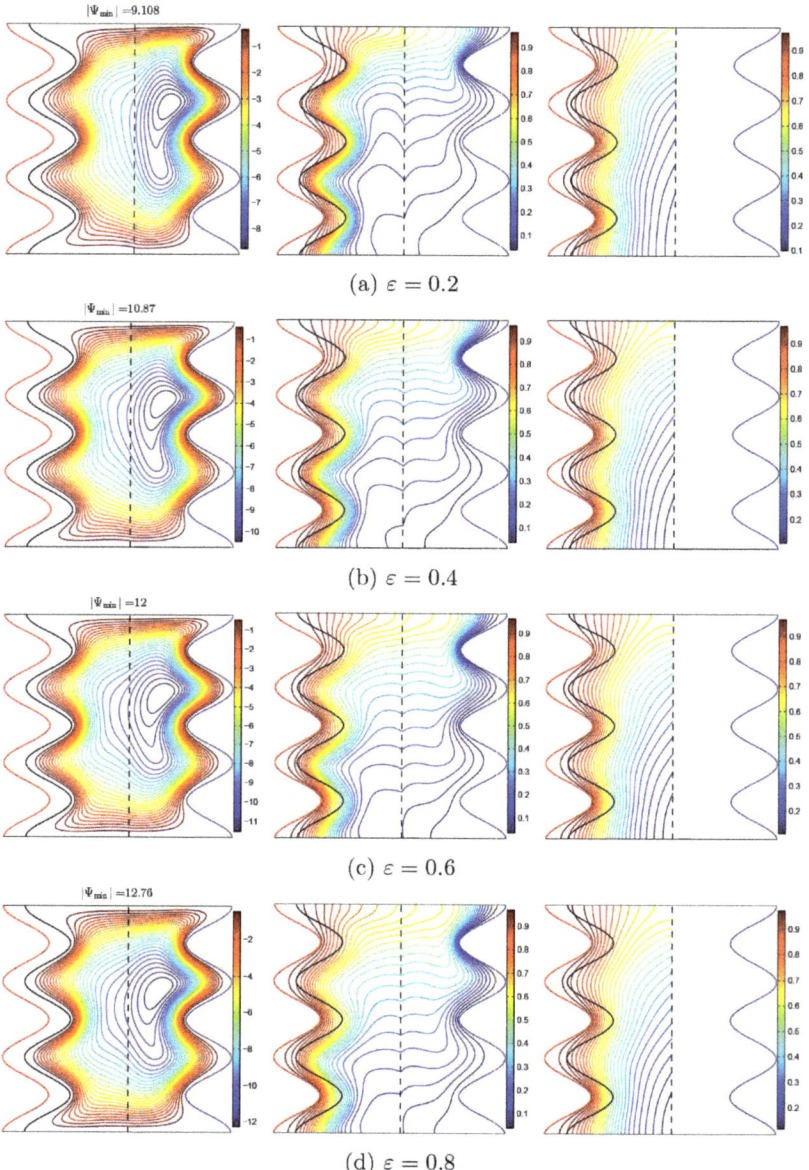

Figure 11. Streamlines (**left**), the isotherms of the nanofluid phase (**middle**), and the isotherms of the solid phase (**right**) with various porosities of the medium (ε); $Da = 10^{-3}$, $\phi = 0.02$, $\gamma = 10$, and $A = 0.1$.

Figure 12a indicates that an improvement in the ε parameter improves the velocity of circulation as well. Figure 12b,c shows that the maximum and minimum of the local heat transfer rates of the nanofluid phase are more extreme with a rise of ε, whereas the rates of the solid-phase heat transfer are not greatly changed by the variations in the porosity magnitude, as the porous media with a large porosity offers more empty spaces to be occupied with the flowing nanofluid.

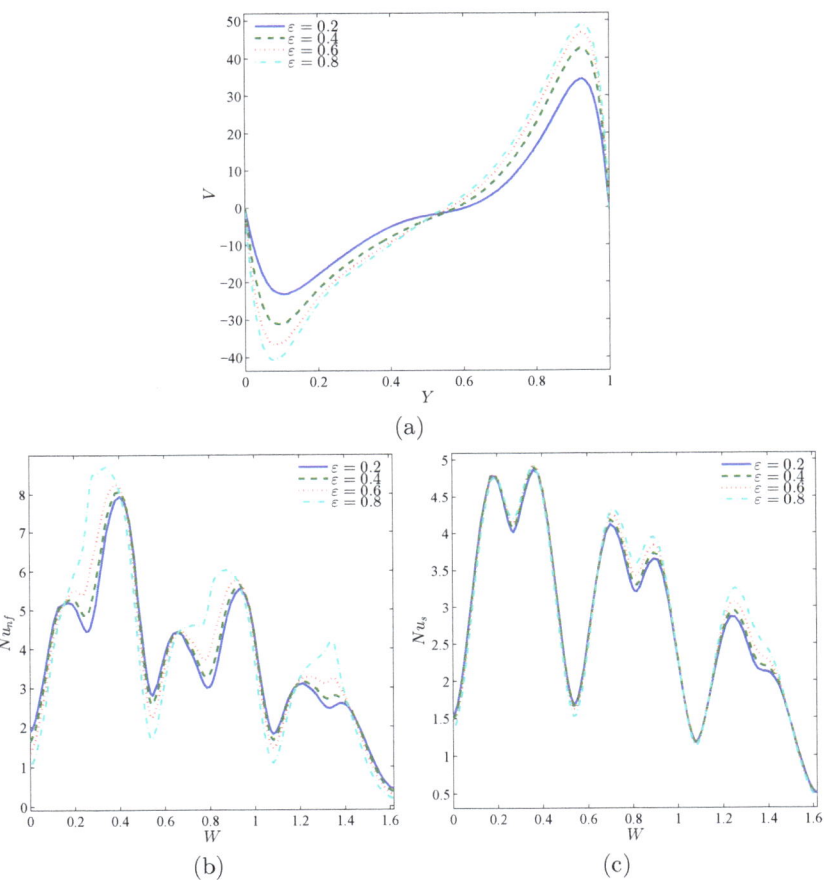

Figure 12. Local velocity (**a**), the local Nusselt number of the nanofluid phase (**b**), and the local Nusselt number of the solid phase (**c**) with the vertical line (Y) for $X = 0.5$ for different ε values; $Da = 10^{-3}$, $\phi = 0.02$, $\gamma = 10$, and $A = 0.1$.

The objective of Figures 13 and 14 is to show the role of the Darcy number (Da) and the modified thermal conductivity ratio (γ) at various nanoparticle concentrations (ϕ) in the average heat transport. It is clear that Da and γ augment the mean Nusselt number of both the nanofluid and the solid phases. The converging values of $\overline{Nu}_{nf}$ and $\overline{Nu}_s$ at lower Da values and higher γ values indicate the local thermal equilibrium situation, as stated earlier. In addition, it is shown that the increasing effect of Da and γ on the mean Nusselt numbers reduces when the Da and γ are higher. Moreover, beyond the value of $\gamma = 100$, the heat transfer rate decreases with γ because the heat transfer through the solid matrix ($\overline{Nu}_s$) is severely limited due to its low thermal conductivity, unlike $\overline{Nu}_{nf}$, which continues to increase with γ due to the improved nanofluid thermal conductivity. It is also evident from the two figures that increasing the nanoparticles' concentration produces increases in both Nusselt numbers for all considered values of Da and γ.

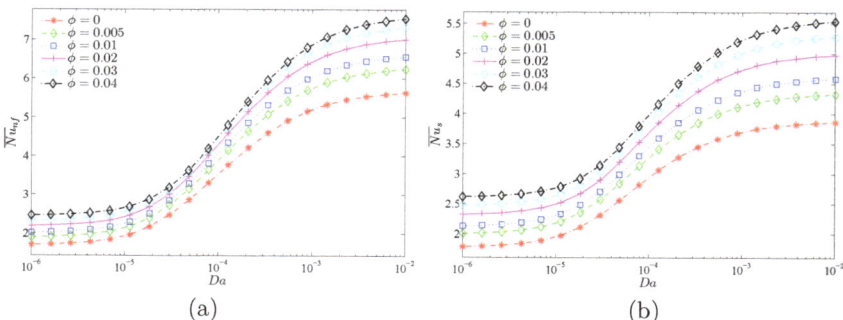

Figure 13. Variations of (**a**) the average Nusselt number of the nanofluid phase and (**b**) the average Nusselt number of the solid phase with Da values for different ϕ values at $\gamma = 10$, $A = 0.1$, and $\varepsilon = 0.5$.

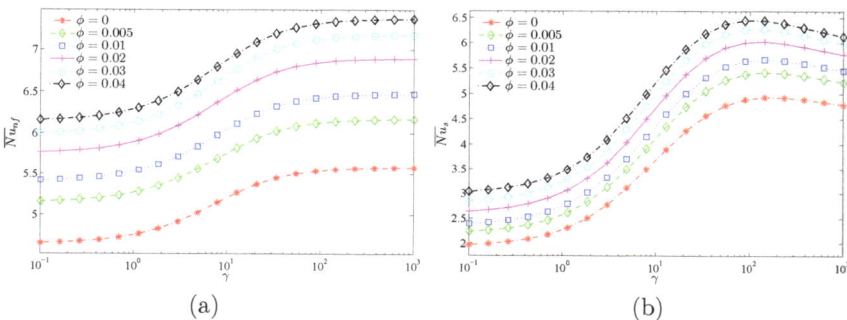

Figure 14. Variations of (**a**) the average Nusselt number of the nanofluid phase and (**b**) the average Nusselt number of the solid phase γ for different ϕ values at $Da = 10^{-3}$, $N = 3$, and $\varepsilon = 0.5$.

Figures 15 and 16 show that the largest global heat transfer for both the solid and nanofluid phases is found for the highest undulation amplitude ($A = 0.2$) for all tested values of Da and ϕ. This can be attributed to the large heat exchange surface of the wavy wall for high values of A. In addition, at a given A, the values of the $\overline{Nu}_{nf}$ and $\overline{Nu}_s$ are found to increase with increasing values of Da and ϕ.

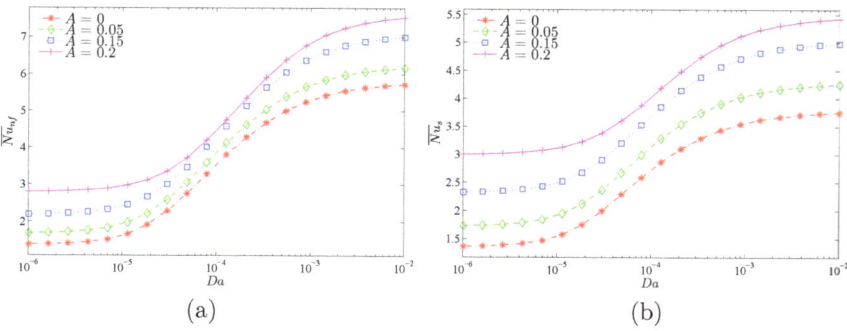

Figure 15. Variations of (**a**) the average Nusselt number of the nanofluid phase and (**b**) the average Nusselt number of the solid phase with Da for different A values at $\phi = 0.02$, $\gamma = 10$, and $\varepsilon = 0.5$.

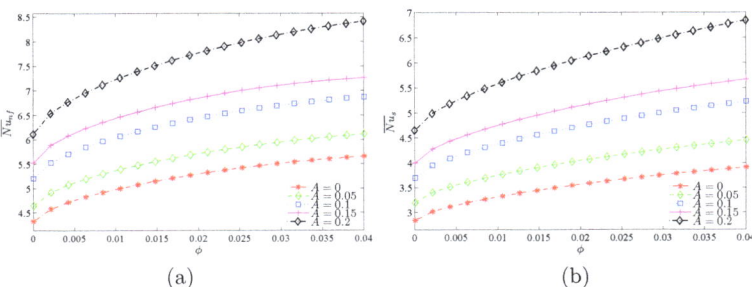

Figure 16. Variations of (**a**) the average Nusselt number of the nanofluid phase and (**b**) the average Nusselt number of the solid phase with ϕ for different A values at $Da = 10^{-3}$, $\gamma = 10$, and $\varepsilon = 0.5$.

Figure 17 aims to examine the role of the porosity (ε) of the porous layer as a function of the Darcy number (Da) in the total heat transfer rates of the nanofluid and solid phases for the case of ε at $\phi = 0.02$, $\gamma = 10$, and $A = 0.1$. At lower Darcy numbers ($Da \leq 10^{-5}$), an increase in the porosity of the porous layer weakens the total heat transfer rates ($\overline{Nu}_{nf}$ and $\overline{Nu}_s$) owing to the fact that high porosity at low permeability contributes to more heat resistance within the porous medium. The opposite is true for high Darcy numbers.

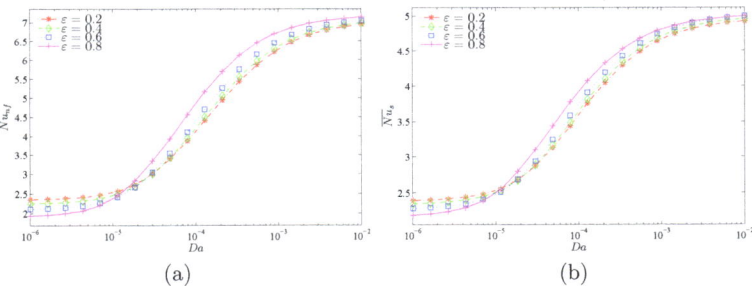

Figure 17. Variations of (**a**) the average Nusselt number of the nanofluid phase and (**b**) the average Nusselt number of the solid phase with Da for different ε values at $\phi = 0.02$, $\gamma = 10$, and $A = 0.1$.

Figure 18 characterizes the variation of $\overline{Nu}_{nf}$ and $\overline{Nu}_s$ with the modified thermal conductivity (γ) for various values of ε. $\overline{Nu}_{nf}$ increases with the increment of γ for all values of ε. Meanwhile, there is a limit in the increase of $\overline{Nu}_s$ with γ after a certain value of ε. This explains that the role of the nanofluid in transferring heat through the porous medium becomes greater than for the solid matrix due to its high thermal conductivity.

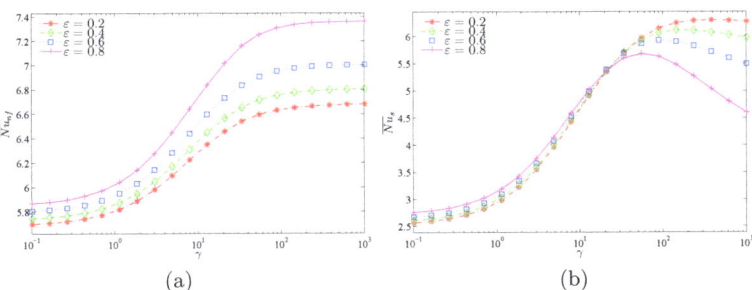

Figure 18. Variations of (**a**) the average Nusselt number of the nanofluid phase and (**b**) the average Nusselt number of the solid phase with γ for different ε values at $Da = 10^{-3}$, $\phi = 0.02$, and $A = 0.1$.

5. Conclusions

The problem of steady thermal-natural convection in a two-dimensional cavity of corrugated vertical walls consisting of three layers–a conducting solid layer of fixed thickness, a porous medium layer filled with a nanofluid, and a third layer filled with a nanofluid–was numerically studied by employing the finite element method. An alumina nanoparticle-water-based nanofluid was used as a working fluid. The LTNE model was considered for the porous medium. The key conclusions of this analysis are listed below:

1. The local thermal non-equilibrium effects are significant for low values of γ and high values of Da.
2. For low values of Da, the flow is almost entirely confined in the nanofluid layer, and the heat transfer is mainly convective in the nanofluid layer and mainly conductive in the other layers.
3. An increase in Da and ε contributes to an increase in the nanofluid circulation rate in the entire cavity, while an increase in γ causes an increase in the flow circulation in the porous region.
4. An increase in A contributes to a decrease in the nanofluid circulation rate in the entire cavity.
5. The best rates of the convective heat transfer through the nanofluid and solid phases are found at high values of A, Da, and ϕ for all other constant parameters.
6. The results show an increase in $\overline{Nu}_{nf}$ with increasing values of γ, while there is a limit in the increase of $\overline{Nu}_s$ with γ, especially for high values of ε.
7. $\overline{Nu}_{nf}$ and $\overline{Nu}_s$ decline as ε boosts for low values of Da and enhance for high values of Da.

Author Contributions: Conceptualization, A.I.A., T.T. and Z.A.S.R.; methodology, A.I.A. and T.T.; software, A.I.A., T.T. and Z.A.S.R.; validation, A.I.A. and T.T.; formal analysis, A.I.A., T.T. and A.S.A.; investigation, A.I.A., T.T., A.S.A. and Z.A.S.R.; resources, I.H. and A.J.C.; writ draft preparation, A.I.A., I.H., T.T., Z.A.S.R., A.J.C. and A.S.A.; writ and editing, A.I.A., I.H., T.T., Z.A.S.R., A.J.C. and A.S.A.; visualization, A.I.A. and T.T.; supervision, I.H. and A.J.C. All authors have read and agreed to the published version of the manuscript.

Funding: The APC was funded by I.H.'s UKM grant GP-2020-K006388.

Acknowledgments: We are grateful for the financial support received from the Malaysian Ministry of Education under the research grant FRGS/1/2019/STG06/UKM/01/2. Moreover, the authors would like to extend their appreciations to the Deanship of Scientific Research at King Khalid University, Abha, Saudi Arabia, for funding this work through the Research Group Project under grant number R.G.P2/144/42. We thank the respected reviewers for their constructive comments, which clearly enhanced the quality of the manuscript.

Conflicts of Interest: The authors declare no conflict of interest.

References

1. Beavers, G.S.; Joseph, D.D. Boundary conditions at a naturally permeable wall. *J. Fluid Mech.* **1967**, *30*, 197–207. [CrossRef]
2. Buretta, R.J.; Berman, A.S. Convective heat transfer in a liquid saturated porous layer. *J. Appl. Mech.* **1976**, *43*, 249–253. [CrossRef]
3. Tong, T.W.; Subramanian, E. Natural convection in rectangular enclosures partially filled with a porous medium. *Int. J. Heat Fluid Flow* **1986**, *7*, 3–10. [CrossRef]
4. Poulikakos, D.; Bejan, A.; Selimos, B.; Blake, K.R. High Rayleigh number convection in a fluid overlaying a porous bed. *Int. J. Heat Fluid Flow* **1986**, *7*, 109–116. [CrossRef]
5. Poulikakos, D. Buoyancy-driven convection in a horizontal fluid layer extending over a porous substrate. *Phys. Fluids* **1986**, *29*, 3949–3957. [CrossRef]
6. Beckermann, C.H.; Viskanta, R.; Ramadhyani, S. Natural convection in vertical enclosures containing simultaneously fluid and porous layers. *J. Fluid Mech.* **1988**, *186*, 257–284. [CrossRef]
7. Sathe, S.B.; Lin, W.Q.; Tong, T.W. Natural convection in enclosures containing an insulation with a permeable fluid-porous interface. *Int. J. Heat Fluid Flow* **1988**, *9*, 389–395. [CrossRef]
8. Hirata, S.C.; Goyeau, B.; Gobin, D. Stability of thermosolutal natural convection in superposed fluid and porous layers. *Transp. Porous Media* **2009**, *78*, 525–536. [CrossRef]

9. Hu, J.T.; Mei, S.J.; Liu, D.; Zhao, F.Y.; Wang, H.Q. Buoyancy driven heat and species transports inside an energy storage enclosure partially saturated with thermal generating porous layers. *Int. J. Therm. Sci.* **2018**, *126*, 38–55. [CrossRef]
10. Mikhailenko, S.A.; Sheremet, M.A.; Mahian, O. Effects of uniform rotation and porous layer on free convection in an enclosure having local heat source. *Int. J. Therm. Sci.* **2019**, *138*, 276–284. [CrossRef]
11. Saleh, H.; Hashim, I.; Jamesahar, E.; Ghalambaz, M. Effects of flexible fin on natural convection in enclosure partially-filled with porous medium. *Alex. Eng. J.* **2020**, *59*, 3515–3529. [CrossRef]
12. Esfe, M.H.; Bahiraei, M.; Hajbarati, H.; Valadkhani, M. A comprehensive review on convective heat transfer of nanofluids in porous media: Energy-related and thermohydraulic characteristics. *Appl. Therm. Eng.* **2020**, *178*, 115487. [CrossRef]
13. Alsabery, A.I.; Chamkha, A.J.; Hussain, S.H.; Saleh, H.; Hashim, I. Heatline visualization of natural convection in a trapezoidal cavity partly filled with nanofluid porous layer and partly with non-Newtonian fluid layer. *Adv. Powder Technol.* **2015**, *26*, 1230–1244. [CrossRef]
14. Al-Zamily, A.M.J. Analysis of natural convection and entropy generation in a cavity filled with multi-layers of porous medium and nanofluid with a heat generation. *Int. J. Heat Mass Transf.* **2017**, *106*, 1218–1231. [CrossRef]
15. Armaghani, T.; Ismael, M.A.; Chamkha, A.J. Analysis of entropy generation and natural convection in an inclined partially porous layered cavity filled with a nanofluid. *Can. J. Phys.* **2017**, *95*, 238–252. [CrossRef]
16. Miroshnichenko, I.V.; Sheremet, M.A.; Oztop, H.F.; Abu-Hamdeh, N. Natural convection of alumina-water nanofluid in an open cavity having multiple porous layers. *Int. J. Heat Mass Transf.* **2018**, *125*, 648–657. [CrossRef]
17. Wang, Y.; Qin, G.; He, W.; Bao, Z. Chebyshev spectral element method for natural convection in a porous cavity under local thermal non-equilibrium model. *Int. J. Heat Mass Transf.* **2018**, *121*, 1055–1072. [CrossRef]
18. Ghalambaz, M.; Tahmasebi, A.; Chamkha, A.; Wen, D. Conjugate local thermal non-equilibrium heat transfer in a cavity filled with a porous medium: Analysis of the element location. *Int. J. Heat Mass Transf.* **2019**, *138*, 941–960. [CrossRef]
19. Mikhailenko, S.A.; Sheremet, M.A. Impacts of rotation and local element of variable heat generation on convective heat transfer in a partially porous cavity using local thermal non-equilibrium model. *Int. J. Therm. Sci.* **2020**, *155*, 106427. [CrossRef]
20. Sivasankaran, S.; Alsabery, A.I.; Hashim, I. Internal heat generation effect on transient natural convection in a nanofluid-saturated local thermal non-equilibrium porous inclined cavity. *Phys. A Stat. Mech. Its Appl.* **2018**, *509*, 275–293. [CrossRef]
21. Tahmasebi, A.; Mahdavi, M.; Ghalambaz, M. Local thermal nonequilibrium conjugate natural convection heat transfer of nanofluids in a cavity partially filled with porous media using Buongiorno's model. *Numer. Heat Transf. Part A Appl.* **2018**, *73*, 254–276. [CrossRef]
22. Izadi, M.; Mehryan, S.A.M.; Sheremet, M.A. Natural convection of CuO-water micropolar nanofluids inside a porous enclosure using local thermal non-equilibrium condition. *J. Taiwan Inst. Chem. Eng.* **2018**, *88*, 89–103. [CrossRef]
23. Mehryan, S.A.M.; Izadi, M.; Sheremet, M.A. Analysis of conjugate natural convection within a porous square enclosure occupied with micropolar nanofluid using local thermal non-equilibrium model. *J. Mol. Liq.* **2018**, *250*, 353–368. [CrossRef]
24. Mehryan, S.A.M.; Ghalambaz, M.; Chamkha, A.J.; Izadi, M. Numerical study on natural convection of Ag–MgO hybrid/water nanofluid inside a porous enclosure: A local thermal non-equilibrium model. *Powder Technol.* **2020**, *367*, 443–455. [CrossRef]
25. Mansour, M.A.; Ahmed, S.E.; Aly, A.M.; Raizah, Z.A.S.; Morsy, Z. Triple convective flow of micropolar nanofluids in double lid-driven enclosures partially filled with LTNE porous layer under effects of an inclined magnetic field. *Chin. J. Phys.* **2020**, *68*, 387–405. [CrossRef]
26. Kumar, B.V.R. A study of free convection induced by a vertical wavy surface with heat flux in a porous enclosure. *Numer. Heat Transf. Part A Appl.* **2000**, *37*, 493–510. [CrossRef]
27. Alsabery, A.I.; Ismael, M.A.; Chamkha, A.J.; Hashim, I.; Abulkhair, H. Unsteady flow and entropy analysis of nanofluids inside cubic porous container holding inserted body and wavy bottom wall. *Int. J. Mech. Sci.* **2021**, *193*, 106161. [CrossRef]
28. Abdulkadhim, A.; Hamzah, H.K.; Ali, F.H.; Abed, A.M.; Abed, I.M. Natural convection among inner corrugated cylinders inside wavy enclosure filled with nanofluid superposed in porous–nanofluid layers. *Int. Commun. Heat Mass Transf.* **2019**, *109*, 104350. [CrossRef]
29. Kadhim, H.T.; Jabbar, F.A.; Rona, A. Cu-Al_2O_3 hybrid nanofluid natural convection in an inclined enclosure with wavy walls partially layered by porous medium. *Int. J. Mech. Sci.* **2020**, *186*, 105889. [CrossRef]
30. Alsabery, A.I.; Mohebbi, R.; Chamkha, A.J.; Hashim, I. Effect of local thermal non-equilibrium model on natural convection in a nanofluid-filled wavy-walled porous cavity containing inner solid cylinder. *Chem. Eng. Sci.* **2019**, *201*, 247–263. [CrossRef]
31. Corcione, M. Empirical correlating equations for predicting the effective thermal conductivity and dynamic viscosity of nanofluids. *Energy Convers. Manag.* **2011**, *52*, 789–793. [CrossRef]
32. Khanafer, K.; Al-Azmi, B.; Marafie, A.; Pop, I. Non-Darcian effects on natural convection heat transfer in a wavy porous enclosure. *Int. J. Heat Mass Transf.* **2009**, *52*, 1887–1896. [CrossRef]
33. Bergman, T.L.; Incropera, F.P. *Introduction to Heat Transfer*, 6th ed.; Wiley: New York, NY, USA, 2011.

Article

1D/2D van der Waals Heterojunctions Composed of Carbon Nanotubes and a GeSe Monolayer

Yuliang Mao [1,*], Zheng Guo [1], Jianmei Yuan [2,*] and Tao Sun [2]

1. Hunan Key Laboratory for Micro–Nano Energy Materials and Devices, School of Physics and Optoelectronic, Xiangtan University, Xiangtan 411105, China; 201821521377@smail.xtu.edu.cn
2. Hunan Key Laboratory for Computation and Simulation in Science and Engineering, School of Mathematics and Computational Science, Xiangtan University, Xiangtan 411105, China; 202021001381@smail.xtu.edu.cn
* Correspondence: ylmao@xtu.edu.cn (Y.M.); yuanjm@xtu.edu.cn (J.Y.)

Abstract: Based on first-principles calculations, we propose van der Waals (vdW) heterojunctions composed of one-dimensional carbon nanotubes (CNTs) and two-dimensional GeSe. Our calculations show that (n,0)CNT/GeSe (n = 5–11) heterojunctions are stable through weak vdW interactions. Among these heterojunctions, (n,0)CNT/GeSe (n = 5–7) exhibit metallic properties, while (n,0)CNT/GeSe (n = 8–11) have a small bandgap, lower than 0.8 eV. The absorption coefficient of (n,0)CNT/GeSe (n = 8–11) in the ultraviolet and infrared regions is around 10^5 cm^{-1}. Specifically, we found that (11,0)CNT/GeSe exhibits type-II band alignment and has a high photoelectric conversion efficiency of 17.29%, which suggests prospective applications in photoelectronics.

Keywords: germanium selenide; carbon nanotubes; heterojunction; photoelectric conversion efficiency

1. Introduction

As a fascinating carbon material, single-wall carbon nanotubes (SWCNTs) [1] have attracted widespread attention due to their unique properties [2–4]. The outstanding physical properties of CNTs make them a good candidate basic material for next-generation electronic devices [3,4]. The accurate prediction of CNTs' electronic properties is very important for their possible applications [5]. For large-diameter CNTs, the CNTs are metals or semiconductors depending on their chiral indices (n,m) [6]. When (n − m) is equal to 3p, where p is an integer, the CNT is a metal. Otherwise, the CNT is a semiconductor [6]. The above criterion is not applicable to CNTs with small diameters due to curvature effects or s-p rehybridization [7–15].

Monolayer germanium selenide (GeSe) is a semiconductor that has a direct bandgap [16]. Few-layer GeSe, including monolayer GeSe, is non-toxic and can exist stably at room temperature [17,18]. Our group successfully prepared a single layer of GeSe using mechanical stripping and laser-thinning technology [19]. First-principles studies, combined with photoluminescence spectra, proved that a direct bandgap exists for less than three layers in a few-layer GeSe [20]. Under conditions of high temperature and high pressure, the GeSe conductivity is higher than that of black phosphorus and graphene [21]. Our previous study showed that monolayer GeSe, with point defect engineering, has a good adsorption effect on toxic gases [22]. Moreover, we found that the bandgap can be tuned by stacking order and external strain in bilayer GeSe [23]. Our designed GeSe/SnSe heterojunction, based on first principles, exhibited a superior photoelectric conversion efficiency (PCE) of 21.47% [24].

Duan et al. [25] recently proposed a state-of-the-art material design called van der Waals (vdW) integration [25]. They suggest combining a two-dimensional (2D) material with materials with other dimensions. For example, 2D and one-dimensional (1D) materials could be combined by the vdW interaction. The literature reported that a 2D/2D GeSe/SnS

heterojunction has stronger optical absorption than GeSe or SnS [24]. A 0D/2D photodiode was proposed by integrating quantum dots or plasma nanoparticles on graphene, which will not damage the original graphene lattice, enhancing the photocurrent [26–28]. Moreover, 1D/2D high-speed transistors are obtained through vdW integration of 1D core-shell nanowires and 2D graphene, which has a high cut-off frequency [29–32]. Based on this progress, in this paper, we propose combining 1D CNTs and 2D GeSe with vdW interaction and explore their electronic properties through first-principles calculations. We aim to provide a theoretical proposal for 1D/2D integration through CNTs and a GeSe monolayer, which has potential applications in the field of optoelectronic devices.

2. Computational Method and Model

Our first-principles calculations used the Vienna ab initio simulation package (VASP) [33]. Based on density functional theory (DFT), a plane wave basis expanded the CNT/GeSe hybrid wave function. To represent the interaction of exchange and correlation between the electrons, the Perdew, Burke, and Ernzerhof (PBE) function in the framework of a generalized gradient approximation was used [34–36]. To ensure sufficient accuracy, we found that a cut-off energy of 450 eV was satisfactory for the convergence standards. The energy convergence was 10^{-6} eV, while a force of 0.01 eV/Å on each atom was sufficient for the calculations. In the structural relaxation and self-consistent calculation, we set a Monkhorst-Pack grid of k points of $8 \times 5 \times 1$ [37] for sampling. When calculating the density of states (DOS) and optical properties, we used a denser Monkhorst-Pack grid of $16 \times 10 \times 1$ k-point sampling. For the simulation of a heterostructure between 1D and 2D materials, vdW interaction is especially important. Our simulations adopted semi-empirical dispersion-corrected D3 (DFT-D3) [38] to represent the weak interaction between CNTs and the GeSe monolayer.

We selected a series of zigzag $(n,0)$CNTs (n = 5–11) with an axial length of 4.26 Å to form a composite structure with monolayer GeSe. To build a reasonable 1D/2D model, we first enlarged the unit cell of monolayer GeSe to a 1×5 supercell (4.25 Å $\times$ 19.95 Å). Based on this supercell, we placed the CNT above the GeSe monolayer along the x-axis. In our model, there was only a 0.2% lattice mismatch. We set 28 Å along the z-axis as the vacuum layer, which avoids interaction between the adjacent supercells. Figure 1 shows the schematic structural model. To better present the schematic model, we enlarged the indicated lattice constant four times along the x-axis.

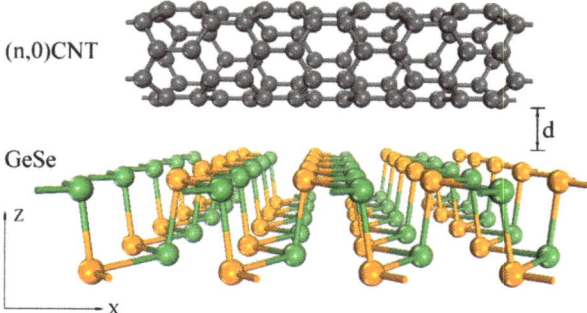

Figure 1. Schematic diagram of the side view of different zigzag CNTs $(n,0)$ (n = 5–11) on monolayer GeSe. C, Ge, and Se atoms are represented by gray, green, and orange spheres, respectively. d is the interfacial spacing between the CNT and monolayer GeSe.

In our optical calculations using the VASP code, the frequency-dependent dielectric matrix after the electronic ground state was determined [39]. As suggested in [39], the imaginary part can be determined by a summation of the empty states using the following formula:

$$\varepsilon_2(\omega) = \frac{4\pi^2 e^2}{\Omega} \lim_{q \to 0} \frac{1}{q^2} \sum_{c,v,k} 2w_k \delta(\varepsilon_{ck} - \varepsilon_{vk} - \omega) \times (u_{ck} + e_{\alpha q}|u_{vk})(u_{ck} + e_{\beta q}|u_{vk})$$

where the indices c and v refer to the conduction and valence band states, respectively. In [39] it is stated that u_{ck} is the periodic cell part of the orbitals at the k-point k, while the k-point weights, w_k, are defined such that their sum is one. In addition, in [39], the real part of the dielectric tensor, $\varepsilon_1(\omega)$, is obtained using the usual Kramers–Kronig transformation:

$$\varepsilon_1(\omega) = 1 + \frac{2}{\pi} P \int_0^\infty \frac{\varepsilon_2(\omega')\omega'}{\omega'^2 - \omega^2 + i\eta} d\omega'$$

where P denotes the principal value while η is an infinitesimal number. In addition, the number of empty bands in the above calculations is twice that of the self-consistent calculations for total energies.

3. Results and Discussion

3.1. Configurations and Stability of 1D/2D Heterostructures

In this work, six types of zigzags $(n,0)$CNTs (n = 5–11) with diameters ranging from 3.92 to 7.83 Å were simulated. It is critical that the 1D CNT and 2D GeSe form a stable composite structure. Figure 2 indicates the optimized configurations of $(n,0)$CNT (n = 5–11) on 2D GeSe. From the perspective of the geometrical structure, CNT/GeSe hybrids maintain their original structure. In $(n,0)$CNT/GeSe (n = 5–11) with optimized structures, the average C–C bond length of the CNT changes little compared to their components, varying between 0.001 to 0.003 Å. Due to compatibility, the average GeSe bond length in GeSe has only minor variations, ranging from 0.006 to 0.009 Å. The interfacial spacing between the top Se atom of monolayer GeSe and the C atom in a CNT ranges from 2.97 to 3.02 Å (see Table 1). According to previous studies [40–42], large interlayer spacing implies weak interaction between GeSe and the CNTs. The calculated formation energy is estimated using the following formula:

$$E_f = E_{CNT/GeSe} - E_{CNT} - E_{GeSe} \quad (1)$$

where $E_{CNT/GeSe}$, E_{CNT}, and E_{GeSe} are the total energies of the CNT/GeSe heterostructure, CNT, and the monolayer GeSe, respectively.

Table 1. Diameter and bandgaps of $(n,0)$ CNTs (E_g) (n = 5–11) and formation energy, E_f, bandgap, $E_g{}^a$, and interfacial spacing, d, of optimized CNT/GeSe hybrids. Bader charge: the positive value indicates gained electrons, while a negative value reveals lost electrons.

Hybrid	Diameter (Å)	E_g (eV)	$E_g{}^a$ (eV)	E_f (eV)	d (Å)	Bader Charge (e)	
						GeSe	CNT
CNT(5,0)/GeSe	3.92	0	0	−3.517	2.97	−0.1098	0.1098
CNT(6,0)/GeSe	4.70	0	0	−3.959	3.02	−0.0490	0.0490
CNT(7,0)/GeSe	5.48	0.1746	0	−4.078	3.02	−0.0280	0.0280
CNT(8,0)/GeSe	6.27	0.5908	0.2112	−4.273	3.00	−0.0204	0.0204
CNT(9,0)/GeSe	7.05	0.1568	0.1643	−4.283	3.02	−0.0159	0.0159
CNT(10,0)/GeSe	7.83	0.7222	0.4669	−4.513	3.01	−0.0280	0.0280
CNT(11,0)/GeSe	8.59	0.9519	0.5924	−4.709	3.01	−0.0174	0.0174

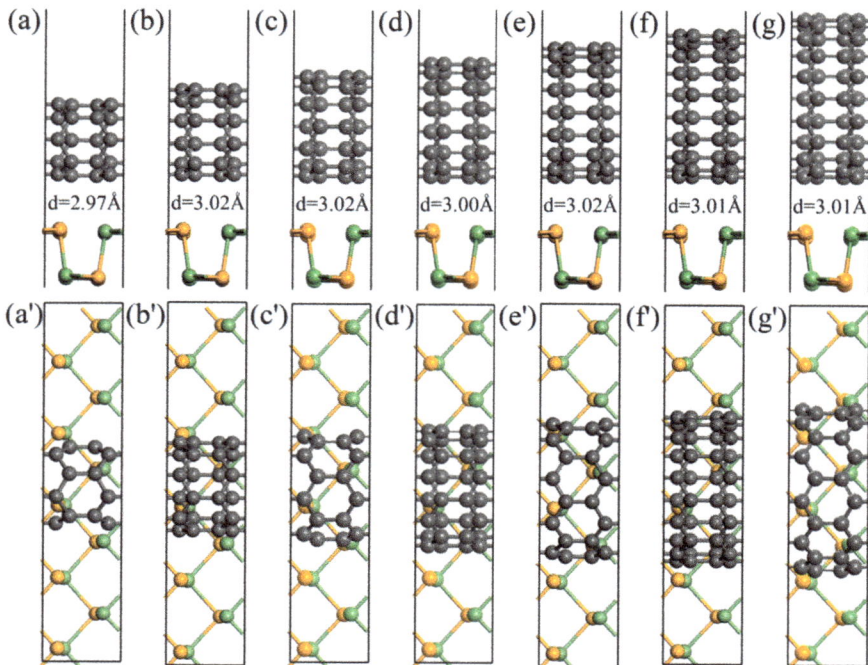

Figure 2. Optimized structures of different CNTs on monolayer GeSe: (**a**–**g**) and (**a**′–**g**′) are side and top view of (*n*,0) CNTs/GeSe (*n* = 5–11), respectively. *d* is the equilibrium spacing between the top Se atomic layer and the annotated wall. Gray, orange, and green spheres represent C, Se, and Ge atoms, respectively.

According to the above definition of formation energy, when E_f is negative, the system tends to be stable. As indicated in Table 1, the negative formation energy of all calculated 1D/2D CNT/GeSe combinations implies the stability of our proposed heterostructures. Our results indicate that the formation energies decrease with increasing CNT diameter. The decreasing formation energy is related to the increased contact area between CNTs and 2D GeSe. In addition, the interlayer interaction of the CNT/GeSe composite structure is reflected by the charge transfer between the CNT and GeSe. We conducted a Bader charge analysis to explore the charge transfer between CNTs and GeSe (see Table 1). A positive Bader charge value indicates that electrons are gained, while a negative value indicates that electrons are lost. Table 1 shows a certain number of electrons are transferred from the 2D GeSe to the CNTs. There are small fluctuations in the amount of charge transfer in (*n*,0)CNT/GeSe (*n* = 5–11), ranging between 0.0159 e to 0.0490 e. The small amount of charge transfer suggests weak interaction between the CNTs and the GeSe monolayer.

Based on the analysis of the charge density difference, the charge transfer and redistribution at the interface in these hybrids can be evaluated (as shown in Figure 3) by the following relationship:

$$\Delta \rho = \rho_{CNT/GeSe} - \rho_{CNT} - \rho_{GeSe} \qquad (2)$$

where $\rho_{CNT/GeSe}$, ρ_{GeSe}, and ρ_{CNT} are the charge densities of CNT/GeSe, CNT, and monolayer GeSe, respectively. As Figure 3 indicates, the redistributed charge is visible due to the interaction between the 1D CNTs and 2D GeSe.

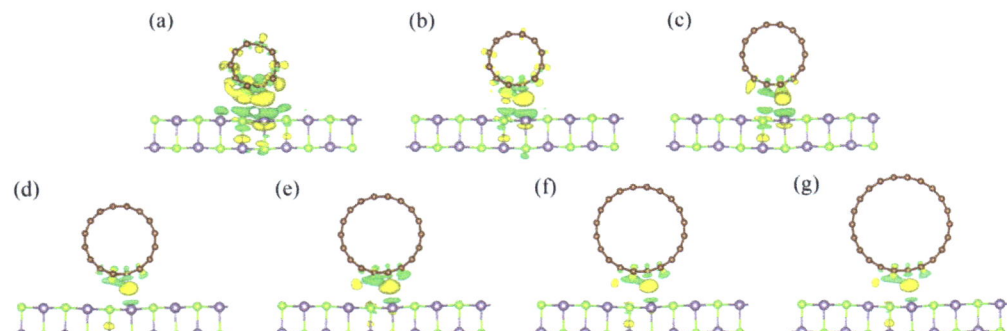

Figure 3. Charge density differences for CNT(n,0)/GeSe with n = (**a**) 5, (**b**) 6, (**c**) 7, (**d**) 8, (**e**) 9, (**f**) 10, and (**g**) 11. The isovalue is set to 0.0015 $e/Å^3$. The yellow and green regions represent charge loss and gain of electrons, respectively.

As shown in Figure 3, at the interface between GeSe and the CNTs, the charge is transferred from the GeSe to the CNTs. Though the amount of charge transfer is very small, the interaction between CNTs and 2D GeSe can be validated. Moreover, the amount of charge transfer in the (5,0)CNT/GeSe is larger than that in other composites. This is because the interlayer distance between the GeSe and (5,0)CNT is 2.97 Å, which is smaller (as shown in Table 1) than in other hybrid structures.

3.2. Band Structure and Density of States

To explore the electronic properties of CNT/GeSe hybrids, we calculated their band structures and DOS. As shown in Table 1, CNT(5,0) and CNT(6,0) are metals while CNT(n,0) (n = 7–11) are semiconductors. Our results using first-principles calculations are consistent with previous reports in the literature [5]. As shown in Figure 4h, the obtained bandgap of monolayer GeSe is 1.14 eV with our PBE calculation, which is the same as a previous report [43]. As shown in Figure 4a–c and Table 1, (n,0)CNT/GeSe (n = 5–7) all have bandgaps of zero, while the bandgaps of (n,0)CNT/GeSe (n = 8–11) are 0.21 eV, 0.16 eV, 0.47 eV, and 0.59 eV. This indicates that CNT/GeSe heterojunctions with small diameter CNTs are metallic. As the diameter of the CNTs increases, the bandgaps of our proposed (n,0)CNT/GeSe (n = 8–11) heterojunctions gradually increase. However, the bandgap of (9,0)CNT/GeSe is smaller than those of (8,0)CNT/GeSe and (10,0)CNT/GeSe. This is because the bandgap of the (9,0)CNTs/GeSe heterojunction is mainly determined by the energy level of the bands in the (9,0)CNT. In contrast, in (8,0)CNT/GeSe and (10,0)CNT/GeSe heterojunctions, the bandgap is a subtraction between the energy level of the conduction band minimum (CBM) in the corresponding CNTs and the energy level of the valence band maximum (VBM) in the GeSe monolayer. As a result, the bandgap of (9,0)CNT/GeSe is smaller than those of (8,0)CNT/GeSe and (10,0)CNT/GeSe. As the diameter of CNTs increases, the bandgaps of our proposed (n,0)CNTs/GeSe (n = 8–11) heterojunctions gradually increase. The projected energy band shown in Figure 4 confirms that the CNTs mainly provide the bands near the Fermi surface in the band structure of a CNT/GeSe heterojunction. In other words, whether (n,0)CNT/GeSe (n = 5–11) heterojunctions composed of (n,0)CNTs (n = 5–11) and 2D GeSe are metals or semiconductors is mainly determined by the conduction bands near the Fermi level. Due to the weak van der Waals interaction between CNTs and 2D GeSe, we can use CNTs with different diameters to obtain suitable bandgaps in CNT/GeSe heterojunctions with the variation of the band structure in CNTs.

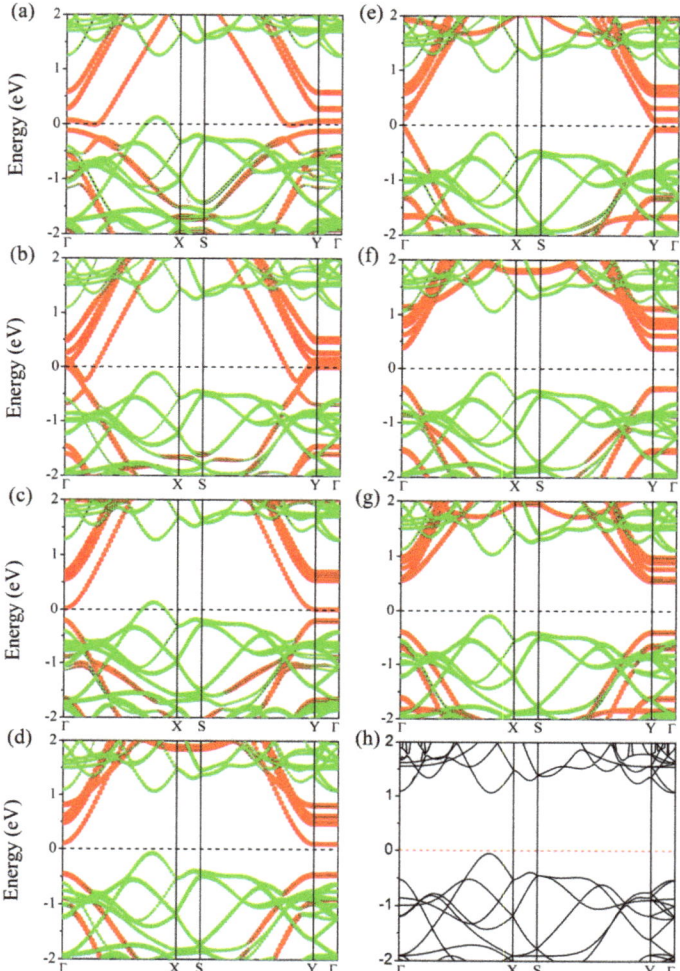

Figure 4. Projected band structures of (n,0)CNT/GeSe hybrids with n = (**a**) 5, (**b**) 6, (**c**) 7, (**d**) 8, (**e**) 9, (**f**) 10, and (**g**) 11. The red and green circle size denotes the weight of the CNT and monolayer GeSe in the CNT/GeSe configuration band structure. (**h**) is the band structure of pristine 2D GeSe calculated for comparison. A horizontal dashed line represents the Fermi level.

In Figure 5, we show the partial DOS (PDOS) of (n,0)CNT/GeSe (n = 5–11) and monolayer GeSe. Compared with the PDOS of monolayer GeSe, the PDOS of GeSe in the heterojunction is the same as that of the monolayer GeSe. The total DOS of the CNT/GeSe heterojunction can be viewed as a superposition of the DOS in the CNT and in the 2D GeSe. This further proves that there is a weak vdW interaction between the CNTs and 2D GeSe. From the perspective of the PDOS, the conduction band near the Fermi surface of the CNT/GeSe heterojunction is provided by the CNT and GeSe together. In contrast, the valence band is mainly provided by the 2p orbital of carbon atoms. Aside from the 2p orbital of carbon atoms, the orbital DOS of the other elements is unchanged. It is the 2p orbital of carbon that determines the top position of the valence bands, thus, affecting the band structure of the heterojunction. As shown in Figure 5a–c, the 2p orbital of carbon in (n,0)CNT/GeSe (n = 5–7) passes through the Fermi level, so those heterojunctions are metallic. The 2*p* orbital of carbon in (8,0)CNT/GeSe and (9,0)CNT/GeSe is closer to the Fermi surface than that in the (10,0)CNT/GeSe and (11,0)CNT/GeSe heterojunctions, which leads to smaller bandgaps in (8,0)CNT/GeSe and (9,0)CNT/GeSe heterojunctions than those in (10,0)CNT/GeSe and (11,0)CNT/GeSe heterojunctions.

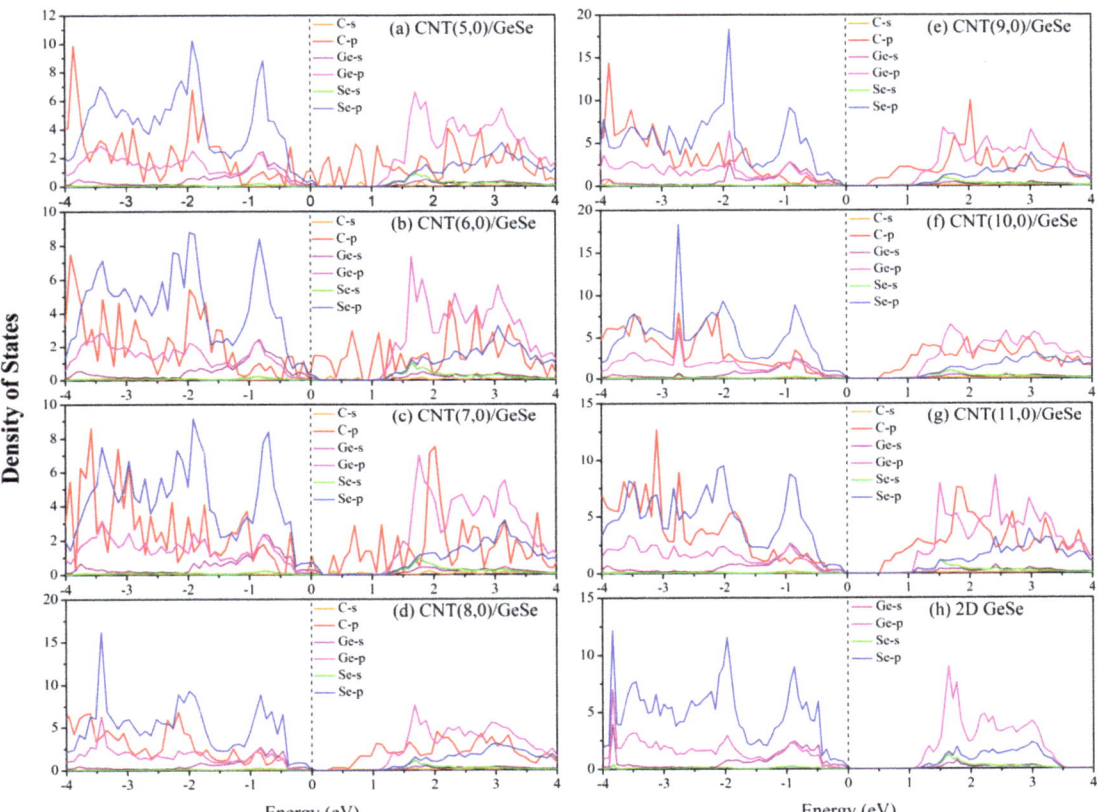

Figure 5. PDOS of CNT(*n*,0)/GeSe hybrids with *n* = (**a**) 5, (**b**) 6, (**c**) 7, (**d**) 8, (**e**) 9, (**f**) 10, (**g**) 11, and a (**h**) monolayer GeSe. The Fermi level is set to zero.

To further confirm the distribution of electronic states near the Fermi surface of the CNT/GeSe, in Figure 6 we show the electronic state distribution of (*n*,0)CNT/GeSe (*n* = 8–11). The CNT and GeSe provide the holes at the top of the valence band, with most holes provided by the CNT. In comparison, the electrons at the bottom of the conduction band are provided by the CNT alone. From the previous discussion of the PDOS, the 2*p* orbital of carbon contributes electrons at the bottom of the conduction band. Therefore, the electronic properties in CNT/GeSe heterojunctions are mainly influenced by CNTs with varying tube diameters. In other words, we found that the bandgap of our studied heterojunctions can be tuned by varying the tube diameter of the CNTs. Our calculated results predict that the bandgap of the heterojunction is smaller than that of monolayer GeSe, which is beneficial for optical absorption.

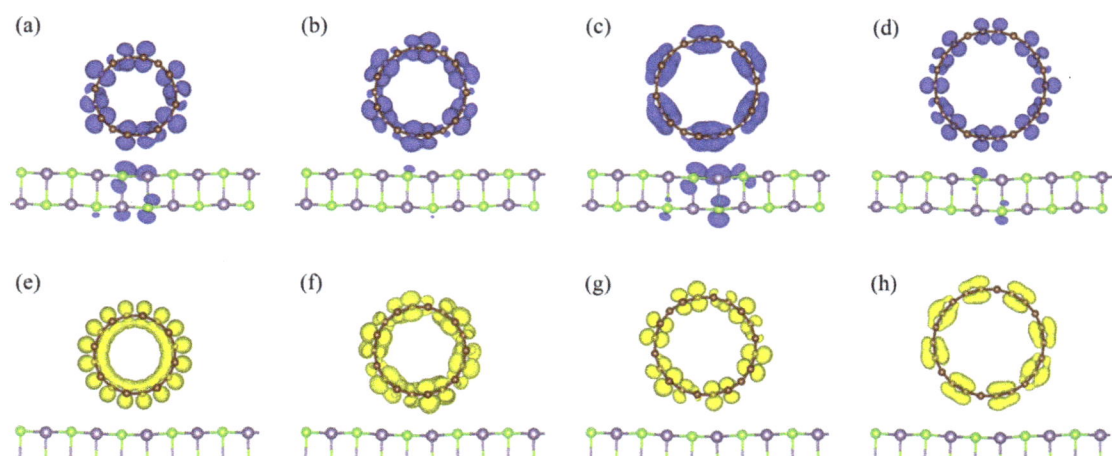

Figure 6. Maps of the hole and electron density distributions for (**a**–**d**) VBM and (**e**–**h**) CBM of CNT(8,0)/GeSe, CNT(9,0)/GeSe, CNT(10,0)/GeSe, and CNT(11,0)/GeSe with an isovalue of 0.007 e/Å³, respectively. Blue and yellow regions denote the hole and electron density distributions of the VBM and CBM, respectively. Brown, cyan, and purple spheres represent C, Se, and Ge atoms, respectively.

3.3. Optical Absorption Properties

To evaluate the optical absorption properties of CNT/GeSe heterojunctions, we utilized the following formula to assess the optical coefficient:

$$\alpha(\omega) = \sqrt{2}\omega \left[\sqrt{\varepsilon_1^2(\omega) + \varepsilon_2^2(\omega)} - \varepsilon_1(\omega) \right]^{\frac{1}{2}} \tag{3}$$

where $\varepsilon_1(\omega)$ and $\varepsilon_2(\omega)$ are the real and imaginary parts of the complex dielectric function, respectively. In Figure 7, we present the calculated optical absorption coefficient, $\alpha(\omega)$ of a monolayer GeSe, pure CNT, and CNT/GeSe hybrids. According to previous studies [16], GeSe reportedly had good optical absorption properties. By combining 2D GeSe with CNTs, the bandgap of the hybrid system is smaller than that of 2D GeSe, which is helpful for the separation of photogenerated electrons and holes. Figure 7 shows the optical absorption of our studied CNT/GeSe hybrids. For comparison, the calculated optical absorption of a GeSe monolayer and CNTs are plotted together. The results indicate that (n,0)CNT/GeSe (n = 8–11) all have good optical absorption in the visible light region, which has a high optical absorption peak of about 6×10^5 cm^{-1}. In the infrared region, the optical absorption coefficients of (n,0)CNT/GeSe (n = 8–11) are significantly enhanced compared to those of GeSe. The optical absorption peaks of (8,0)CNT/GeSe and (11,0)CNT/GeSe in the infrared region reach 2×10^5 cm^{-1}. The optical absorption peak of (10,0)CNT/GeSe is close to 2×10^5 cm^{-1}. In the ultraviolet region, the light absorption of the CNT(n,11)/GeSe (n = 8–11) composite is greatly enhanced compared to the corresponding components of GeSe and CNT. Our results prove that the optical absorption of the combined structure of (n,0) CNTs (n = 8–11) and 2D GeSe is substantially enhanced compared with that of 2D GeSe and (n,0) CNTs (n = 8–11).

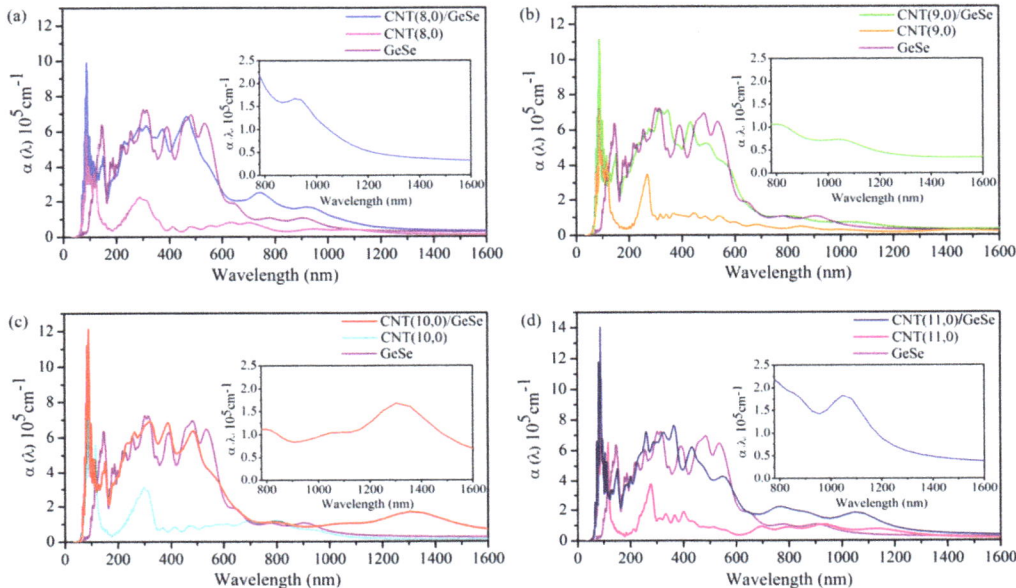

Figure 7. Optical absorption coefficient, α, of (**a**) (8,0)CNT/GeSe, monolayer GeSe, and (8,0)CNT; (**b**) (9,0)CNT/GeSe, monolayer GeSe, and (9,0)CNT; (**c**) (10,0)CNT/GeSe, monolayer GeSe, and (10,0) CNT; and (**d**) (11,0)CNT/GeSe, monolayer GeSe, and (11,0)CNT at the zigzag direction.

According to the theory suggested by Scharber et al., [44], the PCE η of CNT/GeSe can be described as follows:

$$\eta = \frac{J_{SC} V_{OC} \beta_{FF}}{P_{SOLAR}} = \frac{0.65 \left(E_g^d - \Delta E_c - 0.3 \right) \int_{E_g^d}^{\infty} \frac{P(h\omega)}{h\omega} d(h\omega)}{\int_0^{\infty} P(h\omega) d(h\omega)} \quad (4)$$

where the fill factor (β_{FF}) is 0.65. The maximum open-circuit voltage (V_{OC}) is estimated by $\left(E_g^d - \Delta E_c - 0.3 \right)$, where E_g^d is the donor bandgap. ΔE_c is the conduction band offset (CBO) between the donor (GeSe) and acceptor (CNT). $P(h\omega)$ is the AM1.5 solar energy flux at the photon energy ($h\omega$). The integral in the numerator is the short circuit current (J_{SC}) performed by applying an external quantum efficiency limit of 100%, and the integral in the denominator in Equation (4) is the incident solar radiation (P_{SOLAR} = 1000 Wm^{-2}). As mentioned in our analysis of the band structure, the donor layer is the GeSe monolayer with a bandgap of 1.14 eV, while the value of the CBO in the (10,0)CNT/GeSe heterostructure is 0.48 eV. We found that the PCE of the (10,0)CNT/GeSe heterostructure reaches 11.04% following the calculated definition. To achieve a higher PCE by combining CNTs and GeSe, we further obtain the optimized structure of (11,0)/GeSe and the corresponding band structure using the same simulation method. As shown in Figure 8, the (11,0)CNT/GeSe heterostructure has type-II band alignment. The type-II heterostructure facilitates the separation of photogenerated carriers and holes. The value of the CBO in the (11,0)CNT/GeSe heterostructure is 0.22 eV. The PCE of the (11,0)CNT/GeSe heterostructure reaches 17.29%. The obtained high PCE in the (11,0)CNT/GeSe heterostructure is comparable to that in bilayer phosphorene/MoS$_2$ (16%–18%) [45] and GeSe/SnS (18%) [46] heterostructures.

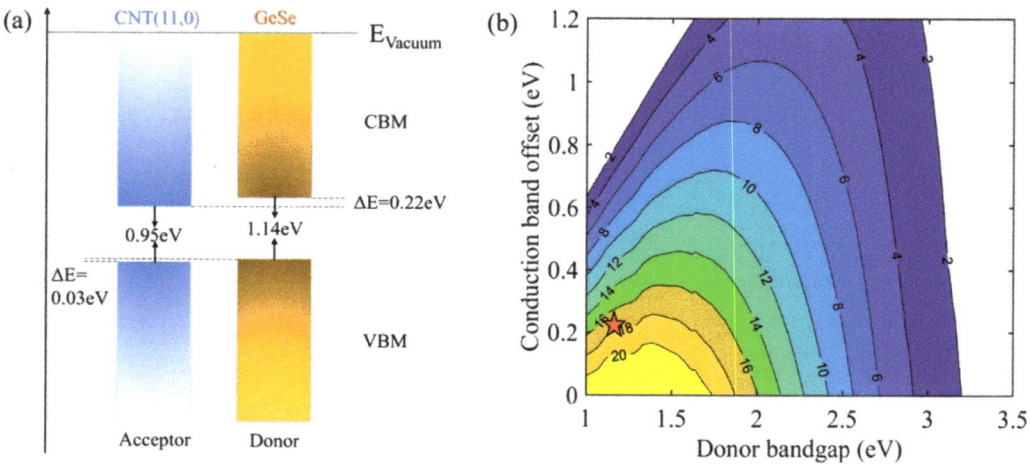

Figure 8. (a) Schematic representation of the band alignment of the (11,0)CNT/GeSe heterostructure. (b) Contour plot of power conversion efficiency (%) for the (11,0)CNT/GeSe heterostructure.

4. Conclusions

We proposed new types of CNT/GeSe heterojunctions by combining a 1D CNT (n = 5–11) and 2D GeSe, and calculated their electronic and optical properties based on DFT. Our calculations show that CNT/GeSe (n = 5–11) are stable through weak vdW interactions. Among the structures, (n,0)CNT/GeSe (n = 5–7) exhibit metallic properties, while (n,0)CNT/GeSe (n = 8–11) have small bandgaps, lower than 0.8 eV. Due to their small bandgaps, (n,0)CNT/GeSe (n = 8–11) have excellent optical absorption properties, especially in ultraviolet and infrared absorption. The absorption coefficient of (n,0)CNT/GeSe (n = 8–11) in the ultraviolet region can reach the order of 10^5 cm^{-1}. In particular, we found that the (11,0)CNT/GeSe heterostructure exhibits type-II band alignment and a high PCE of 17.29%. Our study implies that 1D/2D GeSe/CNT heterostructures have potential applications in photoelectronics and photodetection.

Author Contributions: Y.M. and Z.G. wrote the paper. Y.M. proposed the idea and revised the manuscript. Y.M. and J.Y. provided the research funding. Z.G. did the simulation. J.Y. revised the manuscript. T.S. provided some simulation advices. All authors have read and agreed to the published version of the manuscript.

Funding: This research is funded by the National Natural Science Foundation of China (Grant No. 11471280 and 11374251), by the Natural Science Foundation of Hunan Province, China (Grant No. 2019JJ40280) and the Research and Development Plan of key areas in Hunan Province (2019GK2101).

Data Availability Statement: The data presented in this study are available on request from the corresponding author.

Conflicts of Interest: The authors declare no conflict of interest.

References

1. Iijima, S.; Ichihashi, T. Single-shell carbon nanotubes of 1-nm diameter. *Nat. Cell Biol.* **1993**, *363*, 603–605. [CrossRef]
2. Saito, R.; Dresselhaus, G.; Dresselhaus, M.S. *Physical Properties of Carbon Nanotubes*; World Scientific: Singapore, 1998.
3. Ajayan, P.M.; Zhou, O.Z. Applications of carbon nanotubes. *Carbon Nanotubes* **2001**, *80*, 391–425.
4. Endo, M.; Strano, M.S.; Ajayan, P.M. Potential Applications of Carbon Nanotubes. *Carbon Nanotubes* **2008**, *111*, 13–62.
5. Zólyomi, V.; Kürti, J. First-principles calculations for the electronic band structures of small diameter single-wall carbon nanotubes. *Phys. Rev. B* **2004**, *70*, 085403. [CrossRef]
6. Wilder, J.W.G.; Venema, L.C.; Rinzler, A.G.; Smalley, R.E.; Dekker, C. Electronic structure of atomically resolved carbon nanotubes. *Nat. Cell Biol.* **1998**, *391*, 59–62. [CrossRef]

7. Cabria, I.; Mintmire, J.; White, C.T. Metallic and semiconducting narrow carbon nanotubes. *Phys. Rev. B* **2003**, *67*, 121406. [CrossRef]
8. Blase, X.; Benedict, L.X.; Shirley, E.L.; Louie, S.G. Hybridization Effects and Metallicity in Small Radius Carbon Nanotubes. *Phys. Rev. Lett.* **1994**, *72*, 1878–1881. [CrossRef] [PubMed]
9. Gülseren, O.; Yildirim, T.; Ciraci, S. Systematic ab initio study of curvature effects in carbon nanotubes. *Phys. Rev. B* **2002**, *65*, 153405. [CrossRef]
10. Kanamitsu, K.; Saito, S. Geometries, Electronic Properties, and Energetics of Isolated Single Walled Carbon Nanotubes. *J. Phys. Soc. Jpn.* **2002**, *71*, 483–486. [CrossRef]
11. Sun, G.; Jeno, K.; Miklos, K.; Ray, H. Variations of the Geometries and Band Gaps of Single-Walled Carbon Nanotubes and the Effect of Charge Injection. *J. Phys. Chem. B* **2003**, *107*, 6924–6931. [CrossRef]
12. Li, Z.M.; Tang, Z.K.; Liu, H.J.; Wang, N.; Chan, C.T.; Saito, R.; Okada, S.; Li, G.D.; Chen, J.S.; Nagasawa, N.; et al. Polarized Absorption Spectra of Single-Walled 4 Å Carbon Nanotubes Aligned in Channels of an AlPO4-5 Single Crystal. *Phys. Rev. Lett.* **2001**, *87*, 127401. [CrossRef]
13. Liu, H.J.; Chan, C.T. Properties of 4 Å carbon nanotubes from first-principles calculations. *Phys. Rev. B* **2002**, *66*, 115416. [CrossRef]
14. Machon, M.; Reich, S.; Thomsen, C. Ab initio calculations of the optical properties of 4-Å-diameter single-walled nanotubes. *Phys. Rev. B* **2002**, *66*, 155410. [CrossRef]
15. Kürti, J.; Zólyomi, V.; Kertesz, M.; Sun, G.; Baughman, R.; Kuzmany, H. Individualities and average behavior in the physical properties of small diameter single-walled carbon nanotubes. *Carbon* **2004**, *42*, 971–978. [CrossRef]
16. Xue, D.-J.; Liu, S.-C.; Dai, C.-M.; Chen, S.; He, C.; Zhao, L.; Hu, J.-S.; Wan, L.-J. GeSe Thin-Film Solar Cells Fabricated by Self-Regulated Rapid Thermal Sublimation. *J. Am. Chem. Soc.* **2017**, *139*, 958–965. [CrossRef] [PubMed]
17. Hsueh, H.C.; Vass, H.; Clark, S.J.; Ackland, G.J.; Crain, J. High-pressure effects in the layered semiconductor germanium selenide. *Phys. Rev. B* **1995**, *51*, 16750–16760. [CrossRef]
18. Makinistian, L.; Albanesi, E. Ab initio calculations of the electronic and optical properties of germanium selenide. *J. Phys. Condens. Matter* **2007**, *19*, 186211–186234. [CrossRef]
19. Zhao, H.; Mao, Y.; Mao, X.; Shi, X.; Xu, C.; Wang, C.; Zhang, S.; Zhou, D. Band Structure and Photoelectric Characterization of GeSe Monolayers. *Adv. Funct. Mater.* **2017**, *28*, 1704855–1704864. [CrossRef]
20. Mao, Y.; Xu, C.; Yuan, J.; Zhao, H. Effect of stacking order and in-plane strain on the electronic properties of bilayer GeSe. *Phys. Chem. Chem. Phys.* **2018**, *20*, 6929–6935. [CrossRef] [PubMed]
21. Rohr, F.O.V.; Ji, H.W.; Cevallos, F.A.; Gao, T.; Ong, N.P.; Cava, R.J. High-Pressure Synthesis and Characterization of β-GeSe—A Six-Membered-Ring Semiconductor in an Uncommon Boat Conformation. *J. Am. Chem. Soc.* **2017**, *139*, 2771–2777. [CrossRef] [PubMed]
22. Mao, Y.; Long, L.; Yuan, J.; Zhong, J.; Zhao, H. Toxic gases molecules (NH3, SO2 and NO2) adsorption on GeSe monolayer with point defects engineering. *Chem. Phys. Lett.* **2018**, *706*, 501–508. [CrossRef]
23. Mao, Y.; Xu, C.; Yuan, J.; Zhao, H. A two-dimensional GeSe/SnSe heterostructure for high performance thin-film solar cells. *J. Mater. Chem. A* **2019**, *7*, 11265–11271. [CrossRef]
24. Liu, Y.; Huang, Y.; Duan, X. Van der Waals integration before and beyond two-dimensional materials. *Nat. Cell Biol.* **2019**, *567*, 323–333. [CrossRef] [PubMed]
25. Xia, C.; Du, J.; Xiong, W.; Jia, Y.; Wei, Z.; Li, J. A type-II GeSe/SnS heterobilayer with a suitable direct gap, superior optical absorption and broad spectrum for photovoltaic applications. *J. Mater. Chem. A* **2017**, *5*, 13400–13410. [CrossRef]
26. Jiang, S.; Cheng, R.; Wang, X.; Xue, T.; Liu, Y.; Nel, A.; Huang, Y.; Duan, X. Real-time electrical detection of nitric oxide in biological systems with sub-nanomolar sensitivity. *Nat. Commun.* **2013**, *4*, 2225. [CrossRef]
27. Jia, C.; Famili, M.; Carlotti, M.; Liu, Y.; Wang, P.; Grace, I.; Feng, Z.; Wang, Y.; Zhao, Z.; Ding, M.; et al. Quantum interference mediated vertical molecular tunneling transistors. *Sci. Adv.* **2018**, *4*, eaat8237. [CrossRef]
28. Bediako, D.K.; Rezaee, M.; Yoo, H.; Larson, D.; Zhao, S.Y.F.; Taniguchi, T.; Watanabe, K.; Brower-Thomas, T.L.; Kaxiras, E.; Kim, P. Heterointerface effects in the electrointercalation of van der Waals heterostructures. *Nat. Cell Biol.* **2018**, *558*, 425–429. [CrossRef]
29. Liao, L.; Lin, Y.-C.; Bao, M.; Cheng, R.; Bai, J.; Liu, Y.; Qu, Y.; Wang, K.L.; Huang, Y.; Duan, X. High-speed graphene transistors with a self-aligned nanowire gate. *Nat. Cell Biol.* **2010**, *467*, 305–308. [CrossRef]
30. Liao, L.; Bai, J.; Cheng, R.; Lin, Y.-C.; Jiang, S.; Qu, Y.; Huang, Y.; Duan, X. Sub-100 nm Channel Length Graphene Transistors. *Nano Lett.* **2010**, *10*, 3952–3956. [CrossRef]
31. Cheng, R.; Bai, J.; Liao, L.; Zhou, H.; Chen, Y.; Liu, L.; Lin, Y.-C.; Jiang, S.; Huang, Y.; Duan, X. High-frequency self-aligned graphene transistors with transferred gate stacks. *Proc. Natl. Acad. Sci. USA* **2012**, *109*, 11588–11592. [CrossRef]
32. Cheng, R.; Jiang, S.; Chen, Y.; Liu, Y.; Weiss, N.O.; Cheng, H.-C.; Wu, H.; Huang, Y.; Duan, X. Few-layer molybdenum disulfide transistors and circuits for high-speed flexible electronics. *Nat. Commun.* **2014**, *5*, 1–9. [CrossRef]
33. Kresse, G.; Hafner, J. Ab initio molecular dynamics for liquid metals. *Phys. Rev. B* **1995**, *192*, 222–229. [CrossRef]
34. Kresse, G.; Furthmüller, J. Efficient iterative schemes for ab initio total-energy calculations using a plane-wave basis set. *Phys. Rev. B* **1996**, *5*, 11169–11186. [CrossRef] [PubMed]
35. Ernzerhof, M.; Scuseria, G.E. Assessment of the Perdew–Burke–Ernzerhof exchange-correlation Functional. *J. Chem. Phys.* **1999**, *110*, 5029–5036. [CrossRef]

36. Perdew, J.P.; Burke, K.; Ernzerhof, M. Generalized gradient approximation made simple. *Phys. Rev. Lett.* **1996**, *77*, 3865–3868. [CrossRef] [PubMed]
37. Monkhorst, H.J.; Pack, J.D. Special points for Brillonin-zone integrations. *Phys. Rev. B* **1976**, *13*, 5188–5192. [CrossRef]
38. Grimme, S.; Antony, J.; Ehrlich, S.; Krieg, H. A consistent and accurate ab initio parametrization of density functional dispersion correction (DFT-D) for the 94 elements H-Pu. *J. Chem. Phys.* **2010**, *132*, 154104. [CrossRef] [PubMed]
39. Gajdos, M.; Hummer, K.; Kresse, G.; Uller, J.F.; Bechstedt, F. Linear optical properties in the PAW methodology. *Phys. Rev. B* **2006**, *73*, 045112. [CrossRef]
40. Zhang, Z.; Huang, W.-Q.; Xie, Z.; Hu, W.; Peng, P.; Huang, G.-F. Noncovalent Functionalization of Monolayer MoS2 with Carbon Nanotubes: Tuning Electronic Structure and Photocatalytic Activity. *J. Phys. Chem. C* **2017**, *121*, 21921–21929. [CrossRef]
41. Zhang, Z.; Cheng, M.-Q.; Chen, Q.; Wu, H.-Y.; Hu, W.; Peng, P.; Huang, G.-F.; Huang, W.-Q. Monolayer Phosphorene–Carbon Nanotube Heterostructures for Photocatalysis: Analysis by Density Functional Theory. *Nanoscale Res. Lett.* **2019**, *14*, 1–11. [CrossRef] [PubMed]
42. Zhang, Z.; Huang, W.-Q.; Xie, Z.; Hu, W.; Peng, P.; Huang, G.-F. Simultaneous covalent and noncovalent carbon nanotube/Ag3PO4 hybrids: New insights into the origin of enhanced visible light photocatalytic performance. *Phys. Chem. Chem. Phys.* **2017**, *19*, 7955–7963. [CrossRef] [PubMed]
43. Gomes, L.C.; Carvalho, A. Phosphorene analogues: Isoelectronic two-dimensional group-IV monochalcogenides with orthorhombic structure. *Phys. Rev. B* **2015**, *92*, 085406. [CrossRef]
44. Scharber, M.C.; Mühlbacher, D.; Koppe, M.; Denk, P.; Waldauf, C.; Heeger, A.J.; Brabec, C.J. Design Rules for Donors in Bulk-Heterojunction Solar Cells—Towards 10 % Energy-Conversion Efficiency. *Adv. Mater.* **2006**, *18*, 789–794. [CrossRef]
45. Dai, J.; Zeng, X.C. Bilayer Phosphorene: Effect of Stacking Order on Bandgap and Its Potential Applications in Thin-Film Solar Cells. *J. Phys. Chem. Lett.* **2014**, *5*, 1289–1293. [CrossRef] [PubMed]
46. Cheng, K.; Guo, Y.; Han, N.; Jiang, X.; Zhang, J.; Ahuja, R.; Su, Y.; Zhao, J. 2D lateral heterostructures of group-III monochalcogenide: Potential photovoltaic applications. *Appl. Phys. Lett.* **2018**, *112*, 143902. [CrossRef]

Article

Enhanced Tribological Properties of Vulcanized Natural Rubber Composites by Applications of Carbon Nanotube: A Molecular Dynamics Study

Fei Teng [1], Jian Wu [1,2,*], Benlong Su [1] and Youshan Wang [1,2]

1. Center for Rubber Composite Materials and Structures, Harbin Institute of Technology, Weihai 264209, China; 20S130266@stu.hit.edu.cn (F.T.); subenlong@hit.edu.cn (B.S.); wangsy@hit.edu.cn (Y.W.)
2. National Key Laboratory of Science and Technology on Advanced Composites in Special Environments, Harbin Institute of Technology, Harbin 150001, China
* Correspondence: wujian@hitwh.edu.cn

Abstract: Tribological properties of tread rubber is a key problem for the safety and durability of large aircraft tires. So, new molecular models of carbon nanotube (CNT) reinforced vulcanized natural rubber (VNR) composites have been developed to study the enhanced tribological properties and reveal the reinforced mechanism. Firstly, the dynamic process of the CNT agglomeration is discussed from the perspectives of fractional free volume (FFV) and binding energy. Then, a combined explanation of mechanical and interfacial properties is given to reveal the CNT-reinforced mechanism of the coefficient of friction (COF). Results indicate that the bulk, shear and Young's modulus increase with the increasement of CNT, which are increasement of 19.13%, 21.11% and 26.89% in 15 wt.% CNT/VNR composite compared to VNR; the predicted results are consistent with the existing experimental conclusions, which can be used to reveal the CNT-reinforced mechanism of the rubber materials at atomic scale. It can also guide the design of rubber material prescription for aircraft tire. The molecular dynamics study provides a theoretical basis for the design and preparation of high wear resistance of tread rubber materials.

Keywords: aircraft tire; CNT/VNR composites; friction; MD simulation

1. Introduction

Natural rubber is widely used in industrial products due to its excellent elasticity and mechanical properties, such as tires and seals. However, higher requirements are being placed on the natural rubber due to the harsh working condition of aircraft tire. Carbon-based fillers such as graphene (GE) and carbon nanotube (CNT) have been widely applied in rubber nanocomposites due to the unique structural characteristics, excellent thermodynamic and electromagnetic properties. It has been proved that carbon-based nanofillers can effectively improve the performance of composites, which are suitable for more industrial environments, such as electrical shielding and heating equipment, medical equipment and aircraft tires [1,2].

The properties and potential applications of nanocomposites can be greatly enhanced and expanded by carbon-based nanofillers [3–6]. Well dispersed epoxidized natural rubber/carbon black (ENR/CB) composite with CNT contained was prepared for high-performance flexible sensors [7]. Bokobza et al. [8–13] comprehensively studied the reinforcing effect of CNT on styrene-butadiene rubber (SBR) in mechanical, thermal and electrical properties. The enhancement effect of aminosilane-functionalized carbon nanotube on NR and ENR has been discussed by Shanmugharaj et al. [14,15]. CNT was also used as a model filler for SBR to prepare tightly bound rubber material [16]. The excellent polymer-filler interaction of functionalized CNT was confirmed that it can greatly improve the overall performance including mechanical, thermal and electrical properties

of the NR/CNT and ENR/CNT nanocomposites [17,18]. These fillers are regarded as ideal materials in aircraft tire applications to improve the strength [19], modulus [20] and wear resistance [21] of natural rubber materials. Atieh et al. [22] found that the Young's modulus of the SBR/CNT composite containing 10 wt.% CNT was six times higher than that of the pure SBR due to the excellent strength and interfacial effects of CNT. Compared with other fillers, carbon-based fillers can achieve better reinforcement effects with a smaller dosage, thereby reducing pollution and further improving material performance [23]. Kumar and Lee [24] studied the influence of CB and CNT on the Young's modulus of filled silicone rubbers (SRs). Results showed that the Young's modulus of CNT-filled SR also increased from 272% to 706% when CNT was added from 2 phr to 8 phr, which is much larger than 125% with 10phr CB-filled SR. In addition, the different content of carbon-based fillers also has different reinforcement effects on the rubber matrix. Many research works [7,25–30] have proved that excessive CNT can cause agglomeration of fillers, which leads uneven dispersion and affecting the properties of composite materials.

It is confirmed that cross-linking vulcanization is one of the most important reasons for the excellent elasticity and deformation recovery of rubber materials [31–38]. The degree of crosslinking and cross-linking bond type greatly affect the overall properties of vulcanized rubber. The accelerator is often used to increase the degree of crosslinking of vulcanized rubber, thereby reducing pollution and improving material properties [39]. Sainumsai et al. [40] and Fan et al. [41] both tested vulcanized natural rubbers (VNRs) made by conventional vulcanization (CV), semi-effective vulcanization (SEV) and effective vulcanization (EV) methods. The crosslinking density and the content of monosulfidic, disulfidic and polysulfidic crosslinks were obtained. It was indicated that the distributions of sulfur crosslink types effect the strain-induced crystallization and dynamic mechanical properties of vulcanized rubber.

The enhancing mechanism of rubber composites by CNT, GE and other common reinforcing fillers were explained via molecular dynamics (MD) simulations [42–45]. In particular, it can reveal the micro-reinforcement mechanism of carbon-based fillers and their functionalized products on the polymer matrix [46–52]. The interface interaction between CNT and polymer matrix was studied by molecular simulation [53–55]. It was found that the pull-out force of CNT and the elastic modulus of the matrix are both affected by the diameter of nanotube, however, the shear strength of the interface is mainly affected by the length. In addition, the cross-linked structure of polymers can be developed by MD models, such as epoxy resin and vulcanized rubber [56,57]. Zhang et al. [58] developed a vulcanized SBR molecular model to compare the tribological properties of vulcanized SBR and SBR. It was verified that vulcanization can improve the tribological and interfacial properties of rubber materials at the atomic scale.

In summary, the performance of rubber materials is influenced by the vulcanized crosslinked structure and carbon-based filler. The reinforcement is also affected by crosslinking degree, distribution of crosslinking bond types and dosage of carbon-based fillers. The enhancement mechanism has been studied from the perspective of vulcanization and nanofillers at the atomic level [59,60]; however, it mainly focusses on the oligomers and resin materials rather than rubber materials. In addition, the crosslinking degree, distribution of vulcanization bond types and CNT dosage have not been specific described in the existing studies at atomic scale.

A series of VNR atomic models reinforced by different content CNT were developed to reveal the mechanism of CNT-reinforced VNR. A new VNR model was developed with considering the distribution of sulfur bonds and crosslinking degree. As the main component of road surface, SiO_2 can be regarded as the direct contact material with tires. As results, the interfacial interaction between SiO_2 and VNR plays a significant guiding role in aircraft tire rubber materials design and evaluation. Here, the reinforced CNT/NR models and the CNT/NR-SiO_2 interface models were developed to study the CNT reinforce mechanism on the vulcanized NR and the atomic behaviors of the composites on the friction interface.

2. Molecular Dynamics Model

MD models were developed by Materials Studio Software (MS). Firstly, the NR molecular chain, sulfides for the synthesis of crosslinking bonds, CNT and 45 Å^3 empty periodic cell box were obtained. Then, different numbers of CNT were placed at the center of multiple cell boxes to represent different CNT dosage. The Amorphous Cell Calculation of MS was used to pack 10 NR molecular chains built with 70 repeat units into the periodic cell box by Monte Carlo method. In addition, a variety of sulfides used to generate vulcanized crosslinks were also added to the periodic box in certain proportions and quantities according to the existing experimental measurement results to obtain uncross-linked CNT/VNR models. Next, all the sulfides were used to generate cross-linked bonds by the cross-linking script, then, the cross-linked CNT/VNR composite models were developed. Finally, the silica (quartz glass) model in MS database was imported to build supercell as friction layer. The Cleave surface and the Super cell commands were used for 45 Å × 45 Å × 10.8 Å SiO_2 supercell and interfacial interaction models were constructed by the cross-linked CNT/NR composite and obtained SiO_2 slab for further calculation and analysis.

The degree of polymerization represents the number of repeating units in a single molecular chain in MD simulation. Longer chain can improve the simulation accuracy but reduce the speed. Therefore, the solubility parameters of single molecular chains with different numbers of repeating units were calculated for proper NR chain length. As shown in Figure 1, the solubility parameter of NR chain begins to stabilize while the number of repeat unit larger than 35, which means the number of repeat units should larger than 35. Moreover, the solubility parameter of NR stabilizes at 16.25 $(J/cm^3)^{1/2}$, which is consistent with the experimentally measured value of 16.2~17 $(J/cm^3)^{1/2}$. Here, we built NR chain containing 70 repeat units.

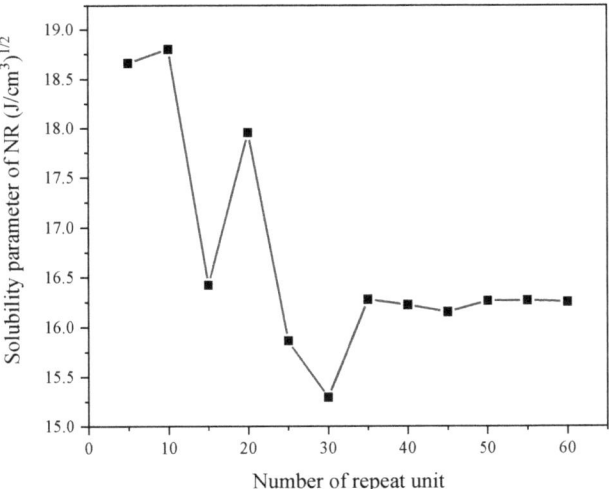

Figure 1. Solubility parameters NR with different degree of polymerization.

The modelling of uncross-linked CNT/VNR is shown in Figure 2. Hydrogen atoms were added at both ends for saturated CNT to eliminate the end-side effects. The size and the corresponding positions of the CNT in different CNT content periodic cells are shown in Figure 2g. The crosslink density is defined by:

$$\rho = v/N_0 \tag{1}$$

where v is the number of crosslinked repeat units and N_0 is the total number of repeat units.

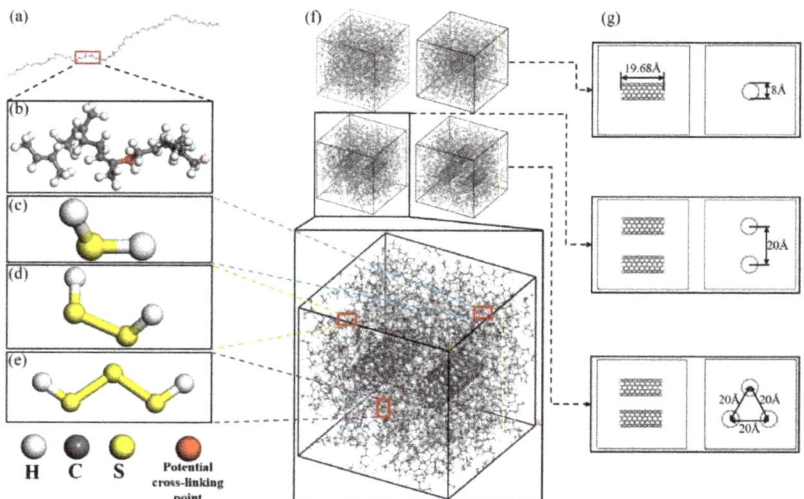

Figure 2. Establishment of un-crosslinked CNT/VNR molecular models. (**a**) NR molecular chain. (**b**) Five repeating units with one potential cross-linking point. (**c**) Monosulfide. (**d**) Disulfide. (**e**) Polysulfide. (**f**) Different CNT contents un-crosslinked CNT/VNR molecular models. (**g**) Size and corresponding positions of the nanotubes.

There is one crosslinked repeat unit in every 50~100 units for conventionally vulcanized rubber materials, which the ρ is about 1~2%. Here, considering the possibility of self-crossing, the number of cross-linking points is determined to be 10 in the model. The ρ of the obtained model is about 2.5%. Experiment results also indicate that the ratio of monosulfidic, disulfidic and polysulfidic crosslinks in the vulcanized natural rubber was about 5:3:2 [40]. Accelerators that promote the cleavage of polysulfide bonds were blended with rubber materials in actual production by EV method, which makes it difficult to form long polysulfide bonds in vulcanized natural rubber. Therefore, vulcanization bonds containing three sulfur atoms were used in this model to characterize polysulfidic crosslinks. Different types of sulfides with the ratio of 5:3:2 and 10 NR molecular chains were introduced to the periodic boxes by the Amorphous Cell Calculation module, which is used to obtain uncross-linked CNT/VNR models with a predefined density of 0.93 g/cm^3.

Uncross-linked CNT/VNR models were further used to generate crosslinked structure by crosslinking Perl script at temperature of 450 K. The flow chart of the programing of vulcanization process was shown in Figure 3. As shown in Figure 2b, potential cross-linking points were set in every five repeating units to avoid two cross-linking points being too close. The carbon atoms on the NR chains and sulfur atoms on the sulfides were used to form carbon–sulfur (C-S) bonds when there was sulfide molecular in the range of 1.5~5.5 Å between two potential cross-linking points and excess hydrogen atoms were removed. The geometric center of the formed bond was located at the center between the two cross-linking points. All sulfides were consumed for the generation of cross-linking bonds at both ends in the cross-linking process. The schematic diagram and chemical formula of the cross-linking principle were shown in Figure 4. Crosslink can be divided into self-crosslinking and crosslinking, which was distinguished by setting each molecular chain as an independent color. Here, the CNT contents of the four models were about 0 wt.%, 5 wt.%, 10 wt.% and 15 wt.%.

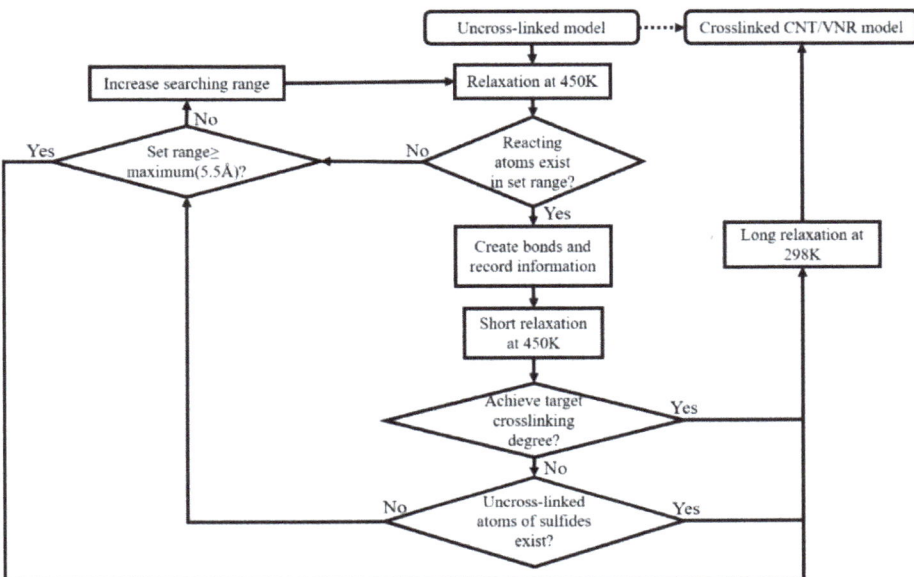

Figure 3. Flow chart of the programing of vulcanization process.

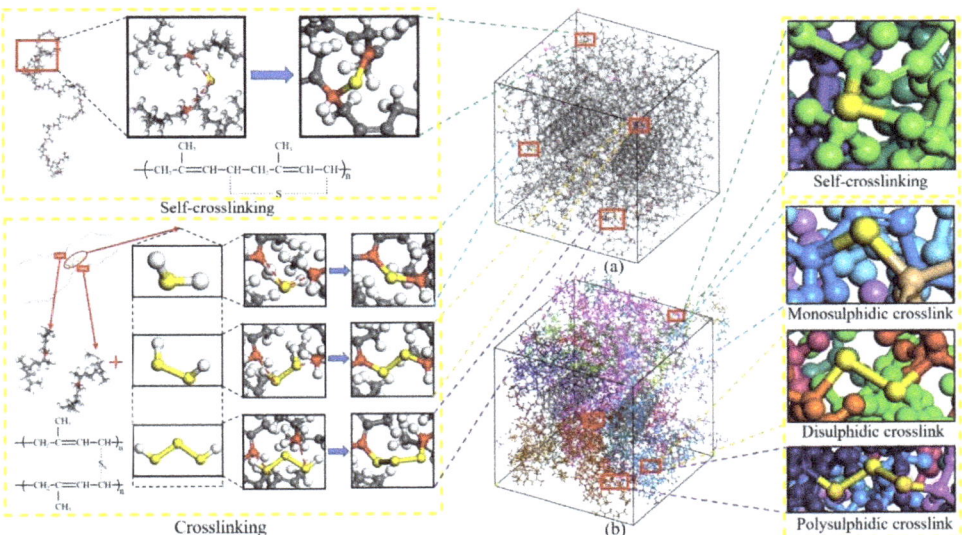

Figure 4. 10 wt.% CNT/VNR molecular model crosslinking process and containing crosslinking bonds. (**a**) Colored by atom types as same as Figure 2. (**b**) Colored by molecular chains.

The geometry optimization was applied to crosslinked CNT/VNR model by conjugate gradient method with the energy and force convergence tolerance of 1×10^{-5} kcal/mol and 5×10^{-4} kcal/mol/Å [61]. Then, a 100ps 5-cycle annealing process was conducted constant volume and temperature (NVT ensemble) from 200 K to 400 K to relax the internal stress of the structure. Finally, a 250 ps dynamic relaxation under constant pressure and temperature (NPT ensemble) of 101kpa and 298 K was performed to obtain the final energy equilibrium model with reasonable vulcanization bond distances and angles.

Double layer and three-layer models were developed to study the interfacial properties and tribological performance. These models were constituted of CNT/VNR model and $45 \times 45 \times 10.8$ Å^3 SiO$_2$ model. The adsorption properties between rubber matrix and SiO$_2$ can be analyzed by the double layer structure as shown in Figure 5. The interfacial energy can be obtained by dynamics calculation for 150 ps under 298 K NVT ensemble with fixed bottom SiO$_2$ layer. The interface interaction energy E_{inter} and the interface van der Waals force energy $E_{inter-vdW}$ can be calculated by Equations (2) and (3), respectively.

$$E_{inter} = E_{Total} - \left(E_{Layer1} + E_{Layer2}\right) \qquad (2)$$

$$E_{inter-vdW} = E_{Total-vdW} - \left(E_{Layer1-vdW} + E_{Layer2-vdW}\right) \qquad (3)$$

where and are the potential energy and van der Waals energy of the entire double-layer model, respectively. and are the potential energy and van der Waals energy of the lower silicon dioxide fixed layer. and are the potential energy and van der Waals energy of the upper EUG/NR composite movable layer.

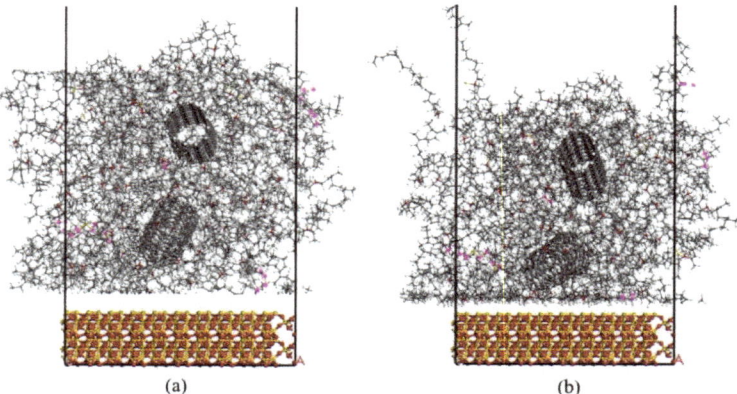

Figure 5. Adsorption double layer structure model of 10 wt.% CNT/VNR composite (Periodic boundary conditions are shown as black lines). (**a**) Initial state. (**b**) Equilibrium state.

Additionally, all atoms of SiO$_2$ layer needs to be unfixed before the calculation.

The frictional simulation was carried out by the confined shear calculation in Forcite module in MS based on the confined nonequilibrium molecular dynamics (NEMD) theory. The SiO$_2$ slabs were initially fixed during relaxation process which including geometry optimization, annealing from 200 K to 400 K and 100 ps dynamic calculation under 298 K NVT ensemble tasks to obtain stable layer structure. After that, the bottom and top SiO$_2$ slabs were unfixed and moved in the opposite direction along the X axis with a speed of 0.2 Å/ps for 250ps. All the friction force, layer pressure and temperature data during friction process were recorded into the trajectory files.

More details have been presented to reproduce the simulation processes. Firstly, the condensed-phase optimized molecular potentials for atomistic simulation studies (COMPASS) force field [62] was used for entire simulation work. The system is considered stable when the energy and density values of the model fluctuate less than 5% during the relaxation process. Secondly, periodic boundary conditions in x and y directions were adopted in the double layer and three-layer models to simulate properties of bulk system. Thirdly, all simulations were conducted with the time step of 1fs, the temperature and pressure controlling methods were Anderson [63] and Berendsen [64] methods, respectively. Finally, the summation methods of energy calculation were Ewald for electrostatic and

Atom based for van der Waals interaction. The accuracy and buffer width of Ewald method were 1×10^{-5} kcal/mol and 0.5 Å. The cutoff distance, spline width and buffer width of van der Waals interaction calculations were 18.5 Å, 1 Å and 0.5 Å [61].

3. Results and Discussion

3.1. Microscopic Inherent Properties Analysis

The fractional free volume (FFV) and mean square displacement (MSD) are calculated in this part to explain and predict the enhancement of CNT on the rubber matrix.

3.1.1. Fractional Free Volume

The total volume (V_T) of solid matrix can be considered as the sum of occupied volume (V_O) and free volume (V_F) based on the free volume theory [65]. The empty space represents the potential area for atoms and chains to move, therefore influencing the mechanical and thermal properties when deformation is applied to material [66]. After the relaxation process, different degrees of distortion occur in different models and it is inaccurate to discuss free volume directly. Therefore, the percentage of fractional free volume is calculated to characterize. The Connolly surface method is adopted for FFVs calculation based on Equation (4). The Connolly radius and Grid interval are 0.1 nm and 0.015 nm, respectively.

$$FFV = \frac{V_F}{V_T} \times 100\% = \frac{V_T - V_O}{V_T} \times 100\% \quad (4)$$

where V_T, V_F and V_O represent the total, free and occupied volume of the models, respectively.

The results of the FFVs of pure VNR and CNT/VNR with different CNT contents are shown in Figure 6. The occupied and free volume are colored by grey and blue, respectively. It is illustrated that the addition of CNT increases the FFV of the composites. Thus, it can be inferred that CNT has an attractive effect on surrounding atoms, which leads to the agglomeration of inside atoms and formation of outside free volume. The pattern of free volume evolution with increasing CNT content can be concluded as generation-growing-convergence. During the increase of CNT content, some small free areas are generated at beginning. Then, those new formed areas become larger due to higher attractive effect from inside CNT. Finally, the grown free areas converge to form larger free volume. The molecular chains inside the concentrated rubber matrix with CNT addition show low tendency of movement, which results in less chance of internal destruction under dynamic stress. This free volume expansion at atom level is in accordance with the experimental results [26].

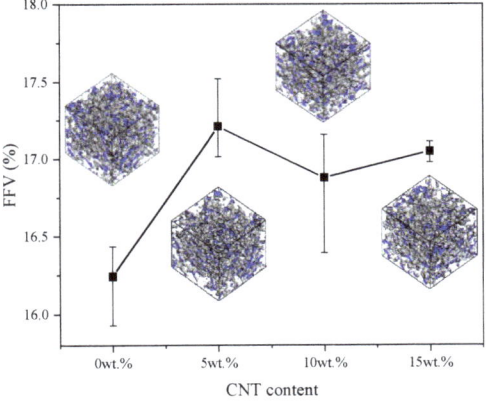

Figure 6. The FFVs and free volume distribution of different CNT contents CNT/VNR composites.

3.1.2. Mean Square Displacement

The diffusion and movement trend of the molecular chains inside the particles can be characterized by the mean square displacement [66,67]. It indicates the statistical square of particle displacement in the system compared to the initial state, which is defined as:

$$MSD = \frac{1}{3N} \sum_{i=0}^{N-1} \left(|R_i(t) - R_i(0)|^2 \right) \quad (5)$$

where $R_i(t)$ and $R_i(0)$ are the displacement vector of atom i at time t and initial time, N is the total number of atoms.

The MSD evolutions of pure VNR and CNT/VNR system during relaxation process are shown in Figure 7. It can be illustrated that the MSD first decrease and then increase with the addition of CNT. It can be conjectured that small amount of CNT is conductive to the aggregation of the matrix and enhancement of the composite strength due to the excellent hardness of CNT. However, agglomeration is caused by excessive nanotubes in the matrix. According to the deformation law of the hard filler-reinforced soft material, the deformation degree of the soft matrix is greater than that of overall material. Similarly, the VNR matrix surrounded by those agglomerated nanotubes shows large deformation and stress concentration, which performs higher tend of molecular movement. As a result, the CNT content should not be too high in order to avoid the agglomeration, which is consistent with related research at different scales [28].

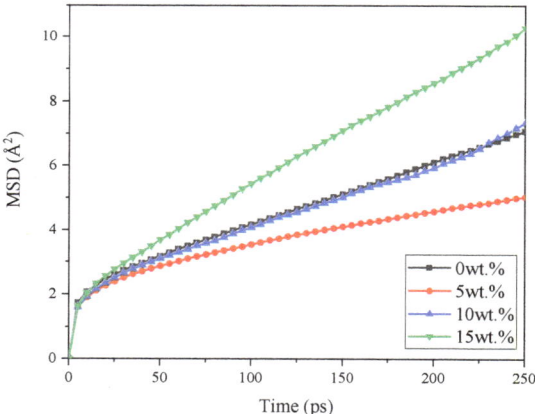

Figure 7. The comparatively MSD of pure VNR and CNT/VNR composites.

3.2. Mechanical Properties Analysis

The elastic mechanical properties of the systems are analyzed by the constant strain method. Relative mechanical properties of the material including Young's, bulk and shear modulus can be obtained by solving the stiffness matrix [68]. In MS, the stiffness matrix can be calculated by applying a series of 0.03% small tensile strains along three axes to the obtained stable models. The stable models used in this section are five independent configurations obtained in last 20 ps of relaxation process. The results are averaged and the error bars are used to express maximum and minimum among these configurations. The elements in the stiffness matrix are expressed by the following equation:

$$C_{ij} = \frac{\partial^2 U}{V \partial \varepsilon_i \partial \varepsilon_j} \quad (6)$$

where U, V and ε represent the second derivative of the deformation, unit volume and strain. For isotropic materials like rubber, the stress-strain relations are completely described by two lame constants λ and μ, which can be expressed by the elements in the stiffness matrix as [51]:

$$\lambda = \frac{1}{3}(C_{11} + C_{22} + C_{33}) - \frac{2}{3}(C_{44} + C_{55} + C_{66}) \tag{7}$$

$$\mu = \frac{1}{3}(C_{44} + C_{55} + C_{66}) \tag{8}$$

The Young's modulus (E), bulk modulus (K) and shear modulus (G) of the systems can be further calculated based on the λ and μ results following equations below [69]:

$$E = \frac{\mu(3\lambda + 2\mu)}{\lambda + \mu} \tag{9}$$

$$K = \lambda + \frac{2}{3}\mu \tag{10}$$

$$G = \mu \tag{11}$$

The calculation results of Young's modulus are separated in three directions. Considering rubber as isotropic material, the modulo is calculated by Equation (12) for quantitative comparison.

$$|E| = \sqrt{E_X^2 + E_Y^2 + E_Z^2} \tag{12}$$

where |E| is the modulo of Young's modulus; E_X, E_Y, E_Z are the Young's modulus in X, Y and Z directions, respectively.

The details of bulk, shear modulus and Young's modulo calculation results are recorded in Table 1. The results are given in the form like: minimum value~maximum value (averaged value). In addition, the increase percentage is calculated according to the averaged value. It can be concluded that the modulus increase with the increase of CNT. The bulk, shear modulus and Young's modulo of CNT/VNR composites with 15 wt.% CNT content are 2.74, 1.09 and 5.19 GPa, which are 19.13%, 21.11% and 26.89% higher than 2.30, 0.90 and 4.09 GPa of pure VNR. This result indicates that the CNT can enhance comprehensive mechanical properties of the rubber matrix, which increases the hardness of the matrix, the resistance to shear deformation [70] and volume change [71]. The addition of CNT also allows the obtained composites to endure larger stress and suit for wider circumstance like aircraft tire production. Moreover, the increasing trend of mechanical properties of CNT/VNR composites at atom level is also conformity with experimental studies [27–29]. The continuous increase of CNT/VNR composites modulus can be explained by the high hardness and modulus of carbon nanotubes.

Table 1. Bulk, Shear modulus and Young's modulo results and increase percentage compared to pure VNR.

CNT Content (wt.%)	Bulk Modulus (GPa)	Increase (%)	Shear Modulus (GPa)	Increase (%)	Young's Modulo (GPa)	Increase (%)
0	2.22~2.34 (2.30)	0	0.88~0.94 (0.90)	0	4.01~4.15 (4.09)	0
5	2.35~2.51 (2.42)	5.22	0.94~1.00 (0.96)	6.67	4.23~4.59 (4.42)	8.07
10	2.50~2.77 (2.61)	13.48	1.01~1.06 (1.03)	14.44	4.70~4.94 (4.83)	18.09
15	2.65~2.85 (2.74)	19.13	1.05~1.12 (1.09)	21.11	4.99~5.31 (5.19)	26.89

3.3. Interfacial Properties Analysis

The interfacial interactions of CNT/VNR composite mainly includes internal interaction of CNT- matrix and external interaction of composite-SiO_2. Adhesion phenomenon between tire and road occurs in friction process. Hence, the double layer structure (seen in Figure 5) was developed to reveal the interface contact mechanism of the adhesion phenomenon. A clear adsorption process between composites and fixed SiO_2 was also observed during dynamic equilibrium. The interfacial energy and atom density between CNT reinforced rubber materials and SiO_2 are calculated in Section 3.3.1. Before dynamical friction happens, a comparatively large friction force occurs in the state of static friction due to the static adsorption. As a result, relatively large deformation occurs in this state, which brings the possibility of material damage. For nanocomposites, the dispersion condition and bonding strength of the filler inside the matrix greatly determine the damage resistance of composites during deformation [25]. In order to characterize the enhancement mechanism of CNT effects on the vulcanized natural rubber, the CNT-matrix binding energy and atom relative concentration inside nanocomposites are discussed in Section 3.3.2.

The system is considered stable when the energy fluctuation is less than 5%. All the results in this section are the average of five independent configurations in the last 20 ps after stabilization. The maximum and minimum values among the calculation results are given in the form of error bars.

3.3.1. Composite-SiO_2 Interface

The interfacial energy calculation of composite-SiO_2 interface is based on the Equations (1) and (2). The interfacial energy and van der Waals energy results of pure VNR and different CNT contents CNT/VNR composites are shown in Figure 8. It can be illustrated that the interfacial energy evolution shows nonlinear trend with the addition of CNT. The largest interfacial energy is obtained in 10 wt.% CNT/VNR composite, which means higher energy barrier needs to be break during the relative movement. This process may lead to higher temperature rise, severer atoms contact and larger static friction force. In addition, the interfacial interaction is mainly caused by the van der Waals interaction according to the contribution of van der Waals energy.

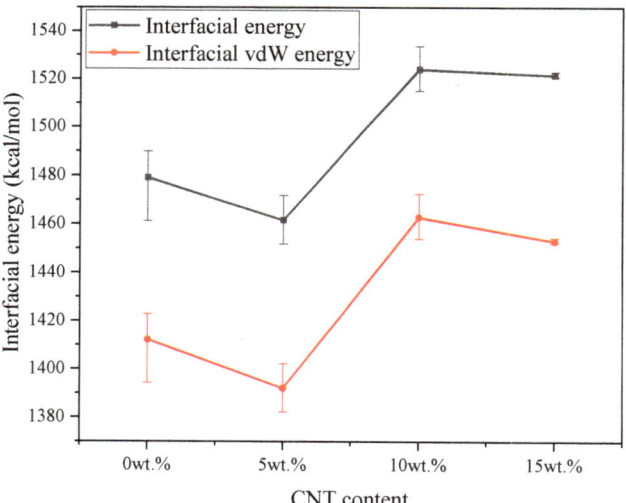

Figure 8. The energy of composite-SiO_2 interface.

The atom relative concentrations along Z direction are shown in Figure 9. It is found that the peaks of atom concentration occur in the range of 12~14 Å along Z direction. These peaks represent the aggregation degree of the atoms at the contact interface between the SiO_2 and rubber matrix, which reflex the severity of relative motion between layers. More concentrated atoms can cause more intense friction process, which is not expected in both micro and macro applications. In the range of 50~60 Å, the relative concentration of the matrix drops rapidly, which can be explained by adsorption effect of SiO_2 slab to the rubber matrix. It can be observed that the concentration value of 10 wt.% CNT/VNR composite in the range of 14~50 Å is higher than that of 0.5 and 15 wt.%. This phenomenon may reflect the superimposition of CNT binding and the binding weakening of excessive nanotubes. The result of interface atom concentration is consistent with the interfacial energy, which indicates that high interfacial energy can attract more internal atoms moving outward to form a high-density shell. This may be a new idea for soft materials coating preparation.

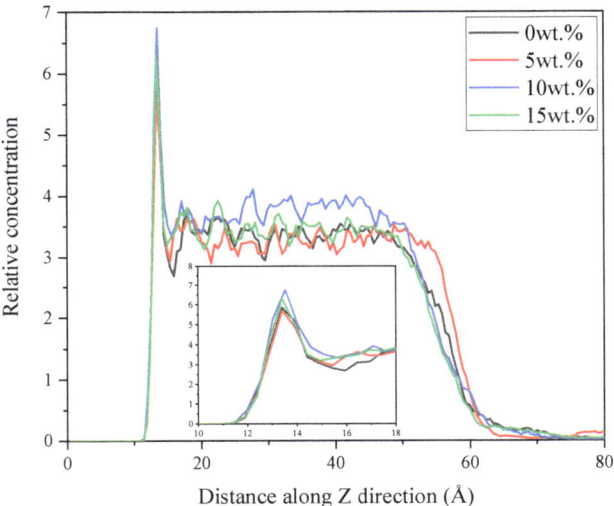

Figure 9. The atom relative concentration of different systems along Z direction.

3.3.2. CNT-Matrix Interface

The interfacial interactions between CNT and rubber matrix are discussed by using the models as same as Section 3.2. The binding energy calculations follow similar pattern of interfacial energy calculations, which is expressed by Equations (13) and (14).

$$E_b = -(E_T - E_r - E_{CNT}) \tag{13}$$

$$E_{b-vdW} = -(E_{T-vdW} - E_{r-vdW} - E_{CNT-vdW}) \tag{14}$$

where E_b, E_T, E_r, E_{CNT} and E_{b-vdW}, E_{T-vdW}, E_{r-vdW}, $E_{CNT-vdW}$ represent the potential and van der Waals energy of binding, total system, rubber and CNT, respectively.

The results of total binding energy and binding energy per CNT are shown in Figure 10. It can be indicated that the binding energy between rubber matrix and CNT increases with the increase of CNT. It is easily to understand that more nanotubes have more contact area with rubber matrix and the superimposition effect of nanotubes binding mentioned before causes greater total binding energy of higher CNT content. However, the binding energy per CNT no longer shows linear growth but decrease when excessive nanotubes exist in the rubber matrix. This binding weakening phenomenon may be caused by the agglomeration of nanotubes. These agglomerated nanotubes begin to compete for atoms located around their geometric center when the superposition effect of CNT exceeded its maximum, which decreases the binding energy per CNT. Macroscopically, it is reflected by the decrease of mechanical properties like tensile strength, elongation at break and so on [7,28,29]. In addition, it can be observed that the van der Waals binding energy constitutes most part of the total binding energy, which proves the van der Waals interaction is the main factor of binding between CNT and rubber matrix.

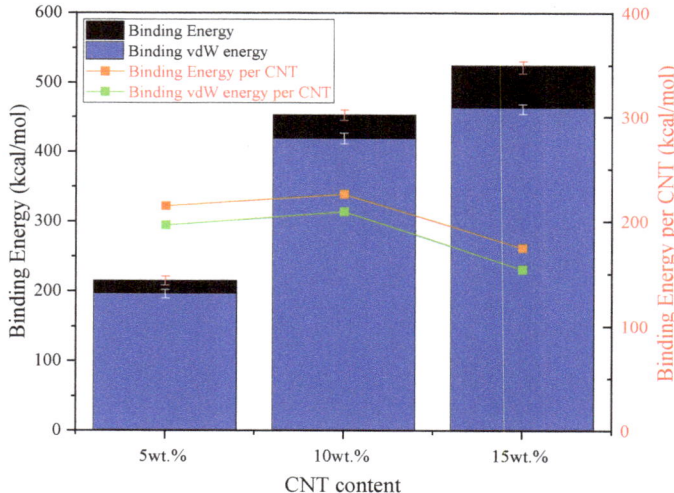

Figure 10. The binding energy of CNT/VNR composites with different CNT contents.

The density distributions of rubber matrix inside the composite are also obtained as shown in Figure 11. Obvious peaks can be observed at the CNT-rubber matrix interface, which indicates the attractive effect of CNT on surrounding atoms. This is consistent with the conclusions obtained in Sections 3.1 and 3.2, which explains the strengthening mechanism of CNT on the rubber matrix. In addition, higher atomic densities are observed between two different nanotubes than that on both sides, which also indicates that the attraction of CNT has a superimposed effect. The relatively low atom density in the 15 wt.%CNT/VNR reflects the compete relationship between excessive nanotubes. This attraction-superposition-competition dynamic changing process caused by the increase of CNT content may be one of the main reasons for the changes of mechanical and tribological properties in actual applications.

In conclusion, both the internal binding and external interaction are influenced by CNT. The dynamic process of attraction-superposition-competition is inferred by molecular simulations. The friction performance is preliminarily predicted based on the interfacial calculation results. In order to verify the rationality of predictions, the tribological properties are discussed below.

3.4. Tribological Properties and Mechanism

The COF is computed in this section for evaluating the tribological properties of the different CNT contents CNT/VNR composites. The details of calculated COF are given in the form like: minimum value~maximum value (averaged value) and listed in the Table 2.

Table 2. COF details of different CNT contents CNT/VNR composites.

CNT Content (wt.%)	COF	Increased Percentage (%)
0	1.017~1.049 (1.033)	0
5	1.037~1.118 (1.078)	4.36
10	1.097~1.192 (1.144)	10.75
15	1.104~1.207 (1.155)	11.81

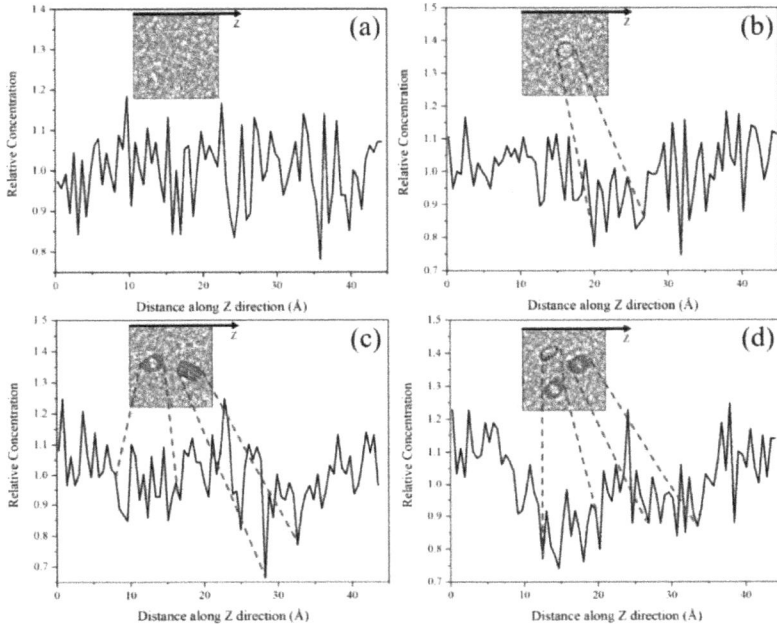

Figure 11. Axial density distribution of CNT/VNR composites with different CNT contents. (a) Pure VNR. (b) 5 wt.%CNT/VNR. (c) 10 wt.%CNT/VNR. (d) 15 wt.%CNT/VNR.

Results indicate that the COF keep rising with the addition of CNT. However, the growth rate drops significantly when CNT content changes from 10 wt.% to 15 wt.%. The discrepancy of COF caused by the surface roughness can be ignored at atomic level in our simulations. The fiction performance of the composite can no longer be simply judged by the interfacial energy but a combined result of mechanical and interfacial properties. On one hand, the increasing mechanical modulus prevent large shear deformation of the composite, which makes the composite become more stable during the friction process. The atoms around the friction surface tend to show lower activity in this stable composite, which achieves better equilibrium adsorption between the composite and the SiO_2 slab. That's the reason why the COF increase with the addition of CNT. On the other hand, the friction performance also effected by the interfacial energy. Although the CNT/VNR composites with lower CNT content are harder to achieve well equilibrium adsorption state during friction process, they still have larger interfacial energy (especially the 10 wt.% CNT/VNR) to offset the inadequate surface interaction caused by insufficient mechanical properties. Thus, the COF of composites are not show a linear trend of increasing like the modulus, but increasing trend of slowing down.

The slip phenomenon is observed during friction process and relative atom concentration along Z direction are further discussed for the tribological properties. During the friction process, the shear deformation of the composite is observed in the beginning. Then, the deformation of the composite reaches its maximum state and the slip phenomenon can be observed afterward. The slip mechanism may be caused by the relative movement, which breaks the energy barrier formed by interfacial interaction. Therefore, the state of the composite and the time of slip phenomenon occurs can be used to reveal the mechanism of the COF evolution. The relative concentration reflects the severity of the friction. More concentrated atoms cause more intense friction process. The state of different CNT/VNR composites and the time when slip phenomenon occurs are shown in Figure 12. The relative atom concentration along Z direction is shown in Figure 13. It can be observed that CNT/VNR composites with higher CNT content have lower shear deformation and the slip phenomenon occurred earlier. This enhanced property prevents the damage of composite during friction and allows the composite to reach stable dynamic friction sooner. The relative concentration indicates that the atoms appear to gather on the rubber matrix rather than friction interface due to the enhanced binding properties of CNT, which can reduce the intensity of friction. The maximum decrease of the atom concentration is observed in the 5 wt.% CNT/VNR composite, which are 14.3% and 13.8% lower than that of pure VNR on different friction interfaces. However, this enhancement can be attenuated by excessive nanotubes, which may be caused by the weakened binding properties. As a result, this impact needs to be noticed in actual prescription design of CNT reinforced rubber materials.

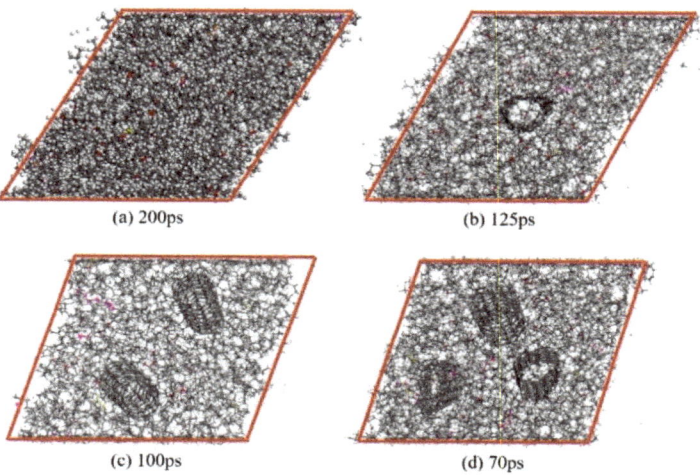

Figure 12. Deformation state of different CNT/VNR composites and time at the beginning of slipping. (**a**) Pure VNR. (**b**) 5 wt.%CNT/VNR composite. (**c**) 10 wt.%CNT/VNR composite. (**d**) 15 wt.%CNT/VNR composite.

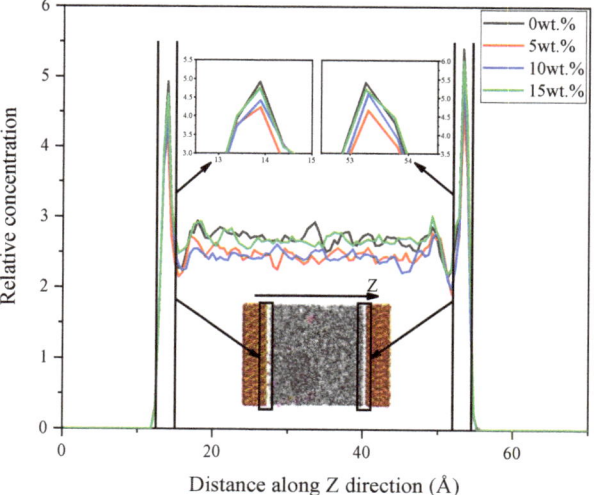

Figure 13. Relative atom concentration of different CNT/VNR composites.

4. Conclusions

MD simulations were performed to reveal the influence of CNT content on the CNT/VNR composites and the enhanced mechanism. The results are consistent with the present research, which can be used to reveal the enhanced mechanism of CNT. The following conclusions are highlighted from the results.

(1) Molecular models for the CNT/VNR composites with different CNT content were developed with considering the distribution of different sulfur bonds. The FFV and MSD were discussed preliminarily. The FFV evolution with CNT addition is summarized as three stages, which are generation, growing and convergence.

(2) The bulk, shear and Young's modulus were calculated by constant strain method to evaluate the mechanical properties of the composites. Approximately linear increase trends were observed in all modulus. The largest bulk, shear and Young's modulus occurred in the 15 wt.% CNT/VNR composite, which were 19.13%, 21.11% and 26.89% higher than that of pure VNR, respectively. The mechanism of these improved mechanical properties can be explained by the high strength of CNT.

(3) The binding energy of CNT-matrix interface and the interfacial energy of composite-SiO_2 interface were obtained, respectively. The largest interfacial energy was obtained in 10 wt.% CNT/VNR composite. Thus, a dynamic attraction-superposition-competition process is concluded to reveal the reinforced mechanism of CNT on the rubber matrix. Both the binding and interfacial interactions are mainly produced by the van der Waals interaction.

(4) The COF and relative concentration at the friction interface were calculated to discuss the tribological properties of CNT reinforced VNR composite. The COF shows a nonlinear trend of increasing. Based on the mechanical and interfacial results, the friction performance is inferred to be a combined consequence of mechanical and interfacial properties. Although the enhanced COF is expected in the production of aircraft tires, the addition of CNT can cause more intense friction process according to the relative concentration results. Therefore, this impact needs to be noticed in the actual design of prescription.

Author Contributions: Conceptualization, J.W.; methodology, F.T.; software, F.T.; formal analysis, F.T.; investigation, J.W. and F.T.; data curation, F.T.; writing—original draft preparation, J.W. and F.T.; writing—review and editing, J.W. and B.S.; visualization, F.T.; supervision, J.W. and Y.W.; project administration, J.W. and Y.W.; funding acquisition, J.W. and Y.W. All authors have read and agreed to the published version of the manuscript.

Funding: This research is funded by National Natural Science Foundation of China (52075119), Major Program of National Natural Science Foundation of China (51790502) and Shandong Provincial Key R & D Program (2019GGX102051).

Institutional Review Board Statement: Not applicable.

Informed Consent Statement: Not applicable.

Data Availability Statement: The data presented in this study are available on request from the corresponding author.

Acknowledgments: This research is funded by National Natural Science Foundation of China (52075119), Major Pro-gram of National Natural Science Foundation of China (51790502) and Shandong Provincial Key R & D Program (2019GGX102051).

Conflicts of Interest: We declare that no known competing financial interests or personal relationships that could have appeared to influence the work reported in this paper.

References

1. Bokobza, L. Multiwall carbon nanotube elastomeric composites: A review. *Polymer* **2007**, *48*, 4907–4920. [CrossRef]
2. Fletcher, A.; Gupta, M.C.; Dudley, K.L.; Vedeler, E. Elastomer foam nanocomposites for electromagnetic dissipation and shielding applications. *Compos. Sci. Technol.* **2010**, *70*, 953–958. [CrossRef]
3. Basheer, B.V.; George, J.J.; Siengchin, S.; Parameswaranpillai, J. Polymer grafted carbon nanotubes—Synthesis, properties, and applications: A review. *Nano Struct. Nano Objects* **2020**, *22*, 100429. [CrossRef]
4. Wang, J.; Jin, X.; Li, C.; Wang, W.; Wu, H.; Guo, S. Graphene and graphene derivatives toughening polymers: Toward high toughness and strength. *Chem. Eng. J.* **2019**, *370*, 831–854. [CrossRef]
5. Tarawneh, M.A.; Ahmad, S.H.; Noum, S.Y.E.; Lau, K.T. Sonication effect on the mechanical properties of MWCNTs reinforced natural rubber. *J. Compos. Mater.* **2012**, *47*, 579–585. [CrossRef]
6. Song, S.; Zhang, Y. Carbon nanotube/reduced graphene oxide hybrid for simultaneously enhancing the thermal conductivity and mechanical properties of styrene-butadiene rubber. *Carbon* **2017**, *123*, 158–167. [CrossRef]
7. Tarawneh, M.A.; Ahmad, S.H.J.; Rasid, R.; Yahya, S.Y.; Shamsul, B.A.R.; Ehnoum, S.; Ka, K.Z.; Seng, L.Y. Mechanical properties of thermoplastic natural rubber (TPNR) reinforced with different types of carbon nanotube. *Sains Malays.* **2011**, *40*, 725–728.
8. Bokobza, L.; Bruneel, J.-L.; Couzi, M. Raman spectra of carbon-based materials (from graphite to carbon black) and of some silicone composites. *C J. Carbon Res.* **2015**, *1*, 77–94. [CrossRef]
9. Bokobza, L. Mechanical and electrical properties of elastomer nanocomposites based on different carbon nanomaterials. *J. Carbon Res.* **2017**, *3*, 10. [CrossRef]
10. Bokobza, L.; Rapoport, O. Reinforcement of natural rubber. *J. Appl. Polym. Sci.* **2002**, *85*, 2301–2316. [CrossRef]
11. Liliane, B.; Mostafa, R. Blends of carbon blacks and multiwall carbon nanotubes as reinforcing fillers for hydrocarbon rubbers. *J. Polym. Sci. Part B Polym. Phys.* **2008**, *46*, 1939–1951. [CrossRef]
12. Bokobza, L.; Bruneel, J.L.; Couzi, M. Raman spectroscopic investigation of carbon-based materials and their composites. Comparison between carbon nanotubes and carbon black. *Chem. Phys. Lett.* **2013**, *590*, 153–159. [CrossRef]
13. Bokobza, L.; Paris, T.; André, L.; Dorothea, K.; Philippe, S. Thermal conductivity and mechanical properties of composites based on multiwall carbon nanotubes and styrene-butadiene rubber. *KGK-Kaut Gummi Kunst.* **2014**, *67*, 45–50. [CrossRef]
14. Shanmugharaj, A.M.; Bae, J.H.; Lee, K.Y.; Noh, W.H.; Lee, S.H.; Ryu, S.H. Physical and chemical characteristics of multiwalled carbon nanotubes functionalized with aminosilane and its influence on the properties of natural rubber composites. *Compos. Sci. Technol.* **2007**, *67*, 1813–1822. [CrossRef]
15. Shanmugharaj, A.M.; Ryu, S.H. Influence of aminosilane-functionalized carbon nanotubes on the rheometric, mechanical, electrical and thermal degradation properties of epoxidized natural rubber nanocomposites. *Polym. Int.* **2013**, *62*, 1433–1441. [CrossRef]
16. Hoshikawa, Y.; Kawaguchi, R.; Nomura, K.; Akahane, H.; Ishii, T.; Ando, M.; Hoshino, N.; Akutagawa, T.; Yamada, H.; Kyotani, T. Quantitative analysis of the formation mechanism of tightly bound rubber by using carbon-coated alumina nanoparticles as a model filler. *Carbon* **2021**, *173*, 870–879. [CrossRef]
17. Peng, Z.; Feng, C.; Luo, Y.; Li, Y.; Kong, L.X. Self-assembled natural rubber/multi-walled carbon nanotube composites using latex compounding techniques. *Carbon* **2010**, *48*, 4497–4503. [CrossRef]
18. Bhattacharyya, S.; Sinturel, C.; Bahloul, O.; Saboungi, M.L.; Thomas, S.; Salvetat, J.P. Improving reinforcement of natural rubber by networking of activated carbon nanotubes. *Carbon* **2008**, *46*, 1037–1045. [CrossRef]
19. Hernández, M.; Bernal, M.; Verdejo, R.; Ezquerra, T.A.; López-Manchado, M.A. Overall performance of natural rubber/graphene nanocomposites. *Compos. Sci. Technol.* **2012**, *73*, 40–46. [CrossRef]
20. Frasca, D.; Schulze, D.; Wachtendorf, V.; Morys, M.; Schartel, B. Multilayer graphene/chlorine-isobutene-isoprene rubber nanocomposites: The effect of dispersion. *Polym. Adv. Technol.* **2016**, *27*, 872–881. [CrossRef]
21. Wu, Y.; Chen, L.; Qin, S.; Li, J.; Zhou, H.; Chen, J. Functionalized graphene-reinforced rubber composite: Mechanical and tribological behavior study. *J. Appl. Polym. Sci.* **2017**, *134*, 1–10. [CrossRef]
22. Atieh, M.A. Effect of functionalized carbon nanotubes with carboxylic functional group on the mechanical and thermal properties of styrene butadiene rubber. *Fuller. Nanotub. Carbon Nanostructures* **2011**, *19*, 617–627. [CrossRef]
23. Xing, W.; Tang, M.; Wu, J.; Huang, G.; Li, H.; Lei, Z.; Fu, X.; Li, H. Multifunctional properties of graphene/rubber nanocomposites fabricated by a modified latex compounding method. *Compos. Sci. Technol.* **2014**, *99*, 67–74. [CrossRef]
24. Kumar, V.; Lee, D.J. Studies of nanocomposites based on carbon nanomaterials and RTV silicone rubber. *J. Appl. Polym. Sci.* **2017**, *134*, 1–9. [CrossRef]
25. Park, J.; Sharma, J.; Monaghan, K.W.; Meyer, H.M.; Cullen, D.A.; Rossy, A.M.; Keum, J.K.; Wood, D.L.; Polizos, G. Styrene-based elastomer composites with functionalized graphene oxide and silica nanofiber fillers: Mechanical and thermal conductivity properties. *Nanomaterials* **2020**, *10*, 1682. [CrossRef] [PubMed]
26. Ponnamma, D.; Ramachandran, R.; Hussain, S.; Rajaraman, R.; Amarendra, G.; Varughese, K.T.; Thomas, S. Free-volume correlation with mechanical and dielectric properties of natural rubber/multi walled carbon nanotubes composites. *Compos. Part A Appl. Sci. Manuf.* **2015**, *77*, 164–171. [CrossRef]
27. Anand, K.A.; Jose, T.S.; Alex, R.; Joseph, R. Natural rubber-carbon nanotube composites through latex compounding. *Int. J. Polym. Mater. Polym. Biomater.* **2010**, *59*, 33–44. [CrossRef]

28. Matchawet, S.; Kaesaman, A.; Bomlai, P.; Nakason, C. Effects of multi-walled carbon nanotubes and conductive carbon black on electrical, dielectric, and mechanical properties of epoxidized natural rubber composites. *Polym. Compos.* **2017**, *38*, 1031–1042. [CrossRef]
29. Kueseng, P.; Sae-Oui, P.; Rattanasom, N. Mechanical and electrical properties of natural rubber and nitrile rubber blends filled with multi-wall carbon nanotube: Effect of preparation methods. *Polym. Test.* **2013**, *32*, 731–738. [CrossRef]
30. Liu, D.; Kong, Q.Q.; Jia, H.; Xie, L.J.; Chen, J.; Tao, Z.; Wang, Z.; Jiang, D.; Chen, C.-M. Dual-functional 3D multi-wall carbon nanotubes/graphene/silicone rubber elastomer: Thermal management and electromagnetic interference shielding. *Carbon* **2021**, *183*, 216–224. [CrossRef]
31. Studebaker, M.L.; Nabors, L.G. Sulfur group analyses in natural rubber vulcanizates. *Rubber Chem. Technol.* **1959**, *32*, 941–961. [CrossRef]
32. Chan, B.L.; Elliott, D.J.; Holley, M.; Smith, J.F. The influence of curing systems on the properties of natural rubber. *J. Polym. Sci. Symp.* **1974**, *48*, 61–86. [CrossRef]
33. Sombatsompop, N. Analysis of cure characteristics on cross-link density and type, and viscoelastic properties of natural rubber. *Polym. Plast. Technol. Eng.* **1998**, *37*, 333–349. [CrossRef]
34. Cunneen, J.; Russell, R. Occurrence and prevention of changes in the chemical structure of natural rubber tire tread vulcanizates during service. *Rubber Chem. Technol.* **1970**, *43*, 1215–1224. [CrossRef]
35. Flory, P.J.; Rabjohn, N.; Shaffer, M.C. Dependence of elastic properties of vulcanized rubber on the degree of cross-linking. *Rubber Chem. Technol.* **1950**, *23*, 9–26. [CrossRef]
36. Mullins, L. Determination of degree of crosslinking in natural rubber vulcanizates. Part I. *Rubber Chem. Technol.* **1957**, *30*, 1–10. [CrossRef]
37. Loo, C.T. High temperature vulcanization of elastomers: 2. Network structures in conventional sulphenamide-sulphur natural rubber vulcanizates. *Polymer* **1974**, *15*, 357–365. [CrossRef]
38. Mukhopadhyay, R.; De, S.K. Effect of vulcanization temperature and vulcanization systems on the structure and properties of natural rubber vulcanizates. *Rubber Chem. Technol.* **1979**, *52*, 263–277. [CrossRef]
39. Saito, T.; Yamano, M.; Nakayama, K.; Kawahara, S. Quantitative analysis of crosslinking junctions of vulcanized natural rubber through rubber-state NMR spectroscopy. *Polym. Test.* **2021**, *96*, 107130. [CrossRef]
40. Sainumsai, W.; Suchiva, K.; Toki, S. Influence of sulphur crosslink type on the strain-induced crystallization of natural rubber vulcanizates during uniaxial stretching by in situ WAXD using a synchrotron radiation. *Mater. Today Proc.* **2019**, *17*, 1539–1548. [CrossRef]
41. Fan, R.; Zhang, Y.; Huang, C.; Zhang, Y.; Fan, Y.; Sun, K. Effect of crosslink structures on dynamic mechanical properties of natural rubber vulcanizates under different aging conditions. *J. Appl. Polym. Sci.* **2001**, *81*, 710–718. [CrossRef]
42. Zhang, X.; Zhao, N.; He, C. The superior mechanical and physical properties of nanocarbon reinforced bulk composites achieved by architecture design—A review. *Prog. Mater. Sci.* **2020**, *113*, 100672. [CrossRef]
43. Li, Y.; Wang, Q.; Wang, S. A review on enhancement of mechanical and tribological properties of polymer composites reinforced by carbon nanotubes and graphene sheet: Molecular dynamics simulations. *Compos. Part B Eng.* **2019**, *160*, 348–361. [CrossRef]
44. Degrange, J.M.; Thomine, M.; Kapsa, P.; Pelletier, J.M.; Chazeau, L.; Vigier, G.; Dudragne, G.; Guerbé, L. Influence of viscoelasticity on the tribological behaviour of carbon black filled nitrile rubber (NBR) for lip seal application. *Wear* **2005**, *259*, 684–692. [CrossRef]
45. Li, Y.; Wang, S.; Arash, B.; Wang, Q. A study on tribology of nitrile-butadiene rubber composites by incorporation of carbon nanotubes: Molecular dynamics simulations. *Carbon* **2016**, *100*, 145–150. [CrossRef]
46. Jin, Y.; Duan, F.; Mu, X. Functionalization enhancement on interfacial shear strength between graphene and polyethylene. *Appl. Surf. Sci.* **2016**, *387*, 1100–1109. [CrossRef]
47. Wang, P.; Qiao, G.; Hou, D.; Jin, Z.; Wang, M.; Zhang, J.; Sun, G. Functionalization enhancement interfacial bonding strength between graphene sheets and calcium silicate hydrate: Insights from molecular dynamics simulation. *Constr. Build. Mater.* **2020**, *261*, 120500. [CrossRef]
48. Liu, F.; Hu, N.; Zhang, J.; Atobe, S.; Weng, S.; Ning, H.; Liu, Y.; Wu, L.; Zhao, Y.; Mo, F.; et al. The interfacial mechanical properties of functionalized graphene-polymer nanocomposites. *RSC Adv.* **2016**, *6*, 66658–66664. [CrossRef]
49. Yu, B.; Fu, S.; Wu, Z.; Bai, H.; Ning, N.; Fu, Q. Molecular dynamics simulations of orientation induced interfacial enhancement between single walled carbon nanotube and aromatic polymers chains. *Compos. Part A Appl. Sci. Manuf.* **2015**, *73*, 155–165. [CrossRef]
50. Cui, J.; Zhao, J.; Wang, S.; Wang, Y.; Li, Y. Effects of carbon nanotubes functionalization on mechanical and tribological properties of nitrile rubber nanocomposites: Molecular dynamics simulations. *Comput. Mater. Sci.* **2021**, *196*, 110556. [CrossRef]
51. Duan, Q.; Xie, J.; Xia, G.; Xiao, C.; Yang, X.; Xie, Q.; Huang, Z. Molecular dynamics simulation for the effect of fluorinated graphene oxide layer spacing on the thermal and mechanical properties of fluorinated epoxy resin. *Nanomaterials* **2021**, *11*, 1344. [CrossRef]
52. Ji, W.M.; Zhang, L.W.; Liew, K.M. Understanding interfacial interaction characteristics of carbon nitride reinforced epoxy composites from atomistic insights. *Carbon* **2021**, *171*, 45–54. [CrossRef]
53. Li, Y.; Liu, Y.; Peng, X.; Yan, C.; Liu, S.; Hu, N. Pull-out simulations on interfacial properties of carbon nanotube-reinforced polymer nanocomposites. *Comput. Mater. Sci.* **2011**, *50*, 1854–1860. [CrossRef]

54. Coto, B.; Antia, I.; Barriga, J.; Blanco, M.; Sarasua, J.R. Influence of the geometrical properties of the carbon nanotubes on the interfacial behavior of epoxy/CNT composites: A molecular modelling approach. *Comput. Mater. Sci.* **2013**, *79*, 99–104. [CrossRef]
55. Yang, S.; Yu, S.; Kyoung, W.; Han, D.S.; Cho, M. Multiscale modeling of size-dependent elastic properties of carbon nanotube/polymer nanocomposites with interfacial imperfections. *Polymer* **2012**, *53*, 623–633. [CrossRef]
56. Nakao, T.; Kohjiya, S. Computer simulation of network formation in natural rubber (NR). In *Chemistry, Manufacture and Applications of Natural Rubber*; Woodhead Publishing: Cambridge, UK, 2014; ISBN 009780857096838.
57. Ma, J.; Nie, Y.; Wang, B. Simulation study on the relationship between the crosslinking degree and structure, hydrophobic behavior for poly (styrene-co-divinylbenzene) copolymer. *J. Mol. Struct.* **2018**, *1173*, 120–127. [CrossRef]
58. Zhang, T.; Huang, H.; Li, W.; Chang, X.; Cao, J.; Hua, L. Vulcanization modeling and mechanism for improved tribological performance of styrene-butadiene rubber at the atomic scale. *Tribol. Lett.* **2020**, *68*, 1–11. [CrossRef]
59. Park, C.; Yun, G.J. Characterization of interfacial properties of graphene-reinforced polymer nanocomposites by molecular dynamics-shear deformation model. *J. Appl. Mech. Trans. ASME* **2018**, *85*, 1–10. [CrossRef]
60. Park, C.; Kim, G.; Jung, J.; Krishnakumar, B.; Rana, S.; Yun, G.J. Enhanced self-healing performance of graphene oxide/vitrimer nanocomposites: A molecular dynamics simulations study. *Polymer* **2020**, *206*, 122862. [CrossRef]
61. Liu, X.; Zhou, X.; Kuang, F.; Zuo, H.; Huang, J. Mechanical and tribological properties of nitrile rubber reinforced by nano-SiO_2: Molecular dynamics simulation. *Tribol. Lett.* **2021**, *69*, 1–11. [CrossRef]
62. Rigby, D.; Sun, H.; Eichinger, B.E. Computer simulations of poly(ethylene oxide): Force field, PVT diagram and cyclization behaviour. *Polym. Int.* **1999**, *44*, 311–330. [CrossRef]
63. Toxvaerd, S. Molecular dynamics at constant temperature and pressure. *Phys. Rev. E* **1993**, *47*, 343–350. [CrossRef] [PubMed]
64. Berendsen, H.J.C.; Postma, J.P.M.; Van Gunsteren, W.F.; Dinola, A.; Haak, J.R. Molecular dynamics with coupling to an external bath. *J. Chem. Phys.* **1984**, *81*, 3684–3690. [CrossRef]
65. Fox, T.G.; Flory, P.J. Second-order transition temperatures and related properties of polystyrene. I. Influence of molecular weight. *J. Appl. Phys.* **1950**, *21*, 581–591. [CrossRef]
66. Zhu, L.; Chen, X.; Shi, R.; Zhang, H.; Han, R.; Cheng, X.; Zhou, C. Tetraphenylphenyl-modified damping additives for silicone rubber: Experimental and molecular simulation investigation. *Mater. Des.* **2021**, *202*, 109551. [CrossRef]
67. Li, K.; Li, Y.; Lian, Q.; Cheng, J.; Zhang, J. Influence of cross-linking density on the structure and properties of the interphase within supported ultrathin epoxy films. *J. Mater. Sci.* **2016**, *51*, 9019–9030. [CrossRef]
68. Zhao, Z.; Zhao, X. Electronic, optical, and mechanical properties of Cu_2ZnSnS_4 with four crystal structures. *J. Semicond.* **2015**, *36*, 083004. [CrossRef]
69. Theodorou, D.N.; Suter, U.W. Atomistic modeling of mechanical properties of polymeric glasses. *Macromolecules* **1986**, *19*, 139–154. [CrossRef]
70. Zhan, Y.; Pang, M.; Wang, H.; Du, Y. The structural, electronic, elastic and optical properties of $AlCu(Se_{1-x}Te_x)_2$ compounds from first-principle calculations. *Curr. Appl. Phys.* **2012**, *12*, 373–379. [CrossRef]
71. Pugh, S.F. XCII. Relations between the elastic moduli and the plastic properties of polycrystalline pure metals. *Lond. Edinb. Dublin Philos. Mag. J. Sci.* **1954**, *45*, 823–843. [CrossRef]

Article

Numerical Analysis of Unsteady Hybrid Nanofluid Flow Comprising CNTs-Ferrousoxide/Water with Variable Magnetic Field

Muhammad Sohail Khan [1], Sun Mei [1,*], Shabnam [1], Unai Fernandez-Gamiz [2], Samad Noeiaghdam [3,4], Said Anwar Shah [5] and Aamir Khan [6]

[1] School of Mathematical Sciences, Jiangsu University, Zhenjiang 212013, China; sohailkhan8688@gmail.com (M.S.K.); shabnam8688@gmail.com (S.)
[2] Nuclear Engineering and Fluid Mechanics Department, University of the Basque Country UPV/EHU, Nieves Cano 12, 01006 Vitoria-Gasteiz, Spain; unai.fernandez@ehu.eus
[3] Industrial Mathematics Laboratory, Baikal School of BRICS, Irkutsk National Research Technical University, 664074 Irkutsk, Russia; noiagdams@susu.ru
[4] Department of Applied Mathematics and Programming, South Ural State University, Lenin Prospect 76, 454080 Chelyabinsk, Russia
[5] Department of Basic Sciences and Islamiat, University of Engineering and Technology Peshawar, Peshawar 25120, Pakistan; anwarshah@uetpeshawar.edu.pk
[6] Department of Pure and Applied Mathematics, University of Haripur, Peshawar 22620, Pakistan; aamir.khan@uoh.edu.pk
* Correspondence: sunm@ujs.edu.cn

Abstract: The introduction of hybrid nanofluids is an important concept in various engineering and industrial applications. It is used prominently in various engineering applications, such as wider absorption range, low-pressure drop, generator cooling, nuclear system cooling, good thermal conductivity, heat exchangers, etc. In this article, the impact of variable magnetic field on the flow field of hybrid nano-fluid for the improvement of heat and mass transmission is investigated. The main objective of this study is to see the impact of hybrid nano-fluid (ferrous oxide water and carbon nanotubes) CNTs-Fe_3O_4, H_2O between two parallel plates with variable magnetic field. The governing momentum equation, energy equation, and the magnetic field equation have been reduced into a system of highly nonlinear ODEs by using similarity transformations. The parametric continuation method (PCM) has been utilized for the solution of the derived system of equations. For the validity of the model by PCM, the proposed model has also been solved via the shooting method. The numerical outcomes of the important flow properties such as velocity profile, temperature profile and variable magnetic field for the hybrid nanofluid are displayed quantitatively through various graphs and tables. It has been noticed that the increase in the volume friction of the nano-material significantly fluctuates the velocity profile near the channel wall due to an increase in the fluid density. In addition, single-wall nanotubes have a greater effect on temperature than multi-wall carbon nanotubes. Statistical analysis shows that the thermal flow rate of (Fe_3O_4-SWCNTs-water) and (Fe_3O_4-MWCNTs-water) rises from 1.6336 percent to 6.9519 percent, and 1.7614 percent to 7.4413 percent, respectively when the volume fraction of nanomaterial increases from 0.01 to 0.04. Furthermore, the body force accelerates near the wall of boundary layer because Lorentz force is small near the squeezing plate, as the current being almost parallel to the magnetic field.

Keywords: ariable magnetic field; magnetic Reynold parameter; hybrid nano-fluid; PCM; BVP4C

1. Introduction

Due to the wide range of engineering and scientific applications, the investigation of heat transfer and fluid movement between surfaces has been one of the most prominent research fields in recent times. Compared to nanofluids, hybrid nanofluids are expected to replace simple nanofluids for a variety of reasons, including wide absorption range, low volatility, low pressure, good thermal conductivity, low-pressure reduction, and low friction losses, solar energy, air conditioning applications, heat pipes, electronic cooling, biomedical engineering, ship, space, automotive industry, transformer cooling, and defence application are the few uses of hybrid nanofluid. In this regard, extensive research has been done to evaluate the heat and flow transfer properties between two plates. Mustafa et al. [1] studied the properties of mass, fluid movement, and heat transfer in two parallel plates. Their results reveal that increasing Schmidt number values reduce the concentration profile and increase the magnitude of the local Sherwood number. In addition, their results signify that the enhancement in the Prandtl number augments the Nusselt number. Dogonchi et al. [2] analyzed the transfer of flow and heat of the MHD graphene oxide/water nanoliquids in the presence of thermal radiation between two parallel plates and an established association between volume fraction, Nusselt number and temperature distribution. In addition, they discovered that the enhancement in the Reynolds number and extension ratio leads to an increase in the skin friction coefficient. Alizadeh et al. [3] investigated the properties of micropolar MHD fluid motion inside a channel loaded by nanoliquids bounded to the thermal radiation. Their results reveal that Nusselt number is a growing function which depends on the volume fraction of nanofluids and radiation parameters. Dib et al. [4] examined the flow and heat transfer of time-dependent nanoliquids between plates and concluded that differents kinds of nanoliquids play an important role in the fortification of heat transmission. Dogonchi et al. [5] studied the heat transfer of transient MHD flow of nanofluid at the surfaces influenced by thermal radiation, and concluded that a growing number of the radiation parameters leads to an increase in the temperature distribution and Nusselt number. Furthermore, they established a relationship between skin friction coefficient and Nusselt number, which increases with the increase in the magnetic parameter and volume fraction of the nanofluid. Sheikholeslami et al. [6] investigated the properties of heat transfer and the flow of nanofluids between two surfaces and in a revolving system. Their results reveal that in both suction and injection cases, the surface heat transfer rate enhances the volume fraction of nanomaterials, Reynolds number, and suction/injection parameter, and this reduces the power of the circulation parameter. In addition, their outcomes reveal that the Nusselt number has a direct relationship with the volume fraction of nanomaterial, suction/injection parameter, and Reynold number, while it possesses an inverse relationship with the power of rotation parameter. Mehmood and Ali [7] analyzed the heat transfer of the viscous MHD fluid flow fluid between two parallel surfaces. They took the influence of viscous dissipation in their analysis and discovered that the injection on the upper plate causes the temperature to rise while the magnetic field decreases the temperature. Furthermore, they reviewed that the Prandtl number with viscous dissipation enhance the temperature profile, while without viscous dissipation, the temperature decreases as the Prandtl number increases.

Fluids play a fundamental role in increasing the rate of heat transfer in many engineering applications, such as heat exchangers, fuel cells, etc. As we know, the regular fluids have a very low thermal conductivity in heat transferring. So, we need unusual, high thermal conductivity fluids to overcome this problem. This special type of fluid is called nanofluids. The practical application of nanofluid was first introduced by Choi [7]. The basic feature of nanofluids is that they have more thermal conductivity in comparison with regular fluids because the particles of metal nanometer-size are put into the liquid, which plays a valuable role in increasing the thermal conductivity. Most of the researchers have focused their analysis on fluid movement and heat transmission using normal fluid or nanofluid. Dogonchi and Ganji [8] studied the behaviour of buoyancy flow and heat transfer of MHD nanofluid on a stretching surface in the light of Brownian motion with thermal radiation. Their research explains that the temperature profile and the velocity of the fluid decrease when the radiation parameter is increased. Furthermore, they explained that the coefficient of skin friction increases with the increase of the magnetic parameter and decreases with the increase in the volume fraction of the nanofluid. Bhatti et al. [9] studied through a permeable extending wall the effect of entropy generation on non-Newtonian

Eyring–Powell nanofluid, and discovered that the greater the effect of the suction parameter, the greater the velocity profile. In addition, their results show that thermophoresis parameters and Brownian motion significantly increase the temperature profile.

Chamkha et al. [10] analyzed the transient conjugate free convection, which applied to a half-circular pot with limited thickness solid walls containing Al_2O_3-Cu-water hybrid nanoliquids. They concluded that only a 5 percent increase in Al_2O_3-Cu nanomaterials indicates an increase in the average Nusselt number ranging from 4.9 to 5.4, while a 5 percent increase in Al_2O_3 nanomaterials increases the average Nusselt number from 4.9 to 5.36. In a square porous wall, MHD heat transfer and free convection flow are cooled and heated by sink or heat source, respectively, and loaded with a Cu-Al_2O_3-water hybrid nanofluid were investigated by Gorla et al. [11]. The results of this investigation signify that the average Nusselt number decreases significantly for hybrid suspension in case of change in the position of heat source. Furthermore, their outcomes reveal that the mean Nusselt number of hybrid suspensions is lower than that of Cu and Al_2O_3. The moving boundaries which produced squeezing flow play a significant role in polymer processing, hydrodynamical machines, and lubrication tools, etc. Jackson [12] studied the relationships between the squeezing fluid flow and loaded earing's performance under the adhesive phenomena. Hayat et al. [13] examined the impact of couple stress fluid flow in the presence of the time-dependent magnetic field. Hayat et al. both [14,15] used the Buongiorno model of nanofluids, which indirectly tested certain nanoparticles. In the meantime, Salehi et al. [16] studied the squeezing hybrid nanofluid which can be formulated by putting the nanomaterial Fe_3O_4-MoS_2 in the base fluid water and ethylene glycol. Acharya [17] analyzed the influence of radiation due to solar energy over Cu-Al_2O_3/water hybrid nanofluid within a channel. Ikram et al. [18] explored another fascinating feature of the hybrid nanofluid movement within a channel. Tayebi and Chamkha [19] studied free convection through the annulus between two elliptical cylinders filled with Cu-Al_2O_3/water hybrid nanofluid. It has been shown that heat transfer rate is more efficient if one uses Cu-Al_2O_3/water hybrid as compared to Al_2O_3/water nanofluid. Tayebi and Chamkha. [20] investigated the free convection of hybrid nanofluid in the eccentric annulus of horizontally cylindrical shape. Bilal and Taseer et al. [14,15] studied the (MHD) hybrid nanofluids (CNT-Fe_3O_4/H_2O) motion between horizontal channels through dilating and squeezing walls with thermal radiation. The flow is transient and laminar. The flow is not symmetric and the lower and upper walls vary in temperature and porosity. The suspension of single and multi-walls carbons nanotubes and Fe_3O_4 in nanofluids is exploited.

The purpose of the analysis was to examine the general quality of local thermal non-equilibrium (LTNE) and focus on the impact of nanomaterials (AA7075 and Cu) in the conventional fluids (methanol and NaAlg) using the Tiwari–Das model. They have investigated the heat transfer properties of nanofluids numerically from an engineering point of view at a stretching plate in the porous medium in [21,22]. Song et al. [23] analyzed the effect of Marangoni convection, thermal radiation, Soret and Dufour effects, viscous dissipation, nonlinear heat sink/source, and activation energy on MHD nanofluids motion produced by revolving the disk. Furthermore, they studied the effect of activation energy on the Darcy–Forchemer movement of Casson fluids, which include the suspension of titanium dioxide and graphene oxide nanomaterials containing 50 percent ethylene glycol as the conventional fluid in a porous medium [23–27]. The Marangoni convection of hybrid nanoliquid has been analyzed by Khan et al. [28,29], which is a combination of (MnZi, Fe_2O, 4-NiZn, Fe_2O_4) nanomaterials and one conventional fluid (H_2O), and the momentum equation updated through the inclusion of Darcy–Forchheimer in the porous medium. Due to heat transfer irreversibility, viscous dissipation irreversibility and mass transfer irreversibility, the entropy generation of the fluid flow is computed and analyzed through pertinent parameters.

A model of PVT unit containing PCM was analyzed by Khodadadi et al. [30]. The impact of different kinds of nanoliquids and NEPCMs at various concentrations on the system efficiency is estimated. The ZnO, SiC, Al_2O_3, MCNT (multi-walled carbon nanotube), Cu, and Ag nanomaterials are used inside the water with phase change particle concentrations of 0, 0.02, 0.04, by Sheikholeslami et al. [31,32]. The solar collector was investigated regarding the collector performance and irreversibility in the presence of the variable solar radiation. Along with these two main factors, performance factors are important functions that must be considered in order to reach the optimal design. Irreversibility is due to a decrease in temperature with an increase in wind speed due to a slight decrease in temperature of different zones. Uddin et al. [33] presented a novel model of bio-nano-transport, the impacts of first and second-order velocity slip, mass slip, heat slip, and gyro-tactic (torque-responsive) microorganism slip of bio-convective nanoliquid motion in the flowing plate with blowing tendency are numerically analyzed. Zohra et al. [34] observed the microfluidic devices based on the microfluidic associated technologies and microelectromechanical processes that have received a warm welcome in the field of science and engineering. This is a mathematical formulation which analyzes the fluid flow of steady forced convective revolving disk put into water-based nanoliquid with microorganisms. Thiyagarajan et al. [35] analyzed pleural outflow as an obstruction of the pleural cavity in the lung wall. The reversal of the lung and chest wall process causes pleural fluid to accumulate in the pleural space. Parietal lymphatic dilation is caused by an increase in pleural fluid. This proposed model has been designed to acquire new outcomes of respiratory tract infections which investigates the response with respect to the injection of transient mixed convection motion of visceral pleural liquid transports between two vertical porous sheets. M. K. Alam et al. [36] studied the behavior of mass and heat transmission on the time-dependent viscous squeezing flow along with changeable magnetic field. They used a revolving channel in their analysis. The proposed model has been updated through the inclusion of energy and variable magnetic field equations. Bilal et al. [37] and Khoshrouye [38] researched that the development of technologies in power engineering and microelectronics requires the improvement of efficient cooling systems. This advancement comprises the use of fins of notably changeable geometry within cavities to increment heat elimination from the heat producing process. It is considered that the fins are playing a significant role in augmenting heat transfer, so the proposed model has studied the impacts of several parameters on the transfer of the heat of embedded fins in cavities. In addition, the impacts range for some parameters on energy transmission.

Motivated by the prescribed literature review, it has been noticed that the analysis of hybrid nanofluids with Fe_3O_4, and SWCNT, MWCNTs, has so far not considered between two parallel porous plates with changeable magnetic fields. In addition, the impacts of changeable magnetic field on mass and heat transfer in rectangular coordinate system is a novel approach in the field. The introduction of hybrid nanofluids flow is very important due to its many applications in industrial and engineering processes. In this article, we are going to investigate the impacts of variable magnetic fields on the flow of hybrid nanofluids for the improvement of heat and mass transmission. In the composition of hybrid nanofluids, the (Fe_3O_4, single-wall carbon nanotubes and multi-wall carbon nanotubes) nanomaterials are used. Thermal radiation is also considered for high-temperature phenomena. The governing equations of the hybrid nanofluid are formulated under certain hypotheses, and solved numerically by (parametric continuation method) in MATLAB. The numerical outcomes of several emerging parameters skin frictions, Nusselt number, etc., are discussed through tables and graphs. In addition, it has been noticed that the thermal flow rate of (Fe_3O_4-SWCNTs-water) nanofluids rises from 1.6336 percent to 6.9519 percent when the volume fraction of nanomaterial increases from 0.01 to 0.04. In the same way, the thermal flow rate of (Fe_3O_4-MWCNTs-water) nanofluids rises from 1.7614 percent to 7.4413 percent when the volume fraction of nano-material increases from 0.01 to 0.04.

2. Formulation

We consider the flow of hybrid nanofluids between two horizontal infinite parallel plates, as depicted in Figure 1. The distance between two plates has taken as $h(t) = l\sqrt{1-at}$; furthermore, the upper plate move towards the lower plate with velocity $v = \frac{dh}{dt}$. We assumed that the temperature T_H is constant at the upper plate, and the flow of hybrid nanofluid (CNTs-Fe_3O_4/H_2O) in the channel is incompressible and viscid. The physical characteristics of hybrid nanofluid flow CNTs-Fe_3O_4/H_2O) rely on time. As a result of the greater influence of variable magnetic fields, this type of behaviour of hybrid nanofluid is thought to be due to their magnetic characteristics. Single and multi-wall carbon nanotubes, on the other hand, have larger thermal conductivity. Subsequently, the hybrid nanofluid (CNTs-Fe_3O_4/H_2O) is developed by suspending a new volume fraction of CNT ($\Phi_2 = 0.5$) into the originally formulated ferrofluid (Fe_3O_4/H_2O). The mathematical formulation of the aforementioned hybrid nanofluids by continuity, momentum, magnetic field and energy conservation equations are as follows:

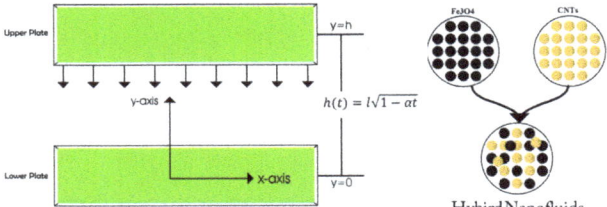

Figure 1. Geometry of the problem with coordinate system.

Continuity equation [15]:

$$\frac{\partial u}{\partial x} + \frac{\partial v}{\partial x} = 0, \qquad (1)$$

Momentum equations [15,21]:

$$\frac{\partial u}{\partial t} + u\frac{\partial u}{\partial x} + v\frac{\partial u}{\partial y} = -\frac{1}{\rho_{hnf}}\frac{\partial P}{\partial x} + \frac{\mu_{hnf}}{\rho_{hnf}}(\frac{\partial^2 u}{\partial x^2} + \frac{\partial^2 u}{\partial y^2}) - \frac{b_2 \sigma_{hnf}}{\rho_{hnf}}(\frac{\partial b_1}{\partial y} - \frac{\partial b_2}{\partial x}) \qquad (2)$$

$$\frac{\partial v}{\partial t} + u\frac{\partial v}{\partial x} + v\frac{\partial v}{\partial y} = -\frac{1}{\rho_{hnf}}\frac{\partial P}{\partial y} + \frac{\mu_{hnf}}{\rho_{hnf}}(\frac{\partial^2 v}{\partial x^2} + \frac{\partial^2 v}{\partial y^2}) - \frac{b_1 \sigma_{hnf}}{\rho_{hnf}}(\frac{\partial b_1}{\partial y} - \frac{\partial b_2}{\partial x}) \qquad (3)$$

Maxwell Equations [21,22]:

$$\frac{\partial b_1}{\partial t} = u\frac{\partial b_2}{\partial y} + b_2\frac{\partial u}{\partial y} - v\frac{\partial b_1}{\partial y} - b_1\frac{\partial v}{\partial y} + \frac{1}{\sigma_{hnf}\mu_e}(\frac{\partial^2 b_1}{\partial x^2} + \frac{\partial^2 b_1}{\partial y^2}) \qquad (4)$$

$$\frac{\partial b_2}{\partial t} = v\frac{\partial b_1}{\partial x} + b_1\frac{\partial v}{\partial x} - u\frac{\partial b_2}{\partial x} - b_2\frac{\partial u}{\partial x} + \frac{1}{\sigma_{hnf}\mu_e}(\frac{\partial^2 b_2}{\partial x^2} + \frac{\partial^2 b_2}{\partial y^2}) \qquad (5)$$

The Energy Equation [15]:

$$\frac{\partial T}{\partial t} + u\frac{\partial T}{\partial x} + v\frac{\partial T}{\partial y} = \frac{\kappa_{hnf}}{(\rho C_p)_{hnf}}(\frac{\partial^2 T}{\partial x^2} + \frac{\partial^2 T}{\partial y^2})$$
$$+ \frac{\mu_{hnf}}{(\rho C_p)_{hnf}}(4(\frac{\partial v}{\partial y})^2 + (\frac{\partial u}{\partial y})^2 + (\frac{\partial v}{\partial x})^2 + 2\frac{\partial u}{\partial y}\frac{\partial v}{\partial x}) \qquad (6)$$

where b_1, b_2 are the components of magnetic field, $(\rho C_p)_{hnf}$ is the heat capacity of the hybrid nanofluid, P is fluid pressure, T is the temperature, ρ_{hnf} is fluid density of hybrid nanofluid, σ_{hnf} is electrical conductivity of hybrid nanofluid, μ_{hnf} is kinematic viscosity of hybrid nanofluid.

Nanofluid are defined as [15]:

$$\nu_{hnf} = \frac{\mu_{hnf}}{\rho_{hnf}}, \quad \frac{\rho_{hnf}}{\rho_f} = (1-\phi_2)((1-\Phi_1 + \frac{\Phi_1 \rho_{MS}}{\rho_f}) + \frac{\Phi_2 \rho_{CNT}}{\rho_f}),$$

$$\frac{\kappa_{hnf}}{\kappa_{bf}} = \frac{\kappa_{CNT} + (m-1)\kappa_{bf} + \phi_2(\kappa_{bf} - \kappa_{CNT})}{\kappa_{CNT} + (m-1)\kappa_{bf} + \phi_2(m-1)(\kappa_{bf} - \kappa_{CNT})}$$

$$\frac{\kappa_{bf}}{\kappa_f} = \left(\frac{\kappa_{MS} + (m-1)\kappa_f + \phi_1(\kappa_f - \kappa_{MS})}{\kappa_{MS} + (m-1)\kappa_f + \phi_1(m-1)(\kappa_f - \kappa_{MS})}\right) \quad (7)$$

$$\frac{(\rho C_p)_{hnf}}{(\rho C_p)_f} = (1-\phi_2)((1-\Phi_1 + \frac{\Phi_1 (\rho C_p)_{MS}}{(\rho C_p)_f}) + \frac{\Phi_2 (\rho C_p)_{CNT}}{(\rho C_p)_f}),$$

$$\frac{\sigma_{hnf}}{\sigma_f} = 1 + \frac{3(\frac{\sigma_p}{\sigma_f} - 1)\phi_1}{(\frac{\sigma_p}{\sigma_f} + 2) - (\frac{\sigma_p}{\sigma_f} - 1)\phi}, \quad \frac{\mu_{hnf}}{\mu_f} = \frac{1}{(1-\phi_1)^{2.5}(1-\phi_2)^{2.5}},$$

with κ_{hnf} is the thermal conductivity of hybrid nanofluid, κ_{bf} is the thermal conductivity of the Fe_3O_4-nanofluid, and ϕ_1, Φ_2 are the volume fraction of CNTs.

3. Boundary Conditions

The boundary conditions of the proposed model are as follows [15]:

$$u = 0, \quad v = -\frac{dh}{dt}, \quad b_1 = \frac{axM_0}{2(1-at)}, \quad b_2 = \frac{axM_0}{2\sqrt{1-at}}, \quad T = T_H \quad at \quad y = h(t)$$
$$u = 0, \quad v = 0, \quad T = 0, \quad b_1 = b_2 = 0, \quad at \quad y = 0 \quad (8)$$

The following similarity transformations have been used for reducing a system of PDEs (1–6) into a non-linear system of ODEs [21],

$$u = \frac{ax}{2(1-at)}f'(\eta), \quad v = -\frac{al}{2\sqrt{1-at}}f(\eta), \quad b_1 = \frac{axM_0}{2(1-at)}K'(\eta),$$
$$b_2 = -\frac{alM_0}{2\sqrt{1-at}}K(\eta), \quad \eta = \frac{y}{l\sqrt{1-at}}, \quad \theta(\eta) = \frac{T}{T_H}, \quad (9)$$

Therefore, Equation (1) of the model has satisfied automatically, and the reduced forms of the remaining Equations (2)–(6) are as follows:

$$f'''' = \frac{\rho_{hnf}}{\rho_f}\frac{\mu_f}{\mu_{hnf}}(S(3f'' + \eta f''' + f'f'' - ff''')) + Ha^2 SRe_m(\frac{\sigma_{hnf}}{\sigma_f})^2\frac{\mu_f}{\mu_{hnf}}(2KK' + fK^2)$$
$$+ \eta Ha^2 S^2 Re_m^2(\frac{\sigma_{hnf}}{\sigma_f})^3\frac{\mu_f}{\mu_{hnf}}(K^2 + \eta KK' - fKK' + f'K^2) \quad (10)$$
$$- Ha^2 S^2 Re_m^2(\frac{\sigma_{hnf}}{\sigma_f})^3\frac{\mu_f}{\mu_{hnf}}(fK^2 + \eta fKK' - f^2 KK' + ff'K^2),$$

$$K'' = Re_m S \frac{\sigma_{hnf}}{\sigma_f}(K + \eta K' - fK' + f'K), \quad (11)$$

$$\theta'' = SPr\frac{(\rho C_p)_{hnf}}{(\rho C_p)_f}\frac{\kappa_f}{\kappa_{hnf}}(\eta\theta' - \theta'f) - \frac{\mu_{hnf}}{\mu_f}\frac{\kappa_f}{\kappa_{hnf}}PrEc(4\delta f'^2 + f''^2), \quad (12)$$

and the boundary conditions in the reduced form as follows,

$$f(0) = 0, \quad f'(0) = 0, \quad K(0) = 0, \quad \theta(0) = 0,$$
$$f(1) = 1, \quad f'(1) = 0, \quad K(1) = 1, \quad \theta(1) = 1, \quad (13)$$

where $Ha^2 = \frac{l^2 M_0^2 a \sigma_f}{\rho_f \nu_f}$ Hartmann number, $S = \frac{al^2}{2\nu_f}$ squeeze number, $Re_m = \sigma_f \nu_f \mu_e$ Rynold's magnetic parameter, $Pr = \frac{\nu_f (\rho C p)_f}{\kappa_f}$ Prandtl number, $Ec = \frac{a^2}{4(Cp)_f T_H (1-at)^2}$ Eckert number, $\delta = \frac{l^2(1-at)}{a^2}$.

Emerging physical parameters in the reduced form of system are the Nusselt number and skin friction coefficient, and can be defined as,

$$C_f = \frac{\mu_{nf}}{\rho_{nf} v_l^2 x}\left(\frac{\partial u}{\partial y}\right)_{y=h(t)}, \quad N_u = -\frac{a\kappa_{nf}\left(\frac{\partial T}{\partial y}\right)_{z=h(t)}}{k_f T_H}, \quad (14)$$

In case of Equation (16), we get

$$R_1^2 C_f \frac{\rho_{hnf}}{\rho_f} \frac{\mu_f}{\mu_{hnf}} = f''(1), \quad -\theta'(1) = \frac{\kappa_f}{\kappa_{hnf}} R_2 N_u. \quad (15)$$

4. Numerical Solution by PCM

In this section, optimal choices of continuation parameters are made through the algorithm of PCM [29] for the solution of non-linear Equations (10)–(12) with boundary conditions in Equation (13):

- **Step 1: First order of ODE**

To transform the Equations (10)–(12) into the first order of ODEs, consider the following

$$\begin{aligned} f = t_1, \quad f' = t_2, \quad f'' = t_3, \quad f''' = t_4 \\ K = t_5, \quad K' = t_6, \quad \theta = t_7, \quad \theta' = t_8 \end{aligned} \quad (16)$$

putting these transformations in Equations (10)–(12), which becomes

$$\begin{aligned} t_4' = & \frac{\rho_{hnf}}{\rho_f} \frac{\mu_f}{\mu_{hnf}} (S(3t_3 + \eta t_3 + t_2 t_3 - t_1 t_4)) + Ha^2 S Re_m \left(\frac{\sigma_{hnf}}{\sigma_f}\right)^2 \frac{\mu_f}{\mu_{hnf}} (2t_5 t_6 + t_1 t_5^2) \\ & + \eta Ha^2 S^2 Re_m^2 \left(\frac{\sigma_{hnf}}{\sigma_f}\right)^3 \frac{\mu_f}{\mu_{hnf}} (t_5^2 + \eta t_5 t_6 - t_1 t_5 t_6 + t_2 t_5^2) \\ & - Ha^2 S^2 Re_m^2 \left(\frac{\sigma_{hnf}}{\sigma_f}\right)^3 \frac{\mu_f}{\mu_{hnf}} (t_1 t_5^2 + \eta t_1 t_5 t_6' - t_1^2 t_5 t_6 + t_1 t_2 t_5^2), \end{aligned} \quad (17)$$

$$t_6' = Re_m S \frac{\sigma_{hnf}}{\sigma_f}(t_5 + \eta t_6 - t_1 t_6 + t_2 t_5), \quad (18)$$

$$t_8' = SPr\frac{(\rho Cp)_{hnf}}{(\rho Cp)_f}\frac{\kappa_f}{\kappa_{hnf}}(\eta t_8 - t_1 t_8) - \frac{\mu_{hnf}}{\mu_f}\frac{\kappa_f}{\kappa_{hnf}}PrEc(4\delta t_2^2 + t_3^2), \quad (19)$$

and the boundary conditions becomes

$$\begin{aligned} t_1(0) = 0, \quad t_1(1) = 1, \quad t_2(0) = 0, \quad t_2(1) = 0, \\ t_5(0) = 0, \quad t_5(1) = 1, \quad t_7(0) = 0, \quad t_7(1) = 1, \end{aligned} \quad (20)$$

- **Step 2: Introducing of parameter p and we obtained ODEs in a p-parameter group**

To get ODE's in a p-parameter group, let we know p-parameter in Equations (17)–(19) and therefore,

$$t'_4 = \frac{\rho_{hnf}}{\rho_f}\frac{\mu_f}{\mu_{hnf}}(S(3t_3 + \eta t_3 + t_2 t_3 - t_1(t_4 - 1)q)) + Ha^2 S Re_m (\frac{\sigma_{hnf}}{\sigma_f})^2 \frac{\mu_f}{\mu_{hnf}}(2t_5 t_6 + t_1 t_5^2)$$
$$+ \eta Ha^2 S^2 Re_m^2 (\frac{\sigma_{hnf}}{\sigma_f})^3 \frac{\mu_f}{\mu_{hnf}}(t_5^2 + \eta t_5 t_6 - t_1 t_5 t_6 + t_2 t_5^2) \tag{21}$$
$$- Ha^2 S^2 Re_m^2 (\frac{\sigma_{hnf}}{\sigma_f})^3 \frac{\mu_f}{\mu_{hnf}}(t_1 t_5^2 + \eta t_1 t_5 t'_6 - t_1^2 t_5 t_6 + t_1 t_2)$$

$$t'_6 = Re_m S \frac{\sigma_{hnf}}{\sigma_f}(t_5 + \eta t_6 - t_1(t_6 - 1)q + t_2 t_5), \tag{22}$$

$$t'_8 = SPr\frac{(\rho C_p)_{hnf}}{(\rho C_p)_f}\frac{\kappa_f}{\kappa_{hnf}}(\eta(t_8 - 1)q - t_1 t_8) - \frac{\mu_{hnf}}{\mu_f}\frac{\kappa_f}{\kappa_{hnf}}PrEc(4\delta t_2^2 + t_3^2), \tag{23}$$

- **Step 3: Differentiation by p, reaches the following system w.r.t the sensitivities to the parameter-p**

  Differentiating the Equations (21)–(23) w.r.t by p

$$d'_1 = h_1 d_1 + e_1 \tag{24}$$

  where h_1 is the coefficient matrix, e_1 is the remainder and $d_1 = \frac{dp_i}{d\tau}, 1 \leq i \leq 8$.

- **Step 3: Cauchy Problem**

$$d_1 = y_1 + a1 v_1, \tag{25}$$

  where y_1, v_1 are vector functions. By resolving the two Cauchy problems for every component. We are satisfied then automatically to ODE's

$$e_1 + h_1(a1 v_1 + y_1) = (a1 v_1 + y_1)' \tag{26}$$

  and left the boundary conditions.

- **Step 4: Using by Numerical Solution**

  An absolute scheme has been used for the resolution of the problem

$$\frac{v_1^{i+1} - v_1^i}{\triangle \eta} = h_1 v_1^{i+1} \tag{27}$$

$$\frac{y^{i+1} - y^i}{\triangle \eta} = h_1 y^{i+1} + e_1 \tag{28}$$

- **Step 5: Taking of the corresponding coefficients**

  As given boundaries are usually applied for p_i, where $1 \leq i \leq 8$, for the solution of ODEs, we required to apply $d_2 = 0$, which seems to be in matrix form as

$$l_1 . d_1 = 0 \quad \text{or} \quad l_1.(a1 v_1 + y_1) = 0 \tag{29}$$

  where $a1 = \frac{-l_1 . y_1}{l_1 . v_1}$

5. Results and Discussions

This paper investigates the hybrid nano-fluid flow between two parallel and squeezing plates under the influence of a variable magnetic field. The heat and mass transfer phenomena are also considered. The governing system of the hybrid nanofluid flow eaquations is converted into a non-linear system of ODEs through similarity transformations. The impacts of various physical parameters, including Hartmann number (Ha), magnetic Reynolds number (Re_m), Prandtl number (Pr), squeezing parameter (S), Eckert number (Ec), and hybrid nano-particle volume fraction (ϕ_1, ϕ_2) have been investigated in the context of heat transfer and fluid flow properties. The statistics in Table 1 provide complete information about the thermophysical properties of nanomaterials. It is mandatory to mention that our proposed model produces a nice results when compared with the results of the models available in the current literature, and the comparison has been shown in Table 2. Tables 3–7 illustrates the numerical outcomes of two important flow parameters skin friction and Nusselt number, which are obtained through two different numerical schemes(BVP4C and PCM). Table 8 displays quantitatively the impact of various flow parameters for different types of hybrid nano-fluid.

Table 1. The thermophysical properties of water base fluid and hybrid nanoparticles.

	ρ	C_p	κ	σ
H_2O	997.1	4179	0.613	5.5×10^{-6}
Fe_3O_4	5200	670	6	9.74×10^6
SWCNT	2600	425	6600	10^6
MWCNT	1600	796	3000	10^7

Table 2. Comparison of the numerical results of $f''(1)$ when $\Phi_1 = \Phi_2 = Re_m = \delta = 0$.

M	S	M. Bilal et al. [15]	Present
0	0.5	4.713254	4.713061
1		4.739148	4.739224
2		4.820361	4.820101
3		4.396271	4.396400
2	0	1.842331	1.842350
	0.3	3.653601	3.653573
	0.6	5.391148	5.391053
	1	7.593006	7.593187

Table 3. Comparison of the numerical results by two methods PCM and BVP4C for Skin friction and Nusselt number, with various physical parameters and $\Phi_1 = \Phi_2 = 0.1$.

	PCM	BVP4C	PCM	BVP4C
S	$f''(1)$	$f''(1)$	$-\theta'(1)$	$-\theta'(1)$
0.1	5.9662	5.9917	2.7615	2.7688
0.2	5.9105	5.9143	2.7369	2.7344
0.3	5.8437	5.8422	2.7103	2.7178
0.4	5.7714	5.7756	2.6831	2.6812
0.5	5.6968	5.6990	2.6557	2.6554
0.6	5.6219	5.6252	2.6278	2.6282
0.7	5.5489	5.5432	2.6057	2.6076
0.8	5.4779	5.4723	2.5822	2.5871
0.9	5.4081	5.4023	2.5538	2.5532

Table 4. Comparison of the numerical results by two methods PCM and BVP4C for skin friction and Nusselt number, with various physical parameters and $\Phi_2 = 0.02$.

	PCM	BVP4C	PCM	BVP4C
Φ_1	$f''(1)$	$f''(1)$	$-\theta'(1)$	$-\theta'(1)$
0.1	5.1032	5.1076	0.6308	0.6366
0.2	5.0736	5.0722	0.6774	0.6732
0.3	5.0801	5.0854	0.7399	0.7321
0.4	5.1222	5.1200	0.7957	0.7960
0.5	5.1989	5.1943	0.7765	0.7790
0.6	5.3095	5.3083	0.4537	0.4502
0.7	5.4558	5.4511	0.3872	0.3845
0.8	5.6343	5.6334	0.2733	0.2712
0.9	5.8421	5.8466	0.1209	0.1255

Table 5. Comparison of the numerical results by two methods PCM and BVP4C for skin friction and Nusselt number, with various physical parameters and $\Phi_1 = \Phi_2 = 0.1$.

Ha	PCM $f''(1)$	BVP4C $f''(1)$	PCM $-\theta'(1)$	BVP4C $-\theta'(1)$
0.1	5.9632	5.9632	2.7584	2.7584
0.2	5.9391	5.9391	2.7486	2.7486
0.3	5.8992	5.8992	2.7325	2.7325
0.4	5.8437	5.8437	2.7103	2.7103
0.5	5.7731	5.7731	2.6826	2.6826
0.6	5.6879	5.6879	2.6499	2.6499
0.7	5.5887	5.5887	2.6128	2.6128
0.8	5.4762	5.4762	2.5721	2.5721
0.9	5.3512	5.3512	2.5283	2.5283

Table 6. Numerical results by parametric continuation method for hybrid nanofluid with various physical parameters.

Φ_2	$H_2O, Fe_3O_4,$ SWCNT $f''(1)$	$-\theta'(1)$	$H_2O, Fe_3O_4,$ MWCNT $f''(1)$	$-\theta'(1)$	H_2O, Fe_3O_4 $f''(1)$	$-\theta'(1)$
0.02	5.6736	0.1161	5.6834	0.1351	5.6997	0.0573
0.03	5.6725	0.1170	5.6822	0.1356	5.6984	0.0564
0.04	5.6717	0.1181	5.6811	0.1361	5.6965	0.0554
0.05	5.6704	0.1196	5.6802	0.1370	5.6953	0.0542
0.06	5.6691	0.1208	5.6793	0.1379	5.6940	0.0529
0.07	5.6679	0.1220	5.6780	0.1387	5.6926	0.0516
0.08	5.6662	0.1236	5.6766	0.1399	5.6911	0.0503
0.09	5.6645	0.1251	5.6752	0.1410	5.6891	0.0488
0.10	5.6633	0.1269	5.6732	0.1422	5.6873	0.0471
0.11	5.6619	0.1282	5.6715	0.1438	5.6858	0.0457
0.12	5.6603	0.1301	5.6690	0.1455	5.6843	0.0441
0.13	5.6593	0.1325	5.6672	0.1471	5.6829	0.0425

Table 7. Numerical results by BVP4C Method for hybrid nanofluid with various physical parameters.

Φ_2	$H_2O, Fe_3O_4,$ SWCNT $f''(1)$	$-\theta'(1)$	$H_2O, Fe_3O_4,$ MWCNT $f''(1)$	$-\theta'(1)$	H_2O, Fe_3O_4 $f''(1)$	$-\theta'(1)$
0.02	5.6731	0.1157	5.6843	0.1343	5.6980	0.0563
0.03	5.6721	0.1164	5.6830	0.1350	5.6971	0.0551
0.04	5.6711	0.1173	5.6821	0.1355	5.6961	0.0540
0.05	5.6698	0.1190	5.6810	0.1363	5.6945	0.0533
0.06	5.6683	0.1201	5.6800	0.1370	5.6930	0.0521
0.07	5.6671	0.1213	5.6791	0.1378	5.6918	0.0502
0.08	5.6653	0.1226	5.6773	0.1391	5.6902	0.0492
0.09	5.6636	0.1242	5.6764	0.1403	5.6887	0.0476
0.10	5.6622	0.1253	5.6750	0.1413	5.6871	0.0462
0.11	5.6610	0.1270	5.6730	0.1428	5.6847	0.0445
0.12	5.6595	0.1292	5.6708	0.1441	5.6828	0.0432
0.13	5.6580	0.1311	5.6691	0.1464	5.6814	0.0413

Table 8. The heat transfer has been calculated percent wise as for the various nanoparticles $Pr = 6.2$, $S = 1.5$, $Ec = 0.4$, using the percentage formula %increase = $\frac{With\ Nanoparticle}{Without\ Nanoparticle} \times 100$ = Result, Result-100 = %enhancement.

Φ_1, Φ_2	$-\theta'(1)\ for\ Fe_3O_4, SWCNT$	$-\theta'(1)\ for\ Fe_3O_4, MWCNT$
0.0	5.5582	5.5582
0.01	5.6490	5.6561
	(1.6336% increase)	(1.7614% increase)
0.03	5.8421	5.8628
	(5.1078% increase)	(5.4802% increase)
0.04	5.9446	5.9718
	(6.9519% increase)	(7.4413% increase)

In Figure 2a, it has been observed that the velocity profile for different values of the squeezing parameter S increases the velocity profile more near the channel due to the rising force in the horizontal direction. This increment in force has a resistance force near the centre, which decreases the horizontal velocity after the central region. Figure 2b depicts that the growing value of squeezing parameter S improves the velocity profile, it is because the velocity gradient is falling near the channel wall for $\eta < 0.5$, while vertical velocity $f'(\eta)$ is rising due to narrowing the channel of the flow for the value $\eta > 0.5$. The velocity profile of the flow is plotted in Figure 3a,b, illustrates that Ha and Re_m are rising. Furthermore, it explains that the increasing value of the Hartmann number raises the velocity profile for $\eta < 0.6$, and reduces it for $\eta > 0.6$. Cross-flow behaviour for the velocity profile has been observed at the centre of the channel wall. The effect of the nano-material volume friction is depicted in Figure 4. It is noticed that a rise in the quantity of nanomaterials significantly fluctuates the velocity profile near the channel wall due to the increase in the density of the nano-materials. Moreover, the velocity of SWCNTs is slightly higher than that of MWCNTs, due to the low-density values of SWCNTs.

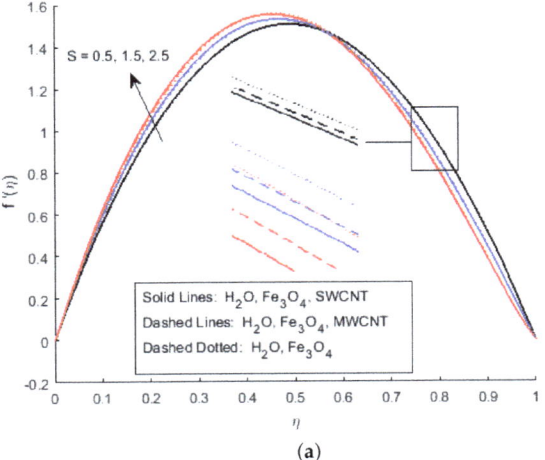

(a)

Figure 2. *Cont.*

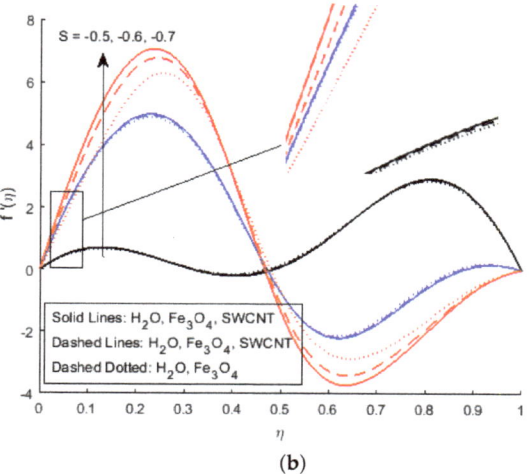

(b)

Figure 2. Effect of $f(\eta)$ and $f'(\eta)$ for (**a**) $S > 0$, (**b**) $S < 0$ and fixed values of $Re_m = 1.5$, $Ha = 0.4, Pr = 6.2, Ec = 0.5, \delta = 0.4, \Phi_1 = 0.02, \Phi_2 = 0.5, m = 0.3$.

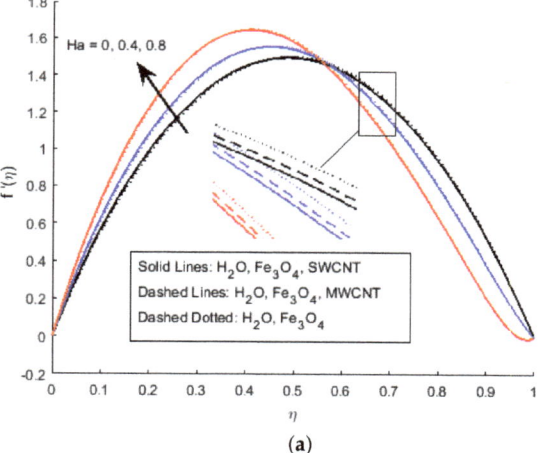

(a)

Figure 3. *Cont.*

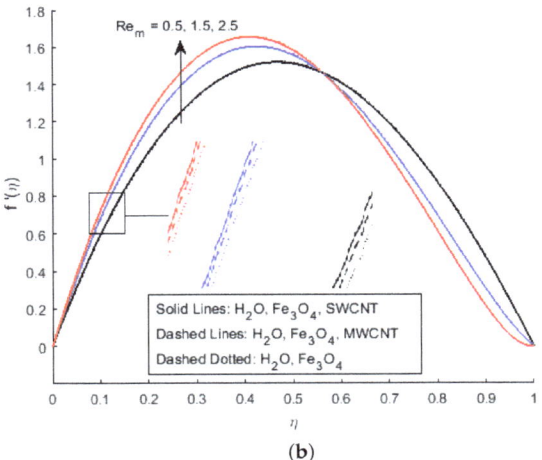

Figure 3. Effect of $f'(\eta)$ for (**a**) Ha, (**b**) Re_m and fixed values of $S = 2.5, Pr = 6.2, Ec = 0.5, \delta = 0.4, \Phi_1 = 0.02, \Phi_2 = 0.5, m = 0.3$.

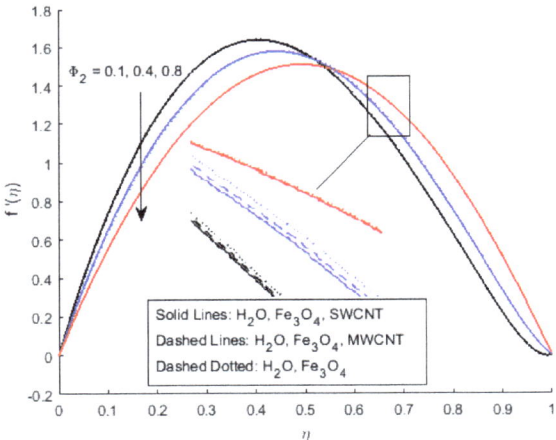

Figure 4. Effect of $f'(\eta)$ for Φ_2 and fixed values of $S = 1.5, Re_m = 2.5, Ha = 0.4, Pr = 6.2, Ec = 0.5, \delta = 0.4, \Phi_1 = 0.02, m = 0.3$.

Figure 5a displays the influence of the squeezing parameter S on the magnetic field profile $G(\eta)$, which illustrates that the magnetic profile becomes parabolic for the various value of the squeezing parameter. Body force is stronger near the bottom of the corresponding main wall because the Lorentz force is smaller near the squeezing plate (because the current is approximately parallel to the magnetic field). Initially, for the x-component of velocity, the velocity decline is identified, but starts augmenting as $\eta \to 1$ maximum value of $f'(\eta)$ is noticed in the middle. In the study of the magnetic profile, we have observed that the magnetic profile $G(\eta)$ is falling in the vicinity of the upper plate when the squeezing parameter increases. Figure 5b shows the falling behaviour of the magnetic Reynold's parameter Re_m, which defines the ratio of fluid flux to the magnetic diffusivity. This parameter therefore is an instrumental in determining the diffusion of magnetic field along streamlines. The variation in Re_m has direct effect on $G(\eta)$. As the increase in magnetic

Reynold's number increases, the opposing force of the axial velocity decreases the axial velocity of the flow. In our study, less importance has been placed on diffusion, and most of the analysis is based on the behaviour of the magnetic field during the flow.

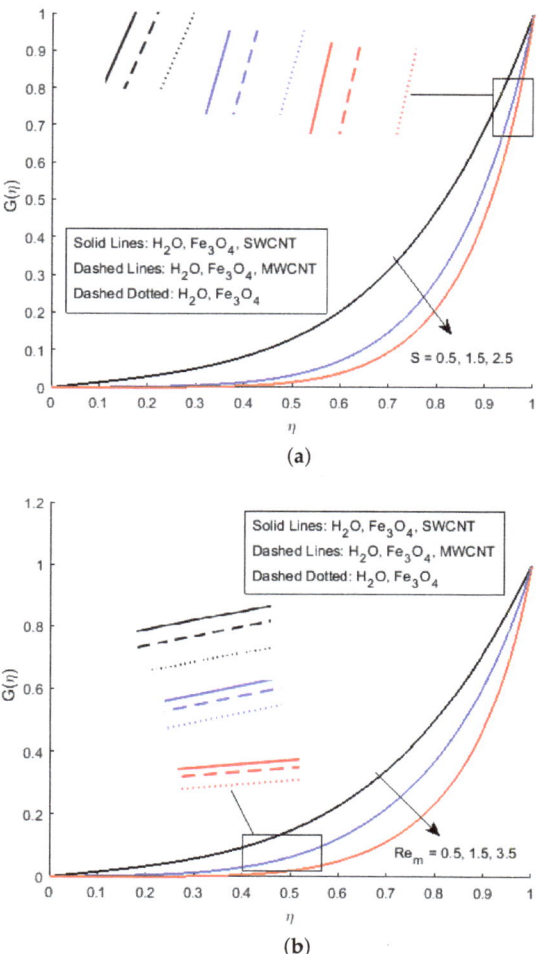

Figure 5. Effect of $G(\eta)$ for (a) S, (b) Re_m and fixed values of $Ha = 1.4, Pr = 6.2, Ec = 0.5, \delta = 0.4, \Phi_1 = 0.02, \Phi_2 = 0.5, m = 0.3$.

Figure 6 illustrates the impact of Φ_2 on the magnetic profile, which signifies that an opposing force is generated due to magnetic fields, as well as an improvement in the fluid viscosity due to the suspension of nanomaterials concentration. This is the basic reason for the fall in the magnetic field. The outcomes reflect that, as the concentration of nanomaterials rises, the magnetic field decreases. Figure 7a explains the impact of growing Eckert number on the temperature profile. The Eckert number is the ratio of kinetic energy to the boundary layer enthalpy difference and is used to explained heat dissipation. An abrupt surge for the temperature profile in the vicinity of the middle line has been noticed, as we know Ec simply the ratio of specific heat to thermal conductivity, so it raises the temperature profile considerably, and $\theta(\eta)$ profile becomes parabolic, which gives the maximum value at the middle of the channel. The heat transmission of (MWCNTs, Fe_3O_4

and H_2O) is slightly more than (SWCNTs, Fe_3O_4 and H_2O). The fluctuation that occurs in temperature profile due to the rising value of nanomaterials volume fraction Φ_2 has been displayed in Figure 8. A rise in the volume fraction of nanomaterials augments heat transmission and generation of heat, which grow the thickness of the thermal layer. In the case of diverging channels, a significant increase in temperature is observed. Nearly the same values for both $MWCNT$ and $SWCNT$ with H_2O and Fe_3O_4 have been observed. The velocity slip parameter significantly affects the temperature profile. Figures 9 and 10 are drawn to investigate the effects of S, Ha, and S, Ec on skin friction and Nusselt number.

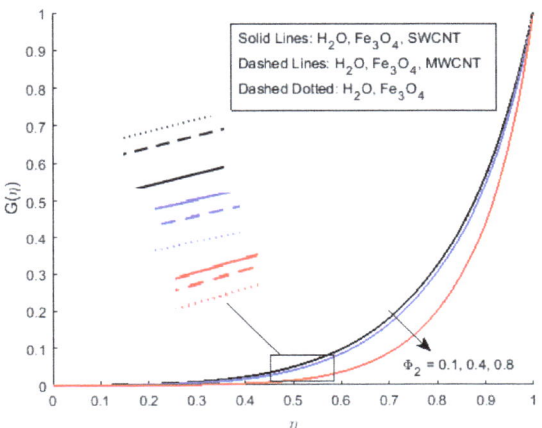

Figure 6. Effect of $G(\eta)$ for Φ_2 and fixed values of $S = 1.5, Re_m = 1.5, Ha = 0.4, Pr = 6.2, Ec = 0.5, \delta = 0.4, \Phi_1 = 0.02, m = 0.3$.

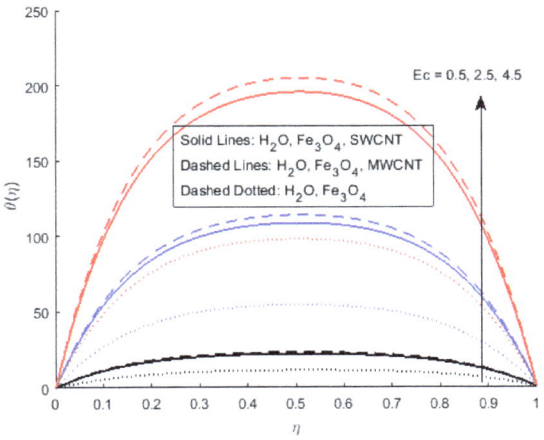

Figure 7. Effect of $\theta(\eta)$ for Ec and fixed values of $S = 0.5, Re_m = 1.5, Pr = 6.2, Ha = 0.4, \delta = 0.4, \Phi_1 = 0.02, \Phi_2 = 0.5, m = 0.3$.

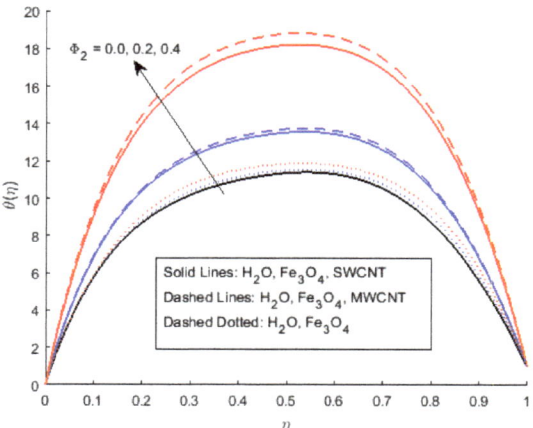

Figure 8. Effect of $\theta(\eta)$ for Φ_2 and fixed values of $S = 0.5, Re_m = 2.5, Ha = 0.4, Pr = 6.2, Ec = 0.5, \delta = 0.4, \Phi_1 = 0.02, m = 0.3$.

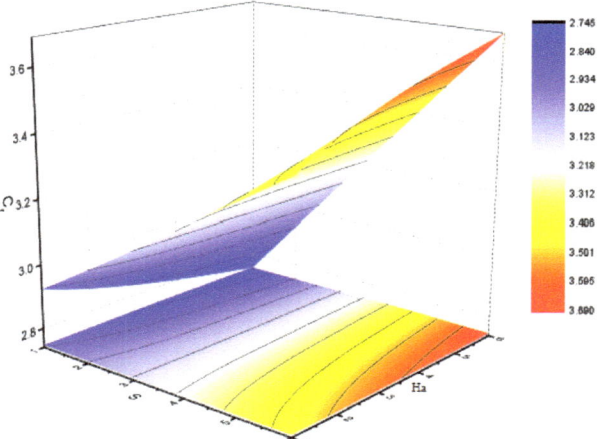

Figure 9. Effect of skin friction for S and Ha and fixed values of $Re_m = 2.5, Pr = 6.2, Ec = 0.5, \delta = 0.4, \Phi_1 = 0.02, m = 0.3$.

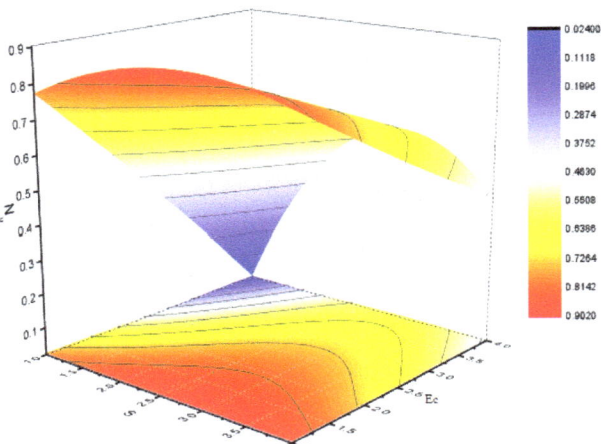

Figure 10. Effect of Nusslet number for S and Ec and fixed values of $Re_m = 2.5, Pr = 6.2, \delta = 0.4, \Phi_1 = 0.02, m = 0.3$.

6. Concluding Remarks

The main objective of this article is to analyze the effects of a changeable magnetic field on the hybrid nanofluids flow between the squeezing parallel plates so as to improve the mass and heat transfer profile. Hybrid nanofluid (ferrous oxide water and carbon nanotubes) CNTs-Fe_3O_4, H_2O have been chosen for this purpose. The governing equation of hybrid nanofluids flow including, momentum equaiton, energy equations, and magnetic field equations have been converted into systems of highly nonlinear ODEs thorugh similarity transformations, and subsequently solved by the parametric continuation method (PCM). For the validity of the numerical solution, the proposed model is also solved by the shooting method.

The main findings of this study are as follows:

- Statistical analysis shows that the thermal flow rate of (Fe_3O_4-SWCNTs-water) rises from 1.6336 percent to 6.9519 percent when the volume fraction of nano-material increases from 0.01 to 0.04 is shown in Table 4.
- It is obvious from Table 3 that the thermal flow rate of (Fe_3O_4-MWCNTs-water) rises from 1.7614 percent to 7.4413 percent when the volume fraction of nano-material increases from 0.01 to 0.04.
- Increase in the volume friction of the value of nano-material significantly fluctuates the velocity profile near the channel wall due to the increase in the fluid density.
- The increasing value of the squeezing parameter comparatively improves the velocity profile near the channel wall more than the center.
- By increasing the squeezing speed of the upper plate improve the velocity profile, it is because the velocity gradient is falling near the channel wall for $\eta < 0.5$, while vertical velocity $f'(\eta)$ is rising, due to narrowing the channel of the flow for the value $\eta > 0.5$.
- Body force is stronger near the bottom of the corresponding main wall, because the Lorentz force is smaller near the squeezing plate (because the current is approximately parallel to the magnetic field). Initially, for the x-component of velocity, the velocity decline is identified, but starts augmenting as $\eta \to 1$ maximum value of $f'(\eta)$ is noticed at the middle.
- It has been noticed that the magnetic Reynolds number increases opposing force of the axial velocity, which decreases the axial velocity of the flow.
- Strengthening the magnetic Reynold's number increases the strength of the opposing force of the axial velocity, which decreases the axial velocity of the flow.

7. Future Work

This problem can be solved in the future in three-dimensional geometry. In addition, the effect of variable magnetic fields on the physical properties of heat and mass transfer and fluids can be interesting work to investigate.

Author Contributions: Writing—original draft preparation, M.S.K.; Supervision, S.M.; Methodology, S.; Investigation, S.A.S.; Resources, S.N.; Writing—review and editing, A.K.; Funding acquisition, U.F.-G. and M.S.K. All authors have read and agreed to the published version of the manuscript.

Funding: The work of U.F.-G. has been supported by the government of the Basque Country for the ELKARTEK21/10 KK-2021/00014 and ELKARTEK20/78 KK-2020/00114 research programs, respectively.

Data Availability Statement: Not Applicable.

Acknowledgments: We acknowledge the financial support provided by the Postdoctoral research support fund of School of Mathematical Sciences, Jiangsu University, Zhenjiang 212013, China.

Conflicts of Interest: The authors declare no conflict of interest.

Nomenclature

The following abbreviations are used in this manuscript:

P	Pressure ($ML^{-1}T^{-2}$)
Φ_1, Φ_2	the volume fraction of CNTs
x, y	Cartesian Coordinates
κ_{bf}	Thermal conductivity of the Fe_3O_4-nanofuid (W/(mK))
u	x-axis velocity (m·s^{-1})
μ_{hnf}	Viscosity of hybrid nanofluid (Ns/m^2)
v	y-axis velocity (m·s^{-1})
κ_{hnf}	Thermal conductivity of hybrid nanofluid (W/(mK))
ρ_{hnf}	Density of hybrid nanofluid (kg/m^3)
H_0	Water
t	time (s)
Ha	Hartmann number
T_H	Upper plate temperature (K)
σ_{hnf}	Electrical conductivity of hybrid nanofluid ($ML^3T^{-3}I^{-2}$)
Re_m	Rynold's Magnetic Parameter
ν_{hnf}	Kinematic viscosity of hybrid nanofluid (m^{2}s^{-1})
Pr	Prandtl number (ν/k)
μ_f	Dynamic viscosity (PaS)
$(\rho Cp)_{hnf}$	Specific Heat of hybrid nanofluid (J/(kgK))
ρ_f	Density (kg/m^3)
S	Squeezing number
Ec	Eckert number
b_1, b_2	Components of Magnetic field
$h(t)$	Distance between parallel plates (L)
$CNTs$	Carbon nanotubes
Fe_3O_4	Ferrous-Oxide
$SWCNTs$	Single-wall carbon nanotubes
$MWCNTs$	Multi-wall carbon nanotubes

References

1. Mustafa, M.; Hayat, T.; Obaidat, S. On heat and mass transfer in the unsteady squeezing flow between parallel plates. *Meccanica* **2012**, *47*, 1581–1589. [CrossRef]
2. Dogonchi, A.S.; Alizadeh, M.; Ganji, D.D. Investigation of MHD Gowater nanofluid flow and heat transfer in a porous channel in the presence of thermal radiation effect. *Adv. Powder Technol.* **2017**, *28*, 1815–1825. [CrossRef]
3. Alizadeh, M.; Dogonchi, A.S.; Ganji, D.D. Micropolar nanofluid flow and heat transfer between penetrable walls in the presence of thermal radiation and magnetic field. *Case Stud. Therm. Eng.* **2018**, *12*, 319–332. [CrossRef]

4. Dib, A.; Haiahem, A.; Bou-Said, B. Approximate analytical solution of squeezing unsteady nanofluid flow. *Powder Technol.* **2015**, *269*, 193–199. [CrossRef]
5. Dogonchi, A.S.; Divsalar, K.; Ganji, D.D. Flow and heat transfer of MHD nanofluid between parallel plates in the presence of thermal radiation. *Comput. Methods Appl. Mech. Eng.* **2016**, *310*, 58–76. [CrossRef]
6. Sheikholeslami, M.; Ashorynejad, H.R.; Domairry, G.; Hashim, I. Flow and heat transfer of Cu-water nanofluid between a stretching sheet and a porous surface in a rotating system. *J. Appl. Math.* **2012**, *2012*, 421320. [CrossRef]
7. Choi, S.U.S. Enhancing thermal conductivity of fluids with nanoparticles. In *Developments and Applications of Non-Newtonian Flows, FED*; Signer, D.A., Wang, H.P., Eds.; ASME: New York, NY, USA, 1995; Volume 231, pp. 99–105.
8. Dogonchi, A.S.; Ganji, D.D. Thermal radiation effect on the Nanofluid buoyancy flow and heat transfer over a stretching sheet considering Brownian motion. *J. Mol. Liq.* **2016**, *223*, 521–527. [CrossRef]
9. Bhatti, M.M.; Abbas, T.; Rashidi, M.M. Entropy generation as a practical tool of optimisation for non-Newtonian nanofluid flow through a permeable stretching surface using SLM. *J. Comput. Des. Eng.* **2017**, *4*, 21–28. [CrossRef]
10. Chamkha, A.J.; Miroshnichenko, I.V.; Sheremet, M.A. Numerical analysis of unsteady conjugate natural convection of hybrid water-based nanofluid in a semicircular cavity. *J. Therm. Sci. Eng. Appl.* **2017**, *9*, 041004. [CrossRef]
11. Gorla, R.S.R.; Siddiqa, S.; Mansour, M.A.; Rashad, A.M.; Salah, T. Heat source/sink effects on a hybrid nanofluid-filled porous cavity. *J. Thermophys Heat Transfer* **2017**. [CrossRef]
12. Jackson, J.D. A study of squeezing flow. *Appl. Sci. Res.* **1963**, *11*, 148–152. [CrossRef]
13. Hayat, T.; Sajjad, R.; Alsaedi, A.; Muhammad, T.; Ellahi, R. On squeezed flow of couple stress nanofluid between two parallel plates. *Results Phys.* **2017**, *7*, 553–561. [CrossRef]
14. Taseer, M.; Hayat, T.; Alsaedi, A.; Qayyum, A. Hydromagnetic unsteady squeezing fow of Jefrey fuid between two parallel plates. *Chin. J. Phys.* **2017**, *55*, 1511–1522.
15. Bilal, M.; Arshad, H.; Ramzan, M.; Shah, Z.; Kumam, P. Unsteady hybrid-nanofuid fow comprising ferrousoxide and CNTs through porous horizontal channel with dilating/squeezing walls. *Nature* **2021**, 12637. [CrossRef]
16. Salehi, S.; Nori, A.; Hosseinzadeh, K.; Ganji, D.D. Hydrothermal analysis of MHD squeezing mixture fluid suspended by hybrid nanoparticles between two parallel plates. *Case Stud. Therm. Eng.* **2020**, *21*, 100650. [CrossRef]
17. Acharya, N. On the flow patterns and thermal behaviour of hybrid nanofluid flow inside a microchannel in presence of radiative solar energy. *J. Therm. Anal. Calorim.* **2020**, *141*, 1425–1442. [CrossRef]
18. Ikram, M.D.; Asjad, M.I.; Akgul, A.; Baleanu, D. Effects of hybrid nanofluid on novel fractional model of heat transfer flow between two parallel plates. *Alex. Eng. J.* **2021**, *60*, 3593–3604. [CrossRef]
19. Tayebi, T.; Chamkha, A.J. Free convection enhancement in an annulus between horizontal confocal elliptical cylinders using hybrid nanofluids. *Numer. Heat Transfer Part A Appl.* **2016**, *70*, 1141–1156. [CrossRef]
20. Tayebi, T.; Chamkha, A.J. Natural convection enhancement in an eccentric horizontal cylindrical annulus using hybrid nanofluids. *Numer. Heat Transfer Part A Appl.* **2017**, *71*, 1159–1173. [CrossRef]
21. Khan, M.S.; Rehan, A.S.; Aamir, K. Effect of variable magnetic field on the flow between two squeezing plates. *Eur. Phys. J. Plus* **2019**, *134*, 219. [CrossRef]
22. Khan, M.S.; Rehan, A.S.; Amjad, A.; Aamir, K. Parametric investigation of the Nernst-Planck model and Maxwell's equations for a viscous fluid between squeezing plates. *Bound. Value Probl.* **2019**, *2019*, 107. [CrossRef]
23. Prasannakumara, B.C. Assessment of the local thermal non-equilibrium condition for nanofluid flow through porous media: A comparative analysis. *Indian J. Phys.* **2021**, 1–9. [CrossRef]
24. Song, Y.Q.; Khan, M.I.; Qayyum, S.; Gowda, R.P.; Kumar, R.N.; Prasannakumara, B.C.; Chu, Y.M. Physical impact of thermo-diffusion and diffusion-thermo on Marangoni convective flow of hybrid nanofluid ($MnZiFe_2O_4$-$NiZnFe_2O_4$-H_2O) with nonlinear heat source/sink and radiative heat flux. *Mod. Phys. Lett. B* **2021**, *35*, 2141006. [CrossRef]
25. Gowda, R.P.; Kumar, R.N.; Prasannakumara, B.C. Two-Phase Darcy-Forchheimer Flow of Dusty Hybrid Nanofluid with Viscous Dissipation Over a Cylinder. *Int. J. Appl. Comput. Math.* **2021**, *7*, 1–18. [CrossRef]
26. Xiong, P.Y.; Khan, M.I.; Gowda, R.P.; Kumar, R.N.; Prasannakumara, B.C.; Chu, Y.M. Comparative analysis of (Zinc ferrite, Nickel Zinc ferrite) hybrid nanofluids slip flow with entropy generation. *Mod. Phys. Lett. B* **2021**, 2150342. [CrossRef]
27. Kumar, R.N.; Gowda, R.P.; Gireesha, B.J.; Prasannakumara, B.C. Non-Newtonian hybrid nanofluid flow over vertically upward/downward moving rotating disk in a Darcy-Forchheimer porous medium. *Eur. Phys. J. Spec. Top.* **2021**, 1–11. [CrossRef]
28. Khan, M.I.; Qayyum, S.; Shah, F.; Kumar, R.N.; Gowda, R.P.; Prasannakumara, B.C.; Kadry, S. Marangoni convective flow of hybrid nanofluid ($MnZnFe_2O_4$-$NiZnFe_2O_4$-H_2O) with Darcy Forchheimer medium. *Ain Shams Eng. J.* **2021**, *12*, 3931–3938. [CrossRef]
29. Kumar, R.V.; Gowda, R.P.; Kumar, R.N.; Radhika, M.; Prasannakumara, B.C. Two-phase flow of dusty fluid with suspended hybrid nanoparticles over a stretching cylinder with modified Fourier heat flux. *SN Appl. Sci.* **2021**, *3*, 1–9. [CrossRef]
30. Khodadadi, M.; Sheikholeslami, M. Heat transfer efficiency and electrical performance evaluation of photovoltaic unit under influence of NEPCM. *Int. J. Heat Mass Transf.* **2021**, *2021*, 122232.
31. Sheikholeslami, M.; Farshad, S.A. Numerical simulation of effect of non-uniform solar irradiation on nanofluid turbulent flow. *Int. Commun. Heat Mass Transf.* **2021**, *129*, 105648. [CrossRef]

32. Said, Z.; Sundar, L.S.; Tiwari, A.K.; Ali, H.M.; Sheikholeslami, M.; Bellos, E.; Babar, H. Recent advances on the fundamental physical phenomena behind stability, dynamic motion, thermophysical properties, heat transport, applications, and challenges of nanofluids. *Phys. Rep.* **2021**. [CrossRef]
33. Uddin, M.J.; Kabir, M.N.; Alginahi, Y.; Beg, O.A. Numerical solution of bio-nano-convection transport from a horizontal plate with blowing and multiple slip effects. Proceedings of the Institution of Mechanical Engineers. *Proc. Inst. Mech. Eng. Part C J. Mech. Eng. Sci.* **2019**, *233*, 6910–6927. [CrossRef]
34. Tuz Zohra, F.; Uddin, M.J.; Basir, M.F.; Ismail, A.I.M. Magnetohydrodynamic bio-nano-convective slip flow with Stefan blowing effects over a rotating disc. *Proc. Inst. Mech. Eng. Part N J. Nanomater. Nanoeng. Nanosyst.* **2020**, *234*, 83–97. [CrossRef]
35. Thiyagarajan, P.; Sathiamoorthy, S.; Santra, S.S.; Ali, R.; Govindan, V.; Noeiaghdam, S.; Nieto, J.J. Free and Forced Convective Flow in Pleural Fluid with Effect of Injection between Different Permeable Regions. *Coatings* **2021**, *11*, 1313. [CrossRef]
36. Alam, M.K.; Bibi, K.; Khan, A.; Noeiaghdam, S. Dufour and Soret Effect on Viscous Fluid Flow between Squeezing Plates under the Influence of Variable Magnetic Field. *Mathematics* **2021**, *9*, 2404. [CrossRef]
37. Bilal, S.; Rehman, M.; Noeiaghdam, S.; Ahmad, H.; Akgul, A. Numerical analysis of natural convection driven flow of a non-Newtonian power-law fluid in a Trapezoidal enclosure with a U-shaped constructal. *Energies* **2021**, *14*, 5355. [CrossRef]
38. Ghiasi, E.K.; Noeiaghdam, S. Truncating the series expansion for unsteady velocity-dependent Eyring-Powell fluid. *Eng. Appl. Sci. Lett.* **2020**, *3*, 28–34. [CrossRef]

Article

Numerical and Experiment Studies of Different Path Planning Methods on Mechanical Properties of Composite Components

Dongli Wang [1], Jun Xiao [1], Xiangwen Ju [1], Mingyue Dou [2], Liang Li [2] and Xianfeng Wang [1,*]

1. College of Material Science and Technology, Nanjing University of Aeronautics and Astronautics, No. 29, Yudao Street, Qinhuai District, Nanjing 210016, China; wdl@nuaa.edu.cn (D.W.); j.xiao@nuaa.edu.cn (J.X.); xiangwen_ju@nuaa.edu.cn (X.J.)
2. Nanjing Chenguang Group, No.1, Zhengxue Street, Qinhuai District, Nanjing 210016, China; Luna20211011@163.com (M.D.); liang1757761210@163.com (L.L.)
* Correspondence: wangxf@nuaa.edu.cn

Citation: Wang, D.; Xiao, J.; Ju, X.; Dou, M.; Li, L.; Wang, X. Numerical and Experiment Studies of Different Path Planning Methods on Mechanical Properties of Composite Components. *Materials* **2021**, *14*, 6100. https://doi.org/10.3390/ma14206100

Academic Editor: Abbas S. Milani

Received: 2 September 2021
Accepted: 8 October 2021
Published: 15 October 2021

Publisher's Note: MDPI stays neutral with regard to jurisdictional claims in published maps and institutional affiliations.

Copyright: © 2021 by the authors. Licensee MDPI, Basel, Switzerland. This article is an open access article distributed under the terms and conditions of the Creative Commons Attribution (CC BY) license (https://creativecommons.org/licenses/by/4.0/).

Abstract: The purpose of this paper is to study the effects of different trajectory planning methods on the mechanical properties of components. The scope of the research includes finite element simulation calculation and experimental tests of the actual structure. The test shall be carried out in the whole load range until the failure of the structure occurs. Taking the composite conical shell as an example, a variable angle initial path generation method of the conical shell surface is proposed, and the parallel offset algorithms based on partition and the circumferential averaging are proposed to fill the surface. Then, finite element analysis is carried out for the paths that satisfy the manufacturability requirements, the analysis results show that the maximum deformation and maximum transverse as well as longitudinal stress of fiber of circumferential averaging variable angle path conical shell are reduced by 16.3%, 5.85%, and 19.76%, respectively, of that of the partition variable angle path. Finally, the strength analysis of conical shells manufactured by different trajectory design schemes is carried out through finite element analysis and actual failure tests. The finite element analysis results are in good agreement with the experimental results of the actual structure. The results show that the circumferential uniform variable angle has good quality, and it is proved that the path planning algorithm that coordinates path planning and defect suppression plays an important role in optimizing placement trajectory and improving mechanical properties of parts.

Keywords: advanced composite; automatic fiber placement; path planning; multi-objective optimization; mechanical properties

1. Introduction

Fiber reinforcement composites have been widely used in aerospace and other high-tech fields due to their light weight, high strength, high modulus, fatigue resistance, corrosion resistance, strong designability, and simplicity to realize automatically integrated manufacturing [1]. Automatic fiber placement (AFP) is one of the most important technologies for manufacturing carbon fiber reinforced plastic (CFRP) components, which can realize the manufacturing of large-size and complex composite structural parts because each slit-tape is driven individually [2–4]. Moreover, it is the development direction of low-cost manufacturing technology of composite materials.

AFP is a manufacturing process in which a numerically controlled robot or CNC machine delivers prepreg tows of thermoset composites at a specific position and orientation through a placement head. It can lay as many as 32 fiber tows simultaneously at a constant angle which is often set as $0°$, $\pm45°$, or $90°$, and the width of a prepreg tow is usually within the range from 3.175 mm up to 25.4 mm. These tows are transported from a creel system through the placement head and deposited onto the mold surface or already existing plies along a specific path [5,6]. Path planning is an extremely important process in AFP [7], which directly determines the quality of the manufactured component and the efficiency of the placement.

Over the past decades, studies on AFP-based path planning were focused on initial path design and optimization [1,8–11], generation of offset path [10,11], and quality control [12–14]. Moreover, models of path deformation were also established to obtain the relationship between placement parameters and layup quality of fiber path [1,14]. Many studies for path generation for open-contoured structures were carried out by Shirinzadeh, Long Yan et al. To sum up, the main construction methods of the initial reference path mainly include [15–18]: forming a fixed angle with an axis [10], along the direction of principal stress distribution or ply bearing information [19], surface–plane intersection strategy [20], projecting the reference line on the mold surface [21], etc. Shirinzadeh et al. [11] proposed a novel path planning algorithm for open-contoured structures, the initial reference path is generated by projecting the major axis on the mold surface, the entire surface is then completely covered by continuously offsetting the initial reference path. However, the placement direction of the parallel path is severely deflected and defects like wrinkles may occur since the steering radius is not controlled, the algorithm is only suitable for simple surfaces. A new path planning algorithm, which uses a surface–plane intersection strategy to construct the initial path, is proposed in [7]. For complex and closed profile parts, the traditional path planning algorithm of isometric offset in fiber placement is difficult to meet both angle deviation and curvature requirements, and it is very easy to cause the angle deviation to exceed the limit. To overcome this limit, Duan [9] proposed an angle control algorithm of trajectory planning, and the algorithm has been successfully validated to plan the path of the airplane's S-shaped inlet. Qu [10] also proposed a path generation method to overcome the contradiction between angle direction and fiber wrinkles on the S-shape inlet surface, and the proposed method can control the angle deviation within 10 degrees and meet the curvature requirements. However, there are still few studies on the effect of carbon fiber ply angle on the mechanical properties of resin matrix composites. Different carbon fiber ply angles will inevitably affect their service properties. Due to the limited deformation of tow, the local steering radius of the placement path must be greater than the minimum steering radius determined by the material properties, otherwise it will produce wrinkles, tears, and other distortions, which will affect the surface quality and result in a decrease in strength as high as 57.7% [22]. This problem is well-described in References [23–26], which presents the results of both experimental testing and numerical calculations regarding the influence of placement angle on the properties of composite plates and profiles under compression. The results show that the carbon fiber ply angle has a significant effect on the buckling and post-buckling characteristics of the test structure. In order to predict, reduce, or even avoid those defects, the optimal path planning method and its influence on component performance must be understood.

To sum up, different path planning schemes will seriously affect the strength of the parts. Resin-rich areas, voids, wrinkles, and other defects are easy to appear in the plies, and there are few related researches on the comprehensive consideration of path manufacturability and part strength, so it is urgent to develop corresponding path planning methods to control the manufacturing quality of the composite component. On the other hand, the research of other scholars on the initial trajectory design is not sufficient, and the influence of path design on the mechanical properties of composite components is rarely studied. Therefore, the objective of the work described in this paper is to study the effects of different trajectory planning methods on the mechanical properties of components. The research scope includes finite element simulation calculation and experimental tests of the actual structure.

The subsequent sections, herein, are organized as follows. The generation of fixed angle, geodesic, and variable angle path of the conical shell are discussed in the automated fiber placement process and the steering radius as well as direction deviation are used to evaluate the path quality in Section 2. Afterwards, using the idea of discretization, the actual placement angle of each discrete element on the conical shell is calculated by using the self-developed path angle calculation software, and the strength analysis is carried out for the conical shell with different path design methods to find the optimal path design scheme in Section 3. The strength analysis of conical shells manufactured by different trajectory design schemes is carried out through finite element analysis and actual failure tests in Section 4. Finally, conclusions are drawn in Section 5.

2. Path Planning for Composites Conical Shell

Composite conical shell structure has been widely used in aircraft, spacecraft, rockets, and missiles, which are frequently subjected to dynamic loads in service [27,28]. However, it is very challenging to design the roller path of composite conical shell components to satisfy both process and structural requirements. In order to further improve the mechanical properties of composite conical shells, a path design method considering multiple constraints needs to be studied.

2.1. Numerical Model

A three-dimensional representation of the geometry of a conical shell is shown in Figure 1. In this mathematical model, the basic parameters that are used to define the cone shell are the semi-vertex angle (α), axial length A = 1160 mm, smaller radius r = 295 mm, and larger radius R = 497 mm, and these parameters have the following mathematical relationship:

$$\tan(\alpha) = \frac{R-r}{A} \quad (1)$$

To define the placement angle, the Frenet Frame at p_i (i is the number of the path) is established, which contains three orthogonal vectors e_1, e_2, and e_3. Here, the unit vectors for the longitudinal and circumferential surface directions, e_1 and e_2, respectively, as well as the surface normal e_3 will be used in the fiber path definitions.

2.2. Trajectory of Conical Shell and Its Manufacturability

The composite material can be treated as a material or a structure, whose one outstanding advantage is the designability of performance. Therefore, composite components with different properties can be obtained by different ply directions and ply forms [29]. The quality of path planning directly affects both the quality of offset paths and the mechanical properties of the components. At present, the commonly used path planning methods mainly include fixed angle, geodesic, and variable angle methods [30]. The fixed angle algorithm meets the requirements of structural design, but it is easy to produce wrinkles [31,32]. Compared with the fixed angle method, the geodesic method has an infinite curvature radius and no wrinkle will occur. However, it is very easy to cause the angle deviation to exceed its upper limit, which results in great difference between the actual performance of the component and the design value [13,33]. The variable angle path planning method is an ideal solution for the placement of composite conical shells at present because it comprehensively considers the angle deviation and path curvature.

2.2.1. Initial Path Generation

The starting point is extremely important to the initial path itself and affects the quality of the offset paths. The starting point is obtained according to the following steps in the paper, as shown in Figure 2.

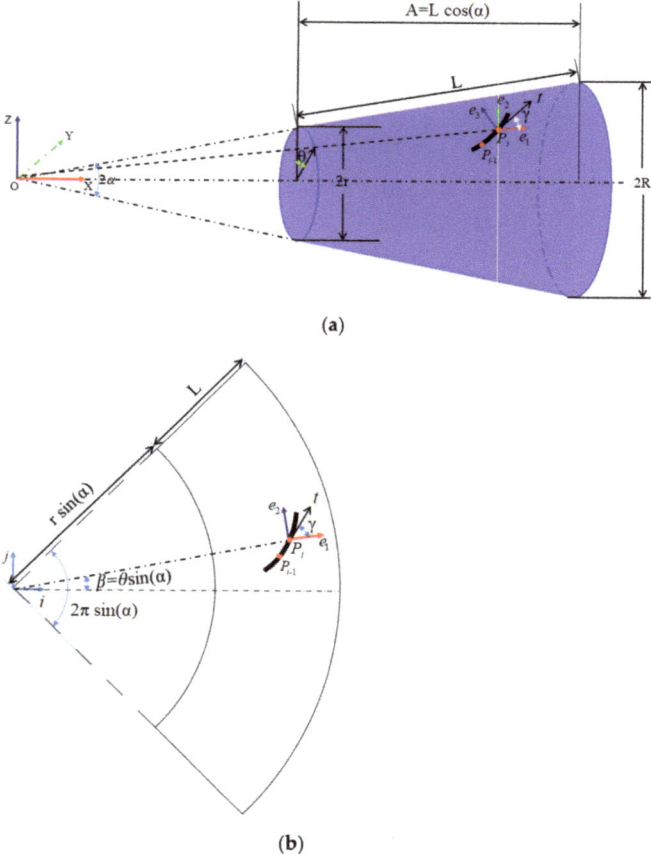

Figure 1. Cone geometry. (**a**) Three-dimensional schematic diagram of the conical shell. (**b**) Developed surface of the conical shell.

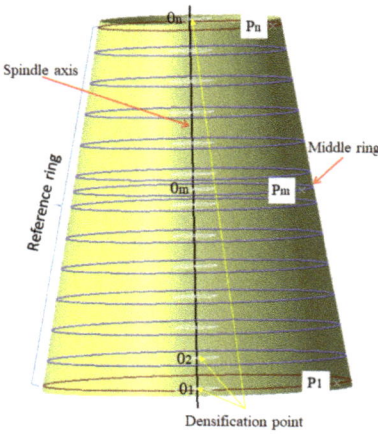

Figure 2. Construction of the initial point.

Step 1. The spindle axis is densified into a series of points O_i (i = 1, 2, ..., n);

Step 2. The slicing ring is formed by the intersection of the mold surface and the normal plane generated by points O_i and perpendicular to the spindle axis;

Step 3. The midpoint of the spindle is projected onto the corresponding ring first to form the initial point in Figure 2. If it cannot meet the uniform angle deviation on either side of the initial path, we iterate the points on both sides of the midpoint in turn.

If the starting point chosen here is not reasonable, it is extremely easy to cause an angle deviation over its upper limit, and cannot generate the required path. Figure 3 shows the geodesic paths at different starting points and their angle deviation. From the Figure 3, we can clearly see that the geodesic path with the starting point P1 (the big end) fails to be generated due to the occurrence of path turn back and unable to reach the boundary of the small end. In addition, even if the angle deviations of the other two geodesic paths are all over the upper limit, their variation laws are completely different.

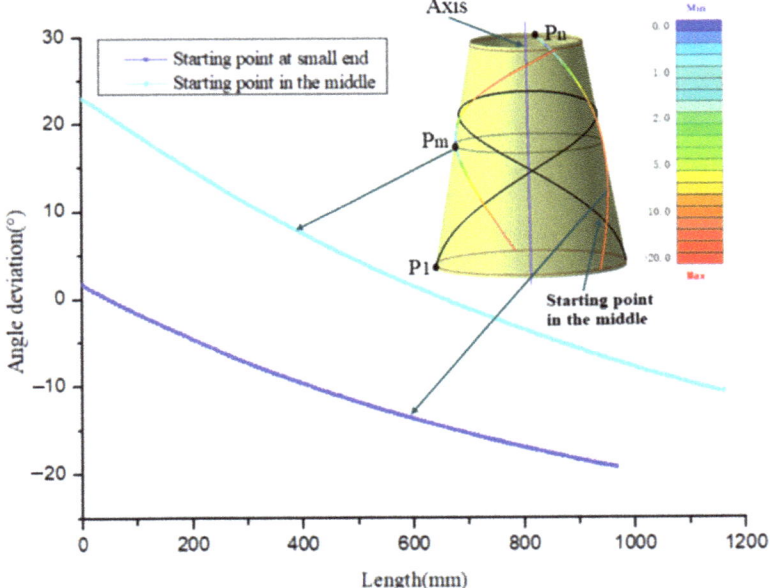

Figure 3. Geodesic trajectory at different starting points.

After the starting point is determined by iterative optimization, the next step is to design the variable angle path on conical shells surfaces, and discrete path points are calculated with a multi constraint model in local Frenet Frames. The flow chart is shown in Figure 4.

As shown in Figure 5, for a given surface S, $T_v(X_v, Y_v, Z_v)$ is the reference direction of the fiber path, P_i is the path point, and Σ is the tangent plane. All path points are obtained according to the following steps.

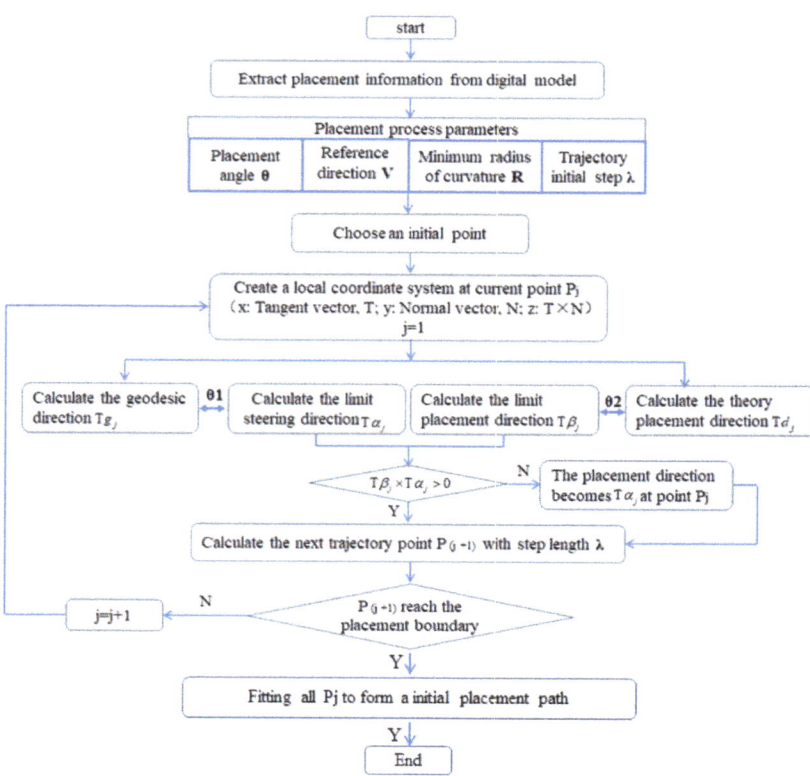

Figure 4. Flow chart of the path generation method.

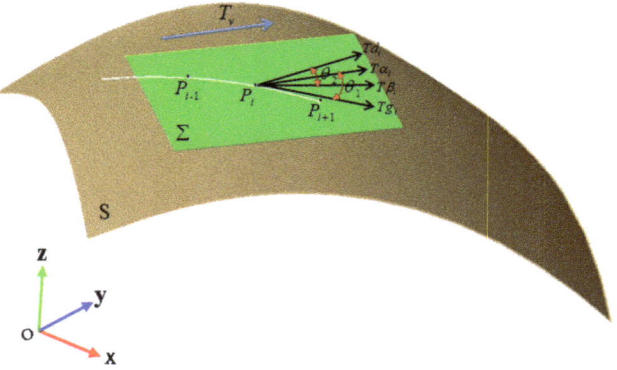

Figure 5. Generation of variable angle trajectory.

Step 1. Select a point P_0 on the mold surface as the starting point through the above methods;

Step 2. Calculate the normal vector N (n_{xi}, n_{yi}, n_{zi}) of the surface at the points $P_i(X_i, Y_i, Z_i)$ and the reference direction T_v is projected onto the tangent plane to obtain the projection vector $T\partial_i$ at the points P_i, the coordinates of $T\partial_i$ are obtained by the following formula:

$$\begin{cases} X'_\partial = -\frac{n_{xi}(n_{xi}X_v + n_{yi}Y_v + n_{zi}Z_v)}{n_{xi}^2 + n_{yi}^2 + n_{zi}^2} + X_i + \lambda X_v \\ Y'_\partial = -\frac{n_{yi}(n_{xi}X_v + n_{yi}Y_v + n_{zi}Z_v)}{n_{xi}^2 + n_{yi}^2 + n_{zi}^2} + Y_i + \lambda Y_v \\ Z'_\partial = -\frac{n_{zi}(n_{xi}X_v + n_{yi}Y_v + n_{zi}Z_v)}{n_{xi}^2 + n_{yi}^2 + n_{zi}^2} + Z_i + \lambda Z_v \end{cases} \quad (2)$$

Step 3. Calculate the P'_{i+1} on the tangent plane with a step λ under the three constraint conditions which including angle direction, steering radius and compaction, then P'_{i+1} is projected onto the mold surface to get the next path point P_{i+1};

Step 4. Repeat Steps 2 and 3 until the path point reaches the boundary of the surface. All the path points are fitted to form the initial path.

Using the above method, the 45-degree variable angle path manufacturability of different starting points is analyzed as shown in Figure 6. It can be seen that the variable angle trajectory with the starting point P1 (at the big end) has a good effect on this conical shells surface. The angle deviation is less than 1 degree and the minimum steering radius is greater than 2000 mm. Therefore, point Pm is selected as the starting point of the 45-degree variable angle trajectory in this paper. The initial trajectory design principle of other angles is the same as 45 degrees.

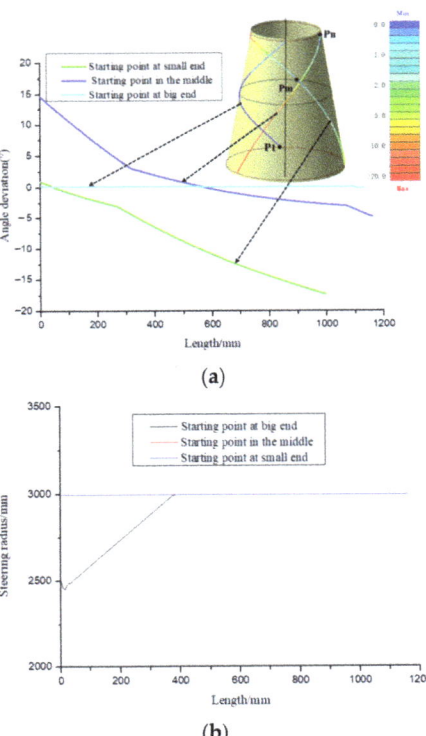

Figure 6. Paths on conical shells surface and its manufacturability. (a) Paths and Angle deviation evaluation. (b) Fiber steering evaluation.

2.2.2. Densification and Coverage Algorithm of the Initial Path

To achieve uniform paths over a mold, parallel equidistant offset algorithm is a common method to densify the path for automated fiber placement path planning. It has good applicability for open free curved surfaces with small and medium curvature. However, if the initial path is always offset for the revolving body, the curvature and angle deviation of the offset path will exceed its limit and produce a larger triangle area. It will produce tows breaking enrichment areas and affect the mechanical properties of components. In order to avoid a series of problems existing in the traditional isometric offset path planning algorithm on the rotating body, the offset paths generation algorithm based on partition and circumferential averaging planning algorithm is adopted in this paper, as shown in Figure 7.

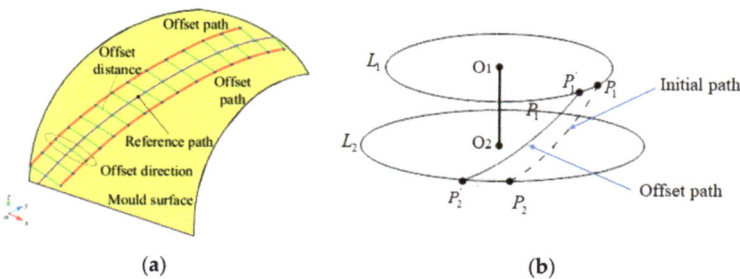

Figure 7. Concept for offsetting the initial path. (**a**) Equidistant offset [7], (**b**) Offset proportionally based on profile perimeter.

As can be seen from Figure 7a, equidistant offset path is that the reference path is offset by a certain distance (tows number multiplied by tow width) along the direction of the bi-normal vector, whose purpose is to formulate a set of paths without gaps and overlaps between subsequent paths. To avoid the transition concentration of triangular region, the conical shell surface is partitioned in this paper, and the number of partitions depends on the quality of offset paths. In this paper, the conical shell surface is divided into the same four regions according to the variable angle trajectory, and other paths are generated by equidistant offset in each region. Those offset paths and its manufacturability evaluations are shown in Figure 8. The minimum steering radius of offset paths all exceeds 2000 mm and the angle deviation ranges from 0 to 15°. It can be seen that the method also has a good effect on the conical shell surface. Other paths are generated using the same method and Figure 9 shows the laying paths of four angles, the red line is the initial path, and the blue line is the offset path.

Different from the equidistant offset paths based on the partition, the placement quality of each circumferential averaging path of the conical shell is the same, and the maximum number of the whole tows is guaranteed. However, it may not meet the technological requirements, which are closely related to the parameters of the cone. Figure 10 shows the circumferential averaging paths based on variable angle and analyzes their manufacturability. From the analysis results, it can be clearly known that the angle deviation of all offset paths is less than 2 degrees, and the minimum steering radius is greater than 2000 mm. Other paths are generated using the same method and Figure 11 shows the laying paths of four angles.

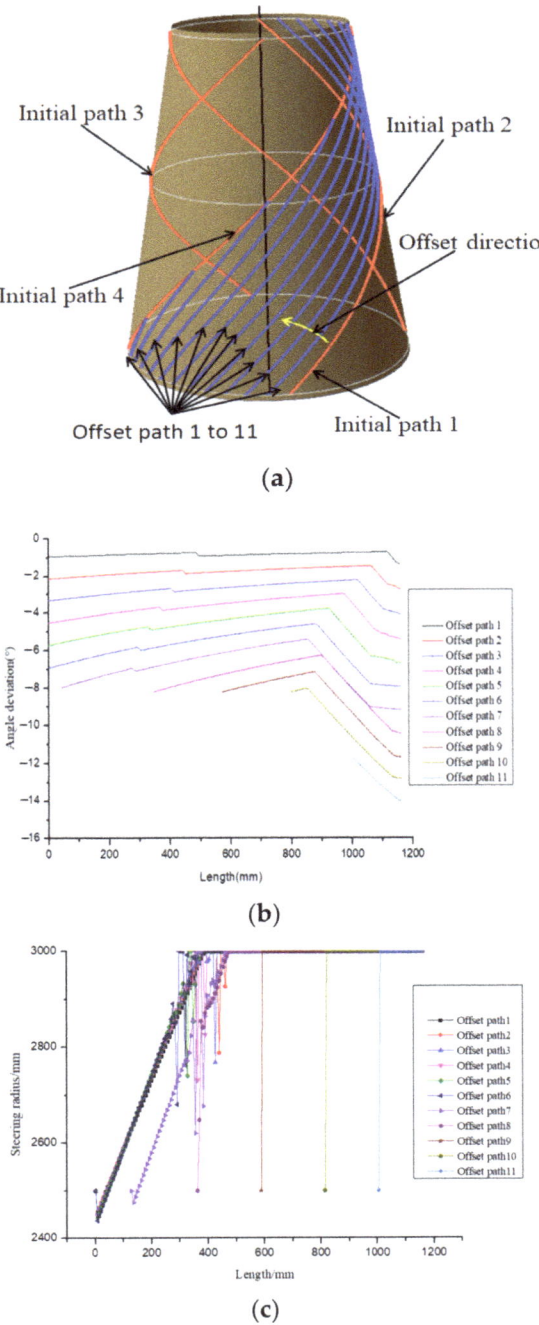

Figure 8. Equidistant offset paths based on partition and its manufacturability. (**a**) Equidistant offset paths based on partition. (**b**) Angular deviation of each offset path. (**c**) Steering radius of each offset path.

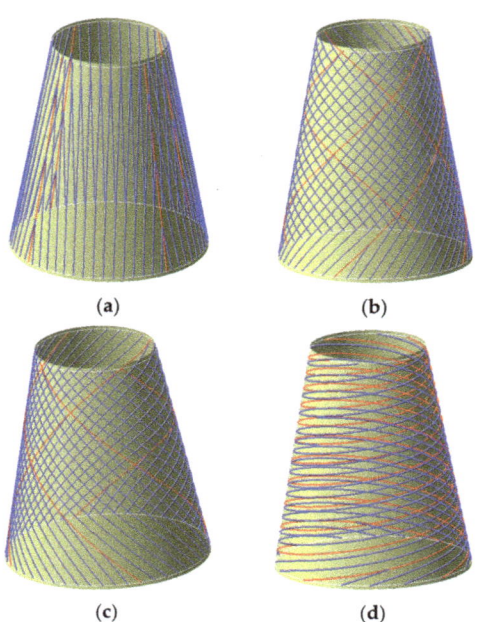

Figure 9. Four typical angle plies are based on partition. (**a**) 0° Ply. (**b**) 45° Ply. (**c**) −45° Ply. (**d**) 80° Ply.

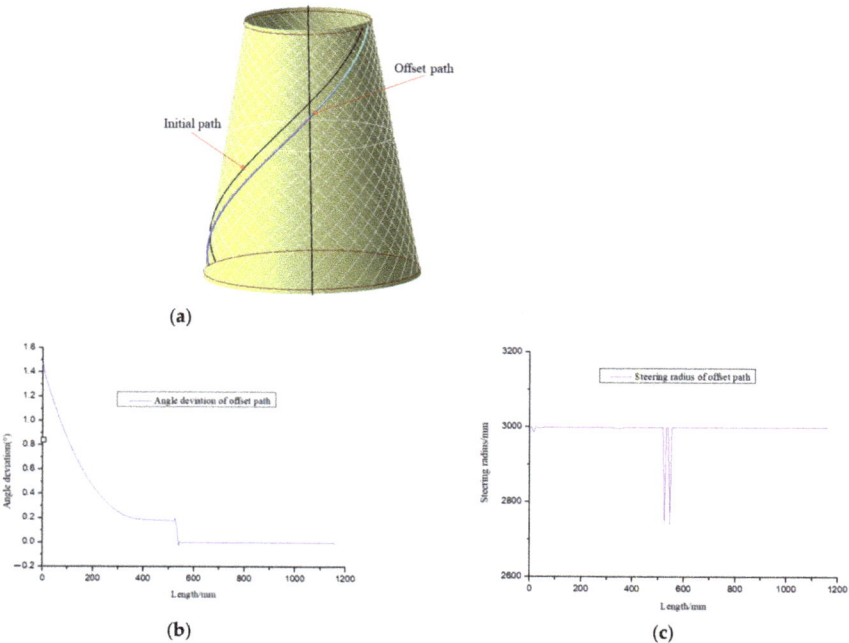

Figure 10. Circumferential averaging planning trajectory and its manufacturability. (**a**) Circumferential averaging path planning. (**b**) Angular deviation of each offset path. (**c**) Steering radius of each offset path.

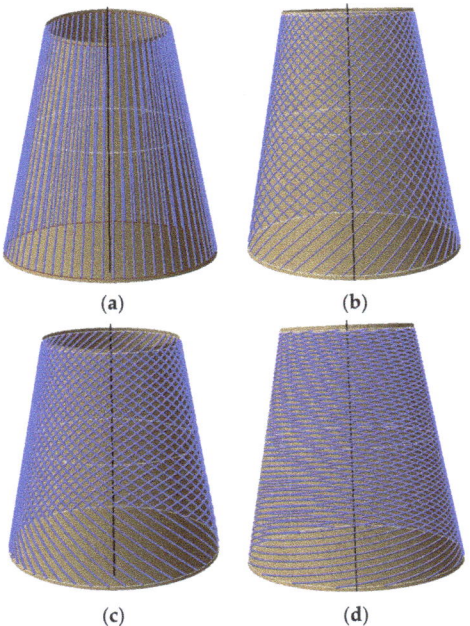

Figure 11. Four typical angle plies based on circumferential averaging. (**a**) 0° Ply. (**b**) 45° Ply. (**c**) −45° Ply. (**d**) 80° Ply.

3. Finite Element Analysis Based on Path Design

From the previous section, we know that there are different path planning methods for the same conical shell surface, and they all meet the technical requirements. However, how to choose the final path planning method to complete the laying still needs to be compared in terms of strength. According to the ply angle for each element data of typical parts, the strength analysis results are obtained by using ABAQUS simulation software, and compared with the traditional ply angle design in ABAQUS software in this paper.

In order to realize finite element analysis based on path design, the path design software is developed by CATIA secondary development technology. The software can design the path of fixed angle, geodesic and variable angle, and can calculate the angle of the center point of each triangular patch. The flow chart in Figure 12 illustrates the detailed process.

3.1. Material Parameters and Ply Settings

The composite prepreg used in the experiment is EH104-HF40-D6/12 made by Jiangsu Hengshen Co., Ltd. of CHINA, and the performance parameters of the material are shown in Table 1. The prepreg tow had a slitting width of 6.35 mm, a nominal thickness of 0.15 mm. In this paper, the conical shell skin adopts the shell element and S3 element. The total number of units is 8528 and the number of nodes is 4368. The layer of the cone shell model created is symmetrical with 24 layers, the total thickness is 3.6 mm, and the layering sequence is designed as $[45/0/-45/90]_{3s}$. The internal pressure of 3 MPa provided by the design department was applied to the external wall of the conical shell to analyze its static mechanical properties. Static pressure loading is shown in Figure 13. In addition, both the big end and the small end are imposed fixed support constraints.

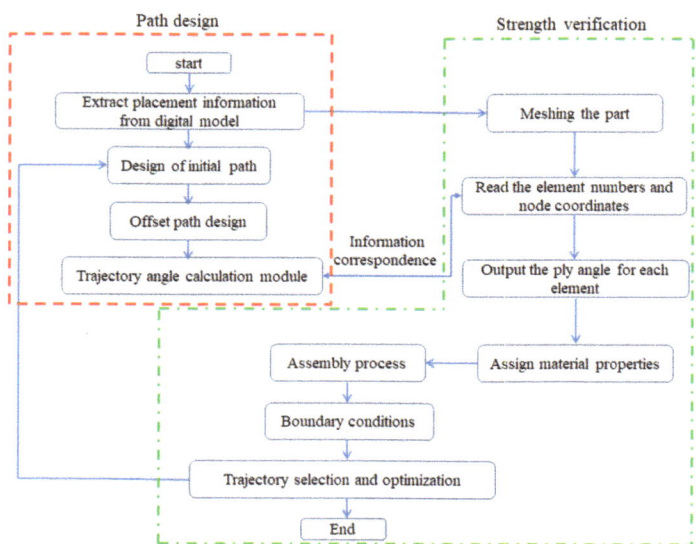

Figure 12. Flow chart of finite element analysis based on trajectory.

Table 1. Basic mechanical property parameters of materials.

Property	E_1/GPa	E_2/GPa	G_{12}/GPa	G_{13}/GPa	G_{23}/GPa	v_{12}
Values	170	8.6	4.8	4.8	3.4	0.2823

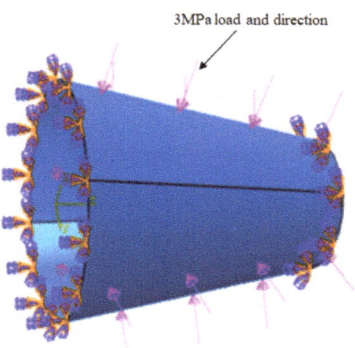

Figure 13. Loading and boundary conditions.

3.2. Simulation Results of Stress Distribution of Composite Conical Shell

The material properties are set according to the same material mechanical properties and the same layup sequence and the same load is applied to analyze the cone shells with two different path design methods (Figures 9 and 11). Table 2 shows the finite element analysis results of the cone shell with two trajectory planning methods of partition equidistant offset and circumferential averaging, which are simulated according to the steps described in Figure 12. From the Table 2, we can clearly see that the distribution of the total displacement nephogram U, fiber transverse stress nephogram S11 and fiber longitudinal stress nephogram S22 of the conical shell under loading, as well as their maximum values as shown in Table 3.

Table 2. Nephogram of displacement and stress distribution.

Fixed Angle Path	Partition Variable Angle Path	Circumferential Averaging Variable Angle Path
(U, Magnitude)	(U, Magnitude)	(U, Magnitude)
(S, S11 Envelope)	(S, S11 Envelope)	(S, S11 Envelope)
(S, S22 Envelope)	(S, S22 Envelope)	(S, S22 Envelope)

As shown in Table 2 and Table 3, different path design schemes based on variable angle affect not only the displacement of the cone shells, but also the stress distribution. Compared with fixed angle trajectory conical shell, the maximum deformation of circumferential averaging variable angle path conical shell increased by 3.01%, the fiber transverse stress decreased by 6.97%, and fiber longitudinal stress increased by 4.42%. If partition variable angle path is applied, these values can be changed to 23.02, 13.62, and 30.13%, respectively. Therefore, the comprehensive evaluation of the performance of conical shells designed by circumferential averaging variable angle trajectory design scheme is better than the partition variable angle path design scheme.

Table 3. Maximum displacement and maximum stress of different path design schemes.

Trajectory Design Scheme	U Max (mm)	S11 Max (MPa)	S22 Max (MPa)
Fixed angle trajectory	2.884	776.3	38.93
Partition variable angle trajectory	3.548	882.0	50.66
Circumferential averaging variable angle trajectory	2.971	830.4	40.65

Based on the results presented above, the possible explanation is that the size of path angle deviation and the position of the overlapping area formed by different path planning methods have a great impact on the mechanical performance. By comparing the circumferential averaging variable angle path design method (Figure 14a) with partition variable angle path design method (Figure 14b) has smaller angle deviation, and the gaps/overlaps position are more evenly distributed on the surface of the conical shell. Therefore, the organic combination of defect suppression, path planning, and strength analysis greatly ensures the rationality of trajectory design, which makes it possible to lay high-quality and efficiency, especially for complex rotating bodies.

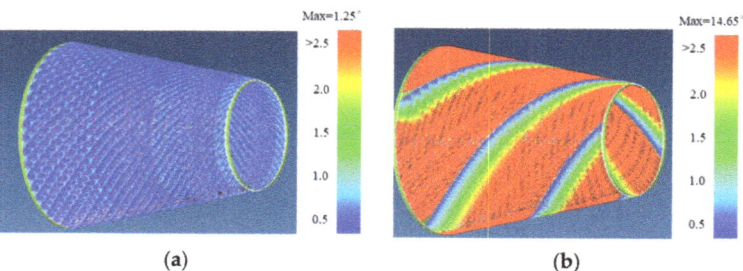

Figure 14. A 45-degree path angle deviation and gaps/overlaps distribution of different path design schemes. (**a**) Circumferential averaging paths. (**b**) Equidistant offset paths based on partition.

4. Experimental and Analysis

To verify the feasibility of the above scheme, the placement experiment of the composite conical shell described in the previous paper is completed by using the circumferential averaging paths and equipment developed independently by our research team. As shown in Figure 15, the automatic fiber placement platform includes fiber placement head, KUKA 6-DOF robot, translation guiderail and spindle, and the guiderail can move the platform in the Y direction and greatly increase the movement capacity in the axis direction of the spindle. The equipment can realize flying cutting/flying sending operation at the laying speed greater than 200 mm/s and can lay up to 8 tows with the width of a single tow is 6.35 mm, simultaneously. Moreover, the prepreg used for laying is EH104-HF40-D6/12, and its corresponding minimum turning radius was 1500 mm which was measured by experiment under the conditions of our common laying technology. Under the process conditions of laying temperature of 35 °C, laying speed of 150 mm/s, belt tension of 4 N and laying forces of 500 N, the laying experiment was carried out with the fiber placement platform.

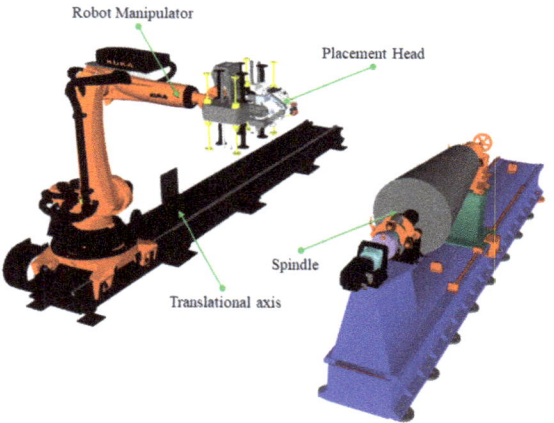

Figure 15. Robot automatic fiber placement platform.

The placement experiment results of four typical plies are shown in Figure 16. It can be seen that the tow fits well with the conical shell surface, and there are neither visible wrinkles nor gaps and overlaps in each ply. This fully shows that the path generation method has strong engineering practicability and can well ensure the quality of tow placements. Therefore, through the path design and optimization based on the laying process and the mechanical properties of the parts, it greatly meets the strength requirements of the part while ensuring the quality of the path laying, and also improves the efficiency of path planning.

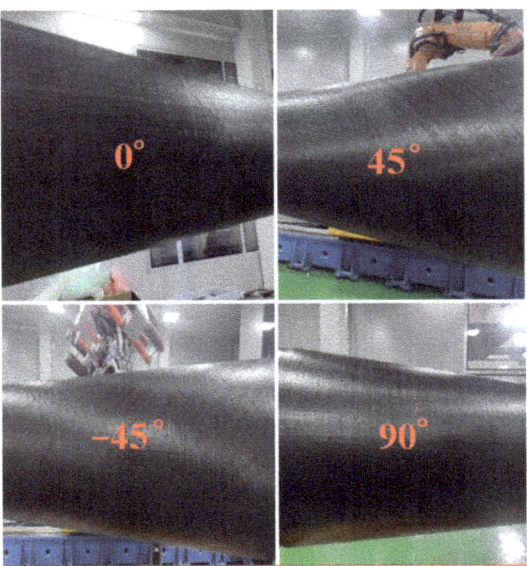

Figure 16. The placement experiment results on the conical shell surface.

In order to verify that the optimized placement path has good quality, the strength of composite components manufactured by two different path design schemes is analyzed. The experimental tests (Figure 17) were conducted in the full range of loading, until the structure's failure. In addition, numerical simulations were conducted the finite element method in ABAQUS, as shown in Figure 18. The total load is 6 MPa. The set load increased by 6/50 for each incremental step, which is reflected in the time step. Increase the time of each incremental step by 1/50 = 0.02 s. The experimental results and simulation analysis are shown in Table 4.

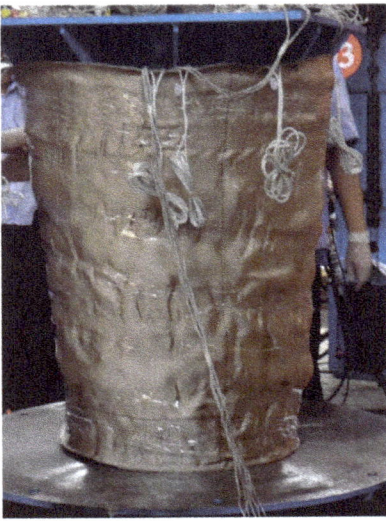

Figure 17. Mechanical properties test on the conical shell surface.

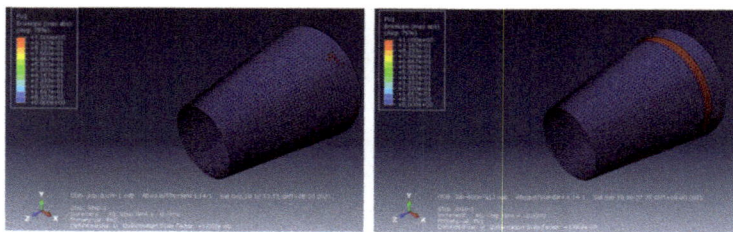

Figure 18. Mechanical properties test on the conical shell surface.

Table 4. Results of both experimental testing and numerical calculations.

Trajectory Design Scheme	Finite Element Simulation	Experimental Test
Partition variable angle trajectory	4.56 MPa	4.32 MPa
Circumferential averaging variable angle trajectory	5.52 MPa	5.38 MPa

From Table 4, we get the following conclusions. The simulation results show that the conical shell manufactured by the partition variable angle path is damaged when the load increases to 4.56 MPa, and the experimental test result is 4.32 MPa. The experimental and simulation results are basically consistent. If the circumferential uniform variation angle path is applied, these values can be changed to 5.52 and 5.38 MPa, respectively. Therefore, both experiments and simulation show that the design of circumferential uniform variation angle path has good quality, as well as proves the rationality of the trajectory design method proposed in this paper. The path planning algorithm that coordinates path planning and defect suppression play an important role in optimizing placement trajectory and improving mechanical properties of parts.

5. Conclusions

The placement path design of conical shell surfaces mainly includes the generation of initial trajectory and offset trajectory. Different path design methods have different mechanical properties of components, and they are very different. In order to optimize and select the optimal placement path, the path planning algorithm of coordinated path planning and defect suppression is studied. The conclusions can be summarized in the following point.

(1) The initial point is extremely important to the initial path and offset paths. If the initial point selection is unreasonable, on the one hand, which will lead to the failure of path generation because the occurrence of path turns back and cannot reach the boundary, on the other hand, when taking the initial point close to the boundary, the angle deviation can reach more than 20°, which cannot meet placement requirement. Therefore, it is necessary to iteratively optimize the initial point to design the initial path to make it have the best laying quality.

(2) The fixed angle algorithm meets the requirements of structural design, but it is easy to produce wrinkles. Compared with the fixed angle method, the geodesic method has an infinite curvature radius and no wrinkles will occur. However, it is very easy to cause the angle deviation to exceed its upper limit. The variable angle trajectory planning method is an ideal solution for the placement of composite conical shells at present because it comprehensively considers the angle deviation and steering radius of the path. For this cone shell, the angle deviation of the variable angle path designed by the path design software developed independently by our research group is less than 2 degrees, and the minimum steering radius of the path is greater than 2000 mm.

(3) By comparing the circumferential averaging variable angle path design method (Figure 13a), the partition variable angle path design method (Figure 14b) has smaller angle deviation, and the gaps/overlaps position are more evenly distributed on the surface of the conical shell. The possible explanation is that the size of path angle deviation and the position of the overlapping area formed by different path planning methods have a great impact on the mechanical performance. Therefore, the organic combination of defect suppression, path planning, and strength analysis greatly ensures the rationality of path design, which makes it possible to lay high-quality and efficient, especially for complex rotating bodies.

(4) According to the finite element analysis results of the cone shell with two trajectory planning methods of partition variable angle path and circumferential averaging variable angle path, it can be observed that the maximum deformation and maximum transverse and longitudinal stress of fiber of circumferential averaging variable angle path conical shell are reduced by 16.3%, 5.85%, and 19.76%, respectively, than that of the partition variable angle trajectory. Considering the manufacturability and strength of all laying paths, the design scheme based on circumferential averaging variable angle path is better than the partition variable angle path design scheme.

(5) The actual placement experiment proves that the proposed path generation method can obtain a good placement effect. This fully shows that the path generation method has strong engineering practicability and can well ensure the quality of tow placement. In addition, both experiments and Finite element simulation results further verify the rationality of path design method proposed in this paper. The path planning algorithm that coordinates path planning and defect suppression plays an important role in optimizing placement trajectory and improving mechanical properties of parts.

Author Contributions: Conceptualization, D.W.; Formal analysis, J.X. and X.W.; Methodology, D.W. and X.W.; Resources, M.D. and L.L.; Software, D.W. and X.W.; Writing—original draft, D.W.; Writing—review & editing, D.W., M.D. and X.J. All authors have read and agreed to the published version of the manuscript.

Funding: This research was funded by Research and demonstration application of key technology system of national quality infrastructure in additive manufacturing industry, grant number 2020YFF0217700 and the scientific and technological innovation project of Shenzhen, grant number CJGJZD20200617102400001.

Institutional Review Board Statement: Not applicable.

Informed Consent Statement: Not applicable.

Data Availability Statement: All data, models, or code that support the conclusions of this study are available from the corresponding author upon reasonable request.

Conflicts of Interest: The authors declare there is no conflict of interest regarding the publication of this paper.

References

1. Shirinzadeh, B.; Aized, T. Robotic fiber placement process analysis and optimization using response surface method. *Int. J. Adv. Manuf. Technol.* **2011**, *55*, 393–404.
2. Wen, L.; Xiao, J.; Wang, X. Research progress of automatic composite placement technology in China. *J. Nanjing Univ. Aeronaut.* **2015**, *47*, 637–649.
3. Asif, M.; Ahmed, I. Advanced composite material for aerospace application—A review. *Int. J. Eng. Manuf. Sci.* **2017**, *7*, 393–409.
4. Marsh, G. Automating aerospace composites production with fiber placement. *Reinf. Plast.* **2011**, *55*, 32–37.
5. Lemaire, E.; Zein, S.; Bruyneel, M. Optimization of composite structures with curved fiber trajectories. *Compos. Struct.* **2015**, *131*, 895–904. [CrossRef]
6. Zhao, C.; Xiao, J.; Li, Y.; Chu, Q.; Xu, T.; Wang, B. An Experimental Study of the Influence of in-Plane Fiber Waviness on Unidirectional Laminates Tensile Properties. *Appl. Compos. Mater.* **2017**, *24*, 1321–1337. [CrossRef]
7. Long, Y.; Chen, Z.; Shi, Y.; Mo, R. An accurate approach to roller path generation for robotic fiber placement of free-form surface composites. *Robot. Comput. Integr. Manuf.* **2014**, *30*, 277–286.

8. Schueler, K.; Miller, J.; Hale, R. Approximate geometric methods in application to the modeling of fiber placed composite structures. *J. Comput. Inf. Sci. Eng.* **2004**, *4*, 251–256. [CrossRef]
9. Duan, Y.; Ge, Y.; Xin, Z. Trajectory planning of fiber placement based on controlled angle and interval. *J. Aeronaut.* **2015**, *25*, 3475–3482.
10. Qu, W.; Gao, J.; Yang, D.; He, R.; Yang, Q.; Cheng, L.; Ke, Y. Automated fiber placement path generation method based on prospective analysis of path performance under multiple constraints. *Compos. Struct.* **2021**, *255*, 112940. [CrossRef]
11. Shirinzadeh, B.; Cassidy, G.; Oetomo, D.; Alici, G.; Ang, M. Trajectory generation for open-contoured structures in robotic fiber placement. *Robot. Comput. Integr. Manuf.* **2007**, *23*, 380–394. [CrossRef]
12. Sonmez, F.; Akbulut, M. Process optimization of tape placement for thermoplastic composites. *Compos. A* **2007**, *38*, 2013–2023. [CrossRef]
13. Brooks, T.; Martins, J. On manufacturing constraints for tow-steered composite design optimization. *Compos. Struct.* **2018**, *204*, 548–559. [CrossRef]
14. Falcóa, O.; Mayugo, J.; Lopes, C.; Gascons, N.; Turon, A.; Costa, J. Variable-stiffness composite panels: As-manufactured modeling and its influence on the failure behavior. *Compos. Part B Eng.* **2014**, *56*, 660–669. [CrossRef]
15. Gao, J.; Qu, W.; Yang, D.; Zhu, W.; Ke, Y. Two-Stage Sector Partition Path Planning Method for Automated Fiber Placement on Complex Surfaces. *Comput. Aided Des.* **2021**, *132*, 102983. [CrossRef]
16. Zhao, C.; Xiao, J.; Huang, W.; Huang, X.; Gu, S. Layup quality evaluation of fiber trajectory based on prepreg tow deformability for automated fiber placement. *J. Reinf. Plast Compos.* **2016**, *35*, 1576–1585. [CrossRef]
17. Zhang, P.; Sun, R.; Zhao, X.; Hu, L. Placement suitability criteria of composite tape for mould surface in automated tape placement. *Chin. J. Aeronaut.* **2015**, *28*, 1574–1581. [CrossRef]
18. Qu, W.; He, R.; Cheng, L.; Yang, D.; Gao, J.; Wang, H.; Yang, Q.; Ke, Y. Placement suitability analysis of automated fiber placement on curved surfaces considering the influence of prepreg tow, roller and AFP machine. *Compos. Struct.* **2021**, *262*, 113608. [CrossRef]
19. Blom, A.; Stickler, P.; Gürdal, Z. Optimization of a composite cylinder under bending by tailoring stiffness properties in circumferential direction. *Compos. Part B* **2010**, *41*, 157–165. [CrossRef]
20. Xiong, W.; Xiao, J.; Wang, X.; Li, J.; Huang, Z. Algorithm of Adaptive Path Planning for automated Placement on Meshed Surface. *Acta Aeronaut. Et Astronaut. Sin.* **2013**, *34*, 434–441.
21. Shirinzadeh, B.; Alici, G.; Foong, C.; Cassidy, G. Fabrication process of open surfaces by robotic fiber placement. *Robot. Comput. Integr. Manuf.* **2004**, *20*, 17–28. [CrossRef]
22. Zhao, C.; Wang, B.; Xiao, J. Macroscopic characterization of fibermicro-buckling and its influence on composites tensile performance. *J. Reinf. Plast. Compos.* **2016**, *36*, 196–205. [CrossRef]
23. Debski, H.; Rozylo, P.; Teter, A. Buckling and limit states of thin-walled composite columns under eccentric load. *Thin Walled Struct.* **2021**, *149*, 106627. [CrossRef]
24. Zimmermann, N.; Wang, P.H. A Review of Failure Modes and Fracture Analysis of Aircraft Composite Materials. *Eng. Fail. Anal.* **2020**, *20*, 104692. [CrossRef]
25. Rozylo, P.; Debski, H.; Wysmulski, P.; Falkowicz, K. Numerical and experimental failure analysis of thin-walled composite columns with a top-hat cross section under axial compression. *Compos. Struct.* **2018**, *204*, 207–216. [CrossRef]
26. Wu, K.; Turpin, J.; Stanford, B. Structural Performance of Advanced Composite tow-steered shells with cutouts. In Proceedings of the 55th AIAA/ASME/ASCE/AHS/ASC Structures, Structural Dynamics, and Materials Conference, National Harbor, MD, USA, 13–17 January 2014; p. 1056.
27. Hu, H.; Ou, S. Maximization of the fundamental frequencies of laminated truncated conical shells with respect to fiber orientations. *Compos. Struct.* **2001**, *52*, 265–275. [CrossRef]
28. Blom, A.; Tatting, B.; Hol, J.; Guerdal, Z. Fiber path definitions for elastically tailored conical shells. *Compos. Part B* **2009**, *40*, 77–84. [CrossRef]
29. Beakou, A.; Cano, M.; Cam, J.; Verney, V. Modelling slit tape buckling during automated prepreg manufacturing: A local approach. *Compos. Struct.* **2011**, *93*, 2628–2635. [CrossRef]
30. Duvaut, G.; Terrel, G.; Lene, F. Optimization of fiber reinforced composites. *Compos. Struct.* **2000**, *48*, 83–89. [CrossRef]
31. Ghiasi, H.; Fayazbakhsh, K.; Pasini, D.; Lessard, L. Optimum stacking sequence design of composite materials Part II: Variable stiffness design. *Compos. Struct.* **2010**, *93*, 1–13. [CrossRef]
32. Wehbe, R.; Tatting, B.; Rajan, S.; Harik, R.; Sutton, M.; Gürdal, Z. Geometrical modeling of tow wrinkles in automated fiber placement. *Compos. Struct.* **2020**, *246*, 112394. [CrossRef]
33. Shinya, H.; Teruki, I.; Narita, Y. Multi-objective optimization of curvilinear fiber shapes for laminated composite plates by using NSGA-II. *Compos. Part B Eng.* **2013**, *45*, 1071–1078.

Article

Soft-Landing Dynamic Analysis of a Manned Lunar Lander Em-Ploying Energy Absorption Materials of Carbon Nanotube Buckypaper

Qi Yuan [1], Heng Chen [2,*], Hong Nie [3], Guang Zheng [4], Chen Wang [5] and Likai Hao [6]

1. State Key Laboratory of Disaster Prevention & Mitigation of Explosion & Impact, Army Engineering University of Chinese People's Liberation Army, Nanjing 210007, China; yuanqi1900@126.com
2. College of Field Engineering, Army Engineering University of PLA, Nanjing 210007, China
3. College of Aerospace Engineering, Nanjing University of Aeronautics and Astronautics, Nanjing 210016, China; hnie@nuaa.edu.cn
4. Faculty of Mechanical Engineering and Mechanics, Ningbo University, Ningbo 315000, China; zhengguang@nbu.edu.cn
5. Key Laboratory of Exploration Mechanism of the Deep Space Planet Surface, Ministry of Industry and Information Technology, Nanjing University of Aeronautics and Astronautics, Nanjing 211100, China; nuaawangchen@nuaa.edu.cn
6. Unit 66133 of Chinese People's Liberation Army, Beijing 100000, China; hao_likai@163.com
* Correspondence: hengchen.123@nuaa.edu.cn; Tel.: +86-14751688128

Citation: Yuan, Q.; Chen, H.; Nie, H.; Zheng, G.; Wang, C.; Hao, L. Soft-Landing Dynamic Analysis of a Manned Lunar Lander Em-Ploying Energy Absorption Materials of Carbon Nanotube Buckypaper. *Materials* **2021**, *14*, 6202. https://doi.org/10.3390/ma14206202

Academic Editors: Simone Morais and Konstantinos Spyrou

Received: 4 September 2021
Accepted: 14 October 2021
Published: 19 October 2021

Publisher's Note: MDPI stays neutral with regard to jurisdictional claims in published maps and institutional affiliations.

Copyright: © 2021 by the authors. Licensee MDPI, Basel, Switzerland. This article is an open access article distributed under the terms and conditions of the Creative Commons Attribution (CC BY) license (https://creativecommons.org/licenses/by/4.0/).

Abstract: With the rapid development of the aerospace field, traditional energy absorption materials are becoming more and more inadequate and cannot meet the requirements of having a light weight, high energy absorption efficiency, and high energy absorption density. Since existing studies have shown that carbon nanotube (CNT) buckypaper is a promising candidate for energy absorption, owing to its extremely high energy absorption efficiency and remarkable mass density of energy absorption, this study explores the application of buckypaper as the landing buffer material in a manned lunar lander. Firstly, coarse-grained molecular dynamics simulations were implemented to investigate the compression stress-strain relationships of buckypapers with different densities and the effect of the compression rate within the range of the landing velocity. Then, based on a self-designed manned lunar lander, buckypapers of appropriate densities were selected to be the energy absorption materials within the landing mechanisms of the lander. For comparison, suitable aluminum honeycomb materials, the most common energy absorption materials in lunar landers, were determined for the same landing mechanisms. Afterwards, the two soft-landing multibody dynamic models are established, respectively, and their soft-landing performances under three severe landing cases are analyzed, respectively. The results depicted that the landers, respectively, adopting the two energy absorption materials well, satisfy the soft-landing performance requirements in all the cases. It is worth mentioning that the lander employing the buckypaper is proved to demonstrate a better soft-landing performance, mainly reflected in reducing the mass of the energy absorption element by 8.14 kg and lowing the maximum center-of-mass overload of the lander by 0.54 g.

Keywords: CNT buckypaper; soft-landing; multibody dynamic models; energy absorption; coarse-grained molecular dynamics simulations

1. Introduction

The moon is not only the closest celestial body to the earth, but also the only natural satellite of the earth. Due to its potential resources and unique space location, the moon has become the preferred target for humanity in carrying out deep space exploration. During the lunar exploration boom from 1958 to 1976, a total of 108 lunar probes were launched by the United States and the former Soviet Union. The most eye-catching one was the "Apollo 11" lunar manned probe, which succeeded in manned soft landing on the moon for the first time [1–4]. After years of silence, China's "Chang'e Project" also successfully realized the soft landing of unmanned lunar landers on the moon three times from 2014 to 2020 [5–7]. Soft landing means that the maximum center-of-mass acceleration of the lander is limited within several g to a dozen g during the whole landing process. The part that lands on the moon's surface is usually called the lander. For both the "Apollo" series of landers and the "Chang'e" series of landers, the landing overloads were controlled within a small range through the soft-landing mechanisms, thereby protecting the integrity of the structure and equipment and the safety of astronauts [8,9]. For all of those landers, impact energy was mainly dissipated by aluminum honeycomb element within the landing mechanisms by means of plastic collapse [10–14]. Aluminum honeycomb, a kind of porous metal material, is a new type of physical engineering material which was developed rapidly in the late 1980s. Due to its excellent physical properties and deformation characteristics of plastic collapse, aluminum honeycomb has been widely used in the fields of shock absorption and energy absorption, and the mass density of energy absorption is about 15–20 J/g [15–19]. With the sharp pace of deep space exploration, such as further manned lunar exploration, the establishment of lunar bases, the development and utilization of lunar resources, etc., the mass of the lander will be greatly increased. Traditional energy absorption materials, even including aluminum honeycomb, are becoming more and more inadequate in meeting the increasing requirements of light weight, high energy absorption efficiency, and high energy absorption density. Therefore, developing novel energy absorption materials is a matter of a great urgency.

In recent years, the randomly distributed CNT network, also called CNT buckypaper, has attracted wide attention [20–27]. In a buckypaper, due to the long range of van der Waals (vdW) interaction between carbon atoms, CNTs gradually aggregate to form bundles or entanglements [28–30]. L. Zhang and S.W. Cranford et al. [31,32] found that the Young's modulus of buckypaper could be adjusted from 0.2 GPa to 3.1 GPa by changing the density and the diameter of CNTs; L.J. Hall et al. [29] found that the Poisson's ratio of buckypaper could be adjusted between positive and negative values by changing the content of multi-walled CNTs. Studies by R.L.D. Whitby et al. [33] showed that a buckypaper usually had a very low density (0.05–0.4 g/cm^3) and high porosity (0.8–0.9); Q. Wu et al. [34] found that the pore size of a buckypaper constructed by single-walled CNTs with diameters of 0.8–1.2 nm and lengths of 100–1000 nm was about 10 nm. What is more noteworthy is that, Y. Li and M. Xu et al. [35–37] revealed the viscoelastic characteristics of buckypaper by experiments and simulations, especially the frequency independence and temperature independence in the temperature range of $-196-1000°$. Additionally, the mechanism of energy dissipation was found to be the zipping-unzipping behavior between CNTs [35,36]. Our previous work [38,39] also indicated that the buckypaper possessed a great capacity of plastic deformation; the energy absorption efficiency could reach up to 98%, the mass density of energy absorption could reach up to 90 J/g (much higher than that of aluminum honeycomb), and by increasing the density of buckypaper, the mass density of energy absorption could be further improved.

All of these results show that buckypaper possesses distinguished energy absorption characteristics and crashworthiness, which makes it a potential candidate for energy absorption materials of future lunar landers [40–42]. Hence, this paper explores the possibility of the application of buckypapers in a manned lunar lander. Firstly, coarse-grained molecular dynamics simulations are implemented to study the uniaxial compression mechanical properties of buckypaper with different densities. Then the corresponding stress-strain relationships are obtained, and the influence of compression rate within the range of landing velocity is analyzed. Secondly, for comparison, suitable buckypapers and aluminum honeycombs are, respectively, determined as the energy absorption materials within the landing mechanisms of a self-designed manned lunar lander. For convenience, the lander adopting buckypapers is called a buckypaper-buffering lander and the lander adopting aluminum honeycomb is called an aluminum honeycomb-buffering lander. Then, soft-landing dynamical models of both the two landers are established and three severe landing cases are selected. Finally, for all the three cases, the soft-landing performances of the two landers are analyzed, respectively.

2. Model and Characteristics of Buckypaper

2.1. Coase-Grained Molecular Dynamic Model

In this work, a buckypaper is built of (5,5) single-walled carbon nanotubes (SWCNTs). A coarse-grained molecular dynamics (CGMD) method is applied to establish large-scale CNT buckypapers. In the CGMD model, each CNT is simplified as a multi-bead chain. For a particular chain, stretching properties can be described by bonds between two adjacent beads, and bending properties can be described by angles among three successive beads. The interaction between different CNTs can be expressed by long-ranged (vdW) interaction between pairs of beads. The total energy of this CGMD model can be expressed as follows:

$$E_{tot} = E_{bond} + E_{angle} + E_{vdW} \tag{1}$$

where $E_{bond} = \sum_{bonds} \frac{1}{2} k_b (r - r_0)^2$ is the inter-chain stretching energy. Buehler and Cranford [43–45] have validated that r_{b0}, θ_0, and σ can be determined by equilibrium conditions, and k_b, k_a, and ε can be determined by the principle of conservation of energy. In detail, the stretching behavior of a CNT can be described by the uniaxial tension tests and then the Young's modulus E can be obtained. Thus k_b can be calculated by EA/r_0, where A is the cross-sectional area of the CNT. Bending tests can determine the bending behavior and bending stiffness EI (I is the bending moment of inertia) of the CNT and thus $k_a = 3EI/2b_0$ can be gained. Simulations of an atomistic assembly of two CNTs can be implemented to calculate the equilibrium distance D between them and their adhesive strength β, and thereby ε and σ can be acquired via $\varepsilon = \beta r_0$ and $\sigma = D/\sqrt[6]{2}$, respectively. The relevant parameters of the coarse-grained model in this work are shown in Table 1.

Table 1. Potential parameters for the coarse-grained model of (5,5) SWCNT.

Parameters	(5,5) SWCNT
Equilibrium bead distance, r_{b0} (Å)	10
Stretching constant, k_b (kcal/(mol·Å^2))	1000
Equilibrium angle, θ_0 (°)	180
Bending constant, k_a (kcal/(mol·rad^2))	14,300
vdW distance, σ (Å)	9.35
vdW energy, ε (kcal/mol)	15.10

2.2. Structure of (5,5) SWCNT Buckypaper

A random walk method is adopted to generate the structure of buckypaper to ensure the randomness and isotropy. Initially, a series of points are put into an orthogonal cell in a spatially uniform manner, serving as the positions for the initial beads of SWCNTs. Each SWCNT grows to the next position for the next bead by a random bond vector with the magnitude equal to the equilibrium bond length (10 Å). Specifically, when the distance between the newly generated bead and its nearest bead is less than a specific distance 2 Å, it is considered that the position of the newly generated bead has been occupied by other beads, so the CNT returns to the position of the previous step, re-generates a random bond vector, and generates a new bead position. The process of coincident inspection needs to be repeated until the new bead position is not occupied by other beads, and then the new bead is formed. When the random walk distance is equal to the desired length of a single CNT, the walk ends, and the current bead is the end bead of this CNT. In this work, all the CNTs in a box have the same length. After the energy minimization and equilibrium process, the equilibrium state of a buckypaper can be obtained, as shown in Figure 1. All three axes of the simulation box are set to be periodic so that a large-scaled buckypaper with continuous mass can be generated. The establishment of the CGMD model of a buckypaper is described in more detail in our published papers [38,39].

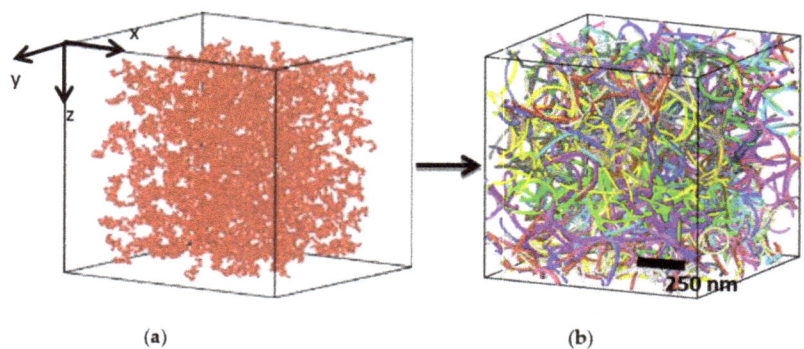

Figure 1. Establishment of a CGMD of buckypaper. (a) Random walk; (b) Buckypaper.

2.3. Compression Characteristics of Buckypaper

The uniaxial compression simulations of buckypapers with different densities are performed on the open source platform LAMMPS [46]. The deformation of the buckypaper is realized by controlling the displacement of the simulation box in the x direction, allowing the y and z directions to fluctuate. The compression ratio is $10^8/s$ and the time step is 10 fs. The compression stress σ_x in the x direction can be calculated by virial formula. The compression strain ε_x is given by $1 - L_x/L_{x0}$, where L_x and L_{x0} are the current length and initial length of the buckypaper in the x direction, respectively. Similarly, the lateral propagation strains are, respectively, defined as $\varepsilon_y = L_y/L_{y0} - 1$ and $\varepsilon_z = L_z/L_{z0} - 1$, where the subscripts y and z represent the corresponding directions, respectively. According to the initial linear stage, the Young's modulus E can be calculated by $E = -\sigma_x/\varepsilon_x$. Compression stress-strain relationships of buckypapers with six different densities are obtained by related compression simulations and demonstrated in Figure 2. It can be seen that the Young's modulus of the buckypaper increases with the increase of its density. Since the vertical landing velocity of the lunar lander is generally less than 4 m/s, in order to understand the influence of compression rate on the stress-strain relationship of the buckypaper, four different compression rates in the 0–4 m/s range (1.141 m/s, 2.282 m/s, 3.423 m/s and 4.564 m/s) are imposed on the buckypaper with a density of 107.6 kg/m^3, and the corresponding relationships between compression stress σ and compression strain ε are depicted in Figure 3. It can be seen that the compression strain rate has no obvious effect on the stress-strain relationship within the specific rate range. Therefore, the influence of the landing velocity on the relationship between cushioning force and cushioning stroke can be ignored. Additionally, for the strain less than 3%, the buckypaper is in the elastic stage and the compression stress increases rapidly with the increase of the compression strain. However, for the strain over 3%, the compression stress of the buckypaper increases slowly with the increase of strain. Generally, the effective cushioning stroke of the buckypaper can reach about 80%. It can be inferred that after the compaction of the buckypaper, the growth of the compression stress will become sharp. When the compression load is removed, the buckypaper will rebound rather slightly with a high level of non-recoverable deformation, resulting in a considerable amount of energy dissipation. The area surrounded by the compression stress-strain curve in Figure 3 is just the energy absorption E_I by the buckypaper per unit volume (that is, $E_I = \int \sigma d\varepsilon$).

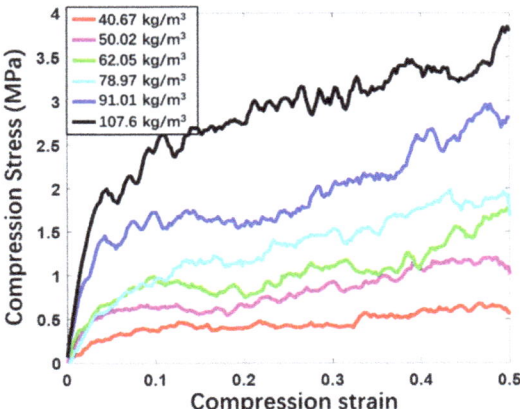

Figure 2. Compression stress as a function of compression strain for buckypapers with different densities.

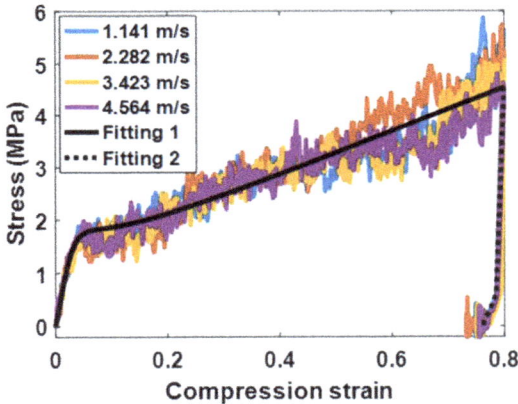

Figure 3. Compression stress as a function of compression strain for the buckypaper with density of 107.6 kg/m^3 under different compression rate.

3. Dynamic Model and Computational Methods of Soft Landing

3.1. Overall Scheme

The overall scheme of a manned lunar lander is shown in Figure 4. It consists of a main body (including an ascent module and a descent module) and four landing mechanisms (also called landing legs). The four landing mechanisms are identical and evenly arranged. In order to improve the computational efficiency, it is necessary to simplify those components that have few influences on landing performance. For example, the main body is considered to be two connected cylinders both with height of 2.5 m, maintaining characteristics of mass and moment of inertia. Each landing gear system is comprised of one primary strut, two auxiliary struts, and one footpad. Both the primary and auxiliary struts contain an outer cylinder and an inner cylinder. The primary strut is attached to the module at the upper end of the outer cylinder by a universal fitting and to the pad at the lower end of the inner cylinder by a ball fitting. Within the primary strut is a compression buffer component which is made of energy absorption materials. The auxiliary strut is attached to the module at the upper end of the outer cylinder by a universal fitting and to the primary strut at the lower end of the inner cylinder by a ball fitting. The auxiliary strut, having a bidirectional buffer capability, contains two buffer components which are made of the same energy absorption materials. Sliding between the outer cylinder and inner cylinder causes the compression of related buffer components, thus absorbing impact energy by plastic deformation. The dish-shaped footpad is allowed to rotate while sliding, preventing the lander from sinking to the soft lunar surface excessively and improving the landing stability in case the initial horizontal velocity is fairly large. The modules, primary and auxiliary struts except buffer components, and footpads can be roughly treated to be rigid.

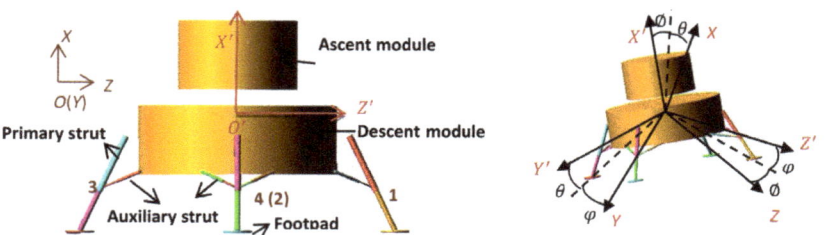

Figure 4. Overall Scheme of a manned lunar lander.

3.2. Definitions of the Coordinate Systems

As shown in Figure 4, two different right-handed Cartesian systems are employed to establish the soft landing dynamic model. They are inertial coordinate system and body coordinate system, which are defined as follows.

(1) In inertial coordinate system $OXYZ$, the origin O is located in initial barycenter of the main body and fixed to the lunar surface, X axis is direct to the opposite of lunar gravity. Z axis is perpendicular to X axis and direct to downhill. Y axis is determined by the rule of right-handed Cartesian system.

(2) In body coordinate system $O'X'Y'Z'$, the origin O' is located in barycenter of the main body. X' axis is perpendicular to the interface of the lunar lander and carrier rocket and point up. Z' axis is perpendicular to X' axis and direct to Leg 1. Y' axis is determined by the rule of right-handed Cartesian system.

Other landing legs are numbered in turn in counterclockwise direction from top view. The auxiliary struts are numbered in form of "i-j". Here, "i" is the number of the corresponding landing leg. "j" equals "1" for the left auxiliary strut and "2" for the right one from outside view.

$$T_{ib} = T_{bi}^T = \begin{bmatrix} \cos\theta\cos\varnothing & -\sin\theta\cos\varphi + \cos\theta\sin\varnothing\sin\varphi & \sin\theta\sin\varnothing + \cos\theta\sin\varnothing\cos\varphi \\ \sin\theta\cos\varnothing & \cos\theta\cos\varphi + \sin\theta\sin\varnothing\sin\varphi & -\cos\theta\sin\varphi + \sin\theta\sin\varnothing\cos\varphi \\ -\sin\varnothing & \cos\varnothing\sin\varphi & \cos\varnothing\cos\varphi \end{bmatrix} \quad (2)$$

3.3. Determination of Key Geometric Parameters

There are two important geometric parameters of landing legs. As shown in Figure 5, One is the initial height from the body's bottom to the local lunar surface, expressed by L_h. The other is the distance between the line connecting the centers of two adjacent footpads and the central axis of the body, expressed by L_v. Since the surface of the moon is strewn with large and small rocks and other bumps, the bottom of the body is necessary to keep a safe distance from the lunar surface during the whole landing process, avoiding contacting the bumps. That means L_h should not be too small. However, L_h should not be too large. Otherwise, the initial distance between the barycenter and the lunar surface, expressed by H_0, could be too large, reducing landing stability. Landing stability is related to the "stability polygon" which is defined as a regular polygon whose vertexes are the centers of four footpads. It can be roughly believed that during the landing process, if the projection of the barycenter of the lander along the direction of gravity is located inside the stability polygon, the landing process is stable, that is, the lander will not tip over. As can be seen from Figure 5, increasing L_v can increase the area of the stability polygon, so as to improve the landing stability of the lander. However, the increase of L_v also means the increase of the length of both primary and auxiliary struts, thus increasing the mass of the landing legs. Therefore, L_h and L_v affect the landing stability and the mass of the landing legs simultaneously. On the premise of ensuring landing stability, in order to reduce the mass of the lander, L_h and L_v should be as small as possible.

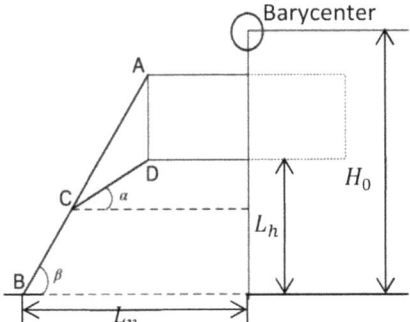

Figure 5. Compression stress as a function of compression strain for the buckypaper with density of 107.6 kg/m³ under different compression rate.

The minimum value of L_h can be estimated by the following formula:

$$L_{h,min} \approx S_h + S_0 + \Delta H \tag{3}$$

where, S_h is the displacement of the primary strut in the vertical direction, S_0 is the safe distance between the bottom of the lander and the lunar surface after landing, and ΔH is the subsidence displacement of the barycenter of the lander relative to the lunar surface during landing.

$$H_0 = L_{h,min} + L_0 \tag{4}$$

In order to ensure that the lander does not tip over during landing, the following condition must be met:

$$W_D \leq W_H \tag{5}$$

where, W_D is the kinetic energy of the lander at the moment of tipping over, which can be approximate to that at the moment of touchdown; W_H refers to the increase of potential energy of the lander with the barycenter of the lander moving from the initial location at touchdown moment to the vertical plane containing centers of any two adjacent footpads; W_D and W_H can be calculated, respectively, by the following formulas:

$$W_D = \frac{1}{2} m_{ld} \left(v_v^2 + v_h^2 \right) \tag{6}$$

$$W_H = m_{ld} g' \Delta h \tag{7}$$

where m_{ld} is the mass of the main body, set as 1.372×10^4 kg; v_v and v_h are the vertical velocity and horizontal velocity of the lander at the initial moment of touchdown respectively; g' is the gravitational acceleration on the lunar surface; and Δh refers to the increase height of the barycenter of the lander corresponding to W_H.

Assuming that the main primary strut has no buffering stroke, Δh can be estimated as follows:

$$\Delta h = \sqrt{H_0^2 + L_v^2} - H_0 \tag{8}$$

According to Formulas (3)–(6), we can get:

$$L_v \geq \frac{1}{2g'}\sqrt{(v_v^2 + v_h^2)^2 + 4g'(v_v^2 + v_h^2)H_0} \qquad (9)$$

Actually, during the process of rolling over, a buffering stroke is produced to absorb partial impact energy, and thus the kinetic energy of the lander is reduced. Therefore, this estimate is conservative.

The impact energy needed to be absorbed can be decomposed into a vertical part W_v and a horizontal part W_h. They can be obtained, respectively, by:

$$W_v = \frac{1}{2}m_{ld}v_v^2 + m_{ld}g'\Delta H \qquad (10)$$

$$W_h = \frac{1}{2}m_{ld}v_h^2 \qquad (11)$$

Since extreme landing and landing surface conditions must be taken into consideration during the design of landing mechanisms, the buffer capacity of each primary strut can be assumed as follows:

$$A_{p\_max} = a_0 W_v \qquad (12)$$

where a_0 is a constant considering landing uncertainties, and its value is generally less than 0.7. Meanwhile, it is assumed that each auxiliary strut is able to absorb impact energy in horizontal direction. Hence, the buffer capacity of each auxiliary strut can be obtained:

$$B_{s\_max} = \frac{1}{2}m_{ld}v_h^2 \qquad (13)$$

Moreover, it is necessary to consider the limitation of the maximum response acceleration of the lander while landing. Thus, we have

$$F \leq m_{ld}a_{max} \qquad (14)$$

where a_{max} is the maximum allowable response acceleration, and F is the buffering force.

According to Formulas (12) and (14), the minimum buffering stroke of the primary strut can be estimated:

$$S_{p\_min} \approx A_{p\_max}/\left(F/n_{lg}\right) \qquad (15)$$

where n_{lg} is the number of landing mechanisms and obviously equals 4 here.

Above all, considering the size, strength and soft landing requirements synthetically, the materials and related parameters of landing mechanisms are as follows. All of the outer and inner cylinders of struts are made of aluminum alloy (7055) with elastic modulus of 70 GPa, poisson's ratio of 0.3, density of 2850 kg/m^3, and yield limit of 550 MPa. The geometric parameters are illustrated in Table 2.

Table 2. Parameters of landing mechanisms.

Parameters of Landing Mechanisms		Value (mm)
	H_0	3318
	L_h	1443
	L_v	5484
Primary struts	Initial length	3506
	Outer diameter of outer cylinders	216
	Outer diameter of inner cylinders	180
	Wall thickness of outer/inner cylinders	3
	Overlap length of outer-inner cylinders	833
Auxiliary strut	Initial length	1774
	Outer diameter of outer cylinders	98
	Outer diameter of inner cylinders	50
	Wall thickness of outer/inner cylinders	3
	Overlap length of outer-inner cylinders	512

3.4. Buffering Forces

The buffering force can be expressed as a function of relative stroke between the outer and inner cylinders. The buffering force of the primary strut is obtained by:

$$F_{P,i} = \begin{cases} c_v \cdot F_{PH,i}(S_{P,i}) - f_{P,i}, & \dot{S}_{P,i} > 0 \text{ and } (S_{P,i} - S_{PR,i}) > 0 \\ -f_{P,i}, & \dot{S}_{P,i} \leq 0 \text{ and } (S_{P,i} - S_{PR,i}) \leq 0 \end{cases} \quad (i = 1,2,3,4) \qquad (16)$$

The buffering force of the auxiliary strut is obtained by:

$$F_{Sj,i} = \begin{cases} c_v \cdot F_{SCj,i}(S_{Sj,i}) - f_{Sj,i}, & \dot{S}_{Sj,i} > 0 \text{ and } (S_{Sj,i} - S_{SCRj,i}) > 0 \\ c_v \cdot F_{STj,i}(S_{Sj,i}) - f_{Sj,i}, & \dot{S}_{Sj,i} < 0 \text{ and } (S_{STRj,i} - S_{Sj,i}) > 0 \\ -f_{Sj,i}, & \text{others} \end{cases} \quad (j = 1,2) \qquad (17)$$

where i denotes the number of landing legs, values 1 and 2 of j, respectively, represent the left auxiliary strut and right auxiliary struts from the outside view, $F_{P,i}$ and $F_{Sj,i}$ are the buffering forces of the main and auxiliary struts, respectively, and c_v is the dynamic load coefficient which takes into account the effect of impact velocity. As is known, the initial impact velocity has little effect on the buckypaper's buffering force as a function of stroke, and thus c_v can be taken as 1. $S_{P,i}$ and $S_{Sj,i}$ represent the buffering stroke of the primary and auxiliary struts, respectively. $\dot{S}_{P,i}$ and $\dot{S}_{Sj,i}$ are the buffering velocity of the primary and auxiliary struts respectively, and they are defined to be positive for compression and negative for tension. $S_{PR,i}$, $S_{SCRj,i}$ and $S_{STRj,i}$ are permanent deformation of the buffer component within the primary strut, the compression buffer component within the auxiliary strut, and the tension buffer component within the auxiliary strut, respectively. $F_{PH,i}(S_{P,i})$ is the compression force as a function of compression displacement of the buffer component within the primary strut. $F_{SCj,i}(S_{Sj,i})$ and $F_{STj,i}(S_{Sj,i})$ are the forces as a function of compression displacement of the buffer component for compression and tension buffer, respectively. $f_{P,i}$ and $f_{Sj,i}$ are the sliding friction between the inner and outer cylinders of the primary and auxiliary struts, respectively.

Since the compression strain rate has no obvious effect on the stress-strain relationship within the range of landing velocity (0–4 m/s), the buffering force as a function of the buffering stroke for a buckypaper-buffer component can be easily obtained based on the compression stress-strain relationship and the physical dimension. The area surrounded by the buffering force-stroke curve is just the energy absorption by the buffer component. Buffer components should meet the limitation of the physical dimensions within the struts and the energy absorption requirements. Meanwhile, buffering force is expected to be as small as possible. By comparing the buckypapers with five densities, the densities

ρ_{bpp} = 52.81 kg/m³ and ρ_{bpf} = 91.01 kg/m³ have been selected for buffering materials of the primary struts and auxiliary struts, respectively. The physical dimensions of the buffer components are described in Table 3. Neglecting the rebound of the buckypaper, the buffering force as a function of the buffering stroke can be obtained based on the compression stress-strain relationship, shown in Figure 6. The mass of buckypaper components within one single landing mechanism can be acquired by:

$$m_{bp} = \rho_{bpp} L_{bpp} \pi D_{bpp}^2/4 + \rho_{bpf} L_{bpf} \pi \left(D_{bpf1}^2 - D_{bpf2}^2\right)/4 \qquad (18)$$

where D_{bpp} and L_{bpp} are the initial diameter and length of the buckypaper component within the primary strut, respectively; and D_{bpf1}, D_{bpf2}, and L_{bpf} are the initial outer diameter, inner diameter, and length of the buckypaper components for both tension and compression buffer. Finally, m_{bp} is equal to 1.76 kg.

Table 3. Parameters of buffer components.

Parameters of Buffer Components (mm)		Buckypaper Components	Aluminum Honeycomb Components		
Buffer components of primary struts	Diameter, D_{bpp}	204		204	204
	Length, L_{bpp}	870	Strong:	580	Weak: 290
	Maximum stroke, $S_{PR,max}$	696		450	225
Tension/compression components of auxiliary struts	Outer diameter, D_{bpf1}	94		94	
	Inner diameter, D_{bpf2}	50		50	
	Length, L_{bpf}	289		289	
	Maximum stroke, $S_{SR,max}$	231		250	

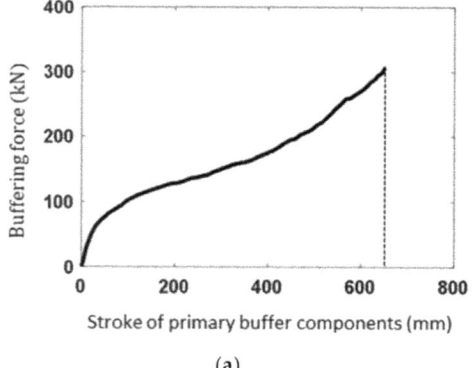
(a)

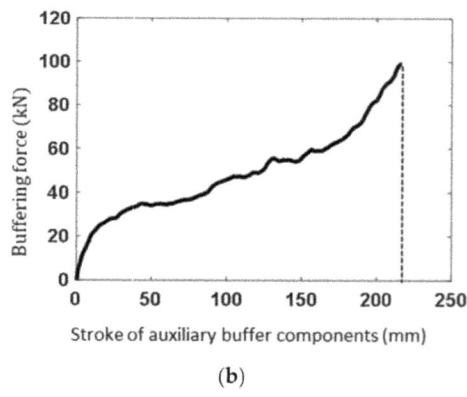
(b)

Figure 6. Buffering characteristics of buckypaper components within (a) primary and (b) auxiliary struts.

As is well known, aluminum honeycomb is the most commonly used energy absorption materials for landing mechanisms of a lunar lander. For the purpose of better understanding the buffering properties of buckypaper, aluminum honeycomb is also selected as an alternative buffering material for the lunar lander. The diagram of aluminum honeycomb and the definition of coordinate system are described in Figure 7. Aluminum honeycomb (abbreviated AH in Table 4) is an anisotropic material with parameters illustrated in Table 4, where E_{xx}, E_{yy} and E_{zz} are elasticity modulus in the corresponding directions, G_{xy}, G_{zx}, and G_{yz} are shear modulus in the corresponding directions, and ρ_{al} is the density. According to the mechanical properties of aluminum honeycomb material, the energy absorption process can be divided into three stages: elastic stage, plastic stage, and elastic stage of matrix material. The elastic stage of aluminum honeycomb material corresponds to the initial compression progress. Upon the compression load, elastic buckling occurs. Before the compression force reaches the peak, both the honeycomb and its matrix material have no plastic deformation. In the plastic stage, the aluminum honeycomb collapses under pressure and the plastic buckling deformation appears, resulting in a long platform of load. When the aluminum honeycomb is completely collapsed and compacted, it enters the elastic stage of the matrix material, causing the corresponding load rising sharply. Energy dissipation mainly depends on the plastic stage and can be obtained by the integral of compression load against compression stroke. The peak load during the elastic phase can be eliminated by applying a preload to the aluminum honeycomb. As is shown in Table 3, the aluminum honeycomb components have the same physical dimensions with the buckypaper ones, yet different available strokes. It is noted that the primary strut has two aluminum honeycomb components with different strengths. The relationship between the buffering force and the buffering stroke of the aluminum honeycomb components can be approximated as depicted in Figure 8.

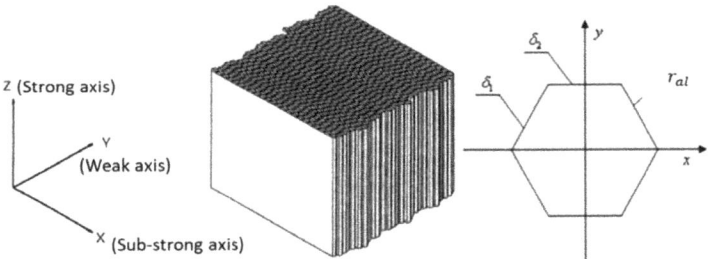

Figure 7. Diagram of aluminum honeycomb and the coordinate system.

Table 4. Parameters of aluminum honeycomb materials.

Aluminum Honeycomb Components	E_{xx} (MPa)	E_{yy} (MPa)	E_{zz} (MPa)	G_{xy} (MPa)	G_{zx} (MPa)	G_{yz} (MPa)	ρ (g/cm^3)
Primary strut-Weak AH	1.01	1.02	2.29	0.59	286	573	0.0912
Primary strut-Strong AH	3.42	3.47	4.59	2.03	429	860	0.1296
Auxiliary strut-Compression AH	3.42	3.47	6.19	2.03	0.75	860	0.1344
Auxiliary strut-Tension AH	11.54	11.70	8.88	6.85	645	1291	0.1984

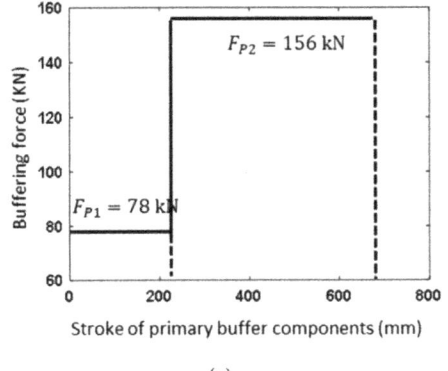

(a)

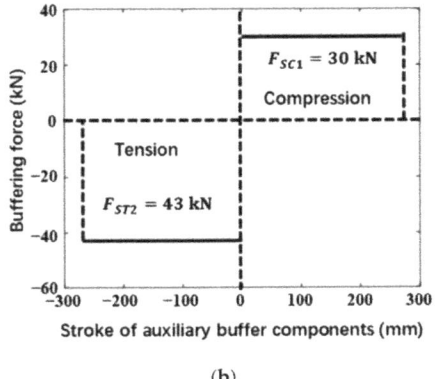

(b)

Figure 8. Buffering characteristics of aluminum honeycomb components within (a) primary and (b) auxiliary struts.

3.5. Forces upon the Main Body

According to Newton's second law, the translational equation of the main body of the lander is:

$$m_{ld} \begin{bmatrix} \ddot{X} \\ \ddot{Y} \\ \ddot{Z} \end{bmatrix} = \begin{bmatrix} F_X - m_{ld}g' \\ F_Y \\ F_Z \end{bmatrix} \quad (19)$$

where X, Y, and Z are, respectively, the components of the displacement of the barycenter of the body in the inertial coordinate system, and F_X, F_Y, and F_Z are, respectively, the components of the resultant force acting on the body induced by the landing mechanisms in the inertial coordinate system.

In the body coordinate system, the rotational Euler equation of the lander is:

$$I_{X'}\dot{\omega}_{X'} - \omega_{Y'}\omega_{Z'}(I_{Y'} - I_{Z'}) = N_{X'} I_{Y'}\dot{\omega}_{Y'} - \omega_{Z'}\omega_{X'}(I_{Z'} - I_{X'}) = N_{Y'} I_{Z'}\dot{\omega}_{Z'} - \omega_{X'}\omega_{Y'}(I_{X'} - I_{Y'}) = N_{Z'} \quad (20)$$

where $I_{X'}$, $I_{Y'}$ and $I_{Z'}$ are, respectively, the rotational inertias of the body about the X' axis, Y' axis, and Z' axes; $\omega_{X'}$, $\omega_{Y'}$ and $\omega_{Z'}$ are, respectively, the corresponding components of angular velocity vector about each axis; and $N_{X'}$, $N_{Y'}$ and $N_{Z'}$ are, respectively, the corresponding components of the moment applied by the landing mechanisms to the body.

The force applied on the body by the landing mechanisms can be expressed in the inertial coordinate system by:

$$\begin{aligned} F_X &= \sum_i (F_{PX,i} + F_{S1X,i} + F_{S2X,i}) \\ F_Y &= \sum_i (F_{PY,i} + F_{S1Y,i} + F_{S2Y,i}) \\ F_Z &= \sum_i (F_{PZ,i} + F_{S1Z,i} + F_{S2Z,i}) \end{aligned} \quad (21)$$

where $i = 1, 2, 3, 4$, denoting the sequence number of the landing mechanisms; $F_{PX,i}$, $F_{PY,i}$ and $F_{PZ,i}$ are, respectively, the components of the force $F_{P,i}$ induced by the primary strut about each axis in the inertial coordinate system; similarly, $F_{S1X,i}$, $F_{S1Y,i}$ and $F_{S1Z,i}$ are respectively the components of the force $F_{S1,i}$ induced by the auxiliary strut with the number of $i-1$; $F_{S2X,i}$, $F_{S2Y,i}$ and $F_{S2Z,i}$ are, respectively, the components of the force $F_{S2,i}$ induced by the auxiliary strut with the number of $i-2$.

The corresponding momentum can be expressed in the body coordinate system by:

$$\begin{aligned} N_{X'} &= \sum_i \left[\left(Y'_{P,i} F_{PZ',i} + Y'_{S1,i} F_{S1Z',i} + Y'_{S2,i} F_{S2Z',i} \right) - \left(Z'_{P,i} F_{PY',i} + Z'_{S1,i} F_{S1Y',i} + Z'_{S2,i} F_{S2Y',i} \right) \right] \\ N_{Y'} &= \sum_i \left[\left(Z'_{P,i} F_{PX',i} + Z'_{S1,i} F_{S1X',i} + Z'_{S2,i} F_{S2X',i} \right) - \left(X'_{P,i} F_{PZ',i} + X'_{S1,i} F_{S1Z',i} + X'_{S2,i} F_{S2Z',i} \right) \right] \\ N_{Z'} &= \sum_i \left[\left(X'_{P,i} F_{PY',i} + X'_{S1,i} F_{S1Y',i} + X'_{S2,i} F_{S2Y',i} \right) - \left(Y'_{P,i} F_{PX',i} + Y'_{S1,i} F_{S1X',i} + Y'_{S2,i} F_{S2X',i} \right) \right] \end{aligned} \quad (22)$$

where i still denotes the sequence number of the landing mechanisms; $F_{PX',i}$, $F_{PY',i}$, and $F_{PZ',i}$ are, respectively, the components of $F_{P,i}$ about each axis of the body coordinate system; $F_{S1X',i}$, $F_{S1Y',i}$, $F_{S1Z',i}$, $F_{S2X',i}$, $F_{S2Y',i}$ and $F_{S2Z',i}$ are, respectively, the components of $F_{S1,i}$ and $F_{S2,i}$ about each axis of the body coordinate system. $X'_{P,i}$, $Y'_{P,i}$ and $Z'_{P,i}$ are the coordinates of the connection point between the primary strut and the body of the lander in the body coordinate system; $X'_{S1,i}$, $Y'_{S1,i}$ and $Z'_{S1,i}$ are the coordinates of the connection point between the auxiliary strut with the number of $i-1$ and the main body in the body coordinate system; and $X'_{S2,i}$, $Y'_{S2,i}$ and $Z'_{S2,i}$ are the coordinates of the connection point between the auxiliary strut with the number of $i-2$ and the main body in the body coordinate system.

3.6. Contact Force between Footpads and the Lunar Surface

The real interaction between the footpads and the lunar surface is rather complicated. In multi-body dynamics, it can be simplified to be a normal force and a tangential force. For the normal force, a nonlinear spring damping model is used to simulate the impact of the footpads on the lunar surface. The tangential force includes the following two parts. One is the frictional force caused by the slipping between the footpads and the lunar surface. And the other is the bulldozing force applied on the side of the footpads by the sinking of the footpads. This frictional force is modeled by coulomb friction and the bulldozing force is a function depending on the depth and velocity of the sinking of the footpads. Hence, the normal force and the tangential force can be calculated by:

$$F_N = K_N \delta_L^{e_N} + C_N \dot{\delta}_L \tag{23}$$

$$F_T = \mu_T F_N + K_T \delta_L^{e_T} \dot{\delta}_L \tag{24}$$

where K_N, e_N and C_N are respectively the stiffness, nonlinear exponent and damping coefficient, all related about the sinking depth δ_L; K_T and e_T are bulldozing coefficients related to δ_L; μ_T is the friction coefficient between the footpads and the lunar surface, which is set as 0.4.

3.7. Initial Landing Conditions and Landing Cases

All the components of the lander except the energy absorption components are approximately regarded to be rigid. The soft landing dynamics simulations of the lander are implemented based on the popular commercial software of MSC. Adams. According to current references [3,4], the initial landing conditions of a lunar lander is summarized as Table 5. To analyze the soft-landing dynamic properties and validate the design reasonability, three severe landing cases are selected based on amounts of simulations, including the high-overloading case (Case 1), the easily overturning case (Case 2) and the long-stroke case (Case 3). According to the reference [4], they are depicted in Table 6 in detail and exhibited in Figure 9.

Table 5. Initial landing conditions.

Requirements	Initial Value
Vertical touchdown velocity(m/s)	−4~0
Horizontal touchdown velicty(m/s)	−1~1
Landing surface slope(deg)	−8~8
Yaw angle(deg)	−45~45
Pitch angle(deg)	−4~4
Depth of concave(mm)	0~200

Table 6. Landing cases.

Requirements	Case 1	Case 2	Case 3
Vertical touchdown velocity V_x (m/s)	−4	−4	−4
Horizontal touchdown velicty V_y (m/s)	0	0	0
Horizontal touchdown velicty V_z (m/s)	1	−1	1
Landing surface slope α (deg)	0	−8	−8
Yaw angle φ (deg)	45	45	0
Pitch angle $\varnothing$ (deg)	0	4	4
Depth of concave d_c (mm)	0	0	200 (Footpads 2, 4)

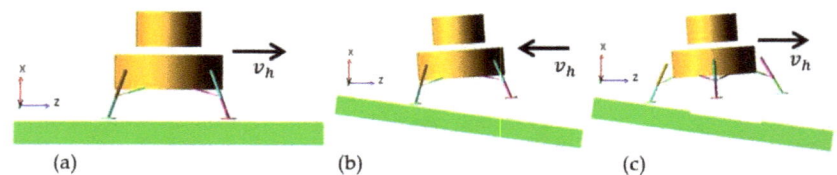

Figure 9. Diagrams of three severe landing case. (**a**) High-overloading case; (**b**) Easily overturning case; (**c**) Long-stroke case.

4. Simulation Results and Discussions

4.1. Overload Response of the Mass Center

In this section, the high-overloading case (Case 1) corresponding to the limitation of the center-of-mass overload is mainly focused on. For better analyzing the overload response, the center-of-mass velocity response is firstly observed. Considering the overload bearing capacity of the astronauts, the center-of-mass acceleration a of the manned lunar lander is generally required to be less than 5 g, that is, $a \leq a_{max} = 5$ g. As is represented in Figure 10, for the two landers with different buffering materials, both the velocities in the X-axis direction (vertical velocity) increase slightly during the short free-fall stage of the initial 0.05 s. After the footpads touching the lunar surface, it decreases to almost zero within 0.2 s. The velocities in Z-axis direction for the two landers are not quite the same. The difference is that the buckypaper-buffering lander needs 1.24 s to mitigate this horizontal velocity completely while the aluminum honeycomb-buffering lander needs 1.3 s. Additionally, the velocity in Y-axis direction maintains zero during the whole landing progress. Above all, the buckypaper-buffering lander reaches the static state earlier than the aluminum-buffering one by 0.06 s.

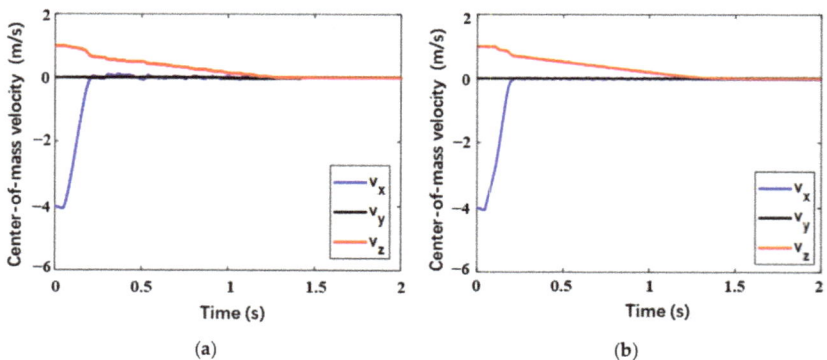

Figure 10. Center-of-mass velocity of lunar lander in Case 1 within (**a**) buckypaper and (**b**) aluminum honeycomb.

In this load case, corresponding to the two buffering materials, the center-of-mass accelerations of the two landers as a function of time are described in Figure 11. It is necessary to declare that all the acceleration responses adopt low passing filter of 80 MHz. It can be seen that the two acceleration responses are analogous. During the initial 0.05 s, the lander is in free fall, therefore, the center-of-mass acceleration is just the local gravitational acceleration of the lunar surface. When the footpads touch the lunar surface, the acceleration in the X-axis direction will rapidly decrease to zero and then points to the opposite direction, followed by a quick increase and a subsequent decrease in the module. Due to the influence of the horizontal velocity, a low acceleration in the Z-axis also appears. In the Y-axis direction, the acceleration is almost zero. It can be seen form Figure 10 that the high overload mainly occurs on the X-axis. The maximum overload of the buckypaper-buffering lander is 3.46 g while that of the aluminum honeycomb-buffering lander is 4.00 g. During the buffering progress, a stepped leapfrog phenomenon appears in the overload response of the aluminum honeycomb-buffering lander. It is due to the fact that there are two stages of aluminum honeycomb components with different strengths in the primary strut, and the strong one begins to work only when the weak one is compacted. However, for the buckypaper-buffering lander, the overload response is relatively smooth. Moreover, the high overload the lander suffers mainly occurs within the initial 0.2 s. Over the period of 0.2 s to 1.3 s, the rest slight accelerations in both X and Y directions will decrease to zero gradually.

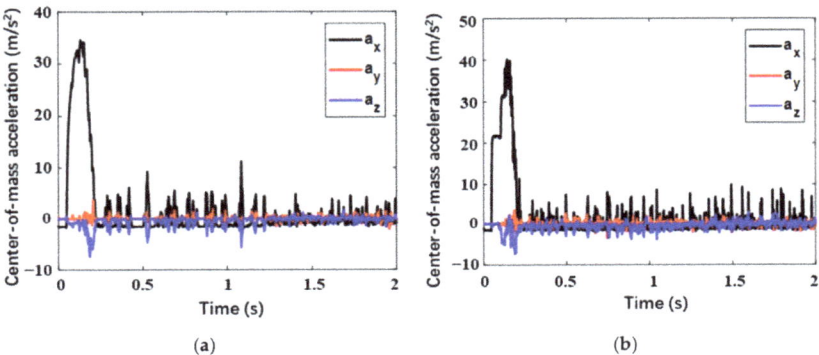

Figure 11. Center-of-mass acceleration of lunar lander in Case 1 within (**a**) buckypaper and (**b**) aluminum honeycomb.

In addition, the important results about the center-of-mass velocity and overload responses for the three cases are demonstrated in Table 7. For both the two landers, the maximum center-of-mass accelerationss are lower than the maximum allowable value of 5 g. Both in Case 1 and Case 2, it spends shorter time finishing the landing process for the buckypaper-buffering lander than the aluminum honeycomb-buffering one, and the overload of the former is lower than that of the latter. However, the conclusions are contrary in Case 3.

Table 7. Important data about the center-of-mass overload responses.

Cases	Landers	Time for the Whole Landing (s)	Maximum Overload (g)
Case 1	Buckypaper	1.24	3.46
	Aluminum honeycomb	1.30	4.00
Case 2	Buckypaper	0.68	2.66
	Aluminum honeycomb	0.92	2.76
Case 3	Buckypaper	3.60	2.71
	Aluminum honeycomb	3.21	2.62

4.2. Overturning Resistance Capability

During the whole landing process, in order to prevent the lander from overturning, the center of mass of the lander should not exceed the "stability walls" which are formed by the vertical planes extended from the "stability polygon". This is to say that the center of mass of the lander should always lies within the space enclosed by the "stability walls". Therefore, it is necessary to monitor the distances between the center of mass of the lander and the four stability walls, respectively. Suppose the centers of all the footpads are B_i ($i = 1, 2, 3, 4$, corresponding to the sequence number of the landing mechanism), the center of mass of the lander is B_0, the projection of the center of mass on the "stability wall ij" which involves centers of two adjacent footpads B_i and B_j is B_0^{ij}, the distance vector from B_0 to B_j is $\boldsymbol{B}_{0j}$, the distance vector from B_i to B_j is $\boldsymbol{B}_{ij}$, and the gravity vector of the lunar is $\boldsymbol{g}'$, the distance B_{0_ij} from the center of mass to the stability wall involving B_i and B_j, also called "stability distance ij", can be calculated by:

$$B_{0\_ij} = \frac{\boldsymbol{B}_{0j} \cdot (\boldsymbol{g}' \times \boldsymbol{B}_{ij})}{|\boldsymbol{g}' \times \boldsymbol{B}_{ij}|} ij = 12, 23, 34, 41 \qquad (25)$$

If $B_{0\_ij} > 0$, the center of mass is located within the space enclosed by the stability walls. If $B_{0\_ij} < 0$, the center of mass is located outside the stability walls. If $B_{0\_ij} = 0$, the center of mass is just located on the stability walls.

In this section, the easily overturning case (Case 2) corresponding to the limitation of the stability distance is mainly focused on. The four stability distances are, respectively, recorded in Figure 12. It can be seen that the simulation results about the stability distances are semblable for the two landers and the landing process has a great influence on stability distance 23 and stability distance 41. During the initial landing stage, only the upslope Leg 2 and Leg 3 touch the lunar soil, and thus Leg 2 and Leg 3 begin to buffer. Due to the influence of horizontal velocity, the stability distance 23 decreases first and then increases until the whole landing is finished. Yet, the stability distance 41 keeps decreasing until Leg 1 and Leg 4 touch the lunar surface, followed by a slight increase as a result of the lateral slipping of the footpads of Leg 1 and Leg 4. Besides, the stability distance 12 and stability distance 34 are nearly identical all the time due to the symmetry of the lander about the xz plane and change slightly over time during the landing process. It also can be seen from the figure that the minimum stability distance is stability distance 41. The corresponding value is 3386 mm and 3349 mm for the buckypaper-buffering lander and aluminum honeycomb-buffering, respectively, indicating a better stability for the former to some extent.

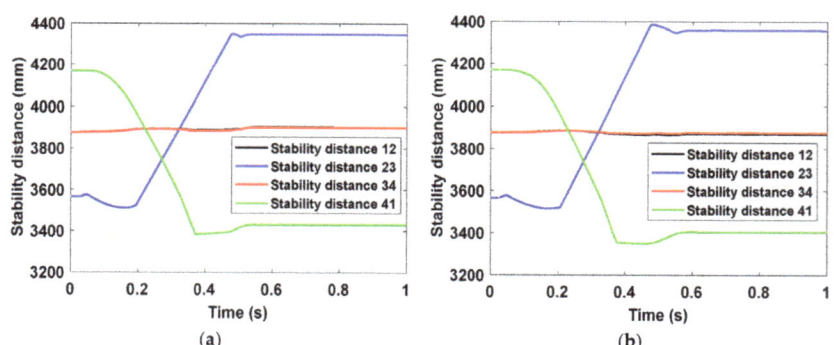

Figure 12. Distances from the center of mass to the stability walls in Case 2 within (**a**) buckypaper and (**b**) aluminum honeycomb.

In addition, the minimum stability distances of the two landers in the three severe cases are listed in Table 8. It can be found that the minimum stability distance of the buckypaper-buffering lander is larger than that of the aluminum honeycomb-buffering lander in both Case 2 and Case 3, while the former is slightly smaller than the latter in Case 1. In addition, considering that if the distance between the bottom of the lander and the lunar surface (also called "bottom-lunar distance" for short) is too small, the bottom structure of the lander is likely to collide with rocks on the lunar surface while landing, probably causing turnover of the lander. Hence, bottom-lunar distance is required to be not less than a certain value D_{cr}, and herein $D_{cr} = 300$ mm. The bottom-lunar distance versus time during the whole landing process can be recorded, and then the minimum values for the two landers in three cases are also summarized in Table 8. It can be found that for the two landers the bottom-lunar distances are adequate in all the cases. In both Case 1 and Case 3, the minimum values for the two landers are quite close, and in Case 2 that of the buckypaper-buffering lander is slightly smaller than the that of the other lander.

Table 8. Important data about key distances related to overturning resistance capability.

Cases	Landers	Minimum Stability Distance (mm)	Minimum Bottom-Lunar Distance (mm)
Case 1	Buckypaper	3789	692
	Aluminum honeycomb	3798	695
Case 2	Buckypaper	3386	694.8
	Aluminum honeycomb	3349	721.5
Case 3	Buckypaper	3384	794.6
	Aluminum honeycomb	3336	794.1

4.3. Buffering Stroke

While landing, the landing mechanisms absorb the impact energy by compressing the buffer components within them, resulting in buffering strokes of the primary struts and the auxiliary struts. The buffering strokes of the primary strut and auxiliary strut, denoted by $S_{P,i}$ and $S_{Sj,i}$ respectively, should not be larger than the designed maximum strokes, denoted $S_{P,max}$ and $S_{S,max}$, respectively. Herein, for aluminum honeycomb buffers, $S_{P,i} \leq S_{P,max} = 675$ mm, and $S_{Sj,i} \leq S_{S,max} = 270$ mm. For buckypaper buffers, $S_{P,i} \leq S_{P,max} = 696$ mm, and $S_{Sj,i} \leq S_{S,max} = 231$ mm. In this section, the long-stroke case (Case 3) is mainly focused on.

In this section, the long stroke case (Case 3) corresponding to the limitation of the buffering stroke is mainly focused on. Figures 13 and 14, respectively, describe the buffering strokes versus time for the primary struts and auxiliary struts. Generally, the change tendencies and laws are approximate for the two landers. Primary Strut 2 and Primary Strut 4 always have the same buffering strokes due to their symmetry. At the initial stage, compression buffering firstly occurs in Primary Strut 3 of upslope Leg 3, and simultaneously the corresponding tension buffering appears in Auxiliary Strut 5 and Auxiliary Strut 6. Due to the concaves below Leg 2 and Leg 3, Leg 1, located downhill, becomes the second landing leg to touch the lunar surface and buffers, quickly followed by Legs 2 and Leg 4. Thereinto, Primary Strut 1 has the largest compression stroke, followed by Primary Sturt 3, and Primary Strut 2 and Primary Strut 4 have the minimum compression strokes with the same value. Besides, Auxiliary Strut 1 and Auxiliary Strut 2 have the largest tension strokes with the same value, followed by Auxiliary Strut 3 and Auxiliary Strut 8 with the second-largest ones, and Auxiliary Strut 1 and Auxiliary Strut 2 with the minimum ones. Especially, a second tension buffering happens on Auxiliary Strut 5 and Auxiliary Strut 6 in Leg 3, during which the tension length is only 1–2 mm owing to the plastic deformation of buffering components. Above all, the maximum compression strokes of the primary struts for the buckypaper-buffering lander and the aluminum honeycomb-buffering lander are, respectively, 471.3 mm and 526.9 mm, the former smaller than the latter by 55.6 mm. The largest tension strokes of the auxiliary struts for the buckypaper-buffering lander and the aluminum honeycomb-buffering lander are, respectively, 115.7 mm and 108.0 mm, the latter slightly larger than the latter. Besides, it can be found that all the auxiliary struts hardly have any compression stroke.

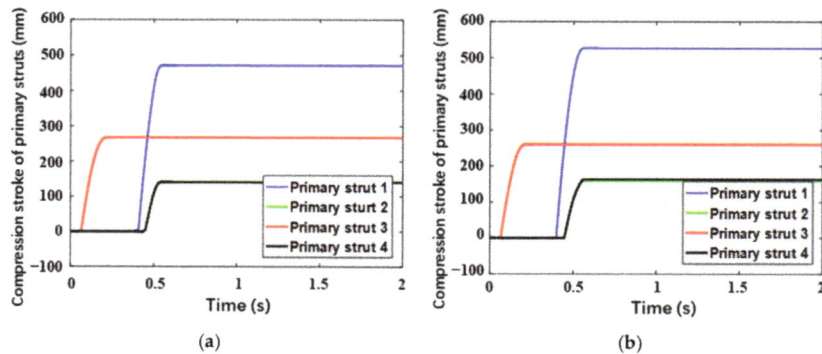

Figure 13. Compression stroke of primary struts versus time for (**a**) buckypaper-buffering lander and (**b**) aluminum honeycomb-buffering lander.

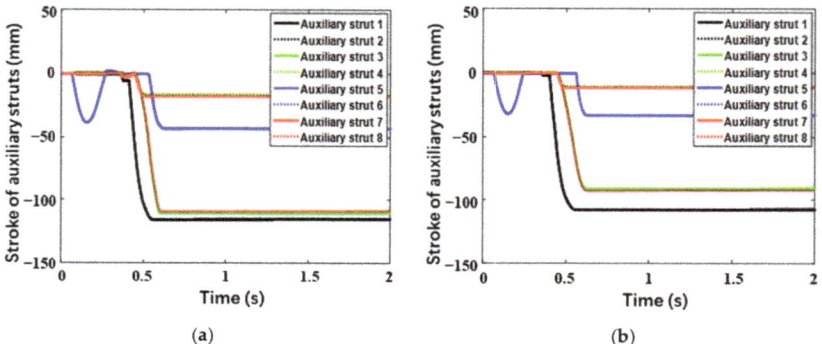

Figure 14. Stroke of auxiliary struts versus time for (**a**) buckypaper-buffering lander and (**b**) aluminum honeycomb-buffering lander.

In addition, Table 9 concludes the maximum buffering stoke of the primary struts and auxiliary struts, respectively, for the two landers in the three cases. It can be observed that all the maximum practical buffering strokes can meet the design requirements. What's more, for each case, the maximum buffering stroke of the primary struts for the buckypaper-buffering lander is always shorter than that of the aluminum honeycomb-buffering lander. Yet, for the maximum buffering strokes of the auxiliary struts, there is little difference between the two landers, only 7.7 mm for the maximum difference.

Table 9. Maximum buffering stroke of the primary struts and auxiliary struts in three cases.

Cases	Landers	Maximum Stoke of Primary Struts (mm)	Maximum Stoke of Auxiliary Struts (mm)
Case 1	Buckypaper	280.5	112.9
	Aluminum honeycomb	289.1	118.6
Case 2	Buckypaper	314.6	110.5
	Aluminum honeycomb	332.6	109.4
Case 3	Buckypaper	471.3	115.7
	Aluminum honeycomb	526.9	108.0

5. Conclusions

In this work, a coarse-grained molecular dynamics method is used to implement a series of simulations about uniaxial compression of buckypapers with different densities. The results indicate that Young's modulus of a buckypaper increases with its density, accompanied with the improvement of its energy absorption capacity. Besides, the compression rate within the range of the landing velocity has little influence on the stress-strain relationship, and thus the influence of the landing velocity on the relationship between the buffering force and buffering stroke can be ignored. That is, the buffering force as a function of the buffering stroke can be substituted by the quasi-static compression stress as a function of compression strain. Two energy absorption materials, buckypaper and aluminum honeycomb, were, respectively, selected for the parametric design of two landing mechanisms, thus, respectively, establishing two soft-landing multibody dynamics models of two landers with a difference only in their energy absorption materials. Then a series of simulations are implemented to explore the landing properties of the two landers under three severe cases. It is found that both of the two landers are able to land on the lunar surface smoothly and safely under all the three cases. In the high-overloading case, during the whole landing process, the highest center-of-mass overloads of the buckypaper-buffering lander and the aluminum honeycomb-buffering lander are, respectively, 3.46 g and 4.00 g, both of which are lower than the allowable value 5 g. What is important is that, compared with the aluminum honeycomb components, the buckypaper components cutdown the highest overload by 0.54 g, which can improve the comfort of astronauts and reduce the loads induced on related structure and devices. In the easily overturning case, the barycenters of both the two landers are always located within the range of the stability walls and keep far away from the walls. The minimum stability distances are 3386 mm and 3349 mm, respectively, for the buckypaper-buffering lander and aluminum honeycomb-buffering lander. Meanwhile the bottoms of the two landers always remain adequate distance from the lunar surface, 471.3 mm, and 526.9 mm, respectively, for the buckypaper-buffering lander and aluminum honeycomb-buffering lander. In the long-stroke case, the maximum compression strokes of the primary struts for both the two landers are within their required stroke ranges, 471.3 mm and 526.9 mm, respectively, for the buckypaper-buffering lander and aluminum honeycomb-buffering lander. The above differences of the results between the two landers are probably due to the difference between the curves of buffering force versus buffering stroke. Remarkably, compared with the aluminum honeycomb components, the buckypaper can reduce the total mass of the buffering components by 52.71%. In other words, the mass of the lander can be decreased by 8.14 kg, indicating less fuel and less cost to a great extent. This is owing to the buckypaper's much lower density than the aluminum honeycomb. In summary, compared to the aluminum honeycomb-buffering lander, the buckypaper-buffering lander has not only better landing performance, but also much lighter mass. In spite of the fact that all of these findings are obtained in silico, this study can shed lights on the superiority of buckypapers in the aspect of crashworthiness while applied in a soft-landing mechanism of a manned lunar lander and is believe that our results will stimulate more relevant practical work.

Author Contributions: Conceptualization, H.C. and Q.Y.; methodology, H.C. and H.N; software, H.C. and G.Z.; validation, H.C., Q.Y., and L.H.; formal analysis, H.C.; investigation, H.C.; resources, H.N.; data curation, Q.Y.; writing—original draft preparation, Q.Y.; writing—review and editing, H.C and H N.; visualization, C.W.; supervision, Q.Y.; project administration, H.C.; funding acquisition, H.C. All authors have read and agreed to the published version of the manuscript.

Funding: This study was supported by the Young Scientists Fund of the National Natural Science Foundation of China (Grant No. 11902157) and the Basic research Program of Jiangsu Province (Natural Science Foundation, Grant No. BK20180417).

Institutional Review Board Statement: Not applicable.

Informed Consent Statement: Not applicable.

Data Availability Statement: The data used to support the findings of this study are included within the article.

Conflicts of Interest: The authors declare no conflict of interest.

References

1. Brooks, G.C.; Grimwood, J.M.; Swenson, L.S. *Chariots for Apollo: The NASA History of Manned Lunar Spacecraft to 1969*; Dover Publications: New York, NY, USA, 2012.
2. Orloff, R.W.; Harland, D.M. Apollo 11: The fifth manned mission: The first lunar landing 16–24 July 1969. In *Apollo: The Definitive Sourcebook*; Springer: New York, NY, USA, 2006; pp. 279–325.
3. Reeves, R. *Russian Robots on the Moon//The Superpower Space Race*; Springer: Boston, MA, USA, 1994; pp. 131–166.
4. Young, A. *Lunar and Planetary Rover: The Wheels of Apollo and the Quest for Mars*; Springer Science Business Media: Berlin/Heidelberg, Germany, 2007.
5. Yang, Y.Z.; Ping, J.S.; Yan, J.G.; Li, J.L. Influence of the layered Moon and Earth's orientation on lunar rotation. *Res. Astron. Astrophys.* **2020**, *20*, 019. [CrossRef]
6. Uhalley, S., Jr. China's Aerospace Prowess Today and Tomorrow. *Am. J. Chin. Stud.* **2018**, *25*, 63–79.
7. Liu, H. An overview of the space robotics progress in China. *System (ConeXpress ORS)* **2014**, *14*, 15.
8. Ji, S.; Liang, S. DEM-FEM-MBD coupling analysis of landing process of lunar lander considering landing mode and buffering mechanism. *Adv. Space Res.* **2021**, *68*, 1627–1643. [CrossRef]
9. Sun, Z. *Ground Test Verification Technology//Technologies for Deep Space Exploration*; Beijing Institute of Technology Press: Beijing, China, 2021; pp. 569–617.
10. Wang, C.; Nie, H.; Chen, J.; Lee, H.P. The design and dynamic analysis of a lunar lander with semi-active control. *Acta Astronaut.* **2019**, *157*, 145–156. [CrossRef]
11. Ramakrishna, S. Microstructural design of composite materials for crashworthy structural applications. *Mater. Des.* **1997**, *18*, 167–173. [CrossRef]
12. Verlinden, B. Severe plastic deformation of metals. *MJoM* **2005**, *11*, 165–182. [CrossRef]
13. Banhart, J. Manufacture, characterisation and application of cellular metals and metal foams. *Prog. Mater. Sci.* **2001**, *46*, 559–632. [CrossRef]
14. Ashby, M.F.; Evans, T.; Fleck, N.A.; Hutchinson, J.; Wadley, H.; Gibson, L. *Metal Foams: A Design Guide A Design Guide*; Elsevier: Amsterdam, The Netherlands, 2000.
15. Zhao, H.; Gary, G. Crushing Behaviour of aluminium honeycombs under impact loading. *Int. J. Impact Eng.* **1998**, *21*, 827–836. [CrossRef]
16. Ruan, D.; Lu, G.; Wang, B.; Yu, T. In-plane dynamic crushing of honeycombs—A finite element study. *Int. J. Impact Eng.* **2003**, *28*, 161–182. [CrossRef]
17. Chen, J.B.; Nie, H.; Zhao, J.C.; Bai, H.M.; Bo, W. Research of the Factors of Buffering Performance in Lunar Lander. *J. Astronaut.* **2008**, *6*, 010.
18. Nemat-Nasser, S.; Kang, W.; McGee, J.; Guo, W.-G.; Isaacs, J. Experimental investigation of energy-absorption characteristics of components of sandwich structures. *Int. J. Impact Eng.* **2007**, *34*, 1119–1146. [CrossRef]
19. Hazizan, M.A.; Cantwell, W.J. The low velocity impact response of an aluminium honeycomb sandwich structure. *Compos. Part B Eng.* **2003**, *34*, 679–687. [CrossRef]
20. Chen, X.; Huang, Y. Nanomechanics modeling and simulation of carbon nanotubes. *J. Eng. Mech.* **2008**, *134*, 211–216. [CrossRef]
21. Delfani, M.R.; Shodja, H.M.; Ojaghnezhad, F. Mechanics and morphology of single-walled carbon nanotubes: From graphene to the elastica. *Philos. Mag.* **2013**, *93*, 2057–2088. [CrossRef]
22. Salvetat, J.P.; Bonard, J.M.; Thomson, N.H.; Kulik, A.J.; Forró, L.; Benoit, W.; Zuppiroli, L. Mechanical properties of carbon nanotubes. *Appl. Phys. A* **1999**, *69*, 255–260. [CrossRef]
23. Wang, C.M.; Zhang, Y.Y.; Xiang, Y.; Reddy, J.N. Recent Studies on Buckling of Carbon Nanotubes. *Appl. Mech. Rev.* **2010**, *63*, 030804. [CrossRef]
24. Becton, M.; Zhang, L.; Wang, X. Molecular Dynamics Study of Programmable Nanoporous Graphene. *J. Nanomech. Micromech.* **2014**, *4*, B4014002. [CrossRef]
25. Zhang, Z.; Wang, X.; Li, J. Simulation of collisions between buckyballs and graphene sheets. *Int. J. Smart Nano Mater.* **2012**, *3*, 14–22. [CrossRef]
26. Zhang, L.; Wang, X. Atomistic insights into the nanohelix of hydrogenated graphene: Formation, characterization and application. *Phys. Chem. Chem. Phys.* **2014**, *16*, 2981–2988. [CrossRef]
27. Zhang, L.; Wang, X. Tailoring Pull-out Properties of Single-Walled Carbon Nanotube Bundles by Varying Binding Structures through Molecular Dynamics Simulation. *J. Chem. Theory Comput.* **2014**, *10*, 3200–3206. [CrossRef] [PubMed]
28. Rinzler, A.G.; Liu, J.; Dai, H.; Nikolaev, P.; Huffman, C.B.; Rodríguez-Macías, F.J.; Boul, P.J.; Lu, A.H.; Heymann, D.; Colbert, D.T.; et al. Large-scale purification of single-wall carbon nanotubes: Process, product, and characterization. *Appl. Phys. A* **1998**, *67*, 29–37. [CrossRef]
29. Hall, L.J.; Coluci, V.R.; Galvao, D.S.; Kozlov, M.E.; Zhang, M.; Dantas, S.O.; Baughman, R.H. Sign change of Poisson's ratio for carbon nanotube sheets. *Science* **2008**, *320*, 504–507. [CrossRef]

30. Thess, A.; Lee, R.; Nikolaev, P.; Dai, H.; Pierre, P.; Robert, J.; Xu, C.; Lee, Y.H.; Kim, S.G.; Rinzler, A.G.; et al. Crystalline ropes of metallic carbon nanotubes. *Sci. AAAS-Wkly. Pap. Ed.* **1996**, *273*, 483–487. [CrossRef]
31. Zhang, L.; Zhang, G.; Liu, C.; Fan, S. High-density carbon nanotube buckypapers with superior transport and mechanical properties. *Nano Lett.* **2012**, *12*, 4848–4852. [CrossRef]
32. Cranford, S.W.; Buehler, M.J. In silico assembly and nanomechanical characterization of carbon nanotube buckypaper. *Nanotechnology* **2010**, *21*, 265706. [CrossRef]
33. Whitby, R.L.D.; Fukuda, T.; Maekawa, T.; James, S.L.; Mikhalovsky, S.V. Geometric control and tuneable pore size distribution of buckypaper and buckydiscs. *Carbon* **2008**, *46*, 949–956. [CrossRef]
34. Wu, Q.; Zhu, W.; Zhang, C.; Liang, Z.; Wang, B. Study of fire retardant behavior of carbon nanotube membranes and carbon nanofiber paper in carbon fiber reinforced epoxy composites. *Carbon* **2010**, *48*, 1799–1806. [CrossRef]
35. Li, Y.; Kroeger, M. Viscoelasticity of carbon nanotube buckypaper: Zipping-unzipping mechanism and entanglement effects. *Soft Matter* **2012**, *8*, 7822–7830. [CrossRef]
36. Xu, M.; Futaba, D.N.; Yamada, T.; Yumura, M.; Hata, K. Carbon nanotubes with temperature-invariant viscoelasticity from–196 to 1000 C. *Science* **2010**, *330*, 1364–1368. [CrossRef]
37. Yang, X.; He, P.; Gao, H. Modeling frequency-and temperature-invariant dissipative behaviors of randomly entangled carbon nanotube networks under cyclic loading. *Nano Res.* **2011**, *4*, 1191–1198. [CrossRef]
38. Chen, H.; Zhang, L.; Chen, J.; Becton, M.; Wang, X.; Nie, H. Effect of CNT length and structural density on viscoelasticity of buckypaper: A coarse-grained molecular dynamics study. *Carbon* **2016**, *109*, 19–29. [CrossRef]
39. Chen, H.; Zhang, L.; Chen, J.; Becton, M.; Wang, X.; Nie, H. Energy dissipation capability and impact response of carbon nanotube buckypaper: A coarse-grained molecular dynamics study. *Carbon* **2016**, *103*, 242–254. [CrossRef]
40. Romo-Herrera, J.M.; Terrones, M.; Terrones, H.; Meunier, V. Guiding electrical current in nanotube circuits using structural defects: A step forward in nanoelectronics. *ACS Nano* **2008**, *2*, 2585–2591. [CrossRef]
41. Li, Y.; Qiu, X.; Yang, F.; Yin, Y.; Fan, Q. Stretching-dominated deformation mechanism in a super square carbon nanotube network. *Carbon* **2009**, *47*, 812–819. [CrossRef]
42. Zsoldos, I.; Laszlo, I. Computation of the loading diagram and the tensile strength of carbon nanotube networks. *Carbon* **2009**, *47*, 1327–1334. [CrossRef]
43. Buehler, M.J. Mesoscale modeling of mechanics of carbon nanotubes: Self-assembly, self-folding, and fracture. *J. Mater. Res.* **2006**, *22*, 2855–2869. [CrossRef]
44. Cranford, S.; Yao, H.; Ortiz, C.; Buehler, M.J. A single degree of freedom 'lollipop' model for carbon nanotube bundle formation. *J. Mech. Phys. Solids* **2010**, *58*, 409–427. [CrossRef]
45. Wang, C.; Buehler Cranford Buehler Cranford Wang, L.; Xu, Z. Enhanced mechanical properties of carbon nanotube networks by mobile and discrete binders. *Carbon* **2013**, *64*, 237–244. [CrossRef]
46. Plimpton, S. Fast Parallel Algorithms for Short-Range Molecular Dynamics. *J. Comput. Phys.* **1995**, *117*, 1–19. [CrossRef]

Article

Titanium Dioxide Nanotubes as Solid-Phase Extraction Adsorbent for the Determination of Copper in Natural Water Samples

Bochra Bejaoui Kefi [1,2,*], Imen Bouchmila [1], Patrick Martin [3,*] and Naceur M'Hamdi [4]

[1] Laboratory of Useful Materials, National Institute of Research and Pysico-Chemical Analysis (INRAP), Technopark of Sidi Thabet, Ariana 2020, Tunisia; bouchmila.imen.ic@gmail.com
[2] Department of Chemistry, Faculty of Sciences of Bizerte, University of Carthage, Zarzouna 7021, Tunisia
[3] Transformations & Agroressources Unity, UR7519, Université d'Artois-UniLaSalle, F-62408 Bethune, France
[4] Research Laboratory of Ecosystems & Aquatic Resources, National Agronomic Institute of Tunisia, Carthage University, Tunis 1082, Tunisia; naceur_mhamdi@yahoo.fr
* Correspondence: bochrabej@yahoo.fr (B.B.K.); patrick.martin@univ-artois.fr (P.M.)

Abstract: To increase the sensitivity of the analysis method of good copper sample preparation is essential. In this context, an analytical method was developed for sensitive determination of Cu (II) in environmental water samples by using TiO_2 nanotubes as a solid-phase extraction absorbent (SPE). Factors affecting the extraction efficiency including the type, volume, concentration, and flow rate of the elution solvent, the mass of the adsorbent, and the volume, pH, and flow rate of the sample were evaluated and optimized. TiO_2 nanotubes exhibited their good enrichment capacity for Cu (II) (~98%). Under optimal conditions, the method of the analysis showed good linearity in the range of 0–22 mg L^{-1} ($R^2 > 0.99$), satisfactory repeatability (relative standard deviation: RSD was 3.16, $n = 5$), and a detection limit of about 32.5 ng mL^{-1}. The proposed method was applied to real water samples, and the achieved recoveries were above 95%, showing minimal matrix effect and the robustness of the optimized SPE method.

Keywords: atomic absorption spectroscopy; copper; hydrothermal treatment; solid phase extraction; TiO_2 nanotubes

1. Introduction

Inorganic contaminants are generally found in trace and ultra-trace levels to a complex matrix. To analyze them, preliminary treatment of samples containing very complex matrix is thus necessary. Indeed, this step is crucial in the analytical process since it affects the quality of the analysis in terms of sensitivity, precision, and repeatability. Several pretreatment methods have been developed, such as liquid–liquid extraction, supercritical phase extraction, microwave extraction, and solid-phase extraction (SPE) [1–9].

Recently, many researchers preferred the SPE method because of its advantages to the enrichment, high recovery, simplicity, rapid, and low organic solvent consumption. SPE is selective, efficient, clean, and can be automated or even used online with several analysis techniques [10–16]. This explains the constant development of this technique through the search for new adsorbents. The most used adsorbents, such as commercial activated carbon and ion-exchange resins [17–20], are characterized by their high production and regeneration costs, which has developed the research toward other adsorbents such as metal-organic frameworks (MOFs), nanocomposites and nanomaterials [10–12,21–31]. The nanometric size of these structures leads to an increase in the proportion of atoms present on their surface, thus allowing a high reactivity and an interesting adsorption capacity [32,33]. The most used of these nanomaterials are carbon nanotubes [10–12,22,24,34,35]. Nevertheless, nanostructured titanium dioxide proves to be a possible alternative to the rather high cost of these nanomaterials since its elaboration is easy to implement and inexpensive. Several morphologies of nanomaterials have been tested, among them the nanotubular one which has an open mesoporous morphology that gives them the properties of adsorbents [36,37]. The typical value of the total volume of specific pores of nanotubes is in the range of 0.60 to 0.85 cm^3 g^{-1}. They have a large specific surface area compared to the starting materials (~50 m^2 g^{-1}) and whose value determined by Brunauer-Emmett-Teller [BET] method varies from 100 to 400 m^2 g^{-1} [38–42]. The values of the total volume and specific surface area depend on the average diameter of the nanotubes. Among these nanotubular structures, titanium dioxide nanotubes can be used for the same applications as those known for the precursor TiO$_2$ before the transformation. Their new morphology is also accompanied by new properties allowing particular and specific applications in catalysis [41–44], medical applications [45,46], environmental applications, and extraction processes of various organic and inorganic contaminants: pesticides, polycyclic aromatic hydrocarbons (PAHs), phthalates, dyes, and heavy metals [47–50].

Among the inorganic contaminants, we are interested in the element copper. Copper is a multipurpose metal, when present in low doses, it is a trace element essential to life, but in higher doses, it is toxic [51–55]. As it is not biodegradable, it can accumulate and eventually reach high levels [53,55]. Its origin in water is very diverse, in addition to the natural contents coming mainly from the deposits, its use by man causes important discharges in the environment [55]. Research in recent years has been based on the identification of new processes of analysis and even good methods for removing copper from water [22,26,27,34].

For Cu^{2+} analysis, atomic absorption spectroscopy (AAS) and inductively coupled plasma are the most widely used [4]. Other more sensitive and selective methods are also used. Qi et al. applied sensor-based methods for rapid detection of Cu^{2+} in water [56].

Therefore, in this work, a solid phase adsorbent, titanium dioxide nanotubes, was synthesized to develop a solid phase extraction (SPE) method for preconcentration/extraction of Cu (II) from natural water.

2. Materials and Methods

2.1. Experimental

Chemical Reagents and Solutions

The TiO$_2$ nanotubes were elaborated using a precursor of titanium dioxide (P25, Degussa-Hüls A.G. (Evonik, Germany), 68% anatase and 32% rutile, SBET = 50 m^2 g^{-1}, non-porous and pHPZC = 5.6), a concentrated solution of caustic soda (Fisher Chemicals, Ohio, USA, Purity: 98.64%) and a diluted solution of hydrochloric acid (Panreac Quimica S.A., Barcelona, Spain, Purity: 37%). The nitric acid was purchased from Scharlau Chemie S.A., (Barcelona, Spain, purity 65%) and the water used throughout the elaboration is ultra-pure purchased from Milli-Q, Millipore (Merck Millipore, MA, USA), 18 MΩ cm. The hydrated copper nitrate Cu(NO$_3$)2 × 3H$_2$O was purchased from the company Panreac Quimica S.A. (Barcelona, Spain) with 99% purity. The different Cu (II) solutions are prepared by dissolving adequate amounts of hydrated copper nitrate in ultra-pure water.

2.2. Apparatus

The copper was determined by a flame atomic absorption spectrometer model Analytik Jena "novAA 400." (Jena, Germany). The pressure and flow rate of the nebulizer were 3 bars and 0.45 L min^{-1}, respectively. The pump speed was 22 rpm and the generator power was maintained at 40.68 MHz. The studied element was identified at the wavelength of 324.754 nm.

2.2.1. Synthesis of Titanium Dioxide Nanotubes

The alkaline hydrothermal method, as described in a previous study [38,39,49] was used to prepare the titanate nanotubes. A commercial TiO$_2$ (P25, 0.50 g) was dispersed in a 15 mL of NaOH solution of 11.25 mol L^{-1} and introduced into a Teflon-lined autoclave with an 80% filling factor. The autoclave was then heated at 130 °C for 20 h to prepare the hydrogenated nanotubes (HNT) samples. This later was washed with one liter of hot ultra-pure water then filtered under vacuum to eliminate the excess of unreacted soda. To exchange the Na$^+$ ions contained in the structure by protons, we proceeded to the neutralization of the product by a solution of hydrochloric acid (0.1 mol L^{-1}). The precipitate was then washed with 0.5 L of hot ultra-pure water to remove NaCl formed in excess. The obtained wet solids were dried in an oven at 80 °C for 24 hr. Finally, HNT was calcined at 500 °C for two hours in an air atmosphere to obtain titanium dioxide nanotubes (TON). The latter were used as adsorbents for Cu (II) removal from water samples.

2.2.2. Characterization of Nanotubes

The structural study of HNT and TON was determined at room temperature with X-ray diffraction (XRD) analyses using an automated "X'Pert PRO MPD, PANalytical Co., Almelo, The Netherlands" diffractometer X-ray diffraction (XRD). Monochromatic Cu Kα-radiation (λ = 1.5418 Å) was obtained with a Ni-filtration and a system of diverging and receiving slides of 0.5° and 0.1 mm, respectively. The diffraction pattern was measured with a voltage of 40 kV and a current of 30 mA over a 2θ range of 3–40° using a step size of 0.02° at a scan speed of 1 s per step.

For a porous solid, the term texture mainly refers to the specific surface area (S$_{BET}$, m^2 g^{-1}), specific pore volume (V$_p$, cm^3 g^{-1}), porosity ε, pore shape, and pore distribution or distribution of pore volumes as a function of pore size. The specific surface areas of HNT and TON, pore volume, and average pore diameters were determined by the BET method and calculated from the nitrogen adsorption-desorption measurements at 77 K using the "Micro metrics ASAP 2000" volumetric apparatus.

Morphology and structure of titanium dioxide nanotubes were characterized by transmission electron microscopy (TEM), purchased from Philadelphia (PA, USA) with an acceleration current of about 200 kV and by high-resolution transmission electron microscopy (HR-TEM) of the "JEOL-2010" type (the acceleration current is 400 kV).

2.2.3. pH of Points Zero Charges (pHpzc)

The pH of zero charge point (pHpzc) of the studied material is an important parameter in adsorption phenomena that depends on electrostatic forces. It gives information on the charge of dominant sites on the solid surface. Indeed, the surface is positively charged at pH values below pHpzc and negatively charged at pH values above pHpzc. To measure the pHpzc values, the salt method was applied [57]. The titration cell is filled with 100 mL of electrolyte NaOH (0.01 mol L^{-1})/NaCl (0.1 mol L^{-1}) and 1 g of the studied material (HNT and TON). The material suspension was equilibrated for 24 h, and then the titration was carried out with 0.1 mol L^{-1} of HCl while measuring the pH for each volume added. Blank titration is done without studied material in the same way as in the presence of the solid with the same concentration of electrolytes.

2.2.4. SPE Process

Empty solid-phase extraction cartridges (0.2 g, 3 mL, polypropylene) are filled with TON. This adsorbent is held in place by the polypropylene upper and lower frits. The packed cartridge was then placed on a vacuum elution apparatus. This solid phase was first conditioned with 10 mL HNO$_3$ (2 M) and 10 mL of ultra-pure water. After that, a volume of ultra-pure water or sample water varying from 100 to 500 mL and spiked with Cu (II) at a concentration of 2.62 mg L^{-1} is percolated through the cartridge at a flow rate to be determined during this work. The washing of the adsorbent, with 10 mL of ultra-pure water is performed only in the case of real water samples. Finally, to ensure complete desorption of Cu (II) three elution solvents of volume 10 mL were tested namely nitric acid, hydrochloric acid, and pure ethanol.

Factors influencing this method (the elution solvent, volume, and concentration, the volume of the sample, the mass of the adsorbent, the pH of the sample, elution, and percolation flow rates) were thus optimized. The extraction yield of Cu (II) cations is calculated according to the following Equation (1):

$$R\% = \frac{C_f - C_i}{C_0} \cdot 100 \quad (1)$$

with:

C_f = concentration of Cu (II) obtained from the spiked sample;
C_i = concentration of Cu (II) obtained from the unspiked sample;
C_0 = concentration of the Cu (II) added to the sample: spiking level: 2.62 mg L^{-1}.

3. Results and Discussion

3.1. Material Characterizations

The change in the crystal structure from HNT to calcined TON was studied by XRD (Figure 1). The comparison of the HNT XRD pattern with the ASTM sheet corresponding to TiO$_2$ (P25), showed that the elaborated nanotubes did not correspond to either anatase or rutile. HNT showed an orthorhombic system with the lattice constants a_0 = 1.926 nm, b_0 = 0.378 nm, and c_0 = 0.300 nm [39,58] and according to the ASTM sheet N° 47–0124, its structure corresponded well to that of hydrogenated nanotubes of type H$_2$Ti$_2$O$_5$ × H$_2$O. Upon calcination of HNT at 500 °C, the XRD pattern of TON showed characteristic peaks of anatase phase TiO$_2$ and additional peaks, which corresponded to H$_{1.2}$Na$_{0.8}$O$_7$Ti$_3$ crystallizing in the monoclinic system.

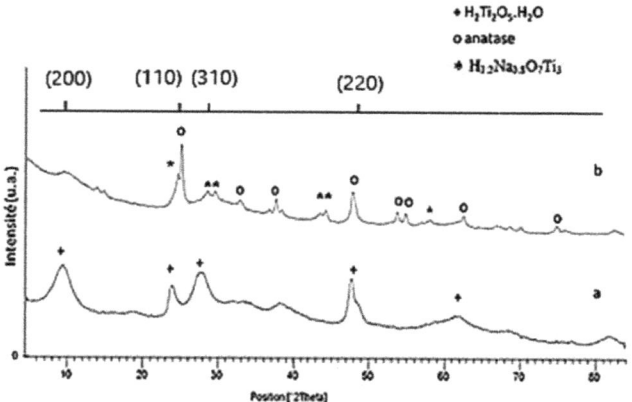

Figure 1. XRD of orthorhombic H$_2$Ti$_2$O$_5$ phase (**a**) and calcined TiO$_2$ nanotubes (TON) (**b**).

The specific surface area (S$_{BET}$), pore volume (V$_p$), as well as an average diameter (d$_p$) of nanotubes before and after calcination are summarized in Table 1. When TiO$_2$ particles (P25) are transformed into hydrogenated nanotubes, a considerable increase in the specific surface area from 50 to 269 m^2 g^{-1} was observed. Indeed, the multi-walled structure of the elaborated nanotubes HNT gives them a high specific surface since nitrogen molecules can intercalate in the interfoliar spaces. When calcined, the specific surface of nanotubes decreased, and the pore diameter increased (from 9 nm for T = 130 °C to 23 nm for T = 500 °C). The decrease in S$_{BET}$ after calcination can be explained by: (i) The disappearance of the multi-walled structure of the nanotubes after calcination, thus the measured S$_{BET}$ corresponds only to the inner and outer surface of the nanotubes; (ii) the decrease of the number of nanotubes after calcination.

Table 1. Textural characteristics of elaborated nanotubes before and after calcination.

T (°C)	S_{BET} (m² g⁻¹)	V_p (cm³ g⁻¹)	d_p (nm)
130	269	0.67	9
500	~100	0.63	23

S_{BET}: specific surface, V_p: porous volume, d_p: average pore diameter.

The analysis of the TEM and HR-TEM images of the HNT nanotubes (Figure 2), shows their hollow and homogeneous nanotubular structure. The outer diameters of tubes were 6–8 nm and inner diameters were 4–6 nm, and the lengths were measured to be several hundred nanometers. The HR-TEM analysis showed that the walls of the nanotubes were amorphous (Figure 2b).

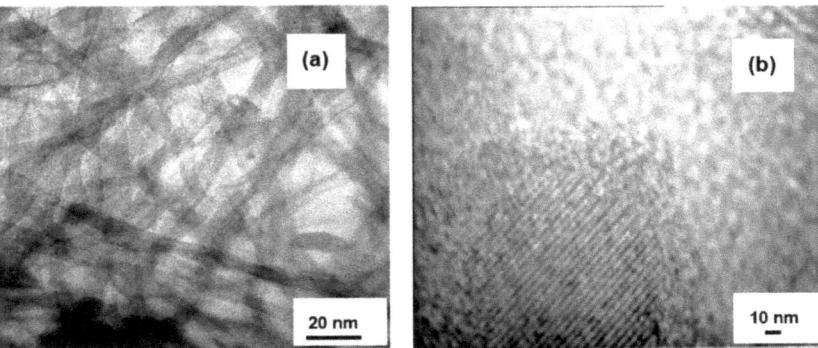

Figure 2. Images of HNT nanotubes with (a) TEM and (b) HR-TEM.

After calcination, the outer and inner diameters of nanotubes decreased to 7–5 nm (Figure 3a). However, HR-TEM analysis was consistent with the XRD results and showed nanotubes walls transformed into crystalline anatase (Figure 3b).

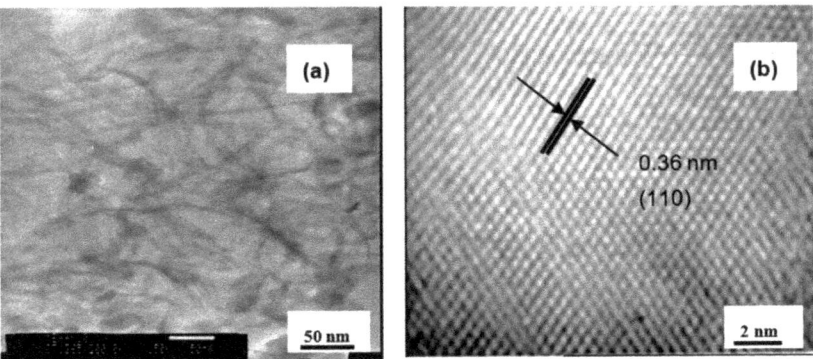

Figure 3. Images of TON nanotubes with (a) TEM and (b) HR-TEM.

The anatase TiO_2 nanotubes (TON) were used as adsorbents for the SPE of Cu (II) from water samples. Factors that may influence the extraction yields of the copper ions namely the type, volume, concentration, and flow rate of elution solvents, were studied. Likewise, the volume, pH, and flow rate of percolated samples and the effect of the mass of the adsorbent were optimized.

3.2. Optimization of SPE Method

3.2.1. Effect of Elution Solvent Type and Volume

The influences of three eluents (nitric acid, hydrochloric acid, and ethanol) on the recoveries of Cu ions from TiO_2 nanotubes were examined (Figure 4).

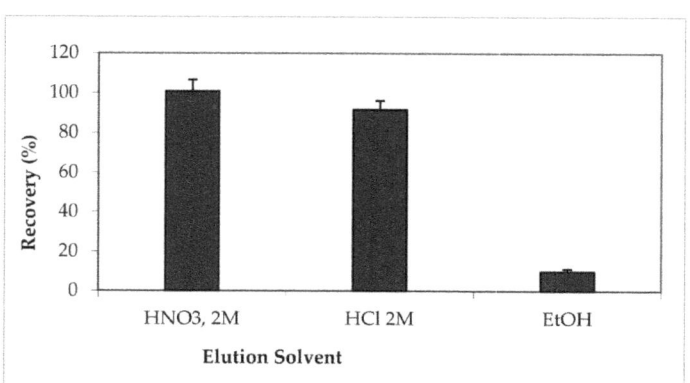

Figure 4. Effect of the elution solvent type on the extraction yield of Cu (II). Solvent volume = 10 mL.

They are the most used solvents for solid-phase extraction of heavy metals [12,25,34,47]. As can be seen in Figure 4, the extraction yield of Cu cations is strongly related to the nature of the elution solvent. The maximum elution is obtained with HNO_3 (2M) [34], therefore it was chosen as elution solvent for the next steps of this optimization. This result is in agreement with that presented by Soylak et al. [34], who studied the extraction of Cu (II) by carbon nanotubes.

The effect of the volume of HNO_3 on the recovery of the Cu (II) was investigated in the range 2–14 mL maintaining its concentration at 2M. The results are shown in Figure 5.

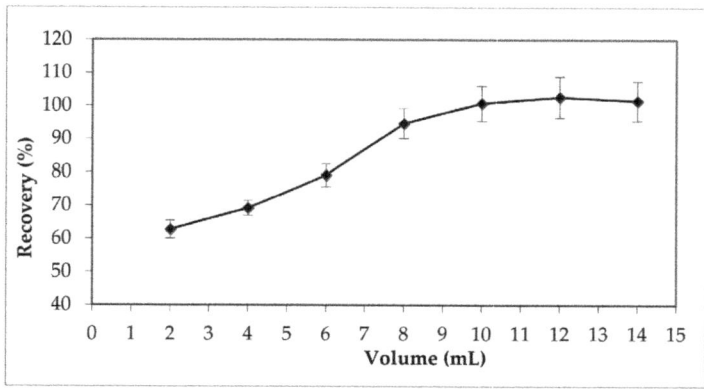

Figure 5. Effect of HNO_3 volume on Cu (II) extraction.

As can be seen, the volume of HNO_3 had a significant effect ($p < 0.001$) on the efficiency of the SPE method, and on the extraction yield of Cu (II) cations. Extraction yield increases progressively with the volume of HNO_3 until reaching a maximum of 10 mL. To ensure a maximum elution of Cu (II) cations, a volume of 12 mL of HNO_3 was used for the next steps of the optimization.

3.2.2. Effect of Eluent Concentration and Flow Rate

Several factors affecting the elution efficiency of analytes, such as eluent concentration, and flow rate were also studied. HNO_3 concentrations between 0.5 and 4 mol L^{-1} were examined, and their effects on the extraction efficiency of Cu (II) cations are shown in Figure 6.

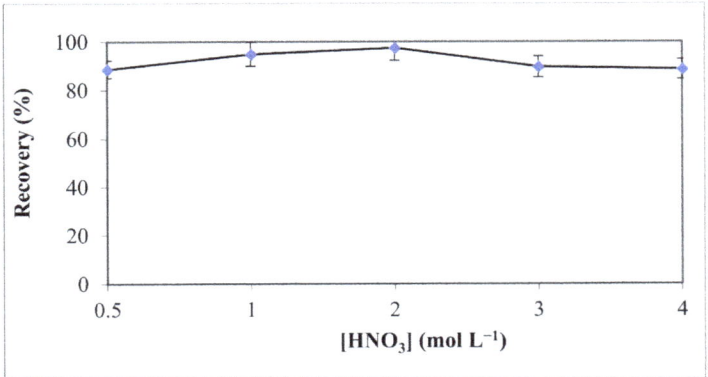

Figure 6. Effect of HNO_3 concentration on Cu (II) extraction (solvent volume = 12 mL).

Obtained results showed insignificant differences in recoveries among the different concentrations of HNO_3. Therefore, the acidic solvent concentration of 2 mol L^{-1} was chosen for the extraction of Cu cations.

Considering the importance of the elution or desorption step in the SPE process, several elution flow rates from 1 to 8 mL min^{-1} were also tested (Figure 7).

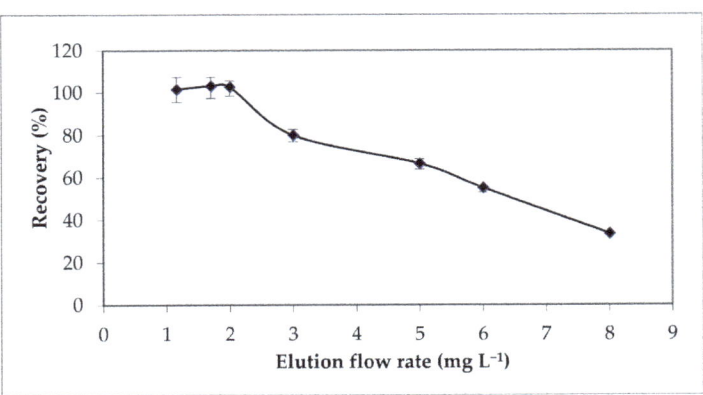

Figure 7. Effect of elution flow rate on the extraction yield of Cu (II). Solvent volume = 12 mL; $[HNO_3]$ = 2 M.

Results show a decrease of the extraction yields when the elution flow rate exceeds 2 mL min^{-1}.

3.2.3. Effect of Sample Volume and Flow Rate

The breakthrough volume in the solid phase extraction was investigated. Different volumes from 100 to 500 mL of ultra-pure water spiked with 2.62 mg L^{-1} of Cu (II) was percolated through the cartridge. Results presented in Figure 8 show a decrease in yields as the percolated volume increased.

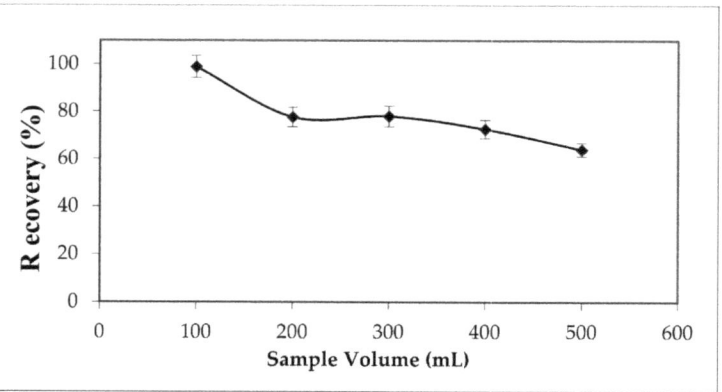

Figure 8. Effect of sample volume on the extraction yield of Cu (II). Solvent volume = 12 mL; [HNO$_3$] = 2 M; the elution flow rate = 2 mL min^{-1}.

Therefore, a sample volume of 100 mL was recommended to obtain the maximum extraction yield.

To ensure better repeatability of results, the flow rate of the sample percolation must be controlled. For this purpose, percolation flow rates ranging from 1 to 8 mL min^{-1} were tested, with the other conditions kept constant. Results are given in Figure 9 and they illustrate maximum extraction yield when the sample was percolated at a flow rate lower than 2 mL min^{-1}.

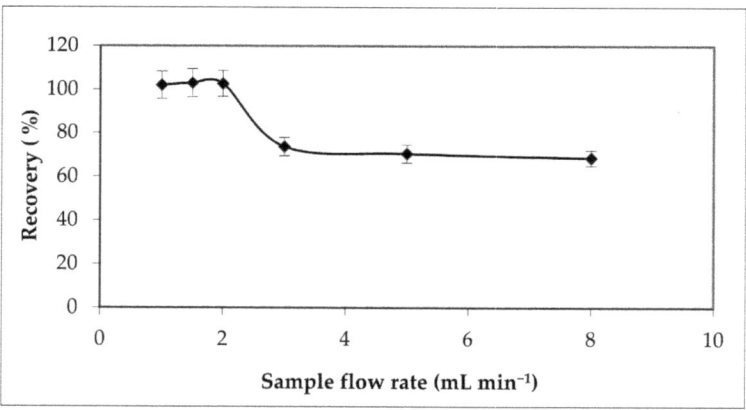

Figure 9. Effect of percolation flow rate on Cu (II) extraction yield. Solvent volume = 12 mL; [HNO$_3$] = 2 mol L^{-1}; the elution flow rate = 2 mL min^{-1}; sample volume = 100 mL.

3.2.4. Effect of Sample pH

The pH is an important factor in the SPE procedure [57,59]. It determines the surface charge of the TiO$_2$ nanotubes, on the one hand, and on the other hand the form of copper in solution. Indeed, the pH values of precipitation of Cu (II) hydroxide, correspond to initial concentrations between 2.62 and 1500 mg L^{-1} and varies from 6.89 to 5.80 respectively. The values of pH at the point of zero charge pHpzc are frequently used to determine the sorption properties of oxides and hydroxides [60]. Results from titration curves (Figure 10) showed pHpzc of 6.7 and 8.3 for HNT and TON, respectively.

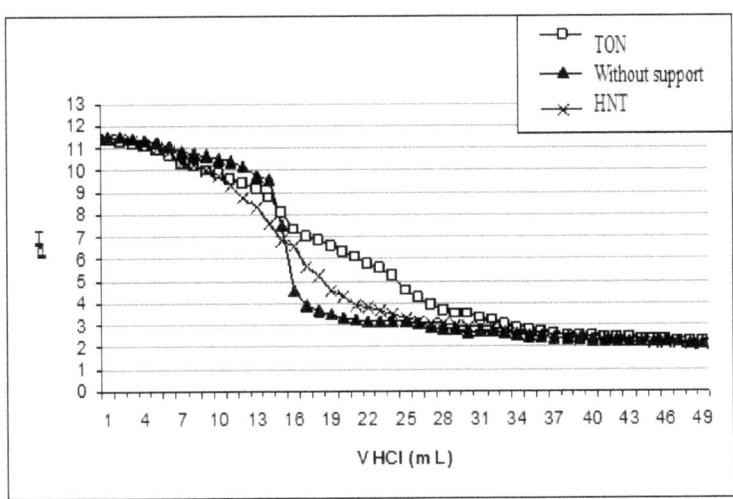

Figure 10. Titration curves of NaOH (0.01 mol L^{-1})/NaCl (0.1 mol L^{-1}) solution by HCl (0.1 mol L^{-1}) in the absence and presence of 1 g of the material (HNT et TON).

The pHpzc of the used adsorbent TON is 8.3, this value shows that the surface is positive for a pH value lower than 8.3 and negative for pH above 8.3.

The effect of the sample pH on recoveries of Cu (II) was examined between 3 and 9. The results presented in Figure 11 show lower adsorption capacities at pH values below 4, the is possibly due to the presence of excess H$^+$ ions competing with Cu (II) ions for the available adsorption sites. Then this adsorption capacity reached its maximum values when the pH was between 5 and 6 [49,59].

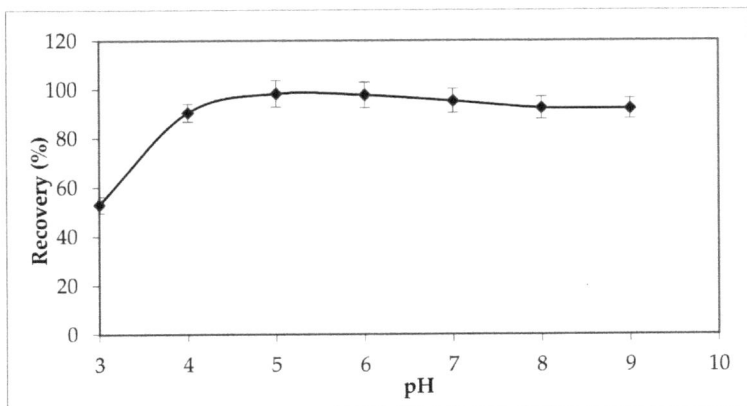

Figure 11. Effect of the sample pH on solid-phase extraction of Cu (II). Solvent volume = 12 mL; [HNO$_3$] = 2 M; the elution flow rate = 2 mL min^{-1}; sample volume = 100 mL; percolation flow rate = 2 mL min^{-1}.

So, we can conclude that the adsorption of copper was more pronounced in the case of positive charges of TON adsorbent.

3.2.5. Effect of the Mass of the Adsorbent

The mass of the adsorbent, which is a determining factor in the solid phase extraction method, is also optimized. TON amounts between 0.1 and 0.5 g were tested for the extraction of copper cations. The results given in Figure 12 show insignificant differences between calculated yields.

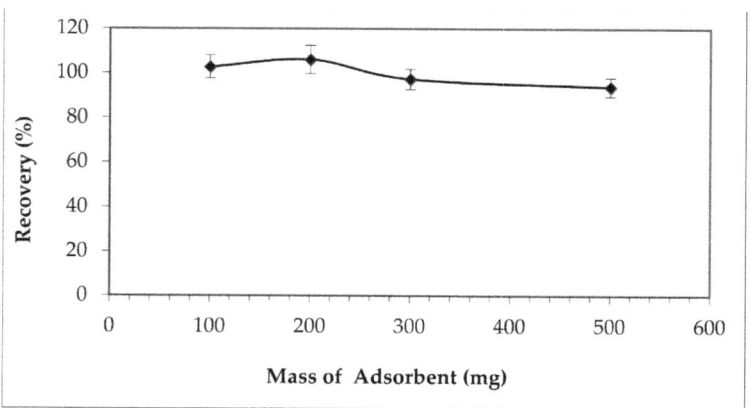

Figure 12. Effect of the mass of the adsorbent on Cu (II) extraction.

So, with a minimum mass of nanotubes, 0.1 g, maximum recovery of ~100 % can be achieved. It is the nanometric size of these structures and the large proportion of atoms present on their surface that favor this high reactivity.

3.3. Optimal Conditions

The SPE procedure following the optimal conditions obtained is therefore summarized as follows: through a cartridge, filled with 0.1 g of TiO$_2$ nanotubes, 10 mL of HNO$_3$ (2 mol L^{-1}), and 10 mL of ultrapure water are first percolated. A sample volume of 100 mL of water is percolated through the adsorbent at a flow rate of 2 mL min^{-1}. The washing of the adsorbent was performed with 10 mL of ultrapure water (in the case of real water samples). Elution of Cu (II) is performed with 12 mL HNO$_3$ (2 M) at a flow rate of 2 mL min^{-1}. These optimum conditions allow an extraction yield higher than 95% and a maximum adsorption capacity of 70 mg g^{-1} of TON. This adsorption capacity is perfect when TiO$_2$ nanotubes are applied as SPE adsorbent. While, for environmental applications and water treatment, higher adsorption capacities are desired [30,61].

3.4. Characteristics of the Method

The analytical parameters such as the sensitivity, detection limits, and reproducibility of the analytical method can be improved with a good sample extraction method. Indeed, it allows a better sensitivity of the technique of analysis by removing interferents from the matrix. The extraction of Cu (II) cations with TiO$_2$ nanotubes is repeated five times according to the optimal conditions obtained. Extraction yields were between 95.81 and 98.85% with satisfactory RSD of 3.16% (lower than 5%). Detection limit of AAS technique was in the range of 32.5 ng mL^{-1}.

In the present work, flame atomic absorption spectrometry was used for the identification of Cu (II) cations. To consider all concentrations of the analyzed copper during the optimization of the SPE method, three calibration ranges are considered (Table 2). The study of linearity showed a linear relationship between the concentration and the response of the technique. Calibration equations and correlation coefficients are illustrated in Table 2 and results show good and linear correlation of the regression method R^2 > 0.998.

Table 2. Calibration equation, correlation coefficient of Cu (II) analyses by SAAF.

Calibrations Range (mg L^{-1})	Calibration Equation	R^2
0–1.65	Y = 0.0869x − 0.0017	0.9982
0–3	Y = 0.0628x − 0.0022	0.9999
0–22	Y = 0.0131x − 0.0015	0.9985

3.5. Application of Optimal Conditions to Real Water Samples

To evaluate the effect of the matrix on the extraction efficiency of the optimized SPE method, the optimal conditions were applied to real water samples (tap water and mineral water). Mineral water was spiked with Cu (II) at a concentration of 2.37 mg L^{-1}, while tap water was spiked at four Cu (II) concentration levels. Analytical results given in Table 3 show the minimal effect of the matrix.

Table 3. Recovery percentages of tap and mineral water samples spiked with Cu cations.

Water Samples	Added Copper (mg L^{-1})	R%
Tap water	2.37	97.04
	1.57	95.23
	0.68	89.54
	0.26	87.21
Mineral water	2.37	96.20

Calculated yields were between 87 and 97% and were not far from those determined with ultra-pure water.

4. Conclusions

In this study, a reliable and efficient method was used for the determination of Cu (II) in water samples. This method consists of using well-characterized titanium dioxide nanotubes TON as an adsorbent in the solid phase extraction procedure of this analyte. TON showed interesting results in terms of calculated extraction yields that are higher than 95% and repeatability since the calculated RSD value is about 3.16%. Furthermore, this method improved the detection limit of the analysis technique of copper (32.5 ng mL^{-1}). The optimal conditions determined during this work were successfully applied for the solid-phase extraction of Cu (II) from real water samples and a minimal matrix effect is observed. The simplicity and relative affordability of the preparation of titanium dioxide nanotubes, as well as their effectiveness as SPE adsorbents, give them great potential in the fields of selective inorganic contaminant separation and sample process.

Author Contributions: Conceptualization, B.B.K. and I.B.; methodology, B.B.K. and I.B.; software, B.B.K. and N.M.; validation, B.B.K., P.M. and I.B.; formal analysis, B.B.K.; data curation, N.M.; writing—original draft preparation, B.B.K.; writing—review and editing, B.B.K., P.M. and N.M.; visualization, B.B.K.; supervision, B.B.K. All authors have read and agreed to the published version of the manuscript.

Funding: This research received no external funding.

Institutional Review Board Statement: Not applicable.

Informed Consent Statement: Not applicable.

Data Availability Statement: All the data is available within the manuscript.

Conflicts of Interest: The authors declare no conflict of interest.

References

1. Özdemir, S.; Yalçın, M.S.; Kılınç, E. Preconcentrations of Ni(II) and Pb(II) from water and food samples by solid-phase extraction using Pleurotus ostreatus immobilized iron oxide nanoparticles. *Food Chem.* **2021**, *336*, 127675. [CrossRef] [PubMed]
2. Buszewski, B.; Szultka, M. Past, Present, and Future of Solid Phase Extraction: A Review. *Crit. Rev. Anal. Chem.* **2012**, *42*, 198–213.
3. Miró, M.; Hansen, E.H. On-line sample processing methods in flow analysis. In *Advances in Flow Analysis*, 1st ed.; Trojanowicz, M., Ed.; Wiley-VCH: Weinheim, Germany, 2008; Volume 11, pp. 291–320.
4. Hoogerstraete, T.V.; Onghena, B.; Binnemans, K. Homogeneous Liquid–Liquid Extraction of Metal Ions with a Functionalized Ionic Liquid. *J. Phys. Chem. Lett.* **2013**, *4*, 1659–1663. [CrossRef] [PubMed]
5. Letellier, M.; Budzinski, H. Microwave assisted extraction of organic compounds. *Analusis* **1999**, *27*, 259–270. [CrossRef]
6. Ventura, S.P.M.; Silva, F.A.E.; Quental, M.V.; Mondal, D.; Freire, M.G.; Coutinho, J.A.P. Ionic-Liquid-Mediated Extraction and Separation Processes for Bioactive Compounds: Past, Present, and Future Trends. *Chem. Rev.* **2017**, *117*, 6984–7052. [CrossRef]
7. Ulla, H.; Wilfred, C.D.; Shaharunn, M.S. Ionic liquid-based extraction and separation trends of bioactive compounds from plant biomass. *Sep. Sci. Technol.* **2019**, *54*, 559–579. [CrossRef]
8. Rochfort, S.; Isbel, A.; Ezernieks, V.; Elkins, A.; Vincent, D.; Deseo, M.A.; Spangenberg, G.C. Utilisation of Design of Experiments Approach to Optimise Supercritical Fluid Extraction of Medicinal Cannabis. *Sci. Rep.* **2020**, *10*, 9124. [CrossRef]
9. Voshkin, A.; Solov'ev, V.; Kostenko, M.; Zakhodyaeva, Y.; Pokrovskiy, O. A Doubly Green Separation Process: Merging Aqueous Two-Phase Extraction and Supercritical Fluid Extraction. *Processes* **2021**, *9*, 727. [CrossRef]
10. Cai, Y.; Jiang, G.; Liu, J.; Zhou, Q. Multiwalled Carbon Nanotubes as a Solid-Phase Extraction Adsorbent for the Determination of Bisphenol A, 4-n-Nonylphenol, and 4-tert-Octylphenol. *Anal. Chem.* **2003**, *75*, 2517–2521. [CrossRef]
11. Wang, W.D.; Huang, Y.M.; Shu, W.Q.; Cao, J. Multiwalled carbon nanotubes as adsorbents of solid-phase extraction for determination of polycyclic aromatic hydrocarbons in environmental waters coupled with high-performance liquid chromatography. *J. Chromatogr. A* **2007**, *1173*, 27–36. [CrossRef]
12. Li, L.; Huang, Y.; Wang, Y.; Wang, W. Hemimicelle capped functionalized carbon nanotubes-based nanosized solid-phase extraction of arsenic from environmental water samples. *Anal. Chim. Acta* **2009**, *631*, 182–188. [CrossRef]
13. Fritz, J.S. *Analytical Solid-Phase Extraction*; WILEY-VCH: Toronto, ON, Canada, 1999.
14. López-Serna, R.; Pérez, S.; Ginebreda, A.; Petrovic, M.; Barceló, D. Fully automated determination of 74 pharmaceuticals in environmental and waste waters by online solid phase extraction–liquid chromatography-electrospray–tandem mass spectrometry. *Talanta* **2010**, *83*, 410–424. [CrossRef]
15. Su, C.K.; Lin, J.Y. 3D-Printed Column with Porous Monolithic Packing for Online Solid-Phase Extraction of Multiple Trace Metals in Environmental Water Samples. *Anal. Chem.* **2020**, *92*, 9640–9648. [CrossRef]
16. Takara, E.; Pasini-Cabello, S.; Cerutti, S.; Gásquez, J.; Martinez, L. On-line preconcentration/determination of copper in parenteral solutions using activated carbon by inductively coupled plasma optical emission spectrometry. *J. Pharm. Biomed. Anal.* **2005**, *39*, 735–739. [CrossRef]
17. Mariana, M.; Khalil, H.P.S.A.; Mistar, E.; Yahya, E.B.; Alfatah, T.; Danish, M.; Amayreh, M. Recent advances in activated carbon modification techniques for enhanced heavy metal adsorption. *J. Water Process Eng.* **2021**, *43*, 102221. [CrossRef]
18. Smith, S.C.; Rodrigues, D. Carbon-based nanomaterials for removal of chemical and biological contaminants from water: A review of mechanisms and applications. *Carbon* **2015**, *91*, 122–143. [CrossRef]
19. Zhang, Z.; Wang, T.; Zhang, H.; Liu, Y.; Xing, B. Adsorption of Pb(II) and Cd(II) by magnetic activated carbon and its mechanism. *Sci. Total Environ.* **2020**, *757*, 143910. [CrossRef]
20. Marin, N.M.; Stanculescu, I. Removal of procainamide and lidocaine on Amberlite XAD7HP resin and of As(V), Pb(II) and Cd(II) on the impregnated resin for water treatment. *Mater. Chem. Phys.* **2021**, *277*, 125582. [CrossRef]
21. Masini, J.C.; Nascimento, F.H.D.; Vitek, R. Porous monolithic materials for extraction and preconcentration of pollutants from environmental waters. *Trends Environ. Anal. Chem.* **2021**, *29*, e00112. [CrossRef]
22. Tolentino, R.; Knaebel, K.S. *Adsorbent Selection*; International Journal of Trend in Research and Development: Dublin, OH, USA, 2004.
23. El Atrache, L.L.; Sghaier, B.R.; Kefi, B.B.; Haldys, V.; Dachraoui, M.; Tortajada, J. Factorial design optimization of experimental variables in preconcentration of carbamates pesticides in water samples using solid phase extraction and liquid chromatography–electrospray-mass spectrometry determination. *Talanta* **2013**, *117*, 392–398. [CrossRef]
24. Ravelo-Pérez, L.M.; Herrera-Herrera, A.V.; Hernandez-Borges, J.; Rodriguez-Delgado, M.A. Review: Carbon nanotubes: Solid-Phase Extraction. *J. Chromatogr. A* **2010**, *1217*, 2618–2641. [CrossRef]
25. Tuzen, M.; Saygi, K.O.; Soylak, M. Solid-phase extraction of heavy metal ions in environmental samples on multiwalled carbon nanotubes. *J. Hazard. Mater.* **2008**, *152*, 632–639. [CrossRef]
26. Chen, S.; Liu, C.; Yang, M.; Lu, D.; Zhu, L.; Wang, Z. Solid-phase extraction of Cu, Co and Pb on oxidized single-walled carbon nanotubes and their determination by inductively coupled plasma mass spectrometry. *J. Hazard. Mater.* **2009**, *170*, 247–251. [CrossRef]
27. Ozcan, S.G.; Satiroglu, N.; Soylak, M. Column solid-phase extraction of iron (III), copper (II), manganese (II) and lead (II) ions food and water samples on multi-walled carbon nanotubes. *Food Chem. Toxicol.* **2010**, *48*, 2401–2406. [CrossRef]
28. Shamspur, T.; Mostafavi, A. Application of modified multiwalled carbon nanotubes as a sorbent for simultaneous separation and preconcentration trace amounts of Au(III) and Mn(II). *J. Hazard. Mater.* **2009**, *168*, 1548–1553. [CrossRef]

29. Heidari, A.; Younesi, H.; Mehraban, Z. Removal of Ni(II), Cd(II), and Pb(II) from a ternary aqueous solution by amino functionalized mesoporous and nano mesoporous silica. *Chem. Eng. J.* **2009**, *153*, 70–79. [CrossRef]
30. Hu, Q.; Xu, L.; Fu, K.; Zhu, F.; Yang, T.; Yang, T.; Luo, J.; Wu, M.; Yu, D. Ultrastable MOF-based foams for versatile applications. *Nano Res.* **2021**, 1–10. [CrossRef]
31. Wu, J.; Xue, S.; Bridges, D.; Yu, Y.; Zhang, L.; Pooran, J.; Hill, C.; Wu, J.; Hu, A. Fe-based ceramic nanocomposite membranes fabricated via e-spinning and vacuum filtration for Cd2+ ions removal. *Chemosphere* **2019**, *230*, 527–535. [CrossRef]
32. Singh, S.; Kapoor, D.; Khasnabis, S.; Singh, J.; Ramamurthy, P.C. Mechanism and kinetics of adsorption and removal of heavy metals from wastewater using nanomaterials. *Environ. Chem. Lett.* **2021**, *19*, 2351–2381. [CrossRef]
33. Kyzas, G.Z.; Matis, K.A. Nanoadsorbents for pollutants removal: A review. *J. Mol. Liq.* **2015**, *203*, 159–168. [CrossRef]
34. Soylak, M.; Ercan, O. Selective separation and preconcentration of copper (II) in environmental samples by the solid phase extraction on multi-walled carbon nanotubes. *J. Hazard. Mater.* **2009**, *168*, 1527–1531. [CrossRef] [PubMed]
35. El Atrache, L.L.; Hachani, M.; Kefi, B.B. Carbon nanotubes as solid-phase extraction sorbents for the extraction of carbamate insecticides from environmental waters. *Int. J. Environ. Sci. Technol.* **2015**, *13*, 201–208. [CrossRef]
36. Kasuga, T. Formation of titanium oxide nanotubes using chemical treatments and their characteristic properties. *Thin Solid Films* **2006**, *496*, 141–145. [CrossRef]
37. Chen, Q.; Zhou, W.; Du, G.; Mao, P.L. Titanate Nanotubes made via a single alkali treatment. In *Advanced Materials*; John Wiley & Son: Hoboken, NY, USA, 2002; Volume 17, pp. 1208–1211.
38. Ma, Y.; Lin, Y.; Xiao, X.; Zhou, X.; Li, X. Sonication–hydrothermal combination technique for the synthesis of titanate nanotubes from commercially available precursors. *Mater. Res. Bull.* **2006**, *41*, 237–243. [CrossRef]
39. Pap, A.L.; Aymes, D.; Saviot, L.; Chassagnon, R.; Heintz, O.; Millot, N. Nanotubes d'oxydes de titane: Synthèse, caractérisations et propriétés. In Proceedings of the Matériaux, Dijon, France, 13–17 November 2006.
40. Hird, M. Transmission of ultraviolet light by films containing titanium pigments—applications in u.v. curing. *Pigment. Resin Technol.* **1976**, *5*, 5–14. [CrossRef]
41. Yuan, Z.-Y.; Su, B.-L. Titanium oxide nanotubes, nanofibers and nanowires. *Colloids Surf. A Physicochem. Eng.* **2004**, *241*, 173–183. [CrossRef]
42. Fujishima, A.; Zhang, X. Titanium dioxide photocatalysis: Present situation and future approaches. *Comptes Rendus. Chim.* **2006**, *9*, 750–760. [CrossRef]
43. Zhang, M.; Jin, Z.; Zhang, J.; Guo, X.; Yang, J.; Li, W.; Wang, X.; Zhang, Z. Effect of annealing temperature on morphology, structure and photocatalytic behavior of nanotubes $H_2Ti_2O_4 \times (OH)_2$. *J. Mol. Catal. A Chem.* **2004**, *217*, 203–210. [CrossRef]
44. Baati, T.; Njim, L.; Jaafoura, S.; Aouane, A.; Neffati, F.; Ben Fradj, N.; Kerkeni, A.; Hammami, M.; Hosni, K. Assessment of Pharmacokinetics, Toxicity, and Biodistribution of a High Dose of Titanate Nanotubes Following Intravenous Injection in Mice: A Promising Nanosystem of Medical Interest. *ACS Omega* **2021**, *6*, 21872–21883. [CrossRef]
45. Alban, L.; Monteiro, W.F.; Diz, F.M.; Miranda, G.M.; Scheid, C.M.; Zotti, E.R.; Morrone, F.; Ligabue, R. New quercetin-coated titanate nanotubes and their radiosensitization effect on human bladder cancer. *Mater. Sci. Eng. C* **2020**, *110*, 110662. [CrossRef]
46. Baati, T.; Kefi, B.B.; Aouane, A.; Njim, L.; Chaspoul, F.; Heresanu, V.; Kerkeni, A.; Neffati, F.; Hammami, M. Biocompatible titanate nanotubes with high loading capacity of genistein: Cytotoxicity study and anti-migratory effect on U87-MG cancer cell lines. *RSC Adv.* **2016**, *6*, 101688–101696. [CrossRef]
47. Niu, H.; Cai, Y.; Shi, Y.; Wei, F.; Mou, S.; Jiang, G. Cetyltrimethylammonium bromide-coated titanate nanotubes for solid-phase extraction of phthalate esters from natural waters prior to high-performance liquid chromatography analysis. *J. Chromatogr. A* **2007**, *1172*, 113–120. [CrossRef]
48. Zhao, X.N.; Shi, Q.Z.; Xie, G.H.; Zhou, Q.X. TiO_2 nanotubes: A novel solid phase extraction adsorbent for the sensitive determination of nickel in environmental wat1er samples. *Chin. Chem. Lett.* **2008**, *19*, 865–867. [CrossRef]
49. Zhou, Q.-X.; Zhao, X.-N.; Xiao, J.-P. Preconcentration of nickel and cadmium by TiO2 nanotubes as solid-phase extraction adsorbents coupled with flame atomic absorption spectrometry. *Talanta* **2009**, *77*, 1774–1777. [CrossRef]
50. Kefi, B.B.; El Atrache, L.L.; Kochkar, H.; Ghorbel, A. TiO2 nanotubes as solid-phase extraction adsorbent for the determination of polycyclic aromatic hydrocarbons in environmental water samples. *J. Environ. Sci.* **2011**, *23*, 860–867. [CrossRef]
51. Keenan, J.; Meleady, P.; O'doherty, C.; Henry, M.; Clynes, M.; Horgan, K.; Murphy, R.; O'sullivan, F. Copper toxicity of inflection point in human intestinal cell line Caco-2 dissected: Influence of temporal expression patterns. *Vitr. Cell. Dev. Biol. Anim.* **2021**, *57*, 1–13. [CrossRef]
52. Kumar, V.; Pandita, S.; Sidhu, G.P.S.; Sharma, A.; Khanna, K.; Kaur, P.; Bali, A.S.; Setia, R. Copper bioavailability, uptake, toxicity and tolerance in plants: A comprehensive review. *Chemosphere* **2021**, *262*, 127810. [CrossRef]
53. Cobine, P.A.; Moore, S.A.; Leary, S.C. Getting out what you put in: Copper in mitochondria and its impacts on human disease. *Biochim. et Biophys. Acta* **2020**, *1868*, 118867. [CrossRef]
54. Kumar, K.S.K.; Dahms, H.-U.; Won, E.-J.; Lee, J.-S.; Shin, K.-H. Microalgae—A promising tool for heavy metal remediation. *Ecotoxicol. Environ. Saf.* **2015**, *113*, 329–352. [CrossRef]
55. Fine, M.; Isheim, D. Origin of copper precipitation strengthening in steel revisited. *Scr. Mater.* **2005**, *53*, 115–118. [CrossRef]
56. Qi, H.; Zhao, M.; Liang, H.; Wu, J.; Huang, Z.; Hu, A.; Wang, J.; Lu, Y.; Zhang, J. Rapid detection of trace Cu 2+ using an l -cysteine based interdigitated electrode sensor integrated with AC electrokinetic enrichment. *Electrophoresis* **2019**, *40*, 2699–2705. [CrossRef] [PubMed]

57. Brunelle, J.P. Preparation of catalysts by metallic complex adsorption on mineral oxides. *Pure Appl. Chem.* **1978**, *50*, 1211–1229. [CrossRef]
58. Kochkar, H.; Lakhdhar, N.; Berhault, G.; Bausach, M.; Ghorbel, A. Optimization of the Alkaline Hydrothermal Route to Titanate Nanotubes by a Doehlert Matrix Experience Design. *J. Phys. Chem. C* **2009**, *113*, 1672–1679. [CrossRef]
59. Zhou, Q.; Ding, Y.; Xiao, J.; Liu, G.; Guo, X. Investigation of the feasibility of TiO2 nanotubes for the enrichment of DDT and its metabolites at trace levels in environmental water samples. *J. Chromatogr. A* **2007**, *1147*, 10–16. [CrossRef]
60. Duran, A.; Tuzen, M.; Soylak, M. Preconcentration of some trace elements via using multiwalled carbon nanotubes as solid phase extraction adsorbent. *J. Hazard. Mater.* **2009**, *169*, 466–471. [CrossRef]
61. Sun, H.; Ji, Z.; He, Y.; Wang, L.; Zhan, J.; Chen, L.; Zhao, Y. Preparation of PAMAM modified PVDF membrane and its adsorption performance for copper ions. *Environ. Res.* **2022**, *204*, 111943. [CrossRef]

Article

Tuning Dielectric Loss of SiO₂@CNTs for Electromagnetic Wave Absorption

Fenghui Cao [1,2], Jia Xu [1], Xinci Zhang [1], Bei Li [1], Xiao Zhang [1,*], Qiuyun Ouyang [1], Xitian Zhang [3] and Yujin Chen [1,4,*]

1. Key Laboratory of In-Fiber Integrated Optics, College of Physics and Optoelectronic Engineering, Harbin Engineering University, Harbin 150001, China; caofenghui@hrbeu.edu.cn (F.C.); xujia110006@hrbeu.edu.cn (J.X.); zhangxinci@hrbeu.edu.cn (X.Z.); 1284034781@hrbeu.edu.cn (B.L.); qyouyang7823@aliyun.com (Q.O.)
2. School of Mechatronic Engineering, Daqing Normal University, Daqing 163712, China
3. Key Laboratory for Photonic and Electronic Bandgap Materials, Ministry of Education, School of Physics and Electronic Engineering, Harbin Normal University, Harbin 150025, China; xtzhangzhang@hotmail.com
4. School of Materials Science and Engineering, Zhengzhou University, Zhengzhou 450001, China
* Correspondence: zhangxiaochn@hrbeu.edu.cn (X.Z.); chenyujin@hrbeu.edu.cn (Y.C.)

Abstract: We developed a simple method to fabricate SiO₂-sphere-supported N-doped CNTs (NC-NTs) for electromagnetic wave (EMW) absorption. EMW absorption was tuned by adsorption of the organic agent on the precursor of the catalysts. The experimental results show that the conductivity loss and polarization loss of the sample are improved. Meanwhile, the impedance matching characteristics can also be adjusted. When the matching thickness was only 1.5 mm, the optimal 3D structure shows excellent EMW absorption performance, which is better than most magnetic carbon matrix composites. Our current approach opens up an effective way to develop low-cost, high-performance EMW absorbers.

Keywords: CNTs; dielectric loss; nitrogen doping; electromagnetic wave absorption

1. Introduction

With the development of science and technology, the rapid rise of artificial intelligence, the popularity of the smart home, and the extensive application of various electrical and electronic products, people's work efficiency and quality of life has improved. However, at the same time, the widespread use of electronic products also hides huge harms: long-term exposure to electromagnetic radiation will damage human health, but also harms other electronic products' electromagnetic interference, affecting their normal work. These hazards have attracted the attention of many countries in the world, and development of efficient electromagnetic absorption and shielding materials has become the main research direction. Therefore, it is necessary to develop a high-performance electromagnetic wave (EMW) absorber. To improve the efficiency of the unitizations, lightweight absorbers with a thin thickness are required. Carbonaceous materials such as graphene, carbon nanotubes (CNTs) and carbon nanofibers have attracted great attention because of their low mass density, good mechanical and chemical stability and high surface areas [1–9]. Carbon nanotubes have received extensive attention and in-depth studies in the field of EMW absorption due to their tubular structure suitable for electron transport, their light weight and good electrical conductivity [10–17]. For example, Lv et al. encapsulated Fe/Fe₃C nanoparticles (NPs) into N-doped CNTs (NCNTs) and obtained the result that the sample had a reflection loss (R_L) of −46.0 dB and a thickness of 4.97 mm at 3.6 GHz [10]. Chang et al. reported Fe₃O₄/PPy/CNT composites with an R_L of −25.9 dB with a thickness of 3.0 mm [11]. The reflection loss of ZnFe₂O₄@CNT/PVDF composite film prepared by Li et al. was −54.5 dB, with a matching thickness of 2.4 mm [12]. Gong et al. reported SiCN(Fe) fibers with an R_L of −47.64 dB and the effective absorption bandwidth of 4.28 GHz [13].

The minimum reflection loss of Fe_3O_4/CNTs prepared by Zeng et al. was −51.0 dB, with a matching thickness of 4.4 mm [14]. Recently, a series of magnetic metal alloys (Fe, Co, Ni, etc.) that encapsulated into NCNTs were designed for EMW absorption [15,16]. However, the impedance matching feature of the composites mentioned above needed to be precisely tuned due to highly conductive magnetic metals and CNTs in these composites [17]. Furthermore, there is still room to improve the EMW absorption property of CNT-based absorbers, such as stronger absorption at a lower filler ratio and thinner matching thickness.

Here, we propose a simple method for SiO_2-sphere-supported NCNTs with embedded Fe_3C/Fe nanoparticles (NPs) (SiO_2@Fe_3C/Fe@NCNT-GT) for EMW absorption. $Fe(OH)_x$ was first coated on the surface of SiO_2 spheres [18], and then the organic solvent (terephthalic acid) was adsorbed on the $Fe(OH)_x$ surface to form relatively larger metal NPs for the growth of NCNTs with moderate diameters. Compared to the counterpart (SiO_2@Fe_3C/Fe@NCNT) without treatment in the organic solvent, the as-prepared SiO_2@Fe_3C/Fe@NCNT showed significantly improved EMW absorption performance. At a filler ratio of 25%, the minimum R_L ($R_{L,\,min}$) and effective bandwidth of the SiO_2@Fe_3C/Fe@NCNT-GT reached −48.43 dB and 4.51 GHz, respectively, while the matching thickness was only 1.5 mm.

2. Materials and Methods

2.1. Materials

Tetraethoxysilane (TEOS, 99 wt%, analytical reagent, A.R.) was purchased from Tianjin Komiou Chemical Reagent Co., Ltd. (Tianjin, China). NH_4OH (25 wt%, A.R.) was purchased from XiLong Scientific Co., Ltd. (Shantou, China). Absolute ethanol (99 wt%, analytical reagent, A.R.) was purchased from Tianjin Fuyu Fine Chemical Co., Ltd. (Tianjin, China). Terephthalic, N,N-Dimethylformamide (DMF) and dicyandiamide were purchased from Tianjin Guangfu Fine Chemical Research Institute (Tianjin, China). Ferric acetylacetonate was purchased from Sinopharm Chemical Reagent Co., Ltd. (Shanghai, China). Paraffin was purchased from Yuyang Wax Industry (Changge, China). It is worth noting that all chemicals were purchased without further treatment before use, and all aqueous solutions were prepared using ultrapure water.

2.2. Characterizations and Electromagnetic Parameter Measurement

The morphology and size of the samples were characterized using scanning electron microscopy (SEM; Hitachi SU70, Tokyo, Japan) and transmission electron microscopy (TEM; FEITecnai-F20, Hillsboro, USA). Energy dispersive X-ray spectroscopy (EDX) was performed to confirm the elemental contents of the samples. X-ray diffraction (XRD) data were measured using a Rigaku D/max 2550 V (Tokyo, Japan) with Cu Kα radiation (λ = 1.5418 Å). X-ray photoelectron spectroscopy (XPS) analyses were carried out by using a spectrometer with Mg Kα radiation (ESCALAB 250, Shanghai, China). Raman spectra were recorded on a Raman spectrometer (Xplora Plus, Paris, France) using a 488 nm He−Ne laser. The Brunauer–Emmett–Teller (BET) surface area and pore volume were tested with a Quantachrome Instruments Autosorb-iQ2-MP (Beijing, China) after the composites were vacuum dried at 200 °C for 10 h. Fourier-transform infrared (FTIR) spectra of samples were collected using a Nicolet FTIR510 spectrometer (KBr pellet method, 4 cm^{-1} resolution, Waltham, MA, USA). The electromagnetic wave absorption properties of the absorbing materials were measured using a vector network analyzer (Anritsu MS4644A Vectorstar, Kanagawa, Japan) in the 2–18 GHz range at room temperature.

2.3. Methods

2.3.1. Synthesis of the SiO_2

Synthesis of the SiO_2 was conducted following the Stöber method. Deionized water (18 mL), absolute ethanol (76 mL) and TEOS (14 mL) were dissolved into NH_4OH (98 mL), and continuously stirred for 4 h. Then, the colloidal solution was centrifuged, and the resultant was placed in an oven at 100 °C for 12 h to obtain silica microspheres [19,20].

2.3.2. Synthesis of the SiO$_2$@Fe(OH)$_x$

SiO$_2$ with diameters of about 400 nm (216 mg) were first dispersed in ethanol (72 mL), and ferric acetylacetonate (270.5 mg) was added to the mixture above, sequentially. Distilled water (3.6 mL) and ammonia (2 mL) were added to the mixture with sonication for 15 min. The as-prepared mixture was sealed in a conical flask and stirred at 80°C for 10 h. The precipitate was washed with distilled water and ethanol several times, then centrifuged and dried in a vacuum oven at 40 °C to obtain the SiO$_2$@Fe(OH)$_x$ [20].

2.3.3. Synthesis of the SiO$_2$@Fe$_3$C/Fe@NCNT

The SiO$_2$@Fe(OH)$_x$ was annealed in Ar (the temperature was 800 °C, the time was 30 min and the ramp rate was 5 °C/min) to obtain SiO$_2$@Fe$_3$C/Fe@NCNT [20].

2.3.4. Synthesis of the SiO$_2$@Fe(OH)$_x$-GT

The SiO$_2$@Fe(OH)$_x$ (200 mg), terephthalic acid (300 mg), pure water (3 mL) and ethanol (3 mL) were added into N,N-Dimethylformamide (DMF) solution (30 mL), and stirred for 30 min. Then, the mixed solution was placed into a 50 mL Teflon container and was treated at a high temperature of 150 °C for 12 h. The precipitate was washed and dried in a 40 °C vacuum oven to obtain the SiO$_2$@Fe(OH)$_x$-GT.

2.3.5. Synthesis of the SiO$_2$@Fe$_3$C/Fe@NCNT-GT

The SiO$_2$@Fe(OH)$_x$-GT was annealed in Ar (the temperature was 800 °C, the time was 30 min, and the ramp rate was 5 °C/min) to obtain the SiO$_2$@Fe$_3$C/Fe@NCNT-GT. The detailed experimental material, structural characterizations and the method are described in the Supplementary Materials.

3. Results and Discussion

The diameter of the prepared SiO$_2$ microspheres were approximately 400 nm. After being coated with an Fe(OH)$_x$ layer on the surface of the SiO$_2$ spheres, the diameter increased from 400 nm to about 500 nm (SiO$_2$@Fe(OH)$_x$). Scanning electron microscopy (SEM) imaging and transmission electron microscopy (TEM) imaging showed that the Fe(OH)$_x$ layer was uniformly coated on the surface of the SiO$_2$ spheres (Figures 1a–c and S1a, Supplementary Materials). In the X-ray diffraction (XRD) pattern, there were two obvious peaks at 2θ 35.02° and 62.72°, indicating that the outmost layer of the SiO$_2$@Fe(OH)$_x$ was mainly composed of weakly crystalline Fe(OH)$_3$, which corresponded to JCPDS card no. 22-0346 (Figure S2a). Energy dispersive X-ray spectroscopy (EDX) element mappings were also confirmed. As shown in Figure 1d, there were Si and O signals in the spherical core region, suggesting that the spherical core region materials were still SiO$_2$ spheres. Fe and O single elements were obviously present in the area outside the sphere, confirming the composition of the SiO$_2$@Fe(OH)$_3$. The XRD pattern of the SiO$_2$@Fe(OH)$_x$-GT indicated that the weakly crystalline Fe(OH)$_3$ (Figure S2a) remained after the treatment, but the rough surface became relatively smooth (Figure S1b). TEM images show that there was a lamellar structure on the surface of the SiO$_2$@Fe(OH)$_3$-GT, different from the SiO$_2$@Fe(OH)$_3$ (Figure 1e–g). The peak at 798 and 1431 cm^{-1} in the Fourier-transform infrared (FTIR) spectra of the SiO$_2$@Fe(OH)$_3$-GT corresponded to the C–H deformation vibration and the C=C stretching vibration (Figure S2b). The peak 1659 cm^{-1} in the spectrum of the SiO$_2$@Fe(OH)$_3$-GT corresponded to the C=O stretching vibration (Figure S2b). Thus, the FTIR results indicated the adsorption of terephthalic acid on the Fe(OH)$_3$ layer. EDX element mappings also confirmed that Fe, O and C single elements were present in the area outside of the sphere, confirming the adsorption of terephthalic acid on the Fe(OH)$_3$ layer (Figure 1h).

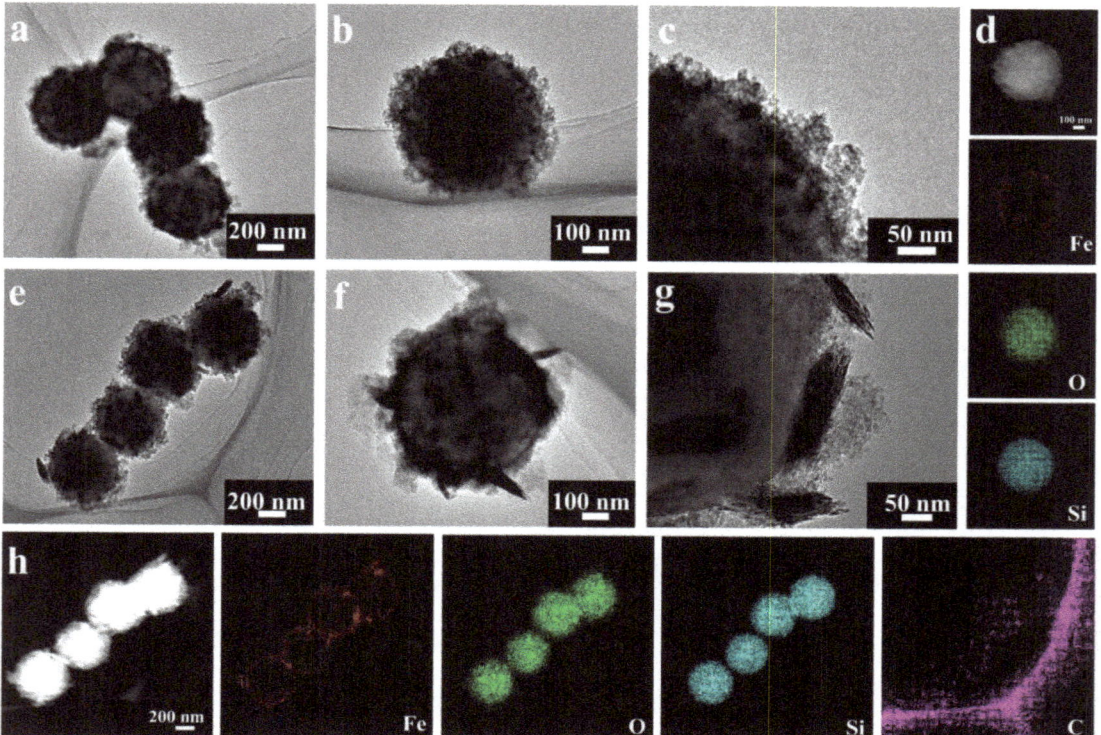

Figure 1. (**a–c**) TEM images of structural characterizations and (**d**) EDX element mappings of the SiO$_2$@Fe(OH)$_x$. (**e–g**) TEM image of structural characterizations and (**h**) EDX element mappings of SiO$_2$@Fe(OH)$_x$-GT.

In order to analyze the composition and valence state of the SiO$_2$@Fe$_3$C/Fe@NCNT-GT, XRD and Raman spectra were performed. As shown in Figure 2a, there was a broad diffraction (2θ) from 10° to 30°, corresponding to amorphous SiO$_2$ in the XRD pattern of the SiO$_2$@Fe$_3$C/Fe@NCNT-GT. In the XRD pattern of the SiO$_2$@Fe$_3$C/Fe@NCNT-GT, the peak at 2θ 44.7° and 42.9° can be indexed to (110) planes of the Fe NPs (JCPDS no. 06-0696) and (211) planes of the Fe$_3$C (JCPDS no. 35-0772), in sequence, while the peak at 26.4° is attributed to the NCNTs (JCPDS no. 41-1487). The Raman spectra of the SiO$_2$@Fe$_3$C/Fe@NCNT-GT showed two distinguishable peaks: one at 1325 cm^{-1} (D band) and the other at 1585 cm^{-1} (G band) (Figure 2b). Their intensity ratios (I_D/I_G) for the SiO$_2$@Fe$_3$C/Fe@NCNT-GT was 1.002, which indicated rich defects in the sample. These defects in the SiO$_2$@Fe$_3$C/Fe@NCNT-GT can greatly improve the polarization relaxation and contribute to enhancing the absorption of electromagnetic waves [21]. As shown in Figure 2c, the N$_2$-sorption isotherms of the SiO$_2$@Fe$_3$C/Fe@NCNT-GT displays type-IV loops, revealing that the mesopores existed in the prepared sample. Furthermore, the Brunauer–Emmett–Teller (BET) surface area of the SiO$_2$@Fe$_3$C/Fe@NCNT-GT was 243.54 m^2 g^{-1}. The illustration in Figure 2c displays pore size distribution diagrams. The pore sizes of the SiO$_2$@Fe$_3$C/Fe@NCNT-GT were centered at 15 nm, and pore volume was 0.568 cm^3 g^{-1}. X-ray photoelectron spectroscopy (XPS) spectra displayed that there were five elements (Fe, N, O, Si and C) in the SiO$_2$@Fe$_3$C/Fe@NCNT-GT (Figure S3). The peaks were at 398.5 (pyridine-N), 399.9 (pyrrolic-N), 401.1 (graphite-Nand) and 404.5 eV (oxide-N) in the XPS spectra of N 1s, respectively (Figure 2d) [22]. The peaks at 709.1 and 721.9 eV in the XPS spectra of Fe 2p can be indexed to metallic Fe. The peaks (717.3 and 733.2 eV) and satellite peaks (712.1 and 724.9 eV) reveal the oxidation state of Fe species in the sample [23–25] (Figure 2e). The binding energies of C–C (284.6 eV), C–N (285.7 eV) and C–O (288.7 eV) were observed on the surface of NCNT (Figure 2f) [26,27]. The carbon atom will tend to form unsaturated covalent bonds with the oxygen anion, increasing the charge state and increasing the band gap. This is due to the electron density shifts from the carbocation to the more electronegative oxygen anion, which in turn affects the electron structure.

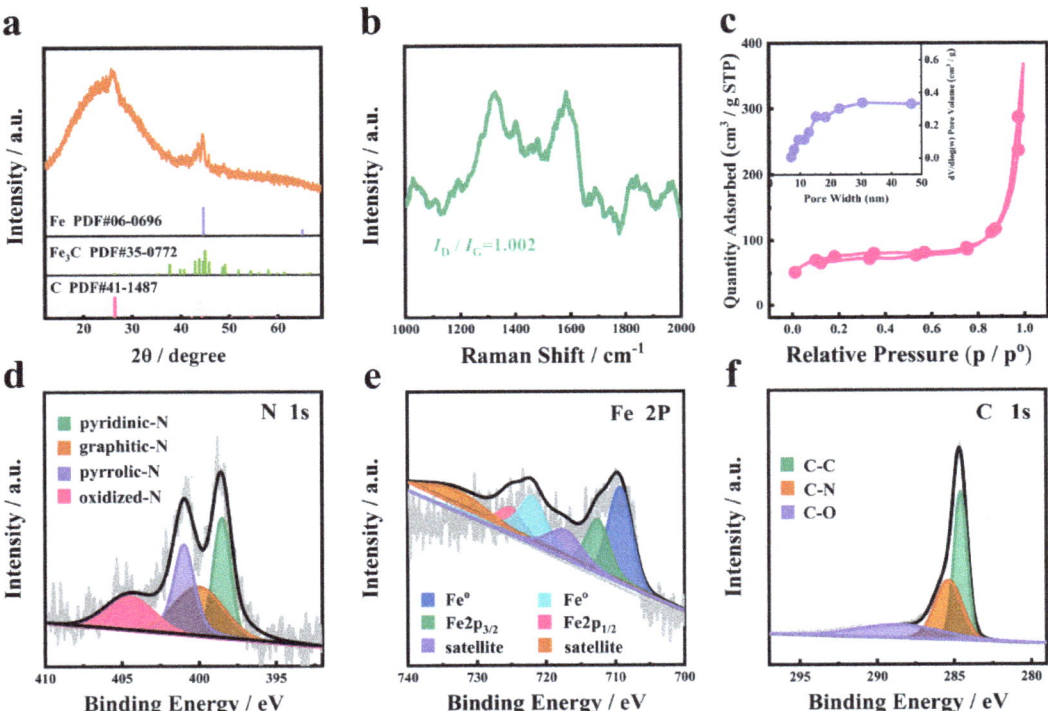

Figure 2. (a) XRD pattern, (b) Raman spectra, (c) pore size distribution and N$_2$-sorption isotherm, (d–f) N 1s, Fe 2p and C 1s XPS spectra of the SiO$_2$@Fe$_3$C/Fe@NCNT-GT.

SEM images indicated that the SiO$_2$@Fe$_3$C/Fe@NCNT-GT exhibited 3D morphology, where the NCNTs were grown on the surface of the SiO$_2$ spheres (Figure S4). Bamboo-like NCNTs in the SiO$_2$@Fe$_3$C/Fe@NCNT-GT are also observed in Figure 3a,b with a length of approximately 1.5 μm. Magnified TEM images show that their average diameter and wall thickness were approximately 51 and 12 nm, respectively (Figures 3b,c and S5). There were some NPs, with an average diameter of about 39 nm, embedded in the bamboo-like NCNTs. Figure 3c shows that the NPs were encapsulated in 25–30 layers of the graphene shell in the high-resolution TEM (HRTEM) images. The *d*-spacing of labeled lattice fringes of 0.20 nm corresponded to the (110) planes of Fe, while the *d*-spacing of 0.35 nm corresponded to the (002) planes of graphite–carbon (Figure 3d). Notably, a mass of defects were present in the NCNT walls and graphene shell of the SiO$_2$@Fe$_3$C/Fe@NCNT-GT. As is shown in Figure 3c, these defects are marked with a yellow frame. Defects including lattice distortion, lattice dislocation and fracture edges are considered to have a positive effect on the absorption property of the SiO$_2$@Fe$_3$C/Fe@NCNT-GT. The distribution of elements in the SiO$_2$@Fe$_3$C/Fe@NCNT-GT was analyzed by EDX element mapping. There were O and Si signals in the spherical core zone, indicating that the spherical core region mediums were still SiO$_2$ spheres (Figure 3e). Fe, C and N single elements were present in the zone of the NCNTs, confirming the composition of NCNTs. Compared to the SiO$_2$@Fe$_3$C/Fe@NCNT [20], the average diameter of the NCNTs in the SiO$_2$@Fe$_3$C/Fe@NCNT-GT increased from 15 nm to 58 nm, the length of bamboo nodes increased from 15 nm to 50 nm, and the wall thickness increased from 3 to 12 nm. The adsorption of terephthalic acid limited the contact of the Fe(OH)$_3$ with the reductive gases, leading to the formation of the larger metal NPs. Consequently, the diameter of NCNTs became larger compared to the counterpart without the adsorption of terephthalic acid.

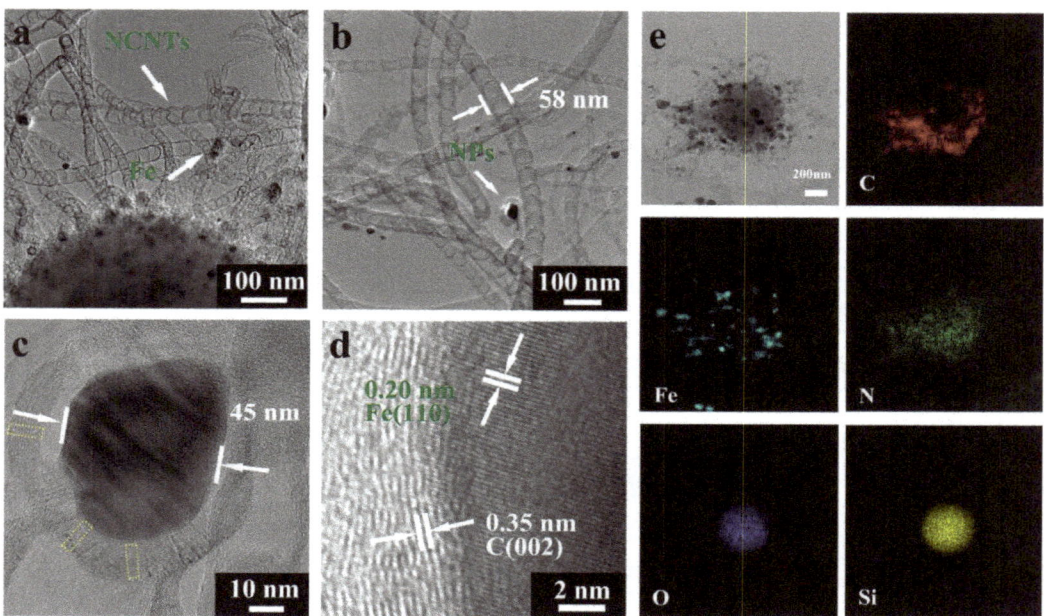

Figure 3. (a–c) TEM images, (d) HRTEM images, (e) TEM image and EDX elemental mappings of SiO$_2$@Fe$_3$C/Fe@NCNT-GT. The defects are marked by yellow dotted square (c).

The factors that may enhance the absorption performance of EMW were investigated through the comparison of electromagnetic parameters of the SiO$_2$@Fe$_3$C/Fe@NCNT-GT and the SiO$_2$@Fe$_3$C/Fe@NCNT, including complex permittivity and permeability. They can be expressed separately by the formula $\varepsilon_r = \varepsilon' - j\varepsilon''$ and $\mu_r = \mu' - j\mu''$. (ε' is the real part of permittivity, ε'' is the imaginary part of permittivity, μ' is the real part of permeability, and μ'' is the imaginary part of permeability) [28–30]. As shown in Figure 4a–c, the ε' values of the SiO$_2$@Fe$_3$C/Fe@NCNT-GT varied in a range of 16.63–9.81, and the ε'' values of the SiO$_2$@Fe$_3$C/Fe@NCNT-GT varied in a range of 6.15–2.44. The permittivity for both samples gradually decreased with the increase in the frequency, which is due to the frequency dispersion effect. The ε' and ε'' values of the SiO$_2$@Fe$_3$C/Fe@NCNT-GT were larger than those of the SiO$_2$@Fe$_3$C/Fe@NCNT. The dielectric loss tangent (tan$\delta_e = \varepsilon''/\varepsilon'$) of the SiO$_2$@Fe$_3$C/Fe@NCNT-GT was also larger. Figure S6 shows that the two samples had very little difference in the real part of permeability, imaginary part of permeability and magnetic loss tangent (tan δ_m), with a value over 2–18 GHz. The saturation magnetization (M_s), remnant magnetization (M_r) and coercivity (H_c) of the SiO$_2$@Fe$_3$C/Fe@NCNT-GT were slightly larger than those of the SiO$_2$@Fe$_3$C/Fe@NCNT (Figure S7). Thus, the magnetic loss of the SiO$_2$@Fe$_3$C/Fe@NCNT-GT is not a determining factor for the EMW performance.

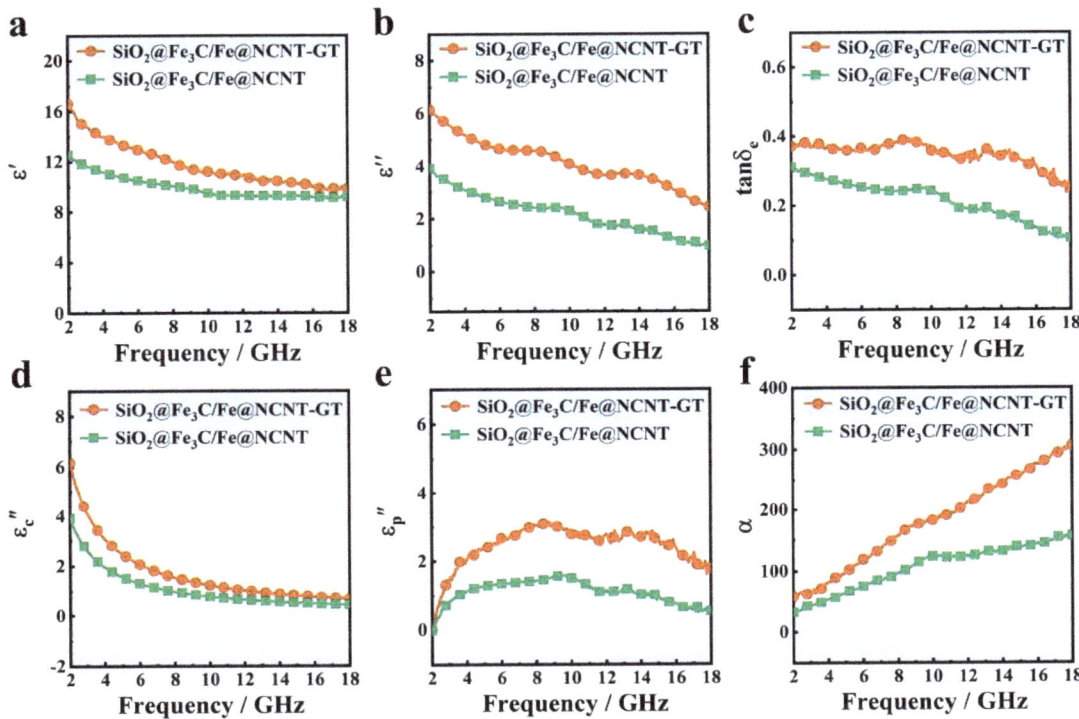

Figure 4. (a) ε'—f curves, (b) ε''—f curves, (c) tanδ_e—f curves, (d) ε_c''—f curves, (e) ε_p''—f curves and (f) α—f curves of SiO$_2$@Fe$_3$C/Fe@NCNT-GT and SiO$_2$@Fe$_3$C/Fe@NCNT.

In general, the dielectric loss of the absorbing material includes the conduction loss and the polarization relaxation loss within the range of gigahertz. The former can be expressed by the formula ($\varepsilon_c'' = \sigma/\varepsilon_0\omega$); the characters σ, ε_0 and ω represent the conductivity, the permittivity in a vacuum and the circular frequency, respectively. The latter is expressed by the formula ($\varepsilon_p'' = \varepsilon'' - \varepsilon_c''$). The experimental results showed that the electrical conductivity of the SiO$_2$@Fe$_3$C/Fe@NCNT-GT was higher than that of the SiO$_2$@Fe$_3$C/Fe@NCNT (Table S1). Therefore, the SiO$_2$@Fe$_3$C/Fe@NCNT-GT had increased conductive loss compared to the SiO$_2$@Fe$_3$C/Fe@NCNT (Figure 4d). Meanwhile, the SiO$_2$@Fe$_3$C/Fe@NCNT-GT also had improved polarization losses compared to the SiO$_2$@Fe$_3$C/Fe@NCNT (Figure 4e) and had a higher attenuation coefficient (Figure 4f). As shown in Figure S8, multiple Cole–Cole semicircles could be found in the curve of the SiO$_2$@Fe$_3$C/Fe@NCNT-GT, confirming the existence of dipole polarization and interfacial polarization relaxation. Therefore, the increased dielectric relaxation loss of the SiO$_2$@Fe$_3$C/Fe@NCNT-GT is relevant to their enhanced dipole and interface polarizations [31–35]. Figure 5 shows the R_L—f curves of the two samples with d of 1.5–5.0 mm over 2–18 GHz. It can be found that the SiO$_2$@Fe$_3$C/Fe@NCNT-GT exhibited a better EMW absorption property than the SiO$_2$@Fe$_3$C/Fe@NCNT. It should be noted that all of the R_L values of SiO$_2$@Fe$_3$C/Fe@NCNT-GT can exceed −20 dB (Figure 5a), where the minimum value was −48.43 dB with d of only 1.5 mm. However, the $R_{L,\,min}$ value for the SiO$_2$@Fe$_3$C/Fe@NCNT was only −16.63 dB with d of 5 mm (Figure 5b). Furthermore, the effective absorption bandwidth (EAB$_{10}$, $R_L \leq -10$ dB) of the SiO$_2$@Fe$_3$C/Fe@NCNT-GT was 4.51 GHz, which is superior to that of the SiO$_2$@Fe$_3$C/Fe@NCNT (2.12 GHz) (Figure 5a,b). Thus, the SiO$_2$@Fe$_3$C/Fe@NCNT-GT showed a significantly enhanced EMW absorption property in the main parameters, including $R_{L,\,min}$, EAB$_{10}$ and d values, showing it has potential applications in practical EMW absorption. In addition, our prepared SiO$_2$@Fe$_3$C/Fe@NCNT-GT had comparable, or better, EMW absorption performance than reported carbon nanotube-based absorbent materials (Figure 5c, Table S2) [36–53]. The M_z—f plot reveals that the SiO$_2$@Fe$_3$C/Fe@NCNT-GT had better impedance matching characteristics compared to the SiO$_2$@Fe$_3$C/Fe@NCNT (Figure S9). Therefore, the increase in diameter of NCNTs may also have a positive effect on the optimization of dielectric loss and impedance matching characteristics, thus enhancing the EMW absorption performance of the SiO$_2$@Fe$_3$C/Fe@NCNT-GT. Overall, compared to the counterpart (SiO$_2$@Fe$_3$C/Fe@NCNT) without treatment in the organic solvent, the as-prepared SiO$_2$@Fe$_3$C/Fe@NCNT-GT showed significantly improved EMW absorption performance. In addition, SiO$_2$@Fe$_3$C/Fe@NCNT-GT exhibited a decreased EMW absorption performance when the filling ratio was 20% or 30% (Figure S10). Therefore, the optimal filler ratio for EMW absorption is 25%.

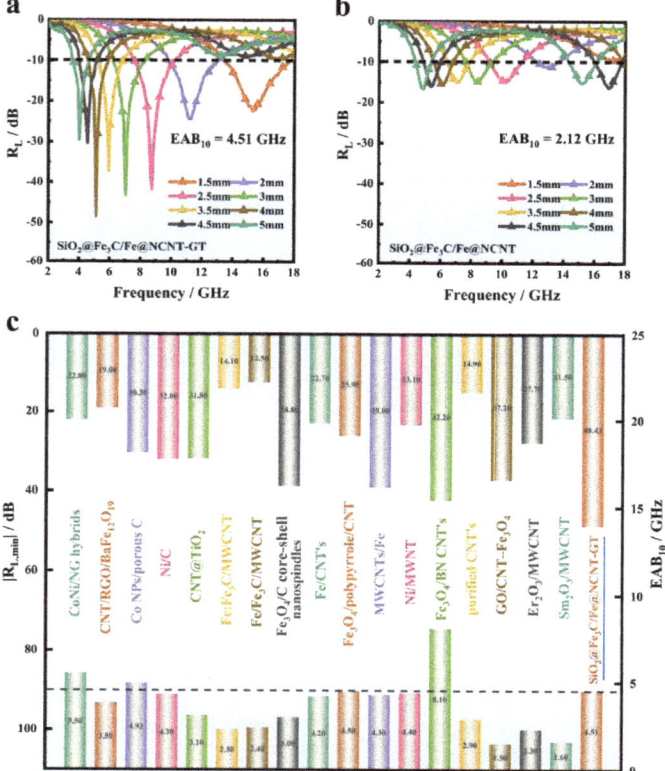

Figure 5. R_L—f curves of (**a**) SiO$_2$@Fe$_3$C/Fe@NCNT-GT and (**b**) SiO$_2$@Fe$_3$C/Fe@NCNT, (**c**) the absorption performance of SiO$_2$@Fe$_3$C/Fe@NCNT-GT with previously reported absorbers.

4. Conclusions

In summary, we fabricated the SiO$_2$@Fe$_3$C/Fe@NCNT-GT with a moderate diameter for EMW absorption. Compared to the counterpart (SiO$_2$@Fe$_3$C/Fe@NCNT) without treatment in the organic solvent, the dielectric loss of the as-prepared SiO$_2$@Fe$_3$C/Fe@NCNT-GT was optimized, the impedance matching characteristics were adjusted, and the absorption performance of EMW was significantly improved. At a filler ratio of 25%, minimum, reflection loss can reach −48.43 dB. In the meantime, effective bandwidth of the SiO$_2$@Fe$_3$C/Fe@NCNT-GT can reach 4.51 GHz, while the matching thickness is only

1.5 mm, which is better than most magnetic carbon matrix composites. Our present approach opens up an effective way to develop low-cost, high-performance EMW absorbers.

Supplementary Materials: The following are available online at https://www.mdpi.com/article/10.3390/nano11102636/s1, Figure S1: SEM images of SiO_2@Fe(OH)$_3$ and SiO_2@Fe(OH)$_3$-GT, Figure S2: XRD patterns and FTIR spectras of SiO_2@Fe(OH)$_3$ and SiO_2@Fe(OH)$_3$-GT, Figure S3: XPS spectra of the SiO_2@Fe$_3$C/Fe@NCNT-GT, Figure S4: SEM image of SiO_2@Fe$_3$C/Fe@NCNT-GT, Figure S5: (a) TEM image, (b) Average diameter of NCNTs and NPs of SiO_2@Fe$_3$C/Fe@NCNT-GT, Figure S6: (a) μ'—f curves, (b) μ''—f curves, and (c) $\tan\delta_m$—f of SiO_2@Fe$_3$C/Fe@NCNT-GT and SiO_2@Fe$_3$C/Fe@NCNT, Figure S7: Magnetization hysteresis loops of the SiO_2@Fe$_3$C/Fe@NCNT-GT and SiO_2@Fe$_3$C/Fe@NCNT, Figure S8: Cole-Cole semicircles of the (a) SiO_2@Fe$_3$C/Fe@NCNT-GT and (b) SiO_2@Fe$_3$C/Fe@NCNT, Figure S9: The M_2—f curves of the (a) SiO_2@Fe$_3$C/Fe@NCNT-GT and (b) SiO_2@Fe$_3$C/Fe@NCNT, Figure S10: R_L—f curves of (a) the SiO_2@Fe$_3$C/Fe@NCNT-GT with a filler ratio of 20 wt.% and (b) 30 wt.%, Table S1: Electrical conductivity of absorbing materials, Table S2: EMW absorption properties of some representative materials.

Author Contributions: Y.C. and X.Z. (Xiao Zhang). conceived and supervised the project. F.C. carried out the experiments, analyzed the experimental data and wrote the first version of the manuscript. Q.O., X.Z. (Xinci Zhang) and X.Z. (Xitian Zhang) evaluated the data and made the intensive discussion. F.C., J.X. and B.L. contributed to electromagnetic parameter measurements. All authors have read and agreed to the published version of the manuscript.

Funding: This research was funded by the NNSF of China (grant number 51972077), the Fundamental Research Funds for the Central Universities (grant number 3072020CF2518, 3072020CFT2505, 3072021CFT2506, 3072021CF2523 and 3072021CF2524), Natural Science Foundation of Daqing Normal University (grant number 19ZR05), Daqing City Directive Science and Technology Plan Project (grant number zd-2020-04) and Heilongjiang Provincial Natural Resources Foundation Joint Guide Project (grant number LH2020E098).

Institutional Review Board Statement: Not applicable.

Informed Consent Statement: Not applicable.

Data Availability Statement: Data is contained within the article or Supplementary Materials.

Conflicts of Interest: The authors declare no conflict of interest.

References

1. Cao, M.; Wang, X.; Zhang, M.; Cao, W.; Fang, X.; Yuan, J. Variable-Temperature Electron Transport and Dipole Polarization Turning Flexible Multifunctional Microsensor beyond Electrical and Optical Energy. *Adv. Mater.* **2020**, *32*, e1907156. [CrossRef] [PubMed]
2. Yakovenko, O.; Matzui, L.; Vovchenko, L.; Lozitsky, O.; Prokopov, O.; Lazarenko, O.; Zhuravkov, A.; Oliynyk, V.; Launets, V.; Trukhanov, S. Electrophysical properties of epoxy-based composites with graphite nanoplatelets and magnetically aligned magnetite. *Mol. Cryst. Liq. Cryst.* **2018**, *661*, 68–80. [CrossRef]
3. Yakovenko, O.S.; Matzui, L.Y.; Vovchenko, L.L.; Oliynyk, V.V.; Trukhanov, A.V.; Trukhanov, S.V.; Borovoy, M.O.; Tesel'ko, P.O.; Launets, V.L.; Syvolozhskyi, O.A.; et al. Effect of magnetic fillers and their orientation on the electrodynamic properties of BaFe$_{12-x}$Ga$_x$O$_{19}$ (x = 0.1–1.2)—epoxy composites with carbon nanotubes within GHz range. *Appl. Nanosci.* **2020**, *10*, 4747–4752. [CrossRef]
4. Arief, I.; Biswas, S.; Bose, S. FeCo-anchored reduced graphene oxide framework-based soft composites containing carbon nanotubes as highly efficient microwave absorbers with excellent heat dissipation ability. *ACS Appl. Mater. Interfaces* **2017**, *9*, 19202–19214. [CrossRef] [PubMed]
5. Vinnik, D.A.; Zhivulin, V.E.; Sherstyuk, D.P.; Starikov, A.Y.; Zezyulina, P.A.; Gudkova, S.A.; Zherebtsov, D.A.; Rozanov, K.N.; Trukhanov, S.V.; Astapovich, K.A.; et al. Electromagnetic properties of zinc–nickel ferrites in the frequency range of 0.05–10 GHz. *Mater. Today Chem.* **2021**, *20*, 100460. [CrossRef]
6. Zdorovets, M.V.; Kozlovskiy, A.L.; Shlimas, D.I.; Borgekov, D.B. Phase transformations in FeCo–Fe$_2$CoO$_4$/Co$_3$O$_4$-spinel nanostructures as a result of thermal annealing and their practical application. *J. Mater. Sci. Mater. Electron.* **2021**, *32*, 16694–16705. [CrossRef]
7. Thakur, A.; Sharma, N.; Bhatti, M.; Sharma, M.; Trukhanov, A.V.; Trukhanov, S.V.; Panina, L.V.; Astapovich, K.A.; Thakur, P. Synthesis of barium ferrite nano-particles using rhizome extract of Acorus Calamus: Characterization and its efficacy against different plant phytopathogenic fungi. *Nano-Struct. Nano-Objects* **2020**, *24*, 100599. [CrossRef]

8. Kozlovskiy, A.; Zdorovets, M. Effect of doping of $Ce^{4+}/^{3+}$ on optical, strength and shielding properties of $(0.5-x)TeO_2$-0.25MoO-0.25Bi_2O_3-xCeO$_2$ glasses. *Mater. Chem. Phys.* **2021**, *263*, 124444. [CrossRef]
9. Wu, Z.; Pei, K.; Xing, L.; Yu, X.; You, W.; Che, R. Enhanced Microwave Absorption Performance from Magnetic Coupling of Magnetic Nanoparticles Suspended within Hierarchically Tubular Composite. *Adv. Funct. Mater.* **2019**, *29*, 1901448. [CrossRef]
10. Lü, Y.; Wang, Y.; Li, H.; Lin, Y.; Jiang, Z.; Xie, Z.; Kuang, Q.; Zheng, L. MOF-Derived Porous Co/C Nanocomposites with Excellent Electromagnetic Wave Absorption Properties. *ACS Appl. Mater. Interfaces* **2015**, *7*, 13604–13611. [CrossRef]
11. Yang, R.B.; Reddy, P.M.; Chang, C.J.; Chen, P.A.; Chen, J.K.; Chang, C.C. Synthesis and characterization of Fe_3O_4/polypyrrole/carbon nanotube composites with tunable microwave absorption properties: Role of carbon nanotube and polypyrrole content. *Chem. Eng. J.* **2016**, *285*, 497–507. [CrossRef]
12. Li, F.; Zhan, W.; Su, Y.; Siyal, S.H.; Bai, G.; Xiao, W.; Zhou, A.; Sui, G.; Yang, X. Achieving excellent electromagnetic wave absorption of $ZnFe_2O_4$@CNT/polyvinylidene fluoride flexible composite membranes by adjusting processing conditions. *Compos. Part A: Appl. Sci. Manuf.* **2020**, *133*, 105866. [CrossRef]
13. Feng, Y.; Guo, X.; Lu, J.; Liu, J.; Wang, G.; Gong, H. Enhanced electromagnetic wave absorption performance of SiCN(Fe) fibers by in-situ generated Fe3Si and CNTs. *Ceram. Int.* **2021**, *47*, 19582–19594. [CrossRef]
14. Zhu, L.; Zeng, X.; Chen, M.; Yu, R. Controllable permittivity in 3D Fe_3O_4/CNTs network for remarkable microwave absorption performances. *RSC Adv.* **2017**, *7*, 26801–26808. [CrossRef]
15. Xu, J.; Zhang, X.; Yuan, H.; Zhang, S.; Zhu, C.; Zhang, X.; Chen, Y. N-doped reduced graphene oxide aerogels containing pod-like N-doped carbon nanotubes and FeNi nanoparticles for electromagnetic wave absorption. *Carbon* **2020**, *159*, 357–365. [CrossRef]
16. Zhang, X.; Xu, J.; Yuan, H.; Zhang, S.; Ouyang, Q.; Zhu, C.; Zhang, X.; Chen, Y. Large-Scale Synthesis of Three-Dimensional Reduced Graphene Oxide/Nitrogen-Doped Carbon Nanotube Heteronanostructures as Highly Efficient Electromagnetic Wave Absorbing Materials. *ACS Appl. Mater. Interfaces* **2019**, *11*, 39100–39108. [CrossRef]
17. Shu, R.; Li, W.; Zhou, X.; Tian, D.; Zhang, G.; Gan, Y.; Shi, J.-J.; He, J. Facile preparation and microwave absorption properties of RGO/MWCNTs/$ZnFe_2O_4$ hybrid nanocomposites. *J. Alloy. Compd.* **2018**, *743*, 163–174. [CrossRef]
18. Yoshiyuki, N.; Kunihiro, S. Application of ferrite to electromagnetic wave absorber and its characteristics. *IEEE Trans. Microw. Theory Tech.* **1971**, *19*, 65–72.
19. Stöber, W.; Fink, A.; Bohn, E. Controlled growth of monodisperse silica spheres in the micron size range. *J. Colloid Interface Sci.* **1968**, *26*, 62–69. [CrossRef]
20. Cao, F.; Yan, F.; Xu, J.; Zhu, C.; Qi, L.; Li, C.; Chen, Y. Tailing size and impedance matching characteristic of nitrogen-doped carbon nanotubes for electromagnetic wave absorption. *Carbon* **2020**, *174*, 79–89. [CrossRef]
21. Wei, H.; Yin, X.; Li, X.; Li, M.; Dang, X.; Zhang, L.; Cheng, L. Controllable synthesis of defective carbon nanotubes/$Sc_2Si_2O_7$ ceramic with adjustable dielectric properties for broadband high-performance microwave absorption. *Carbon* **2019**, *147*, 276–283. [CrossRef]
22. Liu, L.; Yan, F.; Li, K.; Zhu, C.; Xie, Y.; Zhang, X.; Chen, Y. Ultrasmall FeNi$_3$N particles with an exposed active (110) surface an-chored on nitrogen-doped graphene for multifunctional electrocatalysts. *J. Mater. Chem. A.* **2019**, *7*, 1083–1091. [CrossRef]
23. Wu, H.; Yang, T.; Du, Y.; Shen, L.; Ho, G.W. Identification of Facet-Governing Reactivity in Hematite for Oxygen Evolution. *Adv. Mater.* **2018**, *30*, e1804341. [CrossRef]
24. Almessiere, M.A.; Slimani, Y.; Trukhanov, A.V.; Baykal, A.; Gungunes, H.; Trukhanova, E.L.; Kostishin, V.G. Strong correlation between Dy^{3+} concentration, structure, magnetic and microwave properties of the [Ni$_{0.5}$Co$_{0.5}$](Dy$_x$Fe$_{2-x}$)O$_4$ nanosized ferrites. *J. Ind. Eng. Chem.* **2020**, *90*, 251–259. [CrossRef]
25. Kozlovskiy, A.; Egizbek, K.; Zdorovets, M.V.; Ibragimova, M.; Shumskaya, A.; Rogachev, A.A.; Ignatovich, Z.V.; Kadyrzhanov, K. Evaluation of the effi-ciency of detection and capture of manganese in aqueous solutions of FeCeO$_x$ nanocomposites doped with Nb$_2$O$_5$. *Sensors.* **2020**, *20*, 4851. [CrossRef] [PubMed]
26. Quan, B.; Gu, W.; Sheng, J.; Lv, X.; Mao, Y.; Liu, L.; Huang, X.; Tian, Z.; Ji, G. From intrinsic dielectric loss to geometry patterns: Dual-principles strategy for ultrabroad band microwave absorption. *Nano Res.* **2020**, *14*, 1495–1501. [CrossRef]
27. Zhang, Z.; Tan, J.; Gu, W.; Zhao, H.; Zheng, J.; Zhang, B.; Ji, G. Cellulose-chitosan framework/polyailine hybrid aerogel toward thermal insulation and microwave absorbing application. *Chem. Eng. J.* **2020**, *395*, 125190. [CrossRef]
28. Dolmatov, A.; Maklakov, S.; Zezyulina, P.; Osipov, D.; Petrov, D.; Naboko, A.; Polozov, V.; Maklakov, S.; Starostenko, S.; Lagarkov, A. Deposition of a SiO$_2$ Shell of Variable Thickness and Chemical Composition to Carbonyl Iron: Synthesis and Microwave Measurements. *Sensors* **2021**, *21*, 4624. [CrossRef] [PubMed]
29. Sankaran, S.; Deshmukh, K.; Ahamed, M.; Pasha, S.K. Recent advances in electromagnetic interference shielding properties of metal and carbon filler reinforced flexible polymer composites: A review. *Compos. Part A: Appl. Sci. Manuf.* **2018**, *114*, 49–71. [CrossRef]
30. Jacobo, S.E.; Aphesteguy, J.C.; Anton, R.L.; Schegoleva, N.; Kurlyandskaya, G. Influence of the preparation procedure on the properties of polyaniline based magnetic composites. *Eur. Polym. J.* **2007**, *43*, 1333–1346. [CrossRef]
31. Liu, J.; Che, R.; Chen, H.; Zhang, F.; Xia, F.; Wu, Q.; Wang, M. Microwave Absorption Enhancement of Multifunctional Composite Microspheres with Spinel Fe3O4 Cores and Anatase TiO$_2$ Shells. *Small* **2012**, *8*, 1214–1221. [CrossRef]
32. Cao, M.; Song, W.; Hou, Z.; Wen, B.; Yuan, J. The effects of temperature and frequency on the dielectric properties, electro-magnetic interference shielding and microwave-absorption of short carbon fiber/silica composites. *Carbon* **2010**, *48*, 788–796. [CrossRef]

33. Liu, P.; Gao, S.; Zhang, G.; Huang, Y.; You, W.; Che, R. Hollow Engineering to Co@N-Doped Carbon Nanocages via Synergistic Protecting-Etching Strategy for Ultrahigh Microwave Absorption. *Adv. Funct. Mater.* **2021**, *31*, 2102812. [CrossRef]
34. Gu, W.; Cui, X.; Zheng, J.; Yu, J.; Zhao, Y.; Ji, G. Heterostructure design of Fe3N alloy/porous carbon nanosheet composites for efficient microwave attenuation. *J. Mater. Sci. Technol.* **2020**, *67*, 265–272. [CrossRef]
35. Liu, Q.; Cao, Q.; Bi, H.; Liang, C.; Yuan, K.; She, W.; Yang, Y.; Che, R. CoNi@SiO$_2$@TiO$_2$ and CoNi@Air@TiO$_2$ microspheres with strong wideband microwave absorption. *Adv. Mate.* **2016**, *28*, 486–490. [CrossRef]
36. Feng, J.; Pu, F.; Li, Z.; Li, X.; Hu, X.; Bai, J. Interfacial interactions and synergistic effect of CoNi nanocrystals and nitrogen-doped graphene in a composite microwave absorber. *Carbon* **2016**, *104*, 214–225. [CrossRef]
37. Zhao, T.; Ji, X.; Jin, W.; Wang, C.; Ma, W.; Gao, J.; Dang, A.; Li, T.; Shang, S.; Zhou, Z. Direct in situ synthesis of a 3D interlinked amorphous carbon nanotube/graphene/BaFe$_{12}$O$_{19}$ composite and its electromagnetic wave absorbing properties. *RSC Adv.* **2017**, *7*, 15903–15910. [CrossRef]
38. Wang, H.; Xiang, L.; Wei, W.; An, J.; He, J.; Gong, C.; Hou, Y. Efficient and Lightweight Electromagnetic Wave Absorber Derived from Metal Organic Framework-Encapsulated Cobalt Nanoparticles. *ACS Appl. Mater. Interfaces* **2017**, *9*, 42102–42110. [CrossRef]
39. Zhang, X.F.; Dong, X.L.; Huang, H.; Liu, Y.Y.; Wang, W.N.; Zhu, X.G.; Lv, B.; Lei, J.P. Microwave absorption properties of the car-bon-coated nickel nanocapsules. *Appl Phys. Lett.* **2006**, *89*, 053115. [CrossRef]
40. Liu, X.; Cui, X.; Chen, Y.; Zhang, X.-J.; Yu, R.; Wang, G.-S.; Ma, H. Modulation of electromagnetic wave absorption by carbon shell thickness in carbon encapsulated magnetite nanospindles–poly(vinylidene fluoride) composites. *Carbon* **2015**, *95*, 870–878. [CrossRef]
41. Mo, Z.C.; Yang, R.L.; Lu, D.W.; Yang, L.L.; Hu, Q.M.; Li, H.B.; Zhu, H.; Tang, Z.; Gui, X. Lightweight, three-dimensional carbon nanotube@TiO$_2$ sponge with enhanced microwave absorption performance. *Carbon* **2019**, *144*, 433–439. [CrossRef]
42. Wen, F.; Zhang, F.; Liu, Z. Investigation on Microwave Absorption Properties for Multiwalled Carbon Nanotubes/Fe/Co/Ni Nanopowders as Lightweight Absorbers. *J. Phys. Chem. C* **2011**, *115*, 14025–14030. [CrossRef]
43. Zhou, Y.; Miao, Z.; Shen, Y.; Xie, A. Novel porous FexCyNz/N-doped CNT nanocomposites with excellent bifunctions for cat-alyzing oxygen reduction reaction and absorbing electromagnetic wave. *Appl. Surf. Sci.* **2018**, *453*, 83–92. [CrossRef]
44. Wang, L.; Jia, X.; Li, Y.; Yang, F.; Zhang, L.; Liu, L.; Ren, X.; Yang, H. Synthesis and microwave absorption property of flexible magnetic film based on graphene oxide/carbon nanotubes and Fe$_3$O$_4$ nanoparticles. *J. Mater. Chem. A* **2014**, *2*, 14940–14946. [CrossRef]
45. Qi, X.; Xu, J.; Hu, Q.; Deng, Y.; Xie, R.; Jiang, Y.; Zhong, W.; Du, Y. Metal-free carbon nanotubes: Synthesis, and enhanced intrinsic mi-crowave absorption properties. *Sci. Rep.* **2016**, *6*, 28310. [CrossRef]
46. Lin, H.; Zhu, H.; Guo, H.; Yu, L. Investigation of the microwave-absorbing properties of Fe-filled carbon nanotubes. *Mater. Lett.* **2007**, *61*, 3547–3550. [CrossRef]
47. Zhang, T.; Zhong, B.; Yang, J.Q.; Huang, X.X.; Wen, G. Boron and nitrogen doped carbon nanotubes/Fe$_3$O$_4$ composite archi-tectures with microwave absorption property. *Ceram. Int.* **2015**, *41*, 8163–8170. [CrossRef]
48. Zou, T.; Li, H.; Zhao, N.; Shi, C. Electromagnetic and microwave absorbing properties of multi-walled carbon nanotubes filled with Ni nanowire. *J. Alloy. Compd.* **2010**, *496*, L22–L24. [CrossRef]
49. Xu, P.; Han, X.J.; Liu, X.R.; Zhang, B.; Wang, C.; Wang, X.H. A study of the magnetic and electromagnetic properties of γ-Fe$_2$O$_3$–multiwalled carbon nanotubes (MWCNT) and Fe/Fe$_3$C–MWCNT composites. *Mater. Chem. Phys.* **2009**, *114*, 556–560. [CrossRef]
50. Zhang, L.; Zhu, H.; Song, Y.; Zhang, Y.; Huang, Y. The electromagnetic characteristics and absorbing properties of multi-walled carbon nanotubes filled with Er$_2$O$_3$ nanoparticles as microwave absorbers. *Mater. Sci. Eng. B* **2008**, *153*, 78–82. [CrossRef]
51. Zhang, L.; Zhu, H. Dielectric, magnetic, and microwave absorbing properties of multi-walled carbon nanotubes filled with Sm$_2$O$_3$ nanoparticles. *Mater. Lett.* **2009**, *63*, 272–274. [CrossRef]
52. Green, M.; Van Tran, A.T.; Chen, X. Maximizing the microwave absorption performance of polypyrrole by data-driven discovery. *Compos. Sci. Technol.* **2020**, *199*, 108332. [CrossRef]
53. Green, M.; Tran, A.T.; Chen, X. Obtaining strong, broadband microwave absorption of polyaniline through data-driven ma-terials discovery. *Adv. Mater. Interfaces* **2020**, *7*, 2000658. [CrossRef]

Article

Tunable Low Crystallinity Carbon Nanotubes/Silicon Schottky Junction Arrays and Their Potential Application for Gas Sensing

Alvaro R. Adrian [1,2], Daniel Cerda [1,2], Leunam Fernández-Izquierdo [3], Rodrigo A. Segura [4], José Antonio García-Merino [1,2] and Samuel A. Hevia [1,2,*]

[1] Instituto de Física, Pontificia Universidad Católica de Chile, Casilla 306, Santiago 6904411, Chile; aradrian@uc.cl (A.R.A.); ddcerda@uc.cl (D.C.); jose.garcia@uc.cl (J.A.G.-M.)
[2] Centro de Investigación en Nanotecnología y Materiales Avanzados, Pontificia Universidad Católica de Chile, Casilla 306, Santiago 6904411, Chile
[3] Department of Material Science & Engineering, The University of Texas at Dallas, Richardson, TX 75080, USA; lxf180007@utdallas.edu
[4] Instituto de Química y Bioquímica, Universidad de Valparaíso, Avenida Gran Bretaña 1111, Valparaíso 2340000, Chile; rodrigo.segura@uv.cl
* Correspondence: samuel.hevia@fis.puc.cl; Tel.: +56-999986438

Abstract: Highly ordered nanostructure arrays have attracted wide attention due to their wide range of applicability, particularly in fabricating devices containing scalable and controllable junctions. In this work, highly ordered carbon nanotube (CNT) arrays grown directly on Si substrates were fabricated, and their electronic transport properties as a function of wall thickness were explored. The CNTs were synthesized by chemical vapor deposition inside porous alumina membranes, previously fabricated on n-type Si substrates. The morphology of the CNTs, controlled by the synthesis parameters, was characterized by electron microscopies and Raman spectroscopy, revealing that CNTs exhibit low crystallinity (LC). A study of conductance as a function of temperature indicated that the dominant electric transport mechanism is the 3D variable range hopping. The electrical transport explored by I–V curves was approached by an equivalent circuit based on a Schottky diode and resistances related to the morphology of the nanotubes. These junction arrays can be applied in several fields, particularly in this work we explored their performance in gas sensing mode and found a fast and reliable resistive response at room temperature in devices containing LC-CNTs with wall thickness between 0.4 nm and 1.1 nm.

Keywords: low crystallinity carbon nanotubes; anodic aluminum oxide; electric transport; gas sensor; Schottky junction arrays

1. Introduction

Carbon nanotubes (CNTs) remain being considered as promising materials in science and technology owing to their multiple outstanding properties [1–4]. However, these properties strongly depend on the crystallinity or graphitization level of their walls since this determines the electronic structure [5], and consequently, the optical [6], electric [7], and mechanical [8] properties. These nanostructures have proved to be useful as active materials in photo-actuators [9], photodetectors [10], semiconductor electronics [11], and gas sensors [12]. Nevertheless, it is still challenging to produce an array of CNTs with the same morphological characteristics and, therefore, the same physical properties [13]. A particularly suitable method for growing highly ordered nanotube arrays is to use an anodic aluminum oxide (AAO) as a template. One of the significant advantages of using AAO membranes is the cost-efficient synthesis of large area arrays of densely packed nanopores with well-controlled dimensions in the nanometer range. Also, using this dielectric matrix to deposit the carbon nanostructures provides mechanical support to fabricate robust devices [14]. Using the chemical vapor deposition (CVD) process, the diameter, length, wall thickness, and graphitization level of CNT can be controlled [15–18]. The controllability in morphology has an advantage in tuning physical properties [19,20]. The use of these arrays has been reported in electronic systems with better performance than those using non-organized CNTs [21,22]. However, the synthesis of millions of CNTs will hardly be completely crystalline, and there will be amorphous components that affect specific properties. Therein lies the importance of studying and classifying the effects of systems with a high degree of disorder.

The distinctive feature that low-crystalline materials exhibit is that they have localized electronic states; therefore, in a material with a strong structural disorder, an electronic structure with highly localized states arises, exhibiting an electric transport mechanism known as variable range hopping (VRH) [23]. In this case, the conductance tends to zero as the temperature tends to zero due to the hopping process is a phonon-assisted mechanism that transfers an electron from one localized state to another [24]. A good description of this mechanism opens the possibility to tailor the global transport properties in arrays with highly structural defects [25]. The contribution to study CNTs with localized states has a relevant impact since macroscopic samples generally contain several distortions, either in the diameters, chirality, or doping [26]. In this case, rather than making an individualized picture of the properties of the nanotubes, it is necessary to generate approximate models of the average characteristics. For that reason, this work analyzes the electric transport in low-crystallinity CNTs (LC-CNTs) with controllable dimensions in a robust device consisting of Si substrates.

The heterojunction form between carbon allotropes and Si has been reported as a Schottky junction, in which the carrier transport generally follows the thermionic emission (TE) theory at room temperature [27–29]. Moreover, other conduction mechanisms are superposed in complex nanosystems interfaces such as space-charge limited or tunneling [30]. These components are difficult to explore due to several variables, such as substrate doping, CNTs/Si contact, or temperature [31]. It is important to approach the heterojunction as a global system and isolate the crystalline and non-crystalline components from CNTs. For instance, the study of CNTs and Si junctions using AAO and CVD process with synthesis temperature of 950 °C has been reported [31–33]. At this temperature, the CNTs start to lose the defects and behave more as multiwall CNTs with metallic properties [34]. The effects that the impurities in the crystal lattice can induce in electronic transport are not observed in these systems. For that reason, CNTs synthesized at low temperatures are expected to present low crystallinity, and therefore, the effects of interest in the electronic transport mechanism can be observed. Moreover, it is important to analyze the behavior as a function of wall thickness and how this affects the device junction due to dimensionality effects. Furthermore, since these heterojunction devices have been proved to sense gases [34,35], it is appropriate to explore this characteristic in the proposed LC-CNTs. The tunability of electrical transport will allow exploring different morphologies to optimize gas sensing.

This paper presents the synthesis, characterization, and study of the electrical transport properties of LC-CNTs arrays grown directly on Si substrates using AAO as a template. The dominant electric transport mechanism in LC-CNTs is the 3D-VRH and shows a strong dependence on the wall thickness of tubes. Moreover, the LC-CNTs arrays are exposed to reducing gases, and they exhibit a dependence of its electric resistance as a function of gas concentration, which opens the possibilities to use it as gas sensors.

2. Materials and Methods

2.1. Synthesis of Porous Alumina Membranes on Silicon Substrates

Porous alumina membranes (PAMs) were fabricated on Si substrates using a 5 μm layer of Al (99.999% purity). The Al was deposited at a rate of 2 Å/s over polished n-type Si (100) wafers (1–10 Ω·cm) by electron beam evaporation. The Al film was anodized in two steps to obtain PAMs with highly ordered pore patterns [36]. The first anodization step was performed at 40 V in 0.3 M oxalic acid for a period of 40 min, maintaining the temperature of the electrolyte at 5 °C. Furthermore, an aqueous solution with 6.0 wt%

phosphoric acid and 1.8 wt% chromic acid at 60 °C was employed to remove the porous alumina layer produced in the first anodization step, leaving an ordered pattern of pores nucleus in the surface of the mask. Then, the second anodization was performed under the same previously mentioned conditions for 50 min, yielding a homogeneous and ordered membrane with a thickness close to 2.5 µm. Finally, to remove the alumina barrier layer at the bottom of the pores and widen the pores without affecting the membrane order, the samples were subjected to an etching treatment with a 5 wt% phosphoric acid solution at 21 °C for 55 min.

2.2. Low Crystallinity Carbon Nanotubes Synthesis

The synthesis of LC-CNTs inside the pores of PAMs was achieved by CVD method, using a horizontal tube furnace (MTI - OTF 1200X furnace, MTI Corp., Richmond, CA, USA). A piece of PAM/Si (~1 cm^2) was heated at the center of the reactor at a rate of 30 °C/min under an Ar atmosphere (200 sccm) until reaching 650 °C. Then, a flow of C_2H_2 at 25 sccm was introduced as a carbon source, and CNTs were synthesized at different times: from 1 to 30 min. This process promotes the growth of controllable nanotubes inside the pores with tunable wall thickness by keeping the temperature constant [36]. Finally, the whole system LC-CNTs/PAM/Si was cooled down to 21 °C under Ar environment.

2.3. Deposition of the Top Electrode

Top electrodes, consisting of a ~10 mm^2 circular area of 99.9% pure Au, were deposited perpendicular to the top of the LC-CNT/PAM/Si samples at an evaporation rate of 0.4 Å/s until obtaining 100 nm of Au thickness. The deposition was performed in a Balzers evaporator equipped with a Temescal STIH-270-2PT (Ferrotec, Livermore, CA, USA) electron beam source operated at 8 kV and a quartz crystal microbalance to measure the evaporation rate and deposited thickness.

2.4. Characterizations

The structural characterization was carried out through Raman Spectroscopy using a LabRam010 spectrometer (Horiba. Ltd., Kyoto, Japan) at 632.5 nm wavelength. Scanning electron microscopy (SEM) and transmission electron microscopy (TEM) were used to characterize the samples morphologically. SEM analysis was carried out using a FEI Quanta 250 FEG (Thermo Fisher Scientific, Waltham, MA, USA). Standard TEM analysis was performed using a Hitachi HT7700 (Hitachi High Tech Co., Ltd., Tokyo, Japan), and high-resolution TEM (HR-TEM) by using a FEI Tecnai ST-F20 microscope (FEI Company, Hillsboro, OR, USA). In order to perform the TEM measurements, CNTs were released from the PAM by dissolving it in sodium hydroxide solution 3.5 M at 21 °C. Afterward, samples were rinsed with double distilled water and suspended in isopropanol alcohol to obtain a CNTs dispersion.

2.5. Electric Transport Measurements as a Function of Temperature

The electric transport properties were studied in the samples as a function of the wall thickness by analyzing the I–V curves. This measurement assists in the identification of the existing junction between the Si substrate and the LC-CNTs. The I–V curves were measured in the voltage range from −1.5 to 0.8 V to predict a model which fits the electrical characteristics of the junction. Moreover, the conductance was measured around zero bias as a function of temperature from 10 to 300 K to study the dominant transport mechanism. The samples were biased by contacting the top Au-electrode and the Si substrate. The measurements were performed with a Keithley electrometer model 6517B (Keithley Instruments Inc., Cleveland, OH, USA) in a 10 K closed cycle refrigerator system from Janis Research Company (Wilmington, MA, USA) with high vacuum conditions (<10^{-6} Torr). The resistance of the fabricated devices is several orders of magnitude larger than the total resistance of the wires and electrodes; therefore, the errors introduced by using a two-probe measurement are negligible.

2.6. Room Temperature Resistance Measurements in Different Atmospheres

The electrical response of the samples was analyzed in a perturbed atmosphere to test their performances as gas sensors. These were exposed to different cycles of H_2 and C_2H_2 concentrations under an Ar atmosphere at room temperature (21 ± 1 °C), atmospheric pressure, and absence of light. The resistive response (S), shows in Equation (1), is defined as the percentage change in electric resistance when the device is exposed to the analyte ($R(t)$) compared to Ar environment (R_{Ar}). The electric resistance was measured using a Keithley 6487 picoammeter (Tektronix, Portland (Beaverton), OR, USA) around zero bias.

$$S = \frac{R(t) - R_{Ar}}{R_{Ar}} 100\% \qquad (1)$$

The gas mixtures were prepared on the base of a total flow of 100 sccm controlled by Alicat (Alicat Scientific Inc., Tucson, AZ, USA) Mass Flow Controllers (Models MC-5SLM for Ar and MC-10SCCM for H_2 and C_2H_2). The measured response corresponded to changes in current when the device was exposed to a certain amount of analyte by a certain amount of time.

3. Results and discussions

3.1. Morphological and Structural Characterization

Figure 1 shows SEM micrographs of a LC-CNTs/PAM/Si sample, synthesized with 5 min. Figure 1a corresponds to a top view of a PAM with the LC-CNTs grown inside, and Figure 1b shows the same sample after the Au deposition. The incorporation of the top electrode keeps the pores open, but their average diameter is reduced from 43 ± 6 nm to 30 ± 9 nm. Figure 1c shows a lateral view of the sample, in which the electrode thickness (129 ± 23 nm) and the PAM height (2040 ± 2 nm) are determined. Since the Au is deposited with the sample oriented perpendicular to the evaporated metal flow, the Au partially penetrates the nanotubes. Figure 1d shows a backscattered electron image which evidences that the Au average penetration inside the pores is about 320 ± 65 nm. Thus, the Au and Si-electrodes are electrically connected only through the LC-CNTs.

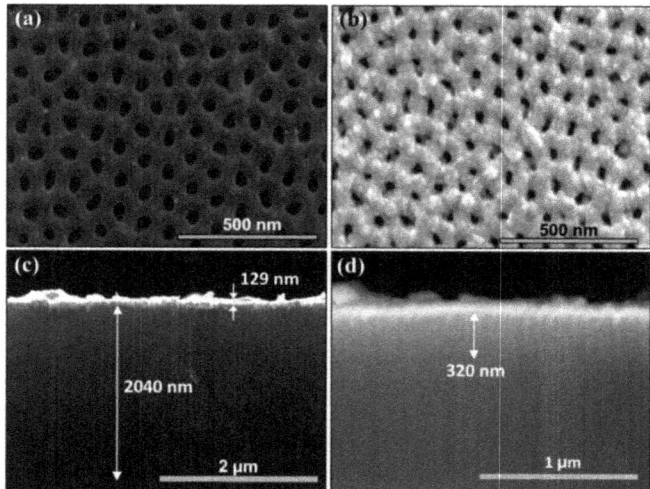

Figure 1. SEM micrographs of a sample synthesized with 5 min of growth time. (**a**) Top view of the pristine sample. (**b**) top view after Au deposition. (**c**) Side view, showing the height of the PAM and the thickness of Au-electrode. (**d**) Backscattered electron image side view, showing the Au penetration inside the nanopores.

Figure 2a–c show TEM images of the LC-CNTs grown with 5, 20, and 30 min of synthesis time, respectively. Moreover, Figure 2d corresponds to a representative HR-TEM image of a nanotube synthesized at 30 min. From this image, it is possible to notice the low degree of crystallinity of the CNTs. The average wall thickness of LC-CNTs with 5 min, 20 min, and 30 min of synthesis time is 0.7 ± 0.4 nm, 1.9 ± 0.4 nm, and 3.2 ± 0.4 nm, respectively. These average wall thicknesses are uniform along the vertical direction of the CNTs. The thickness (w) as a function of time synthesis (ts) is plotted in Figure 2e, and a linear dependence is observed.

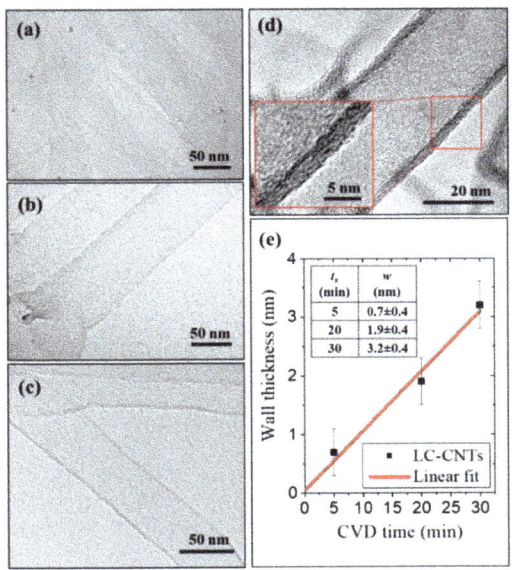

Figure 2. TEM micrographs of LC-CNTs growth with 25 sccm C_2H_2 flow at (**a**) 5 min, (**b**) 20 min, and (**c**) 30 min of synthesis time. (**d**) HR-TEM micrograph of sample growth with 30 min of synthesis time. (**e**) Wall thickness as a function of the time synthesis plot obtained from the HR-TEM images.

Figure 3 shows the first-order Raman spectra between 850 cm^{-1} and 1900 cm^{-1} range of samples with a wall thickness of 0.7 nm, 1.1 nm, 1.9 nm, and 3.2 nm. The spectra showed two main resonances located around 1326 cm^{-1} and 1600 cm^{-1}, which correspond to the G and D bands of carbonaceous materials, respectively [37]. Both resonances are linked to vibrational modes of sp^2 bonded carbon atoms. The G peak involves the bond-stretching motion of C–C atoms (E_{2g} vibrational mode), which occurs even without the presence of the six-fold aromatic rings [38]. The D resonance can be linked to an active A_{1g} breathing mode in amorphous carbon structures [39,40]. In this case, the spectra are characteristic of carbon nanotubes with a low degree of graphitization [37]. Besides, two peaks labeled as $7A_1$ and $5A_1$ shown in Figure 4a need to be considered to fit the data. These resonances, located around 1200 cm^{-1} and 1510 cm^{-1}, can be attributed to the breathing modes of seven and five-member carbon rings, respectively [41].

From the Raman spectra and the plots of fit parameters shown in Figure 3, it is possible to realize that the degree of graphitization of the LC-CNTs does not exhibit significant changes concerning the wall thickness. In all the samples, the Raman shift position of the four resonances, RS($7A_1$), RS(D), RS($5A_1$), and RS(G), is almost constant. Similar behavior is observed for intensity ratios between the peaks, and particularly in the case of D and G peaks, in which the intensity ratio has a value close to 0.9. On the other hand, the full width at half maximum (FWHM) of G and D peaks (FWHM(G) and FWHM(D)) tend to decrease in a bounded range. These values are an indication that the samples have a similar level of graphitization. The low crystallinity of the CNTs observed in TEM micrographs and Raman spectra agree with previous work [15,22,42].

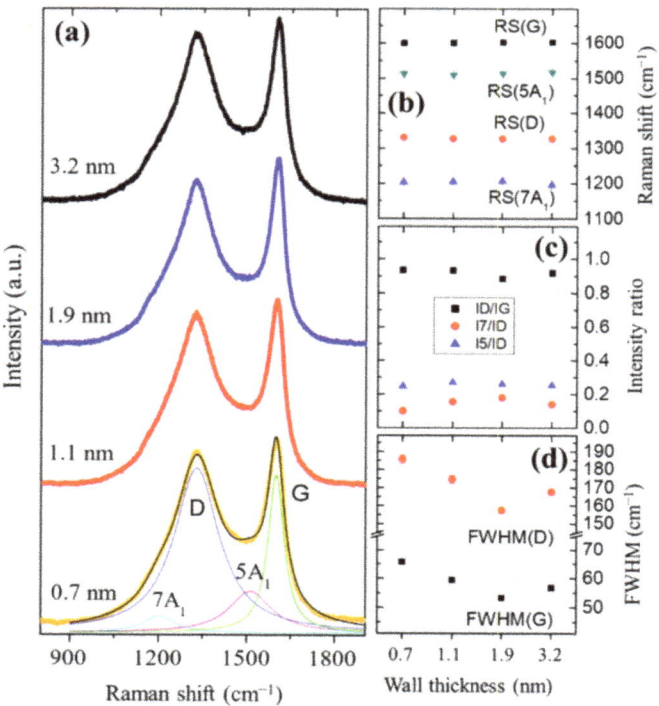

Figure 3. (**a**) Raman spectra of LC-CNTs as a function of wall thickness. (**b**) Peak position of resonances $7A_1$, D, $5A_1$, G. (**c**) Representative ratio of I(D)/I(G), I($7A_1$)/I(D), and I($5A_1$)/I(D). (**d**) FWHM of G and D peaks.

3.2. Study of Conductance as a Function of Temperature

Since there is a considerable degree of disorder in the studied CNTs, a conductance analysis as a function of temperature was used to determine their electric transport properties. Figure 4 shows the conductance characterization of the same samples analyzed by Raman spectroscopy. The electrical behavior of all LC-CNTs exhibits a non-metallic temperature dependence, which can be mainly explained by using the variable range hopping (VRH) model [23]. This transport promotes that charge carriers move by phonon-assisted hopping between localized states. The conductance at zero electric fields can be obtained by Mott's law [43] as follows:

$$G = G_h \exp\left[-(T_0/T)^{1/(d+1)}\right] \quad (2)$$

where G_h is the hopping conductance, T is the absolute temperature, $T_0 = \alpha^3/k_B n(E_f)$ is the characteristic activation temperature and is a measure of the degree of electronic localization, which depends on the parameter α that is related to the spatial decay of the localized electronic state, $n(E_f)$, the density of localized electronic states at the Fermi level, and the Boltzmann constant (k_B). The dimensionality value is $d = 3$, obtained from the best fit of the data, which indicates that the dominant electric transport mechanism is the three-dimensional variable range hopping (3D-VRH). However, as the LC-CNT wall thickness increase, the conductance does not tend to zero when the temperature tends to zero (see insets in Figure 4), as is required by Equation (2). Hence, it is necessary to add a G_m parameter that can be considered as roughly independent of temperature, and which represents the main contribution of a metallic transport mechanism acting in parallel to the 3D-VRH for the low-temperature range (<20 K). It is important to mention that some other transport mechanisms could explain this non-phonon assisted conductivity, such as Bloch Grüneisen [44] or FIT [45]. However, the contribution of G_m is at least two orders of magnitude lower than VRH around 300 K, therefore, for simplicity, the model with the fewest free fit parameters was used for the electric conductance curve fitting. The parameters of the fitting of each curve are presented in Table 1. It is noticed that T_0 decreases when the wall thickness increases. This value changes up to two orders of magnitude from 0.7 nm to 3.2 nm of wall thickness and is in good range for amorphous systems (10^5–10^{12} K) [46]. This difference could be originated by an increment in $n(E_f)$ or a reduction of the localized electronic wave function α. Nevertheless, the magnitude of this difference is expected to mainly arrive from a change in α, due to T_0 has a cubic dependence on this parameter. The previous discussion implies that the wave functions are less localized as the wall thickness increases, which is also consistent with the observed behavior of the parameter G_m.

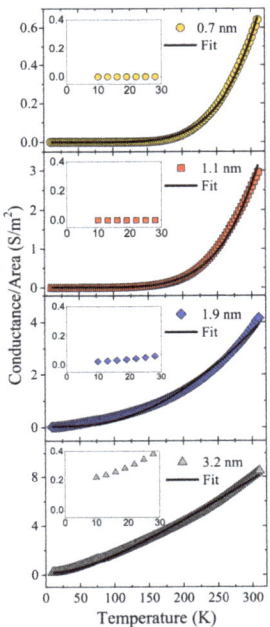

Figure 4. Conductance as a function of temperature and wall thickness of LC-CNTs. Insets show a zoom near 10 K.

Table 1. Fitting parameters for the experimental conductance of Figure 4.

w (nm)	T_0 (K)	G_h (S/m^2)	G_m (S/m^2)
0.7 ± 0.4	8.6 ± 0.2 × 10^7	6.0 ± 0.9 × 10^9	1.0 ± 0.1 × 10^{-4}
1.1 ± 0.4	7.1 ± 0.2 × 10^7	1.0 ± 0.1 × 10^{10}	1.1 ± 0.1 × 10^{-2}
1.9 ± 0.4	1.8 ± 0.1 × 10^6	2.4 ± 0.3 × 10^4	3.0 ± 0.3 × 10^{-2}
3.2 ± 0.4	2.4 ± 0.1 × 10^5	1.6 ± 0.1 × 10^3	2.1 ± 0.2 × 10^{-2}

3.3. Gas Sensing Measurements

The tunable electric transport exhibited by the nanotubes as a function of wall thickness opens the possibility to design devices for a particular application. For instance, a sensor to detect a certain gas atmosphere is feasible since the molecules interacting with the sample are expected to change the electrical parameters [47]. To study this principle, an experiment to measure the electrical resistance was performed in a gas chamber for a set of samples with wall thickness between 0.3 to 3.2 nm. Figure 5a shows the resistive response of a specific device with 0.7 nm wall thickness under different concentrations of analytes (H_2 and C_2H_2) from 1% to 5%. It is observed five representative cycles of the transient responses under H_2 and C_2H_2 analytes. The maximum values reveal a quasi-linear dependence on both gas concentrations (insets), which is an indication that the sensing mechanism is not saturated under these conditions. The resistive response has been tested in different bias voltages, and the results are exposed in Figure 5b. For both analytes, the sample exhibits their maximum response at biased closed to 0.3 V. This nonlinear behavior is consistent with the observed in materials forming junctions in which the maximum sensitivity is related to a potential barrier [48].

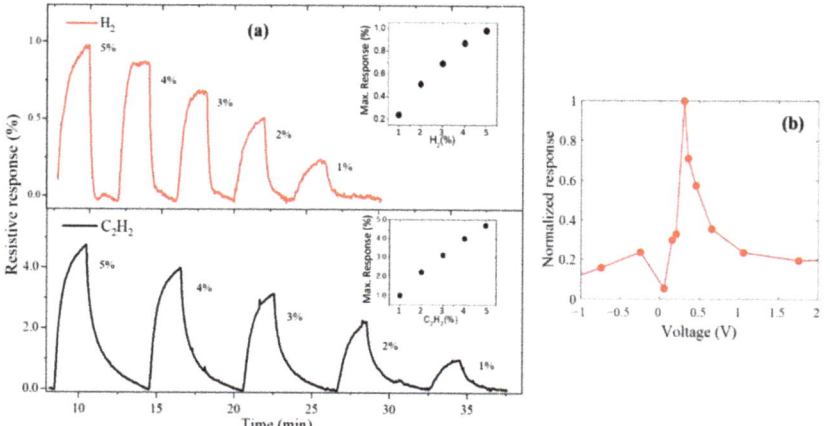

Figure 5. (a) Gas sensing behavior of sample with 0.7 nm wall thickness at different H_2 and C_2H_2 concentrations. Insets show the maximum sensitivity percentage as a function of the analyte concentration. (b) Normalized resistive response as a function of bias voltage.

Additionally, the resistive response of several devices was evaluated at the same concentrations of H_2 and C_2H_2, and a bias voltage around 0.3 V. Particularly, a strong dependence as a function of the wall thicknesses of LC-CNTs was observed. Table 2 summarizes the maximum response measured under H_2 and C_2H_2, and the conductance at zero bias. Only the devices containing LC-CNTs with wall thicknesses between 0.4 and 1.9 nm exhibit a response to the presence of the analytes. In the case of devices with thin walls (<0.4 nm), a high noise was presented in the current measurements; since thinner tubes have a very low conductance (bellow 5×10^{-3} S/m^2), the gas sensing signal was overlapped to the noise. Moreover, for tubes with thicker walls (>1.1 nm) and higher conductance, the sensibility tens to disappear. For both analytes and in all tested concentrations, the response time of the arrays was less than 15 s, a period that can be mainly attributed to the time the analyte takes flowing from the flow controller to the device. The half-maximum time, period which the sensor takes to reach half of the maximum response, was observed between 15 s and 25 s for all cases, being just a few seconds faster for H_2 than C_2H_2.

Previous reports indicate that devices built on the same conditions as these LC-CNTs do not respond electrically to H_2 perturbation by a reaction effect [35]. Besides, the results of Figure 5b point out that the LC-CNTs/Si samples contain a diode-like junction since the electrical response is more representative around 0.3 V bias voltage. This effect is common in porous nanomaterials in which the gas produces a perturbation at the junction interface that gives rise to a change in the electrical signal [49–52]. The physical mechanism behind this is that the gases permeate through the pores until it reaches the contact barriers, producing the electrical difference. Moreover, we performed the gas sensing experiment in self-supported LC-CNTs (without the Si substrate), using Au electrodes in the top and the bottom, and there was no change in the resistivity response. Thus, the heterojunction is expected to promote the sensing gases mechanism and, to explore its nature, an analysis of the electrical transport is carried out below.

Table 2. Maximum resistive response measuring under exposure to 5% of H_2 and C_2H_2 concentration in Ar atmosphere for several LC-CNTs devices.

w (nm)	H_2 Max. Resp. (%)	C_2H_2 Max. Resp. (%)	Conductance/Area (S/m^2)
0.3 ± 0.4	0	0	2.41 ± 0.02 × 10^{-3}
0.4 ± 0.4	2.7 ± 0.1	5.2 ± 0.1	1.62 ± 0.01 × 10^{-1}
0.7 ± 0.4	1.0 ± 0.2	5.7 ± 0.1	1.94 ± 0.02 × 10^{-1}
1.1 ± 0.4	0.4 ± 0.2	1.3 ± 0.2	2.00 ± 0.02 × 10^{0}
1.9 ± 0.4	0	2.2 ± 0.2	2.90 ± 0.03 × 10^{0}
3.2 ± 0.4	0	0	6.83 ± 0.07 × 10^{0}

3.4. Electrical Characterization of the LC-CNTs/Si Junction

To explain the previous results, we study the junction between LC-CNTs and Si, with a focus on the devices that exhibit the highest sensitivity. For that purpose, the dark I–V curves were measured on samples that contain LC-CNTs with 0.4 nm, 0.7 nm, and 1.1 nm of wall thickness, in a voltage range from −1.5 V to 0.7 V at room temperature, connecting the positive terminal to the LC-CNTs (Au-electrode) and the negative to the Si substrate. The results are plotted in Figure 6, in which it is observed that the curves exhibit a rectifying behavior with a high reverse current. In fact, these I–V curves can be modeled by an equivalent circuit consisting of a resistance connecting in parallel with a diode and a resistance in series, as shown in the inset of Figure 6.

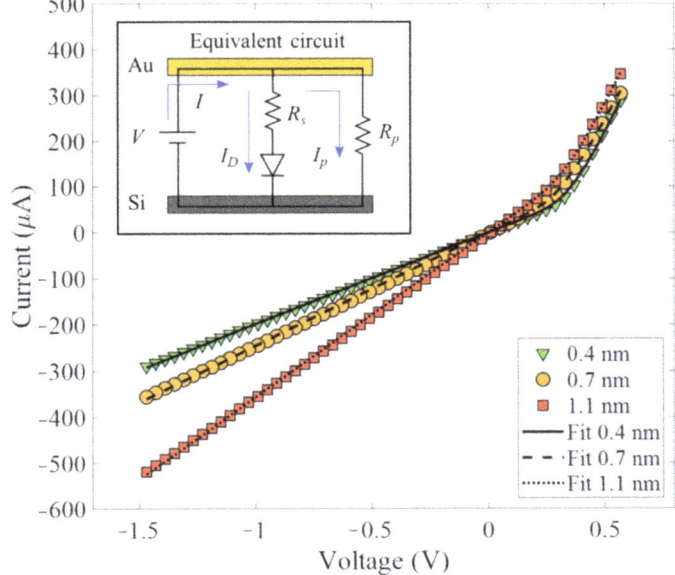

Figure 6. Room temperature dark I–V curves of samples with wall thickness of 0.4 nm, 0.7 nm, and 1.1 nm. Equivalent circuit (inset).

The I–V curves indicate that the arrays present two kinds of junctions attributed to the contacts between LC-CNTs and Si. Since below 0 V, the I–V curves have linear behavior, and above 0 V has a nonlinear tendance, the simplest model that explains this behavior is a parallel system composed of rectifying (Schottky) and non-rectifying (ohmic) junctions. The resistance in parallel (R_p) of the circuit represents the equivalent resistance of all LC-CNTs ohmic-connected with the substrate. Meanwhile, the resistance in series (R_s) represents the equivalent resistance of the LC-CNTs connected to the Si that forming a rectifying junction. Thus, R_s considers the resistance of the LC-CNTs plus the junction resistances. This junction can be modeled as a Schottky barrier [31,53], and the current through the thermionic emission diffusion (TED) theory [54], described by:

$$I_D = I_s \left[\exp\left(\frac{V}{nV_T}\right) - 1 \right] \tag{3}$$

This model initially developed for charge carrier transport across potential barriers in crystalline materials has also been used for non-crystalline systems [49]. Considering the equivalent circuit (inset Figure 6), the total current through it is given by the expression [55]:

$$I = I_D + I_p = I_s \left[\exp\left(\frac{V(1 + R_s/R_p) - IR_s}{nV_T}\right) - 1 \right] + \frac{V}{R_p} \tag{4}$$

where I_D is the current through the diode and R_s, I_p is the current through R_p, I_s corresponds to the saturation current, n is the ideality factor, and $V_T = k_B T$ is the thermal voltage. The analytical solution of this equation can be obtained using the Lambert W function [55].

$$I = \frac{nV_T}{R_s} W \left\{ \frac{I_s R_s}{nV_T} \exp\left(\frac{V + I_s R_s}{nV_T}\right) \right\} - I_s + \frac{V}{R_p} \tag{5}$$

Since the model is dictated by thermionic emission, I_s has the following expression:

$$I_s = A \cdot A^* \cdot T^2 \cdot \exp\left(\frac{-\phi_B}{V_T}\right) \tag{6}$$

where A is the contact area, A^* is the Richardson constant, and ϕ_B is the barrier voltage.

Table 3 shows the parameters of the I–V curves fitting of Figure 6 and other relevant information calculated from the fit parameters. It is observed that n, which is a measure of conformity of the diode behavior to TED theory, has a close value to 1 (ideality), indicating that the model is appropriate to describe the charge transport across the junction. The equivalent resistance R_p decreases as the wall of the CNTs widens. This dependence was expected due to his value is related to the individual resistance of the nanotubes, which are less resistive as their wall thickness increases. The value of R_s slowly decreases as the wall widens; however, in this case, it is not possible to attribute this only to the resistance of the LC-CNTs due the junction resistances are expected to contribute to the equivalent resistance Rs. However, the values of R_s and R_p can be used to get a lower bound of the percentage of CNTs connected through a Schottky junction (SC), because R_s is an upper bound of the equivalent resistance of the CNTs connected through a Schottky junction but without considering the junction resistance contribution. This percentage is given by SC = $100/(1 + R_s/R_p)$, and as we can be observed in Table 3, the number of CNTs connected through a Schottky junction is more than 80% in all the cases.

Table 3. Fitted parameters of I–V curves as a function of wall thickness.

w (nm)	n	R_p (Ω)	R_s (Ω)	I_s (nA)	SC (%)	A (m^2)	ϕ_B (eV)
0.4 ± 0.4	1.09 ± 0.01	11226 ± 171	1093 ± 16	0.11 ± 0.18	>91	5.5 × 10^{-8}	0.34 ± 0.04
0.7 ± 0.4	1.00 ± 0.01	5019 ± 106	1110 ± 28	0.13 ± 0.54	>82	9.0 × 10^{-8}	0.35 ± 0.11
1.1 ± 0.4	1.05 ± 0.01	3891 ± 36	872 ± 14	0.13 ± 0.26	>82	1.4 × 10^{-7}	0.36 ± 0.05

From the currents I_s, whose values are according to the literature [31,56], and based on Equation (6), it is possible to estimate an average barrier voltage ϕ_B of the junctions. For that, first is necessary to compute the contact area A for each sample. Considering that one nanotube has a cross-sectional area given by the expression $A_{CNT} = w\pi(D_p - w)$, where w is wall thickness, and $D_p \approx 50$ nm is the average pore diameter of the bare AAO, is obtain a values of about 6×10^{-17} m^2, 11×10^{-17} m^2, and 17×10^{-17} m^2 for samples with 0.4 nm, 0.7 nm, and 1.1 nm wall thickness, respectively. Then, the total number of CNTs connected in each sample is around 1×10^9, since the area of the top electrode is ~10 mm^2 and the pores cover about 20% of the total area of the electrode (determined from an image processing method of a SEM picture of a bare PAM). Finally, the total contact area A of the Schottky junctions can be calculated considering the percentage of the CNTs connected to the Si through this junction. The values for A are shown in Table 3, together with the average barrier voltage ϕ_B, which has values between 0.34 to 0.36 eV for the three samples.

Figure 7a–c shows the dark I–V curves as a function of temperature (20 K to 300 K) of samples that contain LC-CNTs with a wall thickness of 0.4 nm, 0.7 nm, and 1.1 nm, respectively. The three samples exhibit the same behavior: as the temperature decrease, the slope of both sides of the I–V curves decreases, and the region voltage in which the curve experiences the change in the slope moves to a higher voltage. With the model previously discussed, we fit the I–V curves over the full range of temperatures. The conductance $\sigma_p = 1/R_p$ and $\sigma_s = 1/R_s$ as a function of temperature are plotted in Figure 7d–f for each sample, and the inserts show the ideality factor and the average barrier voltage ϕ_B. These results are consistent with the hopping conduction of the Equation (2) determined for the LC-CNTs, in which σ_p tends to zero and represents the conductance through the nanotubes.

Additionally, the ideality factor increases as the temperature drops since the TE mechanism tends to vanish. This behavior is reported for Schottky junctions in the case of tunneling through the contact barrier [57]. Moreover, the rise of ϕ_B when increasing the temperature in Schottky contacts is related to the temperature-activated charge carriers across the interface. The electrons at low temperatures can surmount the barriers through tunneling, and when the temperature increases, the carriers gain enough energy to reach the higher barrier through thermionic emission. Consequently, the obtained ϕ_B will raise with the increase in temperature and bias voltage [58,59]. These results confirm that the chosen model is satisfactory with the morphology and nature of all samples.

Finally, the existence of Schottky barriers between the LC-CNTs and Si reinforce that the mechanism of gas sensing in the devices is through the permeability of the analytes at the heterojunction. This mechanism is explained to the chemisorption of the analytes on the interface, the layers, or the contact barriers that changes the local density of charge carriers and induces a relatively large difference in the electrical measurements [60–62]. Even though all samples have the same sensing mechanism, there is an optimal CNT morphology in such a way that the conductivity is propitious to transport the charge. For instance, the thinner LC-CNTs have high resistance, and the electrical response cannot be readout. On the other hand, when the tubes have thicker walls, they present lower resistance, and the parallel equivalent circuit suppresses the Schottky contact effect.

Although there are devices with higher gas sensitivity than what is presented here, it is possible to propose the decoration of the internal walls of the LC-CNTs with nanoparticles to enhance their performance [63]. This opens the possibility to adapt the samples in an attractive humidity sensor or multimode analyte analyzer by reading out the electronic response [64]. The highly-order constitution of the nanotubes with perfectly perpendicular orientation opens the possibility to establish a robust platform to develop on-chip devices [65].

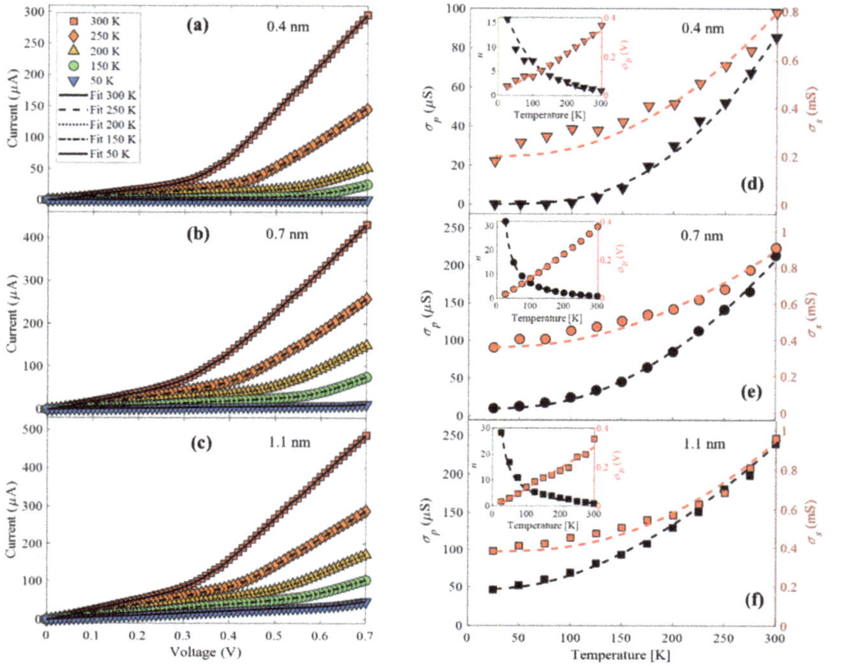

Figure 7. (**a–c**) I–V curves measured at 300 K, 250 K, 200 K, 150 K, and 50 K of samples that contain LC-CNTs with 0.4 nm, 0.7 nm, and 1.1 nm of wall thickness, respectively. (**d–f**) are the plot of the conductance $\sigma_p = 1/R_p$ and $\sigma_s = 1/R_s$ as a function of temperature for each sample, the insert shows the ideality factor and the average barrier voltage ϕ_B as a function of temperature.

4. Conclusions

The fabrication of LC-CNTs arrays with controllable dimensions on Si substrates was successfully achieved using PAMs as templates. The low crystallinity of CNTs was established by Raman spectroscopy and HRTEM. This condition was confirmed by the electrical transport characterization, which shows that the CNTs exhibit a localization of electronic states which depend on their wall thickness, causing 3D-VRH to be the dominant electric transport mechanism. As the wall thickness increases, the electronic wave functions are more delocalized and emerge a metallic transport mechanism parallel to the 3D-VRH.

The arrays containing LC-CNTs with wall thickness in the range of 0.4 to 1.1 nm, exhibit a strong dependence of their resistance as a function of H_2 and C_2H_2 concentrations in an Ar atmosphere. The most representative array was the one containing LC-CNTs with 0.7 nm wall thickness, and it exhibits a maximum resistive response of 5% for C_2H_2 and 1% for H_2 analytes at 5% of concentration. This sample shows a fast response in the gas detection (few seconds) and a short recovery time (few minutes). The origin of this sensing response is related to the existence of a Schottky junction between LC-CNTs and Si. This heterojunction seems to be responsible for gases to permeate and disturb the electrical transport in a specific wall thickness range of LC-CNTs. By engineering the junctions, it is possible to optimize the gas sensing response of these arrays.

Author Contributions: Conceptualization, A.R.A. and S.A.H.; Data curation, D.C.; Formal analysis, D.C. and J.A.G.-M.; Investigation, A.R.A., D.C. and L.F.-I.; Methodology, A.R.A. and S.A.H.; Resources, R.A.S. and S.A.H.; Supervision, R.A.S.; Validation, A.R.A., D.C. and J.A.G.-M.; Writing—original draft, A.R.A.; Writing—review & editing, J.A.G.-M. and S.A.H. All authors have read and agreed to the published version of the manuscript.

Funding: This research was funded by MINECON-Chile through the project Millennium Nucleus MULTIMAT, by the Air Force Office of Scientific Research [FA9550-18-1-0438]; FONDECYT-ANID grant [3190552, 1161614, 1201589], and the Fondequip [EQM150101]; A.R.A. acknowledges a doctoral studies scholarship from CONICYT-ANID.

Conflicts of Interest: The authors declare no conflict of interest.

References

1. Arvand, M.; Hemmati, S. Magnetic nanoparticles embedded with graphene quantum dots and multiwalled carbon nanotubes as a sensing platform for electrochemical detection of progesterone. *Sens. Actuators B Chem.* **2017**, *238*, 346–356. [CrossRef]
2. Marion, S.; Radenovic, A. Towards artificial mechanosensing. *Nat. Mater.* **2020**, *19*, 1043–1044. [CrossRef] [PubMed]
3. Osman, A.I.; Farrell, C.; Al-Muhtaseb, A.H.; Harrison, J.; Rooney, D.W. The production and application of carbon nanomaterials from high alkali silicate herbaceous biomass. *Sci. Rep.* **2020**, *10*, 2563. [CrossRef] [PubMed]
4. Chen, J.; Han, J. Effect of hydroxylated carbon nanotubes on the thermal and electrical properties of derived epoxy composite materials. *Results Phys.* **2020**, *18*, 103246. [CrossRef]
5. Kayang, K.; Nyankson, E.; Efavi, J.; Apalangya, V.; Adetunji, B.; Gebreyesus, G.; Tia, R.; Abavare, E.; Onwona-Agyeman, B.; Yaya, A. A comparative study of the interaction of nickel, titanium, palladium, and gold metals with single-walled carbon nanotubes: A DFT approach. *Results Phys.* **2019**, *12*, 2100–2106. [CrossRef]
6. García-Merino, J.A.; Martínez-González, C.L.; Miguel, C.R.T.S.; Trejo-Valdez, M.; Gutiérrez, H.M.; Torres-Torres, C. Magneto-conductivity and magnetically-controlled nonlinear optical transmittance in multi-wall carbon nanotubes. *Opt. Express* **2016**, *24*, 19552–19557. [CrossRef]
7. Merino, J.A.G.; Martínez-González, C.; Miguel, C.R.T.S.; Trejo-Valdez, M.; Gutiérrez, H.M.; Torres-Torres, C. Photothermal, photoconductive and nonlinear optical effects induced by nanosecond pulse irradiation in multi-wall carbon nanotubes. *Mater. Sci. Eng. B* **2015**, *194*, 27–33. [CrossRef]
8. Cao, G.; Chen, X. The effects of chirality and boundary conditions on the mechanical properties of single-walled carbon nanotubes. *Int. J. Solids Struct.* **2007**, *44*, 5447–5465. [CrossRef]
9. Okazaki, D.; Morichika, I.; Arai, H.; Kauppinen, E.; Zhang, Q.; Anisimov, A.; Varjos, I.; Chiashi, S.; Maruyama, S.; Ashihara, S. Ultrafast saturable absorption of large-diameter single-walled carbon nanotubes for passive mode locking in the mid-infrared. *Opt. Express* **2020**, *28*, 19997–20006. [CrossRef]
10. Kumar, R.; Khan, M.A.; Anupama, A.; Krupanidhi, S.B.; Sahoo, B. Infrared photodetectors based on multiwalled carbon nanotubes: Insights into the effect of nitrogen doping. *Appl. Surf. Sci.* **2021**, *538*, 148187. [CrossRef]
11. Peng, L.-M.; Zhang, Z.; Wang, S. Carbon nanotube electronics: Recent advances. *Mater. Today* **2014**, *17*, 433–442. [CrossRef]
12. Jung, D.; Han, M.; Lee, G.S. Fast-Response Room Temperature Hydrogen Gas Sensors Using Platinum-Coated Spin-Capable Carbon Nanotubes. *ACS Appl. Mater. Interfaces* **2015**, *7*, 3050–3057. [CrossRef]
13. Das, D.; Roy, A. Synthesis of diameter controlled multiwall carbon nanotubes by microwave plasma-CVD on low-temperature and chemically processed Fe nanoparticle catalysts. *Appl. Surf. Sci.* **2020**, *515*, 146043. [CrossRef]
14. Pan, H.; Gao, H.; Lim, S.; Feng, Y.; Lin, J. Highly Ordered Carbon Nanotubes Based on Porous Aluminum Oxide: Fabrication and Mechanism. *J. Nanosci. Nanotechnol.* **2005**, *5*, 277–281. [CrossRef]
15. Segura, R.A.; Contreras, C.; Henriquez, R.; Häberle, P.; Acuña, J.J.S.; Adrian, A.; Alvarez, P.; A Hevia, S. Gold nanoparticles grown inside carbon nanotubes: Synthesis and electrical transport measurements. *Nanoscale Res. Lett.* **2014**, *9*, 207. [CrossRef]
16. Ciambelli, P.; Arurault, L.; Sarno, M.; Fontorbes, S.; Leone, C.; Datas, L.; Sannino, D.; Lenormand, P.; Plouy, S.L.B.D. Controlled growth of CNT in mesoporous AAO through optimized conditions for membrane preparation and CVD operation. *Nanotechnology* **2011**, *22*, 265613. [CrossRef]
17. Sacco, L.; Florea, I.; Châtelet, M.; Cojocaru, C.-S. Electrical and morphological behavior of carbon nanotubes synthesized within porous anodic alumina templates. *J. Phys. Mater.* **2018**, *1*, 015004. [CrossRef]
18. Hou, P.-X.; Yu, W.-J.; Shi, C.; Zhang, L.-L.; Liu, C.; Tian, X.-J.; Dong, Z.-L.; Cheng, H.-M. Template synthesis of ultra-thin and short carbon nanotubes with two open ends. *J. Mater. Chem.* **2012**, *22*, 15221–15226. [CrossRef]
19. Fleming, E.; Du, F.; Ou, E.; Dai, L.; Shi, L. Thermal conductivity of carbon nanotubes grown by catalyst-free chemical vapor deposition in nanopores. *Carbon* **2019**, *145*, 195–200. [CrossRef]
20. Chen, G.; Futaba, D.N.; Sakurai, S.; Yumura, M.; Hata, K. Interplay of wall number and diameter on the electrical conductivity of carbon nanotube thin films. *Carbon* **2014**, *67*, 318–325. [CrossRef]
21. Sohn, J.I.; Kim, Y.-S.; Nam, C.; Cho, B.K.; Seong, T.-Y.; Lee, S. Fabrication of high-density arrays of individually isolated nanocapacitors using anodic aluminum oxide templates and carbon nanotubes. *Appl. Phys. Lett.* **2005**, *87*, 123115. [CrossRef]

22. Hevia, S.A.; Segura, R.; Häberle, P. Low energy electrons focused by the image charge interaction in carbon nanotubes. *Carbon* **2014**, *80*, 50–58. [CrossRef]
23. Mott, N.F. Conduction in non-crystalline materials. *Philos. Mag.* **1969**, *19*, 835–852. [CrossRef]
24. Edwards, P.; Kuznetsov, V.; Slocombe, D.; Vijayaraghavan, R. The Electronic Structure and Properties of Solids. In *Comprehensive Inorganic Chemistry II*, 2nd ed.; Reedikj, J., Poeppelmeier, K., Eds.; Elsevier: Amsterdam, The Netherlands, 2013; pp. 153–176.
25. Qiu, H.; Xu, T.; Wang, Z.; Ren, W.; Nan, H.; Ni, Z.; Chen, Q.; Yuan, S.; Miao, F.; Song, F.; et al. Hopping transport through defect-induced localized states in molybdenum disulphide. *Nat. Commun.* **2013**, *4*, 2642. [CrossRef]
26. Pichler, T.; Knupfer, M.; Golden, M.S.; Fink, J.; Rinzler, A.; Smalley, R.E. Localized and Delocalized Electronic States in Single-Wall Carbon Nanotubes. *Phys. Rev. Lett.* **1998**, *80*, 4729–4732. [CrossRef]
27. Filatzikioti, A.; Glezos, N.; Kantarelou, V.; Kyriakis, A.; Pilatos, G.; Romanos, G.; Speliotis, T.; Stathopoulou, D. Carbon nanotube Schottky type photodetectors for UV applications. *Solid State Electron.* **2019**, *151*, 27–35. [CrossRef]
28. Mohammed, M.; Li, Z.; Cui, J.; Chen, T.-P. Junction investigation of graphene/silicon Schottky diodes. *Nanoscale Res. Lett.* **2012**, *7*, 302. [CrossRef]
29. Tomer, D.; Rajput, S.; Hudy, L.J.; Li, C.H.; Li, L. Carrier transport in reverse-biased graphene/semiconductor Schottky junctions. *Appl. Phys. Lett.* **2015**, *106*, 173510. [CrossRef]
30. Guruprasad, K.; Marappan, G.; Elangovan, S.; Jayaraman, S.V.; Bharathi, K.K.; Venugopal, G.; Di Natale, C.; Sivalingam, Y. Electrical transport properties and impedance analysis of Au/ZnO nanorods/ITO heterojunction device. *Nano Express* **2020**, *1*, 030020. [CrossRef]
31. Kuo, T.-F.; Tzolov, M.B.; Straus, D.A.; Xu, J. Electron transport characteristics of the carbon nanotubes/Si heterodimensional heterostructure. *Appl. Phys. Lett.* **2008**, *92*, 212107. [CrossRef]
32. Kuo, T.-F.; Xu, J. Controlled direct growth of vertical and highly-ordered 'carbon nanotube silicon' heterojunction array. *MRS Proc.* **2005**, *901*. [CrossRef]
33. Tzolov, M.; Chang, B.; Yin, A.; Straus, D.; Xu, J.M.; Brown, G. Electronic Transport in a Controllably Grown Carbon Nanotube-Silicon Heterojunction Array. *Phys. Rev. Lett.* **2004**, *92*, 075505. [CrossRef]
34. Hoa, N.D.; Van Quy, N.; Cho, Y.; Kim, D. An ammonia gas sensor based on non-catalytically synthesized carbon. *Sens. Actuators B* **2007**, *127*, 447–454. [CrossRef]
35. Ding, D.; Chen, Z.; Rajaputra, S.; Singh, V. Hydrogen sensors based on aligned carbon nanotubes in an anodic aluminum oxide template with palladium as a top electrode. *Sens. Actuators B* **2007**, *124*, 12–17. [CrossRef]
36. Hevia, S.; Homm, P.; Cortes, A.; Núñez, V.; Contreras, C.; Ver, J.; Segura, R. Selective growth of palladium and titanium dioxide nanostructures inside carbon nanotube membranes. *Nanoscale Res. Lett.* **2012**, *7*, 342. [CrossRef]
37. Segura, R.A.; Hevia, S.; Häberle, P. Growth of carbon nanostructures using a Pd-based catalyst. *J. Nanosci. Nanotechnol.* **2011**, *11*, 10036–10046. [CrossRef]
38. Ferrari, A.C.; Robertson, J. Interpretation of Raman spectra of disordered and amorphous carbon. *Phys. Rev. B* **2000**, *61*, 14095–14107. [CrossRef]
39. Mapelli, C.; Castiglioni, C.; Zerbi, G.; Müllen, K. Common force field for graphite and polycyclic aromatic hydrocarbons. *Phys. Rev. B* **1999**, *60*, 12710–12725. [CrossRef]
40. Castiglioni, C.; Negri, F.; Rigolio, M.; Zerbi, G. Raman activation in disordered graphites of the A' symmetry forbidden $k \neq 0$ phonon: The origin of the D line. *J. Chem. Phys.* **2001**, *115*, 3769–3778. [CrossRef]
41. Doyle, T.E.; Dennison, J.R. Vibrational dynamics and structure of graphitic amorphous carbon modeled using an embedded-ring approach. *Phys. Rev. B* **1995**, *51*, 196. [CrossRef]
42. García-Merino, J.; Fernández-Izquierdo, L.; Villarroel, R.; Hevia, S. Photo-thermionic emission and photocurrent dynamics in low crystallinity carbon nanotubes. *J. Materiomics* **2021**, *7*, 271–280. [CrossRef]
43. Mott, N.F.; Davis, E.A. *Electronic Processes in Non-Crystalline Materials*, 2nd ed.; Oxford University Press: New York, NY, USA, 1979.
44. Efetov, D.; Kim, P. Controlling Electron-Phonon Interactions in Graphene at Ultrahigh Carrier Densities. *Phys. Rev. Lett.* **2010**, *105*, 256805. [CrossRef]
45. De Nicola, F.; Salvato, M.; Cirillo, C.; Crivellari, M.; Boscardin, M.; Scarselli, M.; Nanni, F.; Cacciotti, I.; De Crescenzi, M.; Castrucci, P. Record efficiency of air-stable multi-walled carbon nanotube/silicon solar cells. *Carbon* **2016**, *101*, 226–234. [CrossRef]
46. Halim, J.; Moon, E.J.; Eklund, P.; Rosen, J.; Barsoum, M.W.; Ouisse, T. Variable range hopping and thermally activated transport in molybdenum-based MXenes. *Phys. Rev. B* **2018**, *98*, 104202. [CrossRef]
47. Zandi, A.; Gilani, A.; Fard, H.G.; Koohsorkhi, J. An optimized resistive CNT-based gas sensor with a novel configuration by top electrical contact. *Diam. Relat. Mater.* **2019**, *93*, 224–232. [CrossRef]
48. Bag, A.; Lee, N.-E. Gas sensing with heterostructures based on two-dimensional nanostructured materials: A review. *J. Mater. Chem. C* **2019**, *7*, 13367–13383. [CrossRef]
49. Potje-Kamloth, K. Semiconductor Junction Gas Sensors. *Chem. Rev.* **2008**, *108*, 367–399. [CrossRef] [PubMed]
50. Ling, Z.; Leach, C.; Freer, R. Heterojunction gas sensors for environmental NO_2 and CO_2 monitoring. *J. Eur. Ceram. Soc.* **2001**, *21*, 1977–1980. [CrossRef]
51. Mirzaei, A.; Hashemi, B.; Janghorban, K. α-Fe_2O_3 based nanomaterials as gas sensors. *J. Mater. Sci. Mater. Electron.* **2016**, *27*, 3109–3144. [CrossRef]

52. Kumar, M.; Bhati, V.S.; Kumar, M. Effect of Schottky barrier height on hydrogen gas sensitivity of metal/TiO$_2$ nanoplates. *Int. J. Hydrogen Energy* **2017**, *42*, 22082–22089. [CrossRef]
53. Uchino, T.; Shimpo, F.; Kawashima, T.; Ayre, G.N.; Smith, D.C.; De Groot, C.H.; Ashburn, P. Electrical transport properties of isolated carbon nanotube/Si heterojunction Schottky diodes. *Appl. Phys. Lett.* **2013**, *103*, 193111. [CrossRef]
54. Rezeq, M.; Ali, A.; Patole, S.; Eledlebi, K.; Dey, R.K.; Cui, B. The dependence of Schottky junction (I–V) characteristics on the metal probe size in nano metal–semiconductor contacts. *AIP Adv.* **2018**, *8*, 055122. [CrossRef]
55. Ortiz-Conde, A.; García Sánchez, F.J.; Muci, J. Exact analytical solutions of the forward non-ideal diode equation with series and shunt parasitic resistances. *Solid-State Electron.* **2000**, *44*, 1861–1864. [CrossRef]
56. Sze, S.M.; Ng, K.K. *Physics of Semiconductor Devices*; John Wiley & Sons: Hoboken, NJ, USA, 2007.
57. Verschraegen, J.; Burgelman, M.; Penndorf, J. Temperature dependence of the diode ideality factor in CuInS2-on-Cu-tape solar cells. *Thin Solid Film.* **2005**, *480-481*, 307–311. [CrossRef]
58. Li, H.; He, D.; Zhou, Q.; Mao, P.; Cao, J.; Ding, L.; Wang, J. Temperature-dependent Schottky barrier in high-performance organic solar cells. *Sci. Rep.* **2017**, *7*, 40134. [CrossRef]
59. Mayimele, M.A.; Diale, M.; Mtangi, W.; Auret, F.D. Temperature-dependent current–voltage characteristics of Pd/ZnO Schottky barrier diodes and the determination. *Mater. Sci. Semicond. Process.* **2015**, *34*, 359–364. [CrossRef]
60. Triet, N.M.; Duy, L.T.; Hwang, B.-U.; Hanif, A.; Siddiqui, S.; Park, K.-H.; Cho, C.-Y.; Lee, N.-E. High-Performance Schottky Diode Gas Sensor Based on the Heterojunction of Three-Dimensional Nanohybrids of Reduced Graphene Oxide–Vertical ZnO Nanorods on an AlGaN/GaN Layer. *ACS Appl. Mater. Interfaces* **2017**, *9*, 30722–30732. [CrossRef]
61. Biswas, R.U.D.; Oh, W.-C. Comparative study on gas sensing by a Schottky diode electrode prepared with graphene–semiconductor–polymer nanocomposites. *RSC Adv.* **2019**, *9*, 11484–11492. [CrossRef]
62. Schierbaum, K.; Kirner, U.; Geiger, J.; Göpel, W. Schottky-barrier and conductivity gas sensors based upon Pd/SnO$_2$ and Pt/TiO$_2$. *Sens. Actuators B* **1991**, *4*, 87–94. [CrossRef]
63. Casanova-Cháfer, J.; Navarrete, E.; Noirfalise, X.; Umek, P.; Bittencourt, C.; Llobet, E. Gas Sensing with Iridium Oxide Nanoparticle Decorated Carbon Nanotubes. *Sensors* **2018**, *19*, 113. [CrossRef]
64. Kim, H.-S.; Kim, J.H.; Park, S.-Y.; Kang, J.-H.; Kim, S.-J.; Choi, Y.-B.; Shin, U.S. Carbon nanotubes immobilized on gold electrode as an electrochemical humidity sensor. *Sens. Actuators B* **2018**, *300*, 127049. [CrossRef]
65. Hung, C.M.; Le, D.T.T.; Van Hieu, N. On-chip growth of semiconductor metal oxide nanowires for gas sensors: A review. *J. Sci. Adv. Mater. Devices* **2017**, *2*, 263–285. [CrossRef]

Article

One-Step Synthesis of SnO₂/Carbon Nanotube Nanonests Composites by Direct Current Arc-Discharge Plasma and Its Application in Lithium-Ion Batteries

Da Zhang, Yuanzheng Tang, Chuanqi Zhang, Qianpeng Dong, Wenming Song and Yan He *

College of Electromechanical Engineering, Qingdao University of Science and Technology, Qingdao 266061, China; qustzd17863928637@163.com (D.Z.); tangyuanzheng@163.com (Y.T.); qustzcq@163.com (C.Z.); 13593574106@163.com (Q.D.); swmpower@163.com (W.S.)
* Correspondence: heyan@qust.edu.cn; Tel.: +86-18660276829

Abstract: Tin dioxide (SnO_2)-based materials, as anode materials for lithium-ion batteries (LIBs), have been attracting growing research attention due to the high theoretical specific capacity. However, the complex synthesis process of chemical methods and the pollution of chemical reagents limit its commercialization. The new material synthesis method is of great significance for expanding the application of SnO_2-based materials. In this study, the SnO_2/carbon nanotube nanonests (SnO_2/CNT NNs) composites are synthesized in one step by direct current (DC) arc-discharge plasma; compared with conventional methods, the plasma synthesis achieves a uniform load of SnO_2 nanoparticles on the surfaces of CNTs while constructing the CNTs conductive network. The SnO_2/CNT NNs composites are applied in LIBs, it can be found that the nanonest-like CNT conductive structure provides adequate room for the volume expansion and also helps to transfer the electrons. Electrochemical measurements suggests that the SnO_2/CNT NNscomposites achieve high capacity, and still have high electrochemical stability and coulombic efficiency under high current density, which proves the reliability of the synthesis method. This method is expected to be industrialized and also provides new ideas for the preparation of other nanocomposites.

Keywords: tin oxide; carbon nanotube; direct current arc-discharge plasma; lithium-ion batteries; anode materials

Citation: Zhang, D.; Tang, Y.; Zhang, C.; Dong, Q.; Song, W.; He, Y. One-Step Synthesis of SnO₂/Carbon Nanotube Nanonests Composites by Direct Current Arc-Discharge Plasma and Its Application in Lithium-Ion Batteries. *Nanomaterials* **2021**, *11*, 3138. https://doi.org/10.3390/nano11113138

Academic Editor: Ilaria Armentano

Received: 27 October 2021
Accepted: 16 November 2021
Published: 21 November 2021

Publisher's Note: MDPI stays neutral with regard to jurisdictional claims in published maps and institutional affiliations.

Copyright: © 2021 by the authors. Licensee MDPI, Basel, Switzerland. This article is an open access article distributed under the terms and conditions of the Creative Commons Attribution (CC BY) license (https:// creativecommons.org/licenses/by/ 4.0/).

1. Introduction

Lithium-ion batteries (LIBs), as the growing popular power sources, have attracted considerable attention in portable electronic devices and are attractive to power electric vehicles [1–5].Therefore, developing new electrode materials and optimizing the preparation process are still hot issues in current research. To circumvent the low theoretical capacity (~372 mAh g^{-1}), low energy and power density of traditional commercial graphite [6] and meet the demands of better performance (higher power and energy density, super-long cycle life and more excellent cycle stability) of LIBs, extensive efforts have been devoted to find new anode materials which have higher theoretical capacity and better cycling performance [7,8], such as silicon [9], metal oxides [10–12], alloys [13,14], carbon-based composite materials [15–18], etc.

Among the above alternative anode materials, in the past few years, SnO_2-based materials have attracted growing research attention due to their suitable charge/discharge voltage range, high theoretical specific capacity (~782 mAh g^{-1}), low toxicity and cost [19], meanwhile, various structures and preparation processes have been explored to remove the biggest bottlenecks that great volume change (>200%) upon the large amount of lithium insertion/extraction into/from SnO_2 and prevent the pulverization of SnO_2 [20,21]. Studies have shown that an effective way to adapt to volume changes and maintain the mechanical integrity of composite electrodes is to uniformly disperse SnO_2 nanoparticles in a conductive matrix (especially carbon materials) [19–21]. Liu et al. [19] synthesized a new type of

SnO$_2$ nanorod structure grown on graphite by hydrothermal method; the results showed that the SnO$_2$/graphite composite maintained higher capacity and better cycling stability than graphite. Wang et al. [20] prepared carbon-coated SnO$_2$/C nanocomposites by a two-step hydrothermal route, which exhibited a markedly improved cycling performance. Chen et al. [22] fabricated the SnO$_2$-reduced graphene oxide-carbon nanotube composites by a facile one-step microwave-assisted method, the electrochemical tests showed that the SnO$_2$-RGO-CNT composite with 60 wt.% SnO$_2$ maintained a maximum capacity of 502 mAh g^{-1} after 50 cycles at 100 mA g^{-1}. Although the above studies have effectively alleviated the volume change upon the large amount of lithium insertion/extraction into/from SnO$_2$, SnO$_2$-based materials have not achieved sustained development in LIBs, the complexity of the chemical preparation process and the contamination of chemical reagents are also important reasons that limit their commercialization besides the discovery of new high-performance electrode materials. Moreover, society's initiatives for green and environmental protection and the urgent desire to accelerate the process of industrialization have driven the preparation of materials to a green and rapid transition. Therefore, developing new material synthesis methods is of great significance to expand the application of SnO$_2$/CNT composites.

In recent years, plasma technology has attracted widespread attention in the preparation and treatment of nanocomposites due to its simple operation, fast synthesis and environmental friendliness [23,24]. Various metal oxides including nano-SnO$_2$ have been successfully prepared by the plasma method [25–27], which greatly avoided these problems including the complexity of chemical methods and the use of chemical reagents. However, the compounding process of SnO$_2$ nanoparticles and conductive matrix is often another process, which poses a higher challenge in the structural design and synthesis process of SnO$_2$-based composite materials.In previous works, the DC arc-discharge plasma has shown excellent effects in realizing the dispersion of CNTs [28,29], providing a basis for the design of CNTs conductive matrix structure, and the CNTs conductive network constructed by plasma showed excellent conductivity in conductive films in our research [30], it is bound to promote its application in LIBs. Besides, instantly loading SnO$_2$ nanoparticles over the surfaces of CNTs followed by the construction of CNTs conductive network structure can not only buffer the volume expansion of SnO$_2$ nanoparticles and realize the synergy of two materials, but promote its industrialization due to the simple and green preparation process.

Herein, the SnO$_2$/CNT nanonests (NNs) composites are first synthesized in one step by DC arc-discharge plasma, in this process, the dispersion of CNTs and the loading of SnO$_2$ nanoparticles are realized simultaneously, which overcomes the complex preparation of traditional chemical methods and the pollution of chemical reagents. Moreover, the nanonest-like conductive structure provides large space for volume change, and also enhances electron transfer between the electrode and SnO$_2$ during lithium ions insertion/extraction process. This method is expected to be industrialized and also provides new ideas for the preparation of other nanocomposites.

2. Materials and Methods

2.1. Synthesis of SnO$_2$/CNTs NNs Composites

The schematic diagram of the synthetic route of SnO$_2$/CNT NNs composites is shown in Figure 1a. First of all, micron-sized Sn particles (99.999% purity), CNTs (GT-400; length: 3–12 μm, diameter: 20–30 nm; Shandong Dazhan Nano Materials co. ltd.) and deionized water (H$_2$O) were initially mixed according to a mass ratio of 8:1:5, and stirred through a glass rod for 30 min until uniform, in this process, additional H$_2$O can be added appropriately to further adjust the viscosity and uniformity of the mixture. Later, with a metal wolfram rod as anode electrode, the prepared mixtures were squeezed into a dense cylindrical electrode as cathode and moved onto a conductive substrate made of copper to keep a certain distance (d) between two electrodes, and d = 2 mm in this experiment. High-voltage direct current (V$_h$ = 10,000 V) was applied between the two electrodes, and the DC arc-discharge plasma was generated while the air was broken down; the mixture as the cathode will produce strong local dispersion under the action of the DC arc-discharge plasma. The hot gas steam pushed the dispersed mixtures upward to the collecting substrate, and finally adhered on the collecting substrate through the electrostatic interaction and significant van der Waals force shown by the nanomaterials [29].

In the micro-composite mechanism of SnO_2/CNTs NNs as shown in Figure 1b, under the DC arc-discharge plasma, the H_2O in the mixture was instantly bumping and the volume expansion force was generated, thereby forming the large pressure gradient between the agglomerated CNTs, which forced the CNTs to be dispersed. Simultaneously, due to the low melting point (232 °C), the micron-sized Sn particles were vaporizing under the action of plasma thermal excitation, and forming the Sn steam zone near the dispersed CNTs. Furthermore, Sn was oxidized to SnO_2 thanks to the oxidizing active substance in the plasma, and uniformly loaded over the surfaces of CNTs, forming the dispersed SnO_2/CNT composites structure. Finally, with the diffusion of SnO_2/CNT and collection on the collecting substrate, secondary agglomeration occurred due to van der Waals force and static electricity [31], and the SnO_2/CNT NNs composites were obtained.

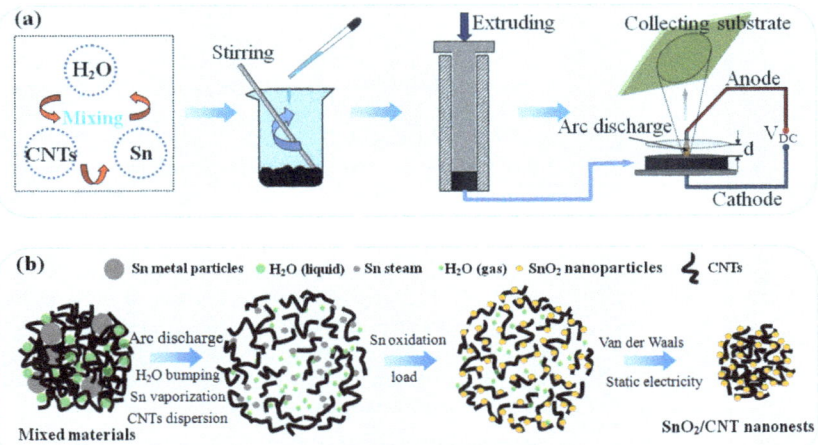

Figure 1. Schematic diagram of the preparation process of SnO_2/CNT NNs composites. (**a**) macro-preparation process; (**b**) micro-composite mechanism.

2.2. Material Characterizations

The morphology was characterized by using a field-emission scanning electron microscopy (SEM) (Hitachi, Tokyo, Japan, SU8010). A transmission electron microscopy (TEM) (Hitachi, Tokyo, Japan, H-8100) was adopted to characterize the further detailed microstructure. Crystallite size determination and phase identification were carried out on an X-ray Diffractometer (XRD) (Rigaku, Tokyo, Japan, Ultima IV) with Cu/Ka radiation (k = 1.5406 Å). The Raman spectroscopy (Renishaw, Shanghai, China) with a 532 nm laser line was applied to characterize the crystallinities of the pristine CNTs and SnO_2/CNT NNs composites obtained by DC arc-discharge plasma. The chemical compositions were further characterized by adopting an X-ray photoelectron spectroscopy (XPS) analysis under ultra-high vacuum using a Thermo ESCALAB 250Xi device employing an Al-Ka (hv = 1253.6 eV) excitation source. Thermogravimetric analysis (TGA) was carried out by using a thermogravimetric analyzer (Netzsch, Selb, Germany, TG 209 F1) with a heating rate of 15 °C min^{-1} in air. In addition, The N_2 adsorption/desorptiontest was estimated by using specific surface and pore size analysis instrument (BET, BSD-PS1, Beijing, China).

2.3. Electrochemical Measurements

For electrochemical measurement, SnO_2/CNT NNs, conductive carbon black, and polyvinylidene fluoride (PVDF), with a weight ratio of 8:1:1, were dissolved in N-methyl pyrrolidinone (NMP) and mixed together thoroughly to form slurry. Then, the resultant slurries were coated onto copper foil substrates. Finally, the working electrodes were dried at 120 °C under vacuum for 12 h. Polypropylene film and Li metals were used as separator and counter anode, respectively, and the 1.15 M $LiPF_6$ electrolyte solution dissolved in a mixture of ethylene carbonate/diethyl carbonate (1:1, vol.%) was electrolyte. The electrochemical measurements were tested using a Battery Testing System (Ningbo baite testing equipment Co., Zhejiang, China). Cyclic voltammetry (CV) curves were collected on a CHI660D electrochemical workstation at 0.2 mV s^{-1} within the voltage range of 0.01–3.00 V and electrochemical impedance spectroscopy (EIS) was performed at 23 °C from 0.01 Hz to 100 KHz with a perturbation amplitude of 5 mV.

3. Results and Discussion

3.1. Microstructure and Morphology of SnO_2/CNT NNs Composites

The microstructure and morphology of SnO_2/CNT NNs composites are shown in Figure 2. Figure 2a shows a typical SEM image of the SnO_2/CNT NNs composites, it can be clearly seen that SnO_2 nanoparticles are uniformly embedded in dispersed CNTs conductive network, which is attributed to the vaporization and oxidation process of metallic Sn and the construction of the dispersed CNTs conductive network under the action of DC arc-discharge plasma. More clearly, Figure 2b,c depicts the TEM images of SnO_2/CNT NNs composites, in which the SnO_2 nanoparticles are densely anchored on the surfaces of CNTs and the average particle size is approximately 5 nm. The overlapping CNTs form a dense nanonest-like conductive network structure, which is conducive to the transmission of electrons, besides, the unique nanonest-like conductive network structure will provide a large void space and mechanical support to relieve the volume change and strain caused upon the alloying/dealloying of SnO_2, thereby preventing the pulverization of SnO_2 nanoparticles. The HRTEM image in Figure 2d shows lattice fringes with a pitch of 0.33 nm, which corresponds to the interplanar distance of the (1 1 0) planes in rutile SnO_2 [32], meanwhile, it can be clearly seen that the lattice fringes of CNTs correspond to the interplanar distance of the (0 0 2) planes.

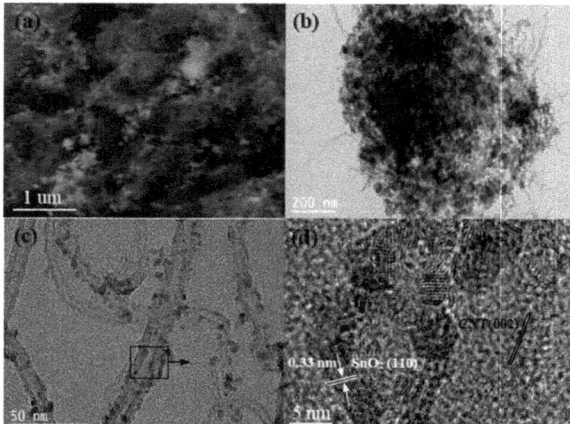

Figure 2. (a) SEM, (b,c) TEM and (d) HRTEM images of SnO_2/CNT NNs composites.

The XRD patterns of bare SnO_2 and SnO_2/CNT NNs composites are shown in Figure 3a. The red line shows the main diffraction peaks of SnO_2, by comparison with the standard values (JCPS No. 21-1272), it is confirmed that the principal diffraction peak has a good correspondence with the tetragonal rutile phase of SnO_2. The black line shows that the peak positions assigned to SnO_2 indexed well with the positions of the bare SnO_2. Besides, the (1 1 0) and (2 1 0) reflection of SnO_2 is overlapped by the (0 0 2) and (1 0 0) reflection of CNTs, respectively.

In order to explore the influence of DC arc-discharge plasma on the structure of CNTs, the structures of the CNTs were analyzed by Raman spectra, as shown in Figure 3b. The Raman spectrum of pristine CNTs were composed of two strong peaks at 1335 cm^{-1} and 1572 cm^{-1}, corresponding to the D and G bands, respectively. The D band constitutes a disordered induction characteristic, which is derived from the vibration of C atoms with dangling bonds, while the G band is derived from the tangential shear mode in C atoms, which corresponds to the tensile mode in the graphite plane [33,34]. The lower intensity of D/G band intensity ratio (I_D/I_G) reflects the higher degree of graphitization; the ratio of the intensities (I_D/I_G) was 1.06 for the pristine CNTs and 1.32 for the SnO_2/CNT NNs, which suggested that there was a certain degree of damage to the CNTs structure in the process of preparing SnO_2/CNT NNs by the DC arc-discharge plasma, which is consistent with our previous researches [31]. The defects on the CNTs walls may cause many cavities and alleyways in the graphite layers, which is beneficial to enhancing the anchoring effect of SnO_2 nanoparticles and CNTs, and also provides more reaction sites for Li$^+$. Besides, CNTs still have a high degree of graphitization and thus retain high electrical conductivity.

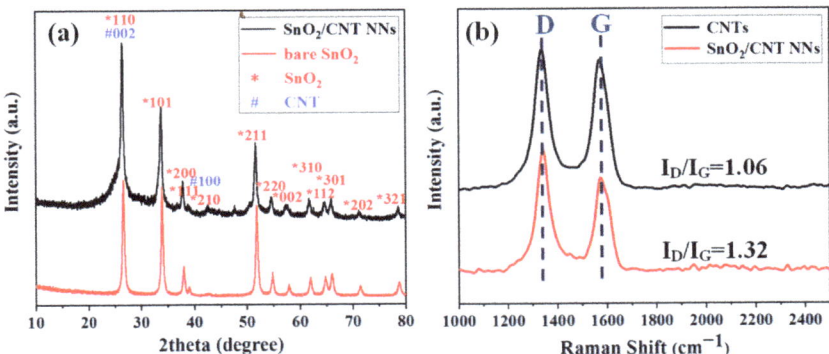

Figure 3. (a) XRD patterns of bare SnO_2, SnO_2/CNT NNs composites and (b) Raman spectra of pristine CNTs, SnO_2/CNT NNs composites.

The XPS studies were carried out to identify the chemical composition of SnO_2/CNT NNs composites. Figure 4a shows the XPS survey spectra of SnO_2/CNT NNs composites, it can be clearly seen that the SnO_2/CNT NNs composites produced peaks corresponding to O 1s, C 1s, Sn 3d as well as Sn 3p, indicating the presence of Sn in the sample besides CNTs and SnO_2, this is due to the fact that a small amount of Sn was not oxidized during the preparing process by plasmas. Figure 4b–d illustrate the spectra for C, O and Sn elements, respectively. The binding energy of 284.8 eV for C 1s mainly corresponds to the carbon atoms in CNTs (Figure 4b). The peaks in Figure 4c correspond to the O spectrum with different chemical states. The peak close to 530 eV could be assigned to O in SnO_2, while peak at around 532.3 eV can be assigned to O in H_2O or adsorbed oxygen. In Figure 4d, the Sn 3d spectrum obtained from the SnO_2/CNT NNs composites exhibited binding energies of 495.3 eV for Sn 3d$_{3/2}$ and 486.9 eV for Sn 3d$_{5/2}$, which confirmed that the rutile SnO_2 nanoparticles were anchored on the surface of CNTs.

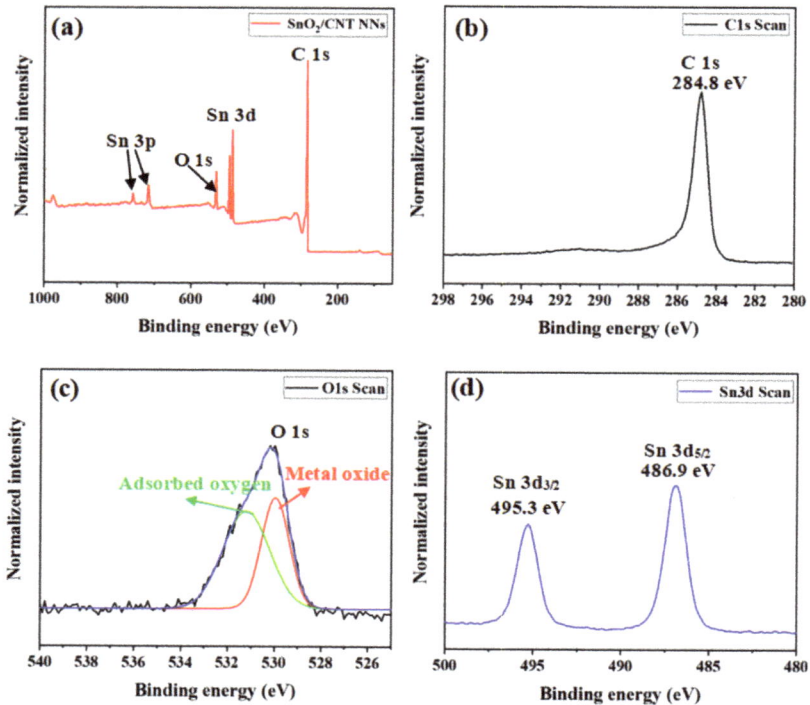

Figure 4. (**a**) XPS spectral survey of the SnO$_2$/CNT NNs composites. (**b**) C 1s, (**c**) O 1s and (**d**) Sn 3d spectrum obtained from the SnO$_2$/CNT NNs composites.

TGA analysis was used to identify the composition and thermal/chemical stability of the SnO$_2$/CNT NNs composites, average of multiple measurements were adopted to ensure the accuracy of SnO$_2$ content, meanwhile, considering the refractory impurities contained in CNTs, the CNTs (without Sn) dispersed by the DC arc-discharge plasma were used as the benchmark to value the content of SnO2, as shown in Figure 5a. It can be seen that the residual content of the same sample under different tests is very stable, the average residual content of SnO$_2$/CNT NNs composites and dispersed CNTs are 47.14% and 2.60%, respectively, thus the content of SnO$_2$ in the sample can be calculated to be 44.54%. Moreover, the SnO$_2$/CNT NNs composites have higher thermal stability than dispersed CNTs.

The porous structure of the SnO$_2$/CNT NNs composites was characterized by N$_2$ adsorption/desorption measurement. As shown in Figure 5b, the adsorption isotherm and pore size distribution analyzed by using the Barrett-Joyner-Halenda (BJH) method. The BET specific surface area of the SnO$_2$/CNT NNs composites is 181.92 m^2 g^{-1}, and the pore volume is 0.89 mL g^{-1}. The average pore diameter of BJH is 16.76 nm. The abundant pore structure and large specific surface are conducive to alleviate strain, enhance electron-electronic contact area and improve the kinetics.

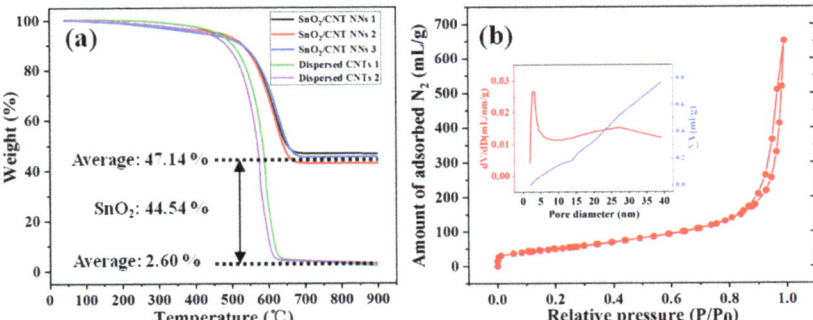

Figure 5. (a) TGA curve of SnO$_2$/CNT NNs composites in air. flow rate 20 mL min^{-1}, heating rate 15 °C min^{-1}, (b) N$_2$ adsorption/desorption isotherm of the SnO$_2$/CNT NNs composites, inset shows the porosity distribution by the Barrett-Joyner-Halenda (BJH) method.

3.2. Electrochemical Performance of SnO$_2$/CNT NNs as Anode Materials in LIBs

The electrochemical behavior of SnO$_2$/CNT NNs composites was evaluated by CV as shown in Figure 6a. The CV curves of SnO$_2$/CNT NNs composites in the first three cycles represents the reaction process of SnO$_2$ and CNTs during the cycle. In the first cycle, the strong reduction peak appears at 0.8 V in the first cycle, which can be attributed to the reduction in SnO$_2$ during the reaction and the formation of a solid electrolyte interphase (SEI) layer [35], and it also can be found with a lower intensity in the second cycle. The peak close to 0.01 V may be attributed to the formation of LiC$_6$ induced by Li intercalation into CNTs, and other reduction peaks (0.01–0.8 V) can be attributed to the formation of Li$_x$Sn [36]. In addition, the peaks at 0.2 V and 0.5 V can be ascribed to deintercalation of LiC$_6$ and the dealloying of Li$_x$Sn, respectively [35], and there is a weak oxidation peak at 1.23 V, which could be attributed to the partly reversible reaction from Sn to SnO$_2$ [37] and the unoxidized Sn within the SnO$_2$/CNT NNs composites confirmed by Figure 4.

Figure 6b compares the charge-discharge cycle performance of bare SnO$_2$ and SnO$_2$/CNT NNs composites, it can be seen that the bare SnO$_2$ particles have an initial discharge and charge capacity of 1914.3 and 1026.7 mAh g^{-1}, respectively. The initial coulomb efficiency is only 53 % comparable with that expected for SnO$_2$ anodes, which is mainly attributed to the formation of Li$_2$O and SEI layer. Although the bare SnO$_2$ exhibits a high initial discharge capacity, the capacity rapidly declines to below 200 mAh g^{-1} after 60 cycles, displaying poor cycle stability of bare SnO$_2$. By contrast, the SnO$_2$/CNT NNs composites shows excellent cycle stability except for the obvious capacity decay in the initial cycle, achieving a capacity of 472 mAh g^{-1} after 200 cycles at 100 mA g^{-1}. Besides, the initial coulomb efficiency of SnO$_2$/CNT NNs composites can reach up to 76 %.

In order to further investigate the rate performances of SnO$_2$/CNT NNs composites, the cycle rate gradually increasing from 100 mA g^{-1} to 1000 mA g^{-1} and then reversing to 100 mA g^{-1}, was adopted as shown in Figure 6c. It can be seen that the SnO$_2$/CNT NNs composites exhibit excellent cycling performance even at the high cycling rate of 1000 mA g^{-1}, the reversible discharge capacity is still preserved at 395 mAh g^{-1} and the coulombic efficiency is around 98.6 % after 40 cycles at different current densities. Although the Coulomb efficiency is slightly reduced (96.5%), when reversing the cycling rate from 1000 mA g^{-1} to 100 mA g^{-1}, the SnO$_2$/CNT NNs composites show strong recovery ability of capacity. The good rate capability can be associated with the nanonest-like structure, the overlapping CNTs provide mechanical support to achieve good electrical contact between the CNTs and SnO$_2$ nanoparticles and can be conducive to the transmission of electrons due to the high electronic contact area ensured by the large specific surface area.

Figure 6d displays the EIS spectra of the bare SnO$_2$ and SnO$_2$/CNT NNs composites, both are composed of a semicircle in high frequencies and a diagonal line in low frequencies. Based on the equivalent circuit, the EIS spectra are fitted as shown in the inset of Figure 6d.

It can be seen that the fitting curves are well consistent with the EIS of both electrodes, respectively. In high frequencies, the kinetic resistance of charge transfer at the electrode–electrolyte interface is represented by Rct, the fitting results show that the R_{ct} of SnO_2/CNT NNs composites is 119.8 Ω, which is lower than the 198.7 Ω of the bare SnO_2, indicating that the introduction of CNTs accelerate electron transport during the electrochemical reaction and it has a higher charge transfer efficiency. Meanwhile, in low frequencies, the slope of the line represents the ionic conductivity of materials, the impedance slope of the SnO_2/CNT NNs composites is greater than that of the bare SnO_2, indicating that the SnO_2/CNT NNs composite has an excellent Li+ diffusion rate. This can explain why SnO_2/CNT NNs composites have better lithium storage characteristics than the bare SnO_2 and exhibit better electrochemical performance.

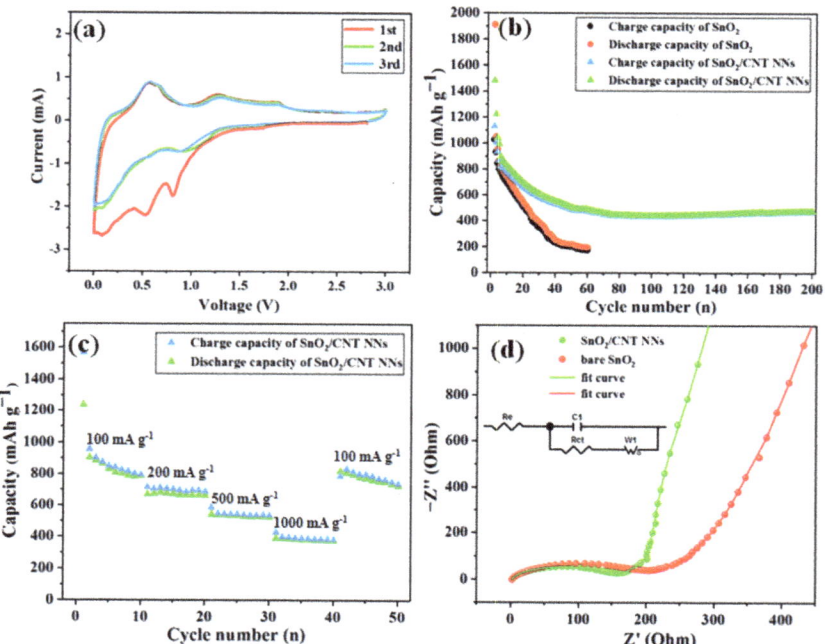

Figure 6. (a) CV curves of the SnO_2/CNT NNs composites at a scanning rate of 0.2 mV s^{-1}, (b) cycling performance at 0.01–3 V and 100 mA g^{-1} of the SnO_2/CNT NNs composites and rare SnO_2, (c) the rate performance of the SnO_2/CNT NNs composites at various current densities, and (d) EIS spectra of the bare SnO_2 and SnO_2/CNT NNs composites at 25 °C from 0.1 to 100 kHz.

Figure 7 exhibits the SEM images after 200 cycles of SnO_2 and SnO_2/CNT NNs electrode, it can be clearly seen that the surface of SnO_2 electrode is rugged and shows serious volume change. In the contrary, the surface of SnO_2/CNT NNs electrode is flat and smooth, and there is no obvious volume change, which is because the unique nanonest-like structure provides adequate room for the volume expansion, this is why the SnO_2/CNT NNs electrode shows excellent electrochemical performance.

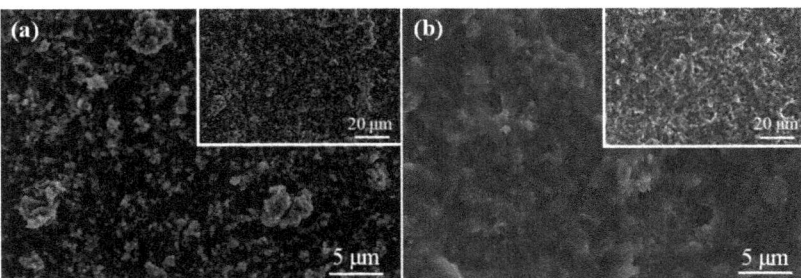

Figure 7. SEM after 200 cycles of (**a**) SnO_2 and (**b**) SnO_2/CNT NNs electrode.

Based on the above results and discussion, the advantages of the plasma one-step synthesis technology and the stable cycle performance of SnO_2/CNT NNs composites are attributed to the following points: (1) plasma one-step synthesis is to achieve a uniform load of SnO_2 nanoparticles while constructing CNTs conductive network, which saves time and energy; (2) this synthesis does not involve any chemicals and it is more environmentally friendly compared with conventional methods; (3) the overlapping CNTs form a dense nanonest-like conductive network structure, which is conducive to the transmission of electrons and ensures the excellent electron contact between Li^+ and SnO_2; (4) the defects on the CNTs walls generated under the action of plasma may create many cavities and channels in graphite layers, providing more reaction sites for Li^+; (5) nanonest-like pore structure provides adequate room for the volume expansion, allowing stable cycle performance by preventing SnO_2 nanoparticles pulverization.

All in all, these results suggest that the SnO_2/CNT NNs composites exhibit high reversible capacity and stable cycle performance. Additionally, the plasma one-step synergy concept can effectively achieve a uniform load of SnO_2 nanoparticles while constructing CNTs conductive network, which is possessed of environmentally friendly, time- and energy-saving advantages. Although SnO_2-based materials are no longer new materials applied in LIBs, experimental results confirm that the DC arc-discharge plasma as a method has exhibited great potential for the synthesis of nanomaterials.

4. Conclusions

In this paper, we successfully synthesized the SnO_2/CNT NNs composites for the first time via DC arc-discharge plasma; in this process, the construction of CNTs conductive network and the loading of SnO_2 nanoparticles were realized simultaneously, this plasma one-step synergy concept is possessed of environmentally friendly, time- and energy-saving advantages compared with chemical synthesis. The SnO_2/CNT NNs composites were applied in LIBs, showing high specific capacity and stable cycle performance. It can achieve a capacity of 472 mAh g^{-1} after 200 cycles at 100 mA g^{-1}, which is due to the fact that the nanonest-like CNT conductive structure provides adequate room for the volume expansion and also helps to transfer the electrons. These results encourage further research in which the DC arc-discharge plasma method can be used for synthesizing energy storage materials.

Author Contributions: Funding acquisition, Y.H.; methodology, D.Z.; supervision, Y.T. and Y.H.; writing—original draft, Q.D.; writing—review and editing, C.Z., W.S. and D.Z. All authors have read and agreed to the published version of the manuscript.

Funding: This research was funded by the National Natural Science Foundation of China (No. 51676103, 52176076), Taishan Scholar Project of Shandong Province (No. ts20190937), Natural Science Foundation of Shandong Province (ZR2019QEE010), and National Science and Technology Development Project of Shandong Province (YDZX20203700003362).

Data Availability Statement: Data underlying the results presented in this paper are not publicly available at this time but may be obtained from the authors upon reasonable request.

Conflicts of Interest: The authors declare no conflict of interest.

References

1. Kang, K.; Meng, Y.S.; Bréger, J.; Grey, C.P.; Ceder, G. Electrodes with high power and high capacity for rechargeable lithium batteries. *Science* **2006**, *311*, 977–980. [CrossRef] [PubMed]
2. Zai, J.T.; Wang, K.X.; Su, Y.Z.; Qian, X.F.; Chen, J.S. High stability and superior rate capability of three-dimensional hierarchical SnS_2 microspheres as anode material in lithium ion batteries. *J. Power Sources* **2011**, *196*, 3650–3654. [CrossRef]
3. Wu, H.B.; Chen, J.S.; Hng, H.H.; Lou, X.W.D. Nanostructured metal oxide-based materials as advanced anodes for lithium-ion batteries. *Nanoscale* **2012**, *4*, 2526–2542. [CrossRef]
4. Liu, C.J.; Huang, H.; Cao, G.Z.; Xue, F.H.; Camacho, R.A.P.; Dong, X.L. Enhanced electrochemical stability of Sn-carbon nanotube nanocapsules as lithium-ion battery anode. *Electrochim. Acta* **2014**, *144*, 376–382. [CrossRef]
5. Deng, W.N.; Chen, X.H.; Liu, Z.; Hu, A.P.; Tang, Q.L.; Li, Z.; Xiong, Y.N. Three-dimensional structure-based tin disulfide/vertically aligned carbon nanotube arrays composites as high-performance anode materials for lithium ion batteries. *J. Power Sources* **2015**, *277*, 131–138. [CrossRef]
6. Xu, Y.H.; Liu, Q.; Zhu, Y.J.; Liu, Y.H.; Langrock, A.; Zachariah, M.R.; Wang, C.S. Uniform nano-Sn/C composite anodes for lithium ion batteries. *Nano Lett.* **2013**, *13*, 470–474. [CrossRef]
7. Zhen, S.; Yi, H.; Chen, Y.; Zhang, X.; Wang, K.; Chen, R. Tin nanoparticle-loaded porous carbon nanofiber composite anodes for high current lithium-ion batteries. *J. Power Sources* **2015**, *278*, 660–667.
8. Zhang, W.J. A review of the electrochemical performance of alloy anodes for lithium-ion batteries. *J. Power Sources* **2011**, *196*, 13–24. [CrossRef]
9. Ashuri, M.; He, Q.R.; Shaw, L. Silicon as a potential anode material for Li-ionbatteries: Where size, geometry and structure matter. *Nanoscale* **2016**, *8*, 74–103. [CrossRef]
10. Shaw, L.; Ashuri, M. Coating—A potent method to enhance electrochemical performance of $Li(Ni_xMn_yCo_z)O_2$ cathodes for Li-ion batteries. *Adv. Mater. Lett.* **2019**, *10*, 369–380. [CrossRef]
11. Li, Y.; Yu, S.L.; Yuan, T.Z.; Yan, M.; Jiang, Y.Z. Rational design of metal oxide nanocomposite anodes for advanced lithium ion batteries. *J. Power Sources* **2015**, *282*, 1–8. [CrossRef]
12. Liang, C.; Gao, M.X.; Pan, H.G.; Liu, Y.F.; Yan, M. Lithium alloys and metal oxides as high-capacity anode materials for lithium-ion batteries. *J. Alloys Compd.* **2013**, *575*, 246–256. [CrossRef]
13. Zhang, C.; Wang, Z.; Cui, Y.; Niu, X.Y.; Chen, M.; Liang, P.; Liu, J.H.; Liu, R.J.; Li, J.C.; He, X. Dealloying-derived nanoporous Cu_6Sn_5 alloy as stable anode materials for lithium-ion batteries. *Materials* **2021**, *14*, 4348. [CrossRef] [PubMed]
14. Chu, D.B.; Li, J.; Yuan, X.M.; Li, Z.L.; Wei, X.; Wan, Y. Tin-based alloy anode materials for lithium ion batteries. *Prog. Chem.* **2012**, *24*, 1466–1476.
15. Jhan, Y.R.; Duh, J.G.; Tsai, S.Y. Synthesis of confinement structure of Sn/C-C (MWCNTs) composite anode materials for lithium ion battery by carbothermal reduction. *Diamond Relat. Mater.* **2021**, *20*, 413–417. [CrossRef]
16. Luo, Z.Y.; Peng, M.L.; Lei, W.X.; Pan, Y.; Zou, Y.L.; Ma, Z.S. Electroplating synthesis and electrochemical properties of CNTs/(Ni-P)/Sn as anodes for lithium-ion batteries. *Mater. Lett.* **2019**, *250*, 1–4. [CrossRef]
17. Huang, L.; Huang, P.; Chen, P.; Ding, Y.L. Metal nanodots anchored on carbon nanotubes prepared by a facile solid-state redox strategy for superior lithium storage. *Funct. Mater. Lett.* **2020**, *13*, 2051039. [CrossRef]
18. Zhong, Y.; Li, X.F.; Zhang, Y.; Li, R.Y.; Cai, M.; Sun, X.L. Nanostructued core–shell Sn nanowires @ CNTs with controllable thickness of CNT shells for lithium ion battery. *Appl. Surf. Sci.* **2015**, *332*, 192–197. [CrossRef]
19. Liu, H.D.; Huang, J.M.; Li, X.L.; Liu, J.; Zhang, Y.X. SnO_2 nanorods grown on graphite as a high-capacity anode material for lithium ion batteries. *Ceram. Int.* **2012**, *38*, 5145–5149. [CrossRef]
20. Wang, F.; Song, X.P.; Yao, G.; Zhao, M.S.; Liu, R.; Xu, M.W.; Sun, Z.B. Carbon-coated mesoporous SnO_2 nanospheres as anode material for lithium ion batteries. *Scripta Mater.* **2012**, *66*, 562–565. [CrossRef]
21. Kuriganova, A.B.; Vlaic, C.A.; Ivanov, S.; Leontyeva, D.V.; Bund, A.; Smirnova, N.V. Electrochemical dispersion method for the synthesis of SnO_2 as anode material for lithium ion batteries. *J. Appl. Electrochem.* **2016**, *46*, 527–538. [CrossRef]
22. Chen, T.Q.; Pan, L.K.; Liu, X.J.; Yu, K.; Sun, Z. One-step synthesis of SnO_2–reduced graphene oxide–carbon nanotube composites via microwave assistance for lithium ion batteries. *RSC Adv.* **2012**, *2*, 11719–11724. [CrossRef]
23. Sadakiyo, M.; Yoshimaru, S.; Kasai, H.; Kato, K.; Takata, M.; Yamauchi, M. A new approach for the facile preparation of metal-organic framework composites directly contacting with metal nanoparticles through arc plasma deposition. *Chem. Commun.* **2016**, *52*, 8385–8388. [CrossRef]
24. Santhosh, N.; Filipič, G.; Tatarova, E.; Baranov, O.; Kondo, H.; Sekine, M.; Hori, M.; Ostrikov, K.; Cvelbar, U. Oriented carbon nanostructures by plasma processing: Recent advances and future challenges. *Micromachines* **2018**, *9*, 565. [CrossRef] [PubMed]
25. Tanaka, M.; Kageyama, T.; Sone, H.; Yoshida, S.; Okamoto, D.; Watanabe, T. Synthesis of lithium metal oxide nanoparticles by induction thermal plasmas. *Nanomaterials* **2016**, *6*, 60. [CrossRef]
26. Guo, B.; Košiček, M.; Fu, J.C.; Qu, Y.Z.; Lin, G.H.; Baranov, O.; Zavašnik, J.; Cheng, Q.J.; Ostrikov, K.; Cvelbar, U. Single-crystalline metal oxide nanostructures synthesized by plasma-enhanced thermal oxidation. *Nanomaterials* **2019**, *9*, 1405. [CrossRef]
27. Wang, C.; Chen, J.Z. Atmospheric-pressure-plasma-jet sintered nanoporous SnO_2. *Ceram. Int.* **2015**, *41*, 5478–5483. [CrossRef]

28. Li, S.L.; He, Y.; Jing, C.G.; Gong, X.B.; Cui, L.L.; Cheng, Z.Y.; Zhang, C.Q.; Nan, F. A novel preparation and formation mechanism of carbon nanotubes aerogel. *Carbon Lett.* **2018**, *28*, 16–23.
29. Li, S.L.; Zhang, C.Q.; He, Y.; Feng, M.; Ma, C.; Cui, Y. Multi-interpolation mixing effects under the action of micro-scale free arc. *J. Mater. Process. Tech.* **2019**, *271*, 645–650. [CrossRef]
30. Li, S.L.; Wang, K.; Feng, M.; Yang, H.L.; Liu, X.Y.; He, Y.; Zhang, C.Q.; Wang, J.Y.; Fu, J.F. Preparation of light-transmissive conductive film by free arc dispersed carbon nanotubes and thermos compression bonding. *Carbon Lett.* **2020**, *30*, 651–656. [CrossRef]
31. Li, S.L.; Ci, Y.D.; Zhang, D.; Zhang, C.Q.; He, Y. Free arc liquid-phase dispersion method for the preparation of carbon nanotube dispersion. *Carbon Lett.* **2020**, *31*, 287–295. [CrossRef]
32. Wang, X.; Fan, H.; Ren, P.; Li, M. Homogeneous SnO_2 core-shell microspheres: Microwave-assisted hydrothermal synthesis, morphology control and photocatalytic properties. *Mater. Res. Bull.* **2014**, *50*, 191–196. [CrossRef]
33. Mouyane, M.; Ruiz, J.M.; Artus, M.; Cassaignon, S.; Jolivet, J.P.; Caillon, G.; Jordy, C.; Driesen, K.; Scoyer, J.; Stievano, L.; et al. Carbothermal synthesis of Sn-based composites as negative electrode for lithium-ion batteries. *J. Power Sources* **2011**, *196*, 6863–6869. [CrossRef]
34. Marcinek, M.; Hardwick, L.J.; Richardson, T.J.; Song, X.; Kostecki, R. Microwave plasma chemical vapor deposition of nano-structured Sn/C composite thin-film anodes for Li-ion batteries. *J. Power Sources.* **2007**, *173*, 965–971. [CrossRef]
35. Kim, J.G.; Nam, S.H.; Lee, S.H.; Choi, S.M.; Kim, W.B. SnO_2 nanorod-planted graphite: An effective nanostructure configuration for reversible lithium ion storage. *Acs Appl. Mater. Int.* **2011**, *3*, 828–835. [CrossRef]
36. Lian, P.C.; Zhu, X.F.; Liang, S.Z.; Li, Z.; Yang, W.S.; Wang, H.H. High reversible capacity of SnO_2/graphene nanocomposite as an anode material for lithium-ion batteries. *Electrochim. Acta.* **2011**, *56*, 4532–4539. [CrossRef]
37. Yao, J.; Shen, X.; Wang, B.; Liu, H.; Wang, G. In situ chemical synthesis of SnO_2-graphene nanocomposite as anode materials for lithium-ion batteries. *Electrochem. Commun.* **2009**, *11*, 1849–1852. [CrossRef]

Article

Design of 3D Carbon Nanotube Monoliths for Potential-Controlled Adsorption

Dennis Röcker *,†, Tatjana Trunzer †, Jasmin Heilingbrunner, Janine Rassloff, Paula Fraga-García and Sonja Berensmeier *

Bioseparation Engineering Group, Department of Mechanical Engineering, Technical University of Munich, Boltzmannstraße 15, 85748 Garching, Germany; t.trunzer@tum.de (T.T.); jheilingbrunner91@gmail.com (J.H.); rassloffj@gmail.com (J.R.); p.fraga@tum.de (P.F.-G.)
* Correspondence: d.roecker@tum.de (D.R.); s.berensmeier@tum.de (S.B.)
† These authors contributed equally to this work.

Abstract: The design of 3D monoliths provides a promising opportunity to scale the unique properties of singular carbon nanotubes to a macroscopic level. However, the synthesis of carbon nanotube monoliths is often characterized by complex procedures and additives impairing the later macroscopic properties. Here, we present a simple and efficient synthesis protocol leading to the formation of free-standing, stable, and highly conductive 3D carbon nanotube monoliths for later application in potential-controlled adsorption in aqueous systems. We synthesized monoliths displaying high tensile strength, excellent conductivity (up to 140 S m^{-1}), and a large specific surface area (up to 177 m^2 g^{-1}). The resulting monoliths were studied as novel electrode materials for the reversible electrosorption of maleic acid. The process principle was investigated using chronoamperometry and cyclic voltammetry in a two-electrode setup. A stable electrochemical behavior was observed, and the synthesized monoliths displayed capacitive and faradaic current responses. At moderate applied overpotentials (± 500 mV vs. open circuit potential), the monolithic electrodes showed a high loading capacity (~20 µmol g^{-1}) and reversible potential-triggered release of the analyte. Our results demonstrate that carbon nanotube monoliths can be used as novel electrode material to control the adsorption of small organic molecules onto charged surfaces.

Keywords: aqueous system; carbon electrodes; electrosorption; maleic acid; setup design; surface oxidation; ultrasonic technology

Citation: Röcker, D.; Trunzer, T.; Heilingbrunner, J.; Rassloff, J.; Fraga-García, P.; Berensmeier, S. Design of 3D Carbon Nanotube Monoliths for Potential-Controlled Adsorption. *Appl. Sci.* **2021**, *11*, 9390. https://doi.org/10.3390/app11209390

Academic Editor: Simone Morais

Received: 20 September 2021
Accepted: 7 October 2021
Published: 10 October 2021

Publisher's Note: MDPI stays neutral with regard to jurisdictional claims in published maps and institutional affiliations.

Copyright: © 2021 by the authors. Licensee MDPI, Basel, Switzerland. This article is an open access article distributed under the terms and conditions of the Creative Commons Attribution (CC BY) license (https://creativecommons.org/licenses/by/4.0/).

1. Introduction

Since the discovery of carbon nanotubes (CNTs) by Ijima [1], research interest in these seamless cylindrical graphene layers has been increasing steadily. Due to their outstanding electrical [2], thermal [3], and mechanical properties [4], CNTs are already commercialized for a variety of applications. For instance, in lithium-ion batteries, small quantities of CNTs provide increased electrical conductivity and mechanical stability leading towards an overall enhanced life cycle [5–8]. Nonetheless, CNTs to date are mainly utilized as additives [5–10]. Thus, focussing the extraordinary properties of individual CNTs into macroscopic 3D assemblies still presents a bottleneck on the route to further technical applications [7].

Over the last years, monolith and aerogel synthesis has proven to be a promising way to tune CNTs into macroscopic functional materials [11–15]. CNT monoliths consist of a continuous particle network while simultaneously maintaining the morphological identity of the CNTs [16]. Moreover, these 3D structures are characterized by a large surface area, hierarchical pores, low density, and many accessible active sites for diverse applications [16]. A common approach for monolith synthesis involves gel formation and the subsequent drying out of a particulate suspension of CNTs. Due to the low crosslinking forces between individual CNTs, the efficient dispersion and induction of gelation play critical roles in monolith synthesis. Therefore, components facilitating the gelation process are often introduced during the synthesis [11–13,15,17–19]. In this context, Kohlmeyer et al. achieved gel formation by using a polymer-based crosslinker [12]. Similarly, Zhang et al. embedded multi-walled carbon nanotubes (MWCNTs) into a polymer matrix [18]. The importance of an initial dispersion of the CNTs has been highlighted by Haghoo et al., who used sodium dodecylbenzene sulfonate (SDBS) for the dispersion of CNTs prior to gelation [15]. However, it has to be considered that these mostly inert additives might alter and impair the later properties of the monolith, as reported by Bryning et al., who reinforced the CNT network with polyvinyl alcohol (PVA) during the monolith synthesis [17]. A first step towards the synthesis of monoliths for technical applications was taken by Shen et al., who focussed on a more economical synthesis protocol using low-cost MWCNTs [11]. While the use of MWCNTs could reduce material costs, the applied protocols in literature still require prolonged time for gelation and special drying techniques such as supercritical fluid drying [11,12,15,17,20]. Thus, the synthesis becomes complex and time-intensive, further reducing the cost-performance ratio of the resulting monoliths [21]. Addressing this research gap, we present a simple and efficient method for synthesizing stable and highly conductive CNT monoliths derived from low-cost bulk MWCNTs. Utilizing ultrasonication for the dispersion and gelation of the CNT suspension and only low amounts of PVA as a stabilizing agent, as well as a simple pressing and heat drying approach, the complexity and time for synthesis can be considerably reduced, enabling a facilitated scale-up and possible technical application of the resulting monoliths.

Many efforts have been made, exploiting the use of CNT-based monoliths as electrode material for supercapacitors or as scaffolds for catalysis [14,22–26]. However, a novel and up-and-coming application alternative presents the targeted electrosorption onto the monoliths' chargeable surface. By applying an electrical potential, the interfacial properties of the monoliths can be tuned in their surface charge and electrical double layer (EDL) structure [27]. Hereby, the adsorption and desorption of ions and molecules can be influenced and steered. In this context, several studies have been published on the use of CNT monoliths for capacitive deionization (CDI) and the removal of inorganic ions from brackish water [24,28,29]. However, the targeted electrosorption of organic molecules onto carbon electrodes is still scarcely studied today.

In a simulative approach, Wagner et al. recently investigated the adsorption equilibria of charged organic molecules at the CNT surface [30]. In view of potential-controlled effects, the electrosorption of small organic molecules on a particulate CNT electrode has been investigated for a chromatographic setup by Trunzer et al. [31,32]. In particular the role of the EDL and the environmental conditions around the solid-liquid interface have been emphasized as critical factors in the electrosorption mechanisms. However, shortcomings of the setup were revealed to be the low electrode capacity and its structural changes throughout the operation time. As a further optimization, the use of a monolithic electrode can increase the mass to volume ratio of the electrode and consequently the number of available binding sites, enable a better and more homogenous potential distribution, and increase the overall structural stability of the electrode.

In the following sections, we characterize: the material properties of untreated and oxidized MWCNTs for monolith synthesis (Section 3.1); the impact of ultrasonication onto the dispersion and gelation of the particle network (Section 3.2); and the resulting monoliths by their macroscopic properties and electrochemically as electrode material (Section 3.3). Moreover, we provide proof of concept for the potential-controlled adsorption and desorption of the small organic molecule maleic acid (Section 3.4). We aim to develop a monolithic electrode for the potential-controlled adsorption of organic molecules as a step ahead to more sustainable and cost-efficient separation processes in biotechnology.

2. Materials and Methods

2.1. Materials and Instruments

MWCNTs were purchased from Future Carbon GmbH (CNT-K, Future Carbon GmbH, Bayreuth, Germany). Prior to monolith synthesis, the particles were washed in 1 M HCl (VWR Chemicals GmbH, Darmstadt, Germany) overnight at 80 °C to reduce the catalytic residue and other contaminants, as suggested by the manufacturer. Surface oxidation of the nanotubes (oxCNTs) was realized in 3 M H_2SO_4 (VWR Chemicals GmbH, Darmstadt, Germany) and HNO_3 (Merck GmbH, Darmstadt, Germany) (ratio 3:1 v/v, 30 min, 80 °C) based on the works of Moraes et al. [33] and Shaffer et al. [34]. After oxidation, the nanotube slurry was filtered and thoroughly washed with deionized water (DI-water) until a neutral pH was reached. PVA (89,000–98,000, 99% hydrolyzed, Sigma Aldrich GmbH, Taufkirchen, Germany) was utilized to reinforce the monoliths' structure [11,17,35]. SDBS was purchased from Sigma Aldrich GmbH. Maleic acid (absolute, Ph. Eur., AppliChem GmbH, Darmstadt, Germany) was used as model adsorbate and electrolyte.

Particle characterization was performed with Fourier-transform infrared spectroscopy (FTIR, ALPHA II, Bruker Co., Billerica, MA, USA), transmission electron microscopy (TEM, 100-CX, JEOL GmbH, Freising, Germany), and light microscopy (Axio 7 observer, Carl Zeiss GmbH, Munich, Germany). CNTs were dispersed and gelled using a Branson Digital Sonifier ultrasonic probe (Wattage 400 W, Branson Ultraschall GmbH, Fürth, Germany). Wet gels were dried in a drying furnace (Heraeus Oven, Thermo Fisher Scientific GmbH, Dreieich, Germany) or in a freeze-dryer (Alpha 1-2LDplus, Martin Christ GmbH, Osterode am Harz, Germany). Monolith characterization was carried out using a Gemini VII surface analyzer (Micromeritics GmbH, Unterschleißheim, Germany), and a Z2.5 tensile strength testing machine (ZwickRoell GmbH, Fürstenfeld, Germany). To measure electrical conductivity, the monoliths were clamped in a 3D printed rig pictured in Figure S1 and contacted by two-point gold pins. Electrochemical experiments were conducted using a Gamry I 1000 E potentiostat (Gamry Instruments, Warminster, PA, USA). For all electrochemical experiments, an asymmetrical two-electrode setup was used. Planar monoliths functioned as working electrodes while a porous steel foam (Alantum Europe GmbH, Munich, Germany) with a pore size of 450 μm was utilized as a counter electrode (see Figure S2). The electrodes were placed parallel in a self-printed test rig (Filament: ABS, Rudolf Wiegand und Partner GmbH, Olching, Germany) (see Figure S2). For potential-controlled adsorption experiments, the amount of adsorbed and desorbed analyte was quantified as triplicates using an Agilent 1100 series high performance liquid chromatography (HPLC) system (Agilent Technologies Inc., Santa Clara, CA, USA).

2.2. Particle Preparation and Characterization

For the determination of the point of zero charge (pHpzc), DI-water was adjusted to pH values from 3 to 10 (pH$initial$) using 10 mM HCl and NaOH. Afterward, dried CNT and oxCNT particles were weighed to concentrations of 0.005 g L^{-1}, 0.010 g L^{-1}, 0.050 g L^{-1}, and 0.100 g L^{-1} and added to the liquid media. The suspensions were incubated at 250 rpm and room temperature (RT) for 24 h before measuring the pH again (pH$final$). All experiments were conducted in triplicates. To study the ultrasonic impact on the gelation of the CNT network, particle concentrations of 10 g L^{-1} and 40 g L^{-1} were ultrasonicated (20%, 6 min, 10 s on, 15 s off) while cooled in an ice bath.

2.3. Monolith Synthesis

Our synthesis protocol is derived from the work of Shen et al. [11]. However, changes were made to the existing protocol as we refrained from using SDBS for the dispersion of the CNTs. Instead, the impact of ultrasonication on the dispersion and gelation of the CNT network was studied. We simplified the process with a shortened incubation time for PVA and a subsequent centrifugation step. For the stabilization of the synthesized CNT monoliths, we developed a simple pressing and heat drying approach. In Figure 1, the synthesis steps for the preparation of CNT monoliths are illustrated.

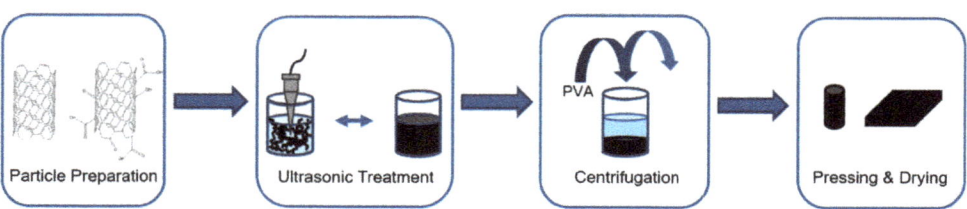

Figure 1. Schematic for the preparation of CNT monoliths.

Wet masses of CNT and oxCNT were dispersed in DI-water. Subsequently, the dispersion was treated ultrasonically for up to 6 min. Afterward, the particles were centrifuged and dispersed in an aqueous 1 wt-% PVA solution. After an incubation time of approx. 5 min and a final centrifugation step, monoliths were formed from the sediment using self-designed pressing molds (pressing weight: 4 kg) for cylindrical and planar monoliths (see Figure 2). For the pressing of monoliths, CNT wet mass is evenly distributed within the pressing mold in a first step. In a second step, the CNT wet mass is compressed with stamps. After pressing, the wet monoliths were dried in a drying furnace (72 h, 60 °C) or by freeze-drying (−58 °C, 0.04 mbar, 48 h). For further monolith characterization and electrochemical experiments, only heat-dried monoliths were utilized.

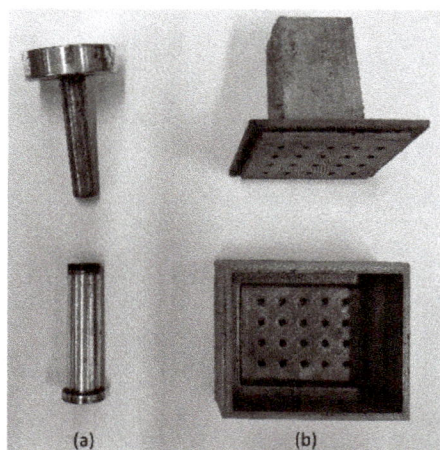

Figure 2. Molds and stamps for the pressing of cylindrical (**a**) ø 10 mm and planar monoliths (**b**) 50 × 40 × 5 mm.

2.4. Monolith Characterization

The BET surface was determined as triplicates at 77 K under nitrogen atmosphere. By exposing the monoliths to predefined weight loads, the mechanical stability of the monoliths was tested. In addition, tensile strength measurements were conducted for planar monoliths. The conductivity was determined by the two point probe method and determined in analytical and technical triplicates. For electrochemical characterization, the potentiostat was operated in surface mode. As electrolyte solution, 6.5 mL of 5 mM maleic acid adjusted to a pH of 7 was utilized. Chronoamperometric measurements were conducted for up to 300 s in ±100 mV steps. Cyclic voltammetry was conducted 3 consecutive times at a scan rate of 1 mV s^{-1}. The operation range was set between −500 mV and +500 mV vs. the open circuit potential (OCP).

2.5. Potential-Controlled Adsorption

Maleic acid was utilized as a model analyte. For potential-controlled adsorption experiments, 6.5 mL of 5 mM maleic acid were adjusted to a pH of 7. A potential of +500 mV vs. OCP was applied during the adsorption phase for 15 min. After the adsorption phase, test rig and counter electrode were rinsed and cleaned extensively using DI-water so that only the effect of the working electrode would be observed and no remaining analyte on the test rig and counter electrode could influence the concentration measurement. DI-water at an adjusted pH of 7 was filled into the test rig to recover the model analyte. A potential of −500 mV vs. OCP was applied for 15 min for the desorption phase. Due to its strong UV absorbance maleic acid can easily be detected spectrophotometrically [31,32,36]. Hence, adsorbed and desorbed amounts of maleic acid were determined as triplicates by measuring the concentration in the supernatant through HPLC runs at a wavelength of 258 nm. Electrosorption experiments were conducted for three subsequent runs with washing steps for the monolith and test rig after each run to test the stability of the electrodes.

3. Results

3.1. Particle Oxidation and Characterization

Surface oxidation is often utilized to improve the dispersive behavior and hydrophilicity of CNTs [37,38]. The increased presence of oxygen-containing surface moieties thus leads to enhanced electrostatic repulsion and stabilization of the nanotube dispersion [34]. Intending to synthesize and stabilize monoliths containing minimal quantities of additives, we compared untreated and oxidized CNT particles as raw material prior to monolith synthesis. It therefore has to be considered that oxidation can lead to structural damage and conductivity loss of CNTs [39–41]. Hence, particular focus was given to developing a mild oxidation protocol (30 min, 3:1 v/v 3M H_2SO_4 to HNO_3, ratio, 80 °C). As expected, no structural damage could be observed through TEM-imaging for our oxidized nanotubes (see Figure S3). In FTIR measurements, an increase in peak intensity for bands associated with oxygen-containing surface groups indicates the successful surface oxidation of the nanotubes (see Figure S4). As a result of functionalization a dominant increase of C-O stretching vibrations visible at 1118 cm^{-1} proves the augmented presence of ester, ether and alcohol groups on the nanotubes surface [42]. Moreover, an increasing peak intensity for bands associated with C=O stretching can be observed between 1700 cm^{-1} and 1500 cm^{-1} and confirms the presence of carboxylic acid and ketone groups [43]. Thus, we observe functionalization of the nanotubes' surface even upon mild oxidation conditions similar to Avilés et al. [37]. Therefore, we expect good electrical properties of the CNTs.

Further, studying the solid-liquid interface is essential to understanding the later adsorption of charged molecules onto the synthesized monoliths in aqueous systems. In Figure 3, pH-shift experiments for untreated (green) and oxidized (blue, grey) CNTs are presented. An increased number of oxidized surface species for oxCNTs could be proven by determining the point of zero charge (pHpzc). The pHpzc is represented by an intersection with the bisecting line in Figure 3 (pH$inital$ = pH$final$). For the untreated nanotubes, a pHpzc of 6.56 was determined. In contrast, a pHpzc between 1.67 and 2.16 was observed for the oxidized samples, resulting from the deprotonation of surface hydroxyl groups [44,45]. Thus, the strong influence of the oxidation onto the surface charge and the nanotubes' dispersibility is illustrated. For increasing mass concentrations of oxCNTs, an explicit dependency on the strength of the pH-shift is visible, resulting from the enlarged number and effect of oxidized surface groups. As also reported by Wagner et al. [30] and Trunzer et al. [31], our experiments highlight the remarkable influence of the nanotubes onto aqueous systems due to changes in the EDL.

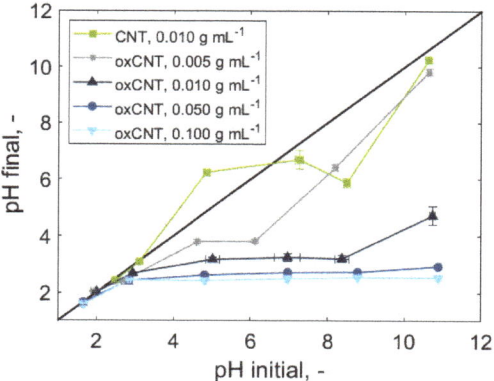

Figure 3. pH-shift experiments for untreated (green) and oxidized CNT-K (blue, grey) particles at differing concentrations. The nanotubes were weighed in as dry mass. Incubation was conducted at RT for 24 h.

3.2. Monolith Stability Dependence on Ultrasonication and Drying

Upon initial dispersion and gelation, carbon aerogels are traditionally dried by supercritical fluid drying or freeze-drying in order to remove the liquid phase without disturbing or collapsing the porous network [46–49]. However, both methods are rather expensive, time-consuming, and hard to handle. With the goal of a simplified and optimized synthesis protocol, we studied the impact of ultrasonication and subsequent heat drying on the monoliths' stability compared to a more traditional dispersant-aided freeze-drying approach. Hereby, we determined ultrasonic treatment prior to pressing and drying of the CNT gel as a critical factor to achieve stable and free-standing monoliths.

Through the high energy impact of sonication onto the CNT suspension and the resulting shear stress, individual agglomerates can be separated, leading to a dispersion of the particles [50,51]. Interestingly, with increasing sonication time, even swelling of the CNT network can be observed, as illustrated in Figure 4a,b. Thus, strong water uptake of the CNT network is indicated, as also discussed by Trunzer et al. [32]. Moreover, with increasing ultrasonic impact, strong gelation of the CNT-water suspension can be triggered. After 6 min of ultrasonic treatment, the CNTs form a gel displaying no free water, as displayed in Figure 4c. However, it has to be considered that similar to harsh oxidation conditions, the high energy impact during ultrasonication can lead to scission and structural damage of the CNTs [50,52,53]. Thus, we additionally studied the impact of sonication-induced shear forces through TEM imaging (see Figure S3), where no significant structural effect on the CNTs could be found upon a six-minute ultrasonic treatment.

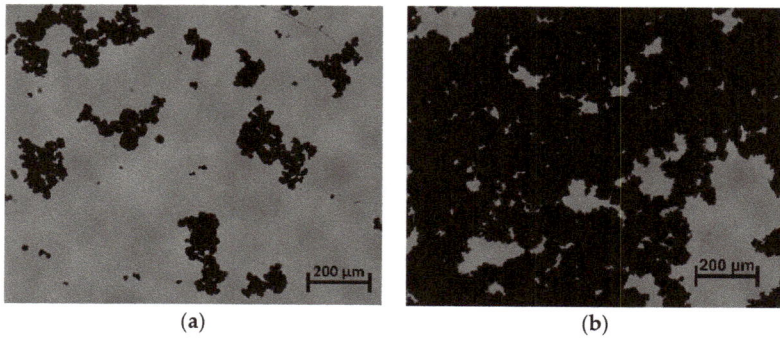

(a)   (b)

Figure 4. *Cont.*

(c)   (d)

Figure 4. Ultrasonic impact on CNT suspensions: (**a**) 0.4 g CNTs in 40 mL DI-water, no ultrasonic impact; (**b**) 0.4 g CNTs in 40 mL DI-water, ultrasonic impact (20%, 6 min, 10 s on, 15 s off) (**c**) 0.4 g in 40 mL DI-water after different times of US (20%, 10 s on, 15 s off): l.t.r.: 0 min (prompt sedimentation), 3 min (more viscous), 6 min (highly viscous, gel-like); (**d**) 0.4 g CNTs in 10 mL 10 g L^{-1} SDBS after different times of US (20%, 10 s on, 15 s off): l.t.r.: 0 min (prompt sedimentation), 3 min (dispersed system), 6 min (dispersed system).

In contrast to an ultrasonically steered approach, the use of dispersing agents has already established itself as a first step in the synthesis of CNT monoliths [11,15–17]. However, utilization of dispersants introduces an additional component to the monoliths' network, further influencing its adsorptive properties, as shown by Li et al. for pristine CNTs [54]. Introducing nanotube gelation through ultrasound, the addition of surfactants disturbed the gelation process in our system (see Figure 4d). Here, the utilized dispersant SDBS strongly accumulated on the surface of the nanotubes during ultrasonication, leading to sterical separation and inhibition of gel formation (see Figure S3d) [55,56]. Consequently, we were able to proceed with our synthesis without additional dispersing agents.

Heat drying the sonicated and pressed CNT-gel, we found that the resulting monoliths remained intact with increasing sonication time even after prolonged water incubation and shaking of the sample container (see Figure 5a). For instance, upon a heat drying approach, the synthesized monoliths showed no sign of disintegration in DI-water even after 6 months of incubation. In contrast, monoliths pressed with no prior sonication treatment dissolved and broke easily upon contact with water. For heat-dried monoliths, a considerable shrinking could be observed during drying as water evaporates from the particle network. Comparison of heat drying to a traditional freeze-drying approach as described by Bryning et al. [17] showed that the initial size of the wet gel is preserved for the most part (see Figure 5b). However, the stability of the monoliths decreases to a great extent as the particle network could not be further stabilized during drying.

 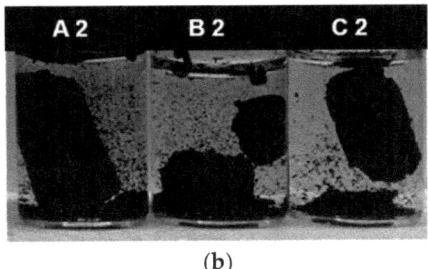

(a)   (b)

Figure 5. Solubility of cylindrical monoliths in DI-water synthesized with different US duration (A: 0 min, B: 3 min, C: 6 min), dried at 60 °C for 72 h (**a**) (A1–C1), and freeze-dried for 48 h (**b**) (A2–C2). Pictures were taken after 1 min of incubation.

3.3. Monolith Structure and Conductivity

For the further characterization of CNT monoliths, PVA was applied as a structural reinforcement as previously described by Bryning et al. [17]. Especially for planar monoliths, utilized for electrosorption and described in Section 3.4, we found that small concentrations of PVA further increased the homogeneity and stability. In the following, we characterize monoliths comprised of untreated and oxidized CNTs to determine their suitability as electrodes in an electrosorption process.

Our synthesized cylindrical monoliths were free-standing and characterized by a remarkably high structural stability. For example, three monoliths with a density of approximately 0.3 g cm^{-3} supported a counterweight of 3000 g (see Figure S5), which roughly translates to over 5000 times their weight. Moreover, the tensile strength for planar monoliths was determined to 4.28 $\pm$ 0.71 MPa, further emphasizing the high structural stability of our CNT monoliths and their suitability as technical electrodes.

The surface area of the untreated nanotubes was determined to 181 $\pm$ 2 m^2 g^{-1}. In this context, Trunzer et al. reported a predominance of structurally closed tube ends for the used CNTs, possibly due to the synthesis process [31,57]. Given that, the oxidation of the CNTs did not influence the tube structure as a specific surface area (SSA) of 175 $\pm$ 2 m^2 g^{-1} was measured for the oxidized sample (see Figure 6). Remarkably, the SSA could be preserved during monolith synthesis with untreated and oxidized nanotubes, as the SSA decreased only by 2%, respectively. This effect can be attributed to our additive-free gelation approach and the scarce utilization of PVA as a binder. Within the development of our synthesis protocol, we changed the incubation conditions of the wet gel in 1 wt-% PVA from several days as described by Bryning et al. [17] to a quick dispersion and centrifugation step with subsequent discard of the supernatant. Thus, we expect less PVA accumulation on the surface of the nanotubes, and therefore preservation of the CNTs' electrical and structural properties. Indeed, we were able to synthesize CNT monoliths with a reduced SSA loss compared to similar synthesis approaches [11,24]. For instance, Shen et al. [11] reported an SSA loss of 26% upon monolith synthesis. Hence, with our synthesis protocol, a vast number of possible binding sites for the later application as an electrode in an electrosorption process should be provided, which will be examined and discussed in Section 3.4.

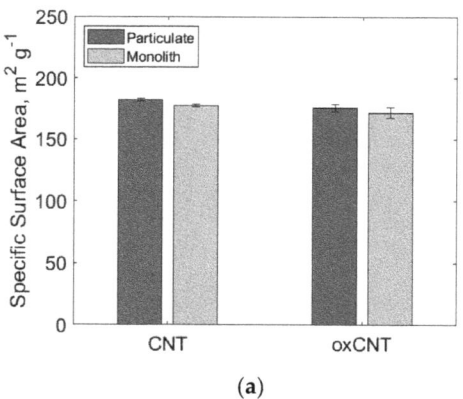 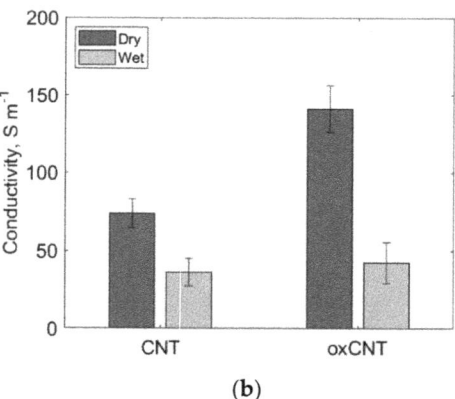

(a) (b)

Figure 6. Specific surface area (a) and electrical conductivity (b) for untreated and oxidized monoliths.

Additionally, we observed the preservation of the CNTs' exciting properties through conductivity measurements illustrated in Figure 6b. The electrical conductivity was found to be strongly dependent on the pretreatment of the nanotubes. For untreated monoliths, a conductivity of 74 $\pm$ 9 S m^{-1} was measured in a dried state. Utilizing monoliths comprised of oxidized nanotubes, the conductivity is almost doubled to 141 $\pm$ 15 S m^{-1}. This indicates an improved particle network formation during the monolith synthesis of the mildly oxidized nanotubes. After incubation in DI-water, both materials experience a considerable reduction of conductivity. Here, DI-water is stored in the porous structure of the monoliths and inhibits the current flow. Monoliths made of untreated CNTs display a loss of conductivity of 51% to 36 ± 9 S m^{-1}, while the oxidized monoliths possess a marginally higher conductivity of 42 ± 13 S m^{-1}. Nonetheless, a high conductivity can be observed for the synthesized monoliths, in a dry and wet state, which is essential for the later electrosorption in aqueous systems. Compared to Shen et al. [11] and Bryning et al. [17], on whom our synthesis protocol builds, we were able to increase the conductivity of MWCNT monoliths by at least one order of magnitude. Here, the pressing and subsequent heat drying of the gelled CNTs might additionally benefit the later conductivity of the resulting monoliths as the particle network becomes more compressed in the process. Simultaneously, we were able to reduce the synthesis steps, as well as the needed equipment and time. Thus, the synthesis and processing costs of the monoliths can be reduced.

3.4. Electrochemical Characterization

Chronoamperometry and cyclic voltammetry (CV) experiments were performed to investigate the suitability for potential-controlled adsorption as well as the electrochemical reactions between the monoliths' surface and the electrolyte environment. Within this framework, maleic acid as a negatively charged electrolyte and target molecule is frequently studied for electrosorption onto carbon electrodes [30–32,36,58]. Brammen et al. [36] and Trunzer et al. [31] already observed reversible maleic acid binding onto a particulate CNT electrode in aqueous systems, thus ensuring its suitability to study the surface-liquid interface of our novel monolithic CNT electrodes.

Figure 7a,b show an initial high current due to the electrodes' double-layer charging. As the double layer extends, the current profile declines until a nearly constant value is achieved. The curves do not reach zero. Hence, a remaining faradaic current is observed for untreated and oxidized monoliths, confirming deviations from an ideal capacitive system. Consequently, the studied monolithic electrodes display a resistance, which can be related to the observed faradaic current. This observed equilibrium current is generated by redox reactions causing constant electron flow on the electrode and is illustrated in Figure 7c. Hereby, an increased current response for monoliths comprised of oxidized CNTs is visible compared to untreated CNT monoliths. The higher faradaic current results from redox reactions between functional groups on the surface of the oxidized nanotubes [59,60]. This assumption can be further confirmed regarding the measured cyclic voltammograms of untreated and oxidized CNT monoliths in Figure 8. As the successive scans remain nearly congruent, no irreversible reactions are expected for the untreated CNTs. Moreover, its smooth shape approximates a capacitive behavior. Monoliths consisting of oxCNTs, on the other hand, behave slightly differently. A greater CV area can be observed, indicating an enlarged capacitance. At around +550 mV, a redox peak is passed through, assigned to oxidized groups on the monolith's surface. Moreover, the voltammetry scans for oxCNT monoliths pictured in Figure 8 behave slightly inconsistent. Thus, further irreversible reactions on the surface-electrolyte-interface might be displayed. While the increased current response of oxCNT demonstrates improved electrosorption capability, indicated irreversible reactions might negatively impact a reversible process. The following section examines how these interrelationships are manifested in a practical electrosorption process.

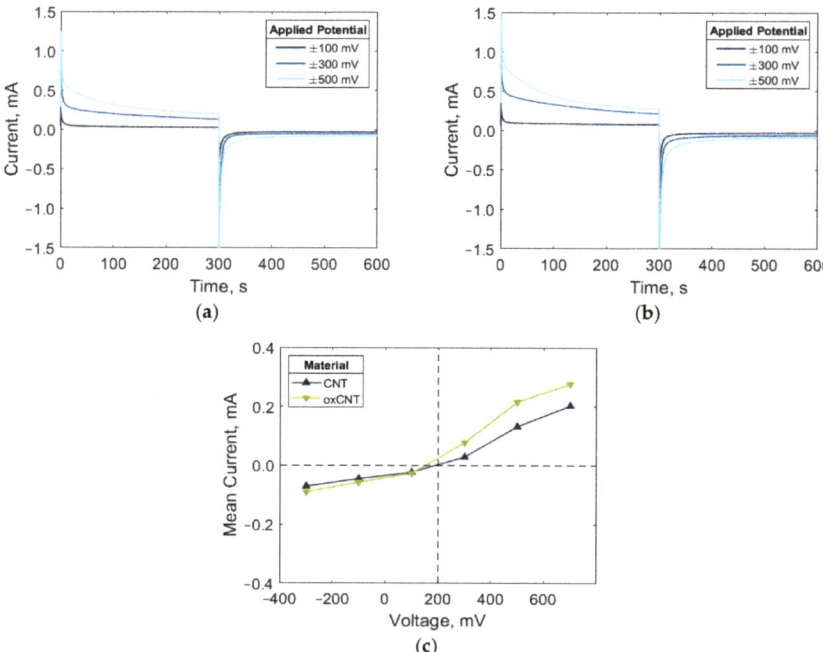

Figure 7. Chronoamperometry experiments for untreated (**a**) and oxidized (**b**) CNT-K monoliths. The calculated resulting mean current after EDL rearrangement for both materials is presented in (**c**).

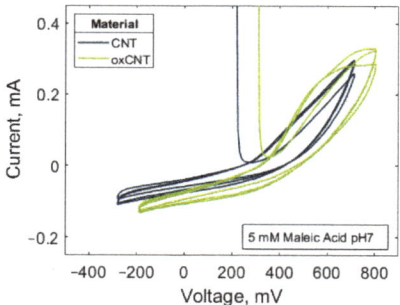

Figure 8. Cyclic voltammetry of untreated and oxidized CNT-K monoliths.

3.5. Potential-Controlled Adsorption on CNT-Monoliths

As a proof of concept, we studied the effect of applied potential on a model analyte's adsorption and desorption behavior, pictured in Figure 9. In potential step experiments, the potential-controlled adsorption behavior onto CNT monoliths could be proven. Untreated CNT monoliths showed initial adsorption of 10.7 ± 4.7 µmol g^{-1}. Upon potential switch, a released amount of 6.2 ± 0.6 µmol g^{-1} could be detected, resulting in a release efficiency of 57%. In the following runs, the amount of adsorbed maleic acid firmly declined for the untreated monoliths. Changes on the monolith's surface and strongly bound analyte, not removed through the washing steps, might impair the repeated use of the electrodes and can lead to deviations of the measured concentrations. The availability of active binding sites is thereby drastically reduced. The released amount of analyte slightly increases during the subsequent runs. The presence of unremoved analyte from the first run is thus confirmed.

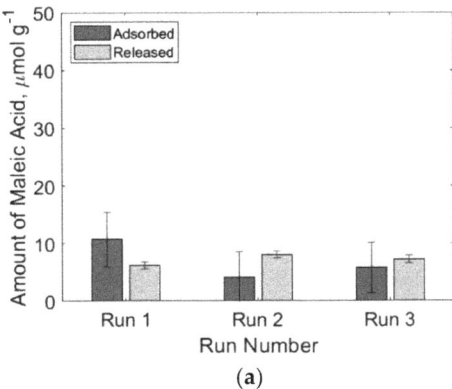
(a)

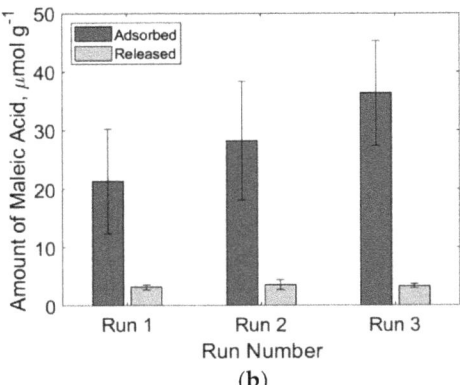
(b)

Figure 9. Potential-controlled adsorption and release experiments for 5 mM maleic acid at pH 7 onto untreated (**a**) and oxidized (**b**) CNT-K monoliths. A potential of +500 mV for the adsorption and −500 mV for the desorption phase was applied for all experiments. Potentials were applied for 15 min during adsorption and release phase.

A major increase in adsorption capacity can be observed during all runs for oxidized monoliths, possibly due to an enlarged number of binding sites and enhanced current response. However, as the amount of released analyte upon potential inversion is reduced, another binding mechanism between the electrode's oxidized surface and the analyte might be revealed. This heterogeneity is often recognized for carbon nanotubes, as different sorption mechanisms, such as π-π bonds, hydrophobic interactions and hydrogen bonds, may act simultaneously [61]. Particularly for monoliths consisting of oxidized CNTs, electrosorption does not take place exclusively on the surface, aggravating the potential-dependent release of the analyte.

Conventional electrosorption processes studied for the adsorption of smaller inorganic ions show similar adsorption capacities compared to our oxidized monoliths [62–64]. However, it has to be considered that the cited studies were conducted in flow-through setups, thus benefiting from an increased mass transport towards the electrode. In this context, the tested monoliths display good electrosorption properties even at weak applied potentials due to their high conductivity and SSA. Nonetheless, further investigation of the monoliths in a flow-through setup might improve the adsorption and desorption behavior due to mass transport effects.

4. Conclusions and Outlook

This study presents a simple and cost-efficient synthesis protocol, preserving the outstanding structural and electrical properties of carbon nanotubes during monolith synthesis and leading to stable and free-standing monoliths. Moreover, our work combines a broad material characterization of the synthesized monoliths with a subsequent investigation of interfacial effects occurring on the monoliths' surface, as well as an integrated electrosorption study. Thus, it paves the way towards a future application of monolithic CNT electrodes in an electrochemically driven separation process.

For the stability of the synthesized monoliths, we determined the dispersion and gelation of the particles by ultrasonication during the synthesis process to be a key factor. Thus, ultrasonic treatment led tow strong gelation as drying in a heating chamber further stabilized the CNT network. We utilized untreated and mildly oxidized CNTs for monolith synthesis. Both materials enable the formation of stable monoliths with a comparable specific surface area. However, by using oxidized CNTs the conductivity of the monoliths could be strongly increased, indicating a more homogenous and pronounced particle network. Upon electrochemical characterization, the presence of capacitive and faradaic currents could be determined for both materials. However, pronounced faradaic currents were detected with oxidized monoliths, leading to an overall stronger current response. The potential-controlled adsorption of maleic acid was investigated as a proof of concept for an electrosorption process on CNT monoliths. Untreated and oxidized monoliths showed electrosorption behavior. While the adsorption capacity could be improved through oxidation, strong heterogeneous interactions still complicate a potential-triggered release of the model analyte. In this respect, the utilization of electro-active composites might be a possible way to control the reversible electrosorption onto CNT monoliths. Furthermore, the interfacial interactions guiding the reversible electrosorption of organic molecules should be further examined. With the goal of developing a potential-controlled process for biotechnological separations, we continue working on a preparative setup for the targeted electrosorption onto CNT monolith electrodes.

Supplementary Materials: The following are available online at https://www.mdpi.com/article/10.3390/app11209390/s1, Figure S1: Test rig for conductivity measurements on cylindrical monoliths. Figure S2: Electrode contacting and test rig for potential-controlled adsorption experiments. Figure S3: Agglomerates of CNTs after oxidative and ultrasonic treatment in DI-Water and SDBS. Figure S4: FTIR spectra of untreated and oxidized CNT-K particles. Figure S5: Mechanical strength of cylindrical and planar monoliths.

Author Contributions: T.T., P.F.-G. and S.B. designed the project, D.R., T.T., J.H., J.R., performed the experiments. All authors assisted in evaluating the results. D.R. wrote the manuscript. T.T., P.F.-G. and S.B. assisted in reviewing and editing the manuscript. Financial acquisition and project administration by P.F.-G. and S.B. All authors have read and agreed to the published version of the manuscript.

Funding: The research was funded by the Federal Ministry of Education and Research under the name S3kapel (grant number 13XP5038A). This work was supported by the Technical University of Munich (TUM) through the TUM Open Access Publishing Fund.

Institutional Review Board Statement: Not applicable.

Informed Consent Statement: Not applicable.

Data Availability Statement: The original data related to this research can be asked at any time to the corresponding author's email: d.roecker@tum.de.

Acknowledgments: We thank Sebastian Schwaminger and Chiara Turrina for the support and advice regarding TEM imaging. In addition, we thank Tim Mader for his assistance in the laboratory. Moreover, we want to thank several groups of the TUM for their support and the use of their equipment: the Institute for Machine Tools and Industrial Management for their assistance regarding the 3D printing of the setup and the pressing molds; the Chair of Medical Materials and Implants for the support in tensile testing of the monoliths and the Associate Professorship of Electron Microscopy for TEM imaging.

Conflicts of Interest: The authors declare no conflict of interest. The funders had no role in the design of the study; in the collection, analyses, or interpretation of data; in the writing of the manuscript, or in the decision to publish the results.

References

1. Iijima, S. Helical microtubules of graphitic carbon. *Nature* **1991**, *354*, 56–58. [CrossRef]
2. Ebbesen, T.W.; Lezec, H.J.; Hiura, H.; Bennett, J.W.; Ghaemi, H.F.; Thio, T. Electrical conductivity of individual carbon nanotubes. *Nat. Cell Biol.* **1996**, *382*, 54–56. [CrossRef]
3. Pop, E.; Mann, D.; Wang, Q.; Goodson, K.; Dai, H. Thermal conductance of an individual single-wall carbon nanotube above room temperature. *Nano Lett.* **2006**, *6*, 96–100. [CrossRef] [PubMed]
4. Treacy, M.M.J.; Ebbesen, T.W.; Gibson, J.M. Exceptionally high Young's modulus observed for individual carbon nanotubes. *Nat. Cell Biol.* **1996**, *381*, 678–680. [CrossRef]
5. Evanoff, K.; Khan, J.; Balandin, A.A.; Magasinski, A.; Ready, W.J.; Fuller, T.F.; Yushin, G. Towards ultrathick battery electrodes: Aligned carbon nanotube-enabled architecture. *Adv. Mater.* **2012**, *24*, 533–537. [CrossRef] [PubMed]

6. Sotowa, C.; Origi, G.; Takeuchi, M.; Nishimura, Y.; Takeuchi, K.; Jang, I.Y.; Kim, Y.J.; Hayashi, T.; Kim, Y.A.; Endo, M.; et al. The reinforcing effect of combined carbon nanotubes and acetylene blacks on the positive electrode of lithium-ion batteries. *ChemSusChem* **2008**, *1*, 911–915. [CrossRef] [PubMed]
7. De Volder, M.F.L.; Tawfick, S.; Baughman, R.H.; Hart, A.J. Carbon nanotubes: Present and future commercial applications. *Science* **2013**, *339*, 535–539. [CrossRef]
8. Park, M.-Y.; Kim, C.-G.; Kim, J.-H. CNT-coated quartz woven fabric electrodes for robust lithium-ion structural batteries. *Appl. Sci.* **2020**, *10*, 8622. [CrossRef]
9. Comba, F.N.; Romero, M.R.; Garay, F.S.; Baruzzi, A.M. Mucin and carbon nanotube-based biosensor for detection of glucose in human plasma. *Anal. Biochem.* **2018**, *550*, 34–40. [CrossRef]
10. Cai, H.; Cao, X.; Jiang, Y.; He, P.; Fang, Y. Carbon nanotube-enhanced electrochemical DNA biosensor for DNA hybridization detection. *Anal. Bioanal. Chem.* **2003**, *375*, 287–293. [CrossRef]
11. Shen, Y.; Du, A.; Wu, X.-L.; Li, X.-G.; Shen, J.; Zhou, B. Low-cost carbon nanotube aerogels with varying and controllable density. *J. Sol Gel Sci. Technol.* **2016**, *79*, 76–82. [CrossRef]
12. Kohlmeyer, R.R.; Lor, M.; Deng, J.; Liu, H.; Chen, J. Preparation of stable carbon nanotube aerogels with high electrical conductivity and porosity. *Carbon* **2011**, *49*, 2352–2361. [CrossRef]
13. Zou, J.; Liu, J.; Karakoti, A.; Kumar, A.; Joung, D.; Li, Q.; Khondaker, S.I.; Seal, S.; Zhai, L. Ultralight multiwalled carbon nanotube aerogel. *ACS Nano* **2010**, *4*, 7293–7302. [CrossRef] [PubMed]
14. You, B.; Wang, L.; Yao, L.; Yang, J. Threedimensional N-doped graphene–CNT networks for supercapacitor. *Chem. Commun.* **2013**, *49*, 5016–5018. [CrossRef]
15. Haghgoo, M.; Yousefi, A.A.; Mehr, M.J.Z.; Celzard, A.; Fierro, V.; Léonard, A.; Job, N. Characterization of multi-walled carbon nanotube dispersion in resorcinol–formaldehyde aerogels. *Microporous Mesoporous Mater.* **2014**, *184*, 97–104. [CrossRef]
16. Du, R.; Zhao, Q.; Zhang, N.; Zhang, J. Macroscopic carbon nanotube-based 3D monoliths. *Small* **2015**, *11*, 3263–3289. [CrossRef]
17. Bryning, M.B.; Milkie, D.E.; Islam, M.; Hough, L.A.; Kikkawa, J.M.; Yodh, A.G. Carbon nanotube aerogels. *Adv. Mater.* **2007**, *19*, 661–664. [CrossRef]
18. Zhang, X.; Liu, J.; Xu, B.; Su, Y.; Luo, Y. Ultralight conducting polymer/carbon nanotube composite aerogels. *Carbon* **2011**, *49*, 1884–1893. [CrossRef]
19. Du, R.; Zhang, N.; Xu, H.; Mao, N.; Duan, W.; Wang, J.; Zhao, Q.; Liu, Z.; Zhang, J. CMP aerogels: Ultrahigh-surface-area carbon-based monolithic materials with superb sorption performance. *Adv. Mater.* **2014**, *26*, 8053–8058. [CrossRef]
20. Maghfirah, A.; Yudianti, R.; Sinuhaji, P.; Hutabarat, L.G. Preparation of poly(vinyl) alcohol—Multiwalled carbon nanotubes nanocomposite as conductive and transparent film using casting method. *J. Phys. Conf. Ser.* **2018**, *1116*, 032017. [CrossRef]
21. Singh, R.; Singh, M.; Bhartiya, S.; Singh, A.; Kohli, D.; Ghosh, P.C.; Meenakshi, S.; Gupta, P. Facile synthesis of highly conducting and mesoporous carbon aerogel as platinum support for PEM fuel cells. *Int. J. Hydrog. Energy* **2017**, *42*, 11110–11117. [CrossRef]
22. Chen, W.; Rakhi, R.B.; Alshareef, H.N. High energy density supercapacitors using macroporous kitchen sponges. *J. Mater. Chem.* **2012**, *22*, 14394–14402. [CrossRef]
23. Li, Y.; Kang, Z.; Yan, X.; Cao, S.; Li, M.; Guo, Y.; Huan, Y.; Wen, X.; Zhang, Y. A three-dimensional reticulate CNT-aerogel for a high mechanical flexibility fiber supercapacitor. *Nanoscale* **2018**, *10*, 9360–9368. [CrossRef]
24. Zhou, Y.; Hu, X.; Guo, S.; Yu, C.; Zhong, S.; Liu, X. Multi-functional graphene/carbon nanotube aerogels for its applications in supercapacitor and direct methanol fuel cell. *Electrochim. Acta* **2018**, *264*, 12–19. [CrossRef]
25. Tesfaye, R.M.; Das, G.; Park, B.J.; Kim, J.; Yoon, H.H. Ni-Co bimetal decorated carbon nanotube aerogel as an efficient anode catalyst in urea fuel cells. *Sci. Rep.* **2019**, *9*, 1–9. [CrossRef] [PubMed]
26. Du, R.; Zhang, N.; Zhu, J.; Wang, Y.; Xu, C.; Hu, Y.; Mao, N.; Xu, H.; Duan, W.; Zhuang, L.; et al. Nitrogen-doped carbon nanotube aerogels for high-performance ORR catalysts. *Small* **2015**, *11*, 3903–3908. [CrossRef] [PubMed]
27. Biesheuvel, P.M.; Porada, S.; Levi, M.; Bazant, M. Attractive forces in microporous carbon electrodes for capacitive deionization. *J. Solid State Electrochem.* **2014**, *18*, 1365–1376. [CrossRef]
28. Wang, L.; Wang, M.; Huang, Z.-H.; Cui, T.; Gui, X.; Kang, F.; Wang, K.; Wu, D. Capacitive deionization of NaCl solutions using carbon nanotube sponge electrodes. *J. Mater. Chem.* **2011**, *21*, 18295–18299. [CrossRef]
29. Wang, X.Z.; Li, M.G.; Chen, Y.W.; Cheng, R.M.; Huang, S.M.; Pan, L.K.; Sun, Z. Electrosorption of NaCl solutions with carbon nanotubes and nanofibers composite film electrodes. *Electrochem. Solid State Lett.* **2006**, *9*, e23–e26. [CrossRef]
30. Wagner, R.; Bag, S.; Trunzer, T.; Fraga-García, P.; Wenzel, W.; Berensmeier, S.; Franzreb, M. Adsorption of organic molecules on carbon surfaces: Experimental data and molecular dynamics simulation considering multiple protonation states. *J. Colloid Interface Sci.* **2021**, *589*, 424–437. [CrossRef]
31. Trunzer, T.; Stummvoll, T.; Porzenheim, M.; Fraga-García, P.; Berensmeier, S. A carbon nanotube packed bed electrode for small molecule electrosorption: An electrochemical and chromatographic approach for process description. *Appl. Sci.* **2020**, *10*, 1133. [CrossRef]
32. Trunzer, T.; Fraga-García, P.; Tschuschner, M.-P.A.; Voltmer, D.; Berensmeier, S. The electrosorptive response of a carbon nanotube flow-through electrode in aqueous systems. *Chem. Eng. J.* **2022**, *428*, 13100. [CrossRef]
33. Moraes, R.A.; De Matos, C.F.; Castro, E.G.; Schreiner, W.H.; Oliveira, M.M.; Zarbin, A.J.G. The effect of different chemical treatments on the structure and stability of aqueous dispersion of iron- and iron oxide-filled multi-walled carbon nanotubes. *J. Braz. Chem. Soc.* **2011**, *22*, 2191–2201. [CrossRef]

34. Shaffer, M.; Fan, X.; Windle, A. Dispersion and packing of carbon nanotubes. *Carbon* **1998**, *36*, 1603–1612. [CrossRef]
35. Vigolo, B.; Pénicaud, A.; Coulon, C.; Sauder, C.; Pailler, R.; Journet, C.; Bernier, P.; Poulin, P. Macroscopic fibers and ribbons of oriented carbon nanotubes. *Science* **2000**, *290*, 1331–1334. [CrossRef] [PubMed]
36. Brammen, M.; Fraga-García, P.; Berensmeier, S. Carbon nanotubes-A resin for electrochemically modulated liquid chromatography. *J. Sep. Sci.* **2017**, *40*, 1176–1183. [CrossRef] [PubMed]
37. Avilés, F.; Cauich-Rodríguez, J.; Moo-Tah, L.; May-Pat, A.; Vargas-Coronado, R.F. Evaluation of mild acid oxidation treatments for MWCNT functionalization. *Carbon* **2009**, *47*, 2970–2975. [CrossRef]
38. Farbod, M.; Tadavani, S.K.; Kiasat, A. Surface oxidation and effect of electric field on dispersion and colloids stability of multiwalled carbon nanotubes. *Colloids Surf. A Physicochem. Eng. Asp.* **2011**, *384*, 685–690. [CrossRef]
39. Wepasnick, K.A.; Smith, B.A.; Schrote, K.E.; Wilson, H.K.; Diegelmann, S.R.; Fairbrother, D.H. Surface and structural characterization of multi-walled carbon nanotubes following different oxidative treatments. *Carbon* **2011**, *49*, 24–36. [CrossRef]
40. Kim, Y.-T.; Mitani, T. Competitive effect of carbon nanotubes oxidation on aqueous EDLC performance: Balancing hydrophilicity and conductivity. *J. Power Sour.* **2006**, *158*, 1517–1522. [CrossRef]
41. Schwaminger, S.P.; Brammen, M.W.; Zunhammer, F.; Däumler, N.; Fraga-García, P.; Berensmeier, S. Iron oxide nanoparticles: Multiwall carbon nanotube composite materials for batch or chromatographic biomolecule separation. *Nanoscale Res. Lett.* **2021**, *16*, 1–10. [CrossRef]
42. Țucureanu, V.; Matei, A.; Avram, M.A. FTIR spectroscopy for carbon family study. *Crit. Rev. Anal. Chem.* **2016**, *46*, 502–520. [CrossRef]
43. Stobinski, L.; Lesiak, B.; Kövér, L.; Tóth, J.; Biniak, S.; Trykowski, G.; Judek, J. Multiwall carbon nanotubes purification and oxidation by nitric acid studied by the FTIR and electron spectroscopy methods. *J. Alloy. Compd.* **2010**, *501*, 77–84. [CrossRef]
44. Morales-Torres, S.; Silva, T.L.S.; Pastrana-Martínez, L.M.; Brandão, A.T.S.C.; Figueiredo, J.L.; Silva, A.M.T. Modification of the surface chemistry of single- and multi-walled carbon nanotubes by HNO_3 and H_2SO_4 hydrothermal oxidation for application in direct contact membrane distillation. *Phys. Chem. Chem. Phys.* **2014**, *16*, 12237–12250. [CrossRef]
45. Gatabi, M.P.; Moghaddam, H.M.; Ghorbani, M. Point of zero charge of maghemite decorated multiwalled carbon nanotubes fabricated by chemical precipitation method. *J. Mol. Liq.* **2016**, *216*, 117–125. [CrossRef]
46. Fu, R.; Zheng, B.; Liu, J.; Dresselhaus, M.S.; Dresselhaus, G.; Satcher, J.H.; Baumann, T.F. The fabrication and characterization of carbon aerogels by gelation and supercritical drying in isopropanol. *Adv. Funct. Mater.* **2003**, *13*, 558–562. [CrossRef]
47. Pekala, R.W.; Farmer, J.C.; Alviso, C.T.; Tran, T.D.; Mayer, S.T.; Miller, J.M.; Dunn, B. Carbon aerogels for electrochemical applications. *J. Non Cryst. Solids* **1998**, *225*, 74–80. [CrossRef]
48. Tamon, H.; Ishizaka, H.; Yamamoto, T.; Suzuki, T. Influence of freeze-drying conditions on the mesoporosity of organic gels as carbon precursors. *Carbon* **2000**, *38*, 1099–1105. [CrossRef]
49. Thongprachan, N.; Nakagawa, K.; Sano, N.; Charinpanitkul, T.; Tanthapanichakoon, W. Preparation of macroporous solid foam from multi-walled carbon nanotubes by freeze-drying technique. *Mater. Chem. Phys.* **2008**, *112*, 262–269. [CrossRef]
50. Huang, Y.Y.; Knowles, T.P.J.; Terentjev, E.M. Strength of nanotubes, filaments, and nanowires from sonication-induced scission. *Adv. Mater.* **2009**, *21*, 3945–3948. [CrossRef]
51. Yamamoto, T.; Miyauchi, Y.; Motoyanagi, J.; Fukushima, T.; Aida, T.; Kato, M.; Maruyama, S. Improved bath sonication method for dispersion of individual single-walled carbon nanotubes using new triphenylene-based surfactant. *Jpn. J. Appl. Phys.* **2008**, *47*, 2000–2004. [CrossRef]
52. Park, H.J.; Park, M.; Chang, J.Y.; Lee, H. The effect of pre-treatment methods on morphology and size distribution of multi-walled carbon nanotubes. *Nanotechnology* **2008**, *19*, 335702. [CrossRef]
53. Huang, Y.Y.; Terentjev, E.M. Dispersion of carbon nanotubes: Mixing, sonication, stabilization, and composite properties. *Polymers* **2012**, *4*, 275–295. [CrossRef]
54. Li, H.; Wei, C.; Zhang, D.; Pan, B. Adsorption of bisphenol A on dispersed carbon nanotubes: Role of different dispersing agents. *Sci. Total. Environ.* **2019**, *655*, 807–813. [CrossRef]
55. Niyogi, S.; Boukhalfa, S.; Chikkannanavar, S.B.; McDonald, T.J.; Heben, A.M.J.; Doorn, S.K. Selective aggregation of single-walled carbon nanotubes via salt addition. *J. Am. Chem. Soc.* **2007**, *129*, 1898–1899. [CrossRef]
56. Clark, M.D.; Subramanian, S.; Krishnamoorti, R. Understanding surfactant aided aqueous dispersion of multi-walled carbon nanotubes. *J. Colloid Interface Sci.* **2011**, *354*, 144–151. [CrossRef] [PubMed]
57. Frackowiak, E.; Metenier, K.; Bertagna, V.; Beguin, F. Supercapacitor electrodes from multiwalled carbon nanotubes. *Appl. Phys. Lett.* **2000**, *77*, 2421–2423. [CrossRef]
58. Wagner, R.; Winger, S.; Franzreb, M. Predicting the potential of capacitive deionization for the separation of pH-dependent organic molecules. *Eng. Life Sci.* **2021**, 1–18. [CrossRef]
59. Lee, S.W.; Gallant, B.M.; Lee, Y.; Yoshida, N.; Kim, D.Y.; Yamada, Y.; Noda, S.; Yamada, A.; Shao-Horn, Y. Self-standing positive electrodes of oxidized few-walled carbon nanotubes for light-weight and high-power lithium batteries. *Energy Environ. Sci.* **2012**, *5*, 5437–5444. [CrossRef]
60. Lee, S.W.; Gallant, B.M.; Byon, H.R.; Hammond, P.T.; Shao-Horn, Y. Nanostructured carbon-based electrodes: Bridging the gap between thin-film lithium-ion batteries and electrochemical capacitors. *Energy Environ. Sci.* **2011**, *4*, 1972–1985. [CrossRef]
61. Pan, B.; Xing, B. Adsorption mechanisms of organic chemicals on carbon nanotubes. *Environ. Sci. Technol.* **2008**, *42*, 9005–9013. [CrossRef] [PubMed]

62. Suss, M.E.; Porada, S.; Sun, X.; Biesheuvel, P.M.; Yoon, J.; Presser, V. Water desalination via capacitive deionization: What is it and what can we expect from it? *Energy Environ. Sci.* **2015**, *8*, 2296–2319. [CrossRef]
63. Peng, Z.; Zhang, D.; Yan, T.; Zhang, J.; Shi, L. Three-dimensional micro/mesoporous carbon composites with carbon nanotube networks for capacitive deionization. *Appl. Surf. Sci.* **2013**, *282*, 965–973. [CrossRef]
64. Pan, L.; Wang, X.; Gao, Y.; Zhang, Y.; Chen, Y.; Sun, Z. Electrosorption of anions with carbon nanotube and nanofibre composite film electrodes. *Desalination* **2009**, *244*, 139–143. [CrossRef]

Article

Comprehensive Characterization of Structural, Electrical, and Mechanical Properties of Carbon Nanotube Yarns Produced by Various Spinning Methods

Takayuki Watanabe [1], Satoshi Yamazaki [2], Satoshi Yamashita [2], Takumi Inaba [1], Shun Muroga [1], Takahiro Morimoto [1], Kazufumi Kobashi [1] and Toshiya Okazaki [1,*]

1. CNT-Application Research Center, National Institute of Advanced Industrial Science and Technology, Tsukuba 305-8565, Japan; takayuki.watanabe.r@gmail.com (T.W.); takumi.inaba@aist.go.jp (T.I.); muroga-sh@aist.go.jp (S.M.); t-morimoto@aist.go.jp (T.M.); kobashi-kazufumi@aist.go.jp (K.K.)
2. Research Association of High-Throughput Design and Development for Advanced Functional Materials (ADMAT), Tsukuba 305-8565, Japan; satoshi.mayama.yamazaki@furukawaelectric.com (S.Y.); satoshi.yamashita@furukawaelectric.com (S.Y.)
* Correspondence: toshi.okazaki@aist.go.jp

Abstract: A comprehensive characterization of various carbon nanotube (CNT) yarns provides insight for producing high-performance CNT yarns as well as a useful guide to select the proper yarn for a specific application. Herein we systematically investigate the correlations between the physical properties of six CNT yarns produced by three spinning methods, and their structures and the properties of the constituent CNTs. The electrical conductivity increases in all yarns regardless of the spinning method as the effective length of the constituent CNTs and the density of the yarns increase. On the other hand, the tensile strength shows a much stronger dependence on the packing density of the yarns than the CNT effective length, indicating the relative importance of the interfacial interaction. The contribution of each physical parameter to the yarn properties are quantitatively analyzed by partial least square regression.

Keywords: carbon nanotubes; yarn; electrical conductivity; tensile strength; nanotube length

1. Introduction

Carbon nanotubes (CNTs) display very high electrical conductivities, thermal conductivities, as well as mechanical strengths, and sorption abilities [1–5]. Because they also have lower densities than metals such as copper and steel [6], CNTs have potential as alternating-current power cables and wires. In fact, their estimated electrical conductivity is as high as 900,000 S/cm, which is much larger than that of copper (600,000 S/cm) [7]. Additionally, the tensile strength of CNT bundles is ~80 GPa, whereas that of steel is ~1 GPa [2].

To preserve superior properties in macroscopic CNT-based structures, the production of yarns composed of CNTs offers a potential for high-strength and lightweight materials that are also thermally and electrically conductive [8–12]. Macroscopic CNT yarns show an electrical conductivity and tensile strength of 10,900 S/cm and 9.6 GPa, respectively [13–15]. Their values are within a factor of those of aluminum and commercially available carbon fibers, respectively. CNT yarns exhibit a strength similar to carbon fibers and an electrical conductivity similar to metal.

Currently, CNT yarns are mainly produced by three methods: dry spinning from multi-walled CNT (MWCNT) forests, direct spinning from a CVD furnace, and wet-spinning of CNTs (Scheme 1). Elucidating their structures, properties, and relationships are critical to realize practical applications of CNT yarns.

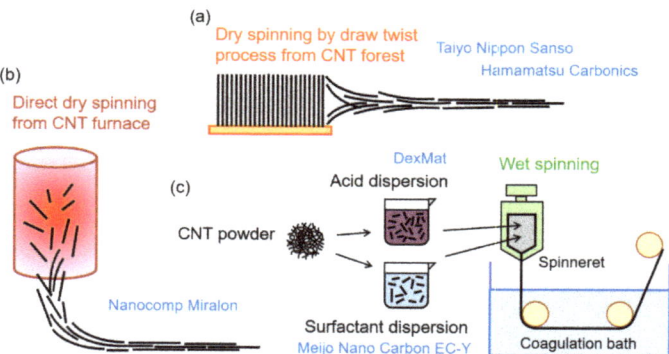

Scheme 1. Schematic representation of spinning methods for CNTs. (**a**) Dry spinning by draw twist process from CNT forest, (**b**) dry spinning from CNT furnace, and (**c**) wet-spinning.

We previously investigated the electrical and mechanical properties of wet-spun CNT yarns using far-infrared (FIR) spectroscopy. These properties are related to the effective CNT length [16–18]. The observed FIR peak can be explained by the one-dimensional plasmon model. Consequently, the estimated length can be ascribed to that of the clean and straight CNT portion between defects or kinks (effective CNT length) [19–22]. FIR spectroscopy is applicable to single-walled CNTs (SWCNTs) and MWCNTs because the resonant frequency is insensitive to the CNT diameter [19,21]. On the other hand, the G-band/D-band ratio in the Raman spectrum strongly depends on the CNT diameter and CNT types.

Here, we systematically investigate the structure of various CNT yarns and the properties of the constituent CNTs. Our study evaluated yarns fabricated by three different spinning methods using various CNTs. We observed the yarn structures and cross-sections by scanning electron microscopy (SEM). We also estimated the average CNT diameters and wall numbers by transmission electron microscopy (TEM), the CNT alignment by wide angle X-ray diffraction (WAXD), the G/D ratios and the radial breathing modes by Raman spectroscopy, and the effective CNT lengths by FIR spectroscopy.

Then we evaluated the electrical conductivities and the tensile strengths of the yarns, and their correlation with the yarn and CNT properties. The electrical conductivity of the yarns is significantly correlated with the effective CNT length, yarn density, and CNT alignment in the yarns. In contrast, the tensile strength depends more on the yarn density and the CNT alignment. To quantitatively evaluate the contribution of each physical parameter to the properties of CNT yarn, the measurement data was analyzed using the partial least square (PLS) regression [23–25].

2. Materials and Methods

The six commercially available CNT yarns were studied here (Scheme 1). Two CNT yarns by dry spinning from CNT forests were purchased from Hamamatsu Carbonics Corporation (Shizuoka, Japan) and Taiyo Nippon Sanso Corporation (Tokyo, Japan). The CNT yarn by direct spinning from a CVD furnace (Miralon CNT yarn) was obtained from Nanocomp Technologies Inc. (Merrimack, NH, USA). The wet-spinning CNT yarns from surfactant-assisted aqueous dispersion and strong acid dispersion were purchased from Meijo Nano Carbon Co., Ltd. (Aichi, Japan) (EC-Y type I and II) and DexMat Inc. (Houston, TX, USA), respectively.

The morphologies of the CNT yarns were observed by SEM microscopy (Hitachi SU-8200, Tokyo, Japan). The diameter and wall number of the CNTs were estimated by TEM microscopy (EM002B, Topcon, Tokyo, Japan), while WAXD (AichiSR, beam line BL8S3, Aichi, Japan) assessed the CNT alignment in the yarns. The photon energy was 13.48 KeV (λ = 0.092 nm). The sample-to-detector distance was 0.208 m. The WAXD spectra were measured using a two-dimensional detector (R-AXIS VII++). The scattering wave vector $q = 4\pi\sin\theta/\lambda$, where θ is the scattering angle. The Raman spectra were measured with inVia (Renishaw, Wotton-under-Edge, England, UK) with an excitation wavelength of 532 nm. The effective length of each CNT was estimated from its FIR optical absorption spectrum [20,22]. The FIR measurements were acquired using Vertex 80v (Bruker Optics, Billerica, MA, USA) and TR-1000 (Otsuka Electronics, Osaka, Japan). The average structural CNT lengths were estimated from the AFM measurements (Shimadzu SFT-4500, Kyoto, Japan).

The electrical resistance of the CNT yarn was measured using the four-terminal method with a probe system (Summit12000, Cascade Microtech, Beaverton, OR, USA) and a semiconductor device analyzer (B1500A, Keysight, Santa Rosa, CA, USA). Three samples were measured for each CNT yarn type. The distance between the two terminals of the differential voltage measurement system was 30 mm. The obtained values were averaged for, at least, three samples for each CNT yarn.

The breaking strengths of the CNT yarns were measured in tensile tests with a micro strain tester (MST-I, Shimadzu, Kyoto, Japan). The test pieces and the test conditions followed the Japanese standard, JIS R 7606, which is a standard protocol for carbon fiber tensile testing. The gauge length was fixed to 25 mm, and the head speed was fixed to 1 mm/min. Three samples were measured for each CNT yarn type.

Thermogravimetric analysis (TGA) was used to estimate the carbonaceous purity of the CNT yarns with TGAQ500 (TA instruments, New Castle, DE, USA).

PLS models were constructed from the covariance between the standardized response variables (electrical conductivity or tensile strength) and the standardized explanatory variables (physical parameters). Details for select measurement procedures are described in the Supplementary Materials.

3. Results and Discussion

3.1. Structural Characterizations of Commercially Available CNT Yarns and Their Constituent CNTs

The structural and physical properties of CNT yarn strongly depend on the spinning method [8–12,26]. Figures 1 and 2 show SEM images of the CNT yarns evaluated in this study. Hamamatsu and Taiyo Nippon Sanso yarns were made by dry spinning from CNT forests [27]. Their estimated diameters by laser micrometer are 59 and 31 μm, respectively (Table 1, see the Supporting Information). This method spins CNTs from a CNT forest by twisting. The twisted structure is visible from the side views (Figure 1). The roundness of the cross-sections exceeds those of the other yarns, reflecting the high controllability of the dry spinning methods (Figure 2a). The network structures in the high-magnification images show the different diameters of the constituent CNTs (Figure S1 in the Supporting Information).

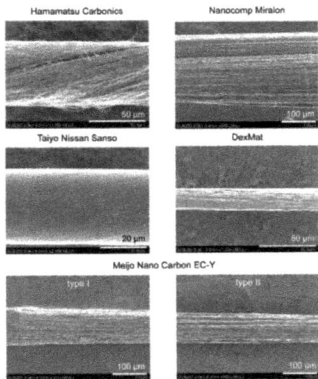

Figure 1. Side-view SEM images of CNT yarns.

Nanocomp Miralon yarn is produced by direct spinning from a CVD furnace [28]. It has a unique structure among the CNT yarns, including conventional direct spun CNT yarns [29–31]. The low-magnification image shows non-negligible voids in the yarn (Figure 2a), whereas the CNTs seem well packed and sub micrometer voids rarely appear in the high-magnification image (Figure 2b). The estimated yarn diameter is 230 μm, which is the largest among the yarns in this study (Table 1).

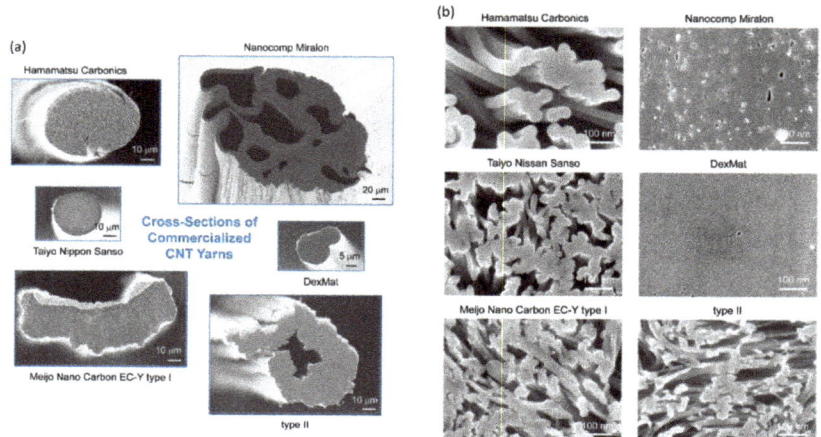

Figure 2. Cross-sectional SEM images of CNT yarns. (**a**) Low-magnification images. (**b**) High-magnification images.

Table 1. Physical properties of the CNT yarns in this study.

CNT Yarns	Hamamatsu Carbonics	Taiyo Nippon Sanso	Nanocomp Miralon	Meijo EC-Y Type I	Meijo EC-Y Type II	DexMat
Yarn diameter (μm)	59	31	230	90	76	20
Average diameter of CNTs (nm)	45 ± 1	12 ± 0.2	5.3 ± 0.1	1.8 ± 0.05	1.7 ± 0.04	2.2 ± 0.05
Average wall number	50 ± 2	8.5 ± 0.1	1.9 ± 0.03	1.4 ± 0.05	1.4 ± 0.05	1.6 ± 0.05
CNT type	MWCNT	MWCNT	FWCNT	SWCNT	SWCNT	FWCNT
WAXD (002) FWHM (degree)	42.8	23.7	9.4	21.4	38.4	7.7
Herman orientation factor	0.86	0.89	0.94	0.82	0.78	0.94
G/D ratio	2.2 ± 0.03	1.2 ± 0.03	3.3 ± 0.3	50 ± 5	98 ± 9	31 ± 2
CNT effective length (nm)	270	130	470	2300	2300	1800
CNT length by AFM (μm)	11 ± 2 [a]	1.1 ± 0.07 [a]	3.9 ± 0.4 [b]	3.2 ± 0.2 [b]	1.7 ± 0.1 [b]	1.3 ± 0.1 [b]
Yarn Density (mg/cm^3)	450	1111	985	353	575	1670
Electrical conductivity (S/cm)	462	588	16,627	1725	3334	70,659
Tensile strength (MPa)	151	716	1193	69	111	1595

[a] Individual MWCNTs; [b] Bundles of SW or FWCNTs.

Meijo CNT yarns are wet-spun in a surfactant-assisted aqueous dispersion [32]. These yarns have a ribbon shape, and the CNT bundles are sparsely entangled, rather than aligned (Figure 2). DexMat yarn is produced by wet-spinning CNTs dispersed in a strong acid [10]. DexMat yarn has a rounded cross-section, and CNTs are well packed similar to Nanocomp Miralon (Figure 2). Despite its high density indicated by the cross-section, the side view clearly shows a bundled structure (Figure S1). DexMat has thicker bundles but its average diameter is smaller (20 μm) than those of Meijo yarns (90 μm and 76 μm for type I and type II, respectively) (Table 1).

The diameters and the wall numbers of the constituent CNTs were estimated from TEM observations (Figure 3 and Figure S2, Table 1). The average values of the CNT diameters are widely distributed from 1.4 to 45 (Figure S2). Based on the observed wall numbers, Hamamatsu Carbonics and Taiyo Nippon Sanso yarns consist of MWCNTs. The other yarns, Nanocomp Miralon, Meijo, and DexMat, contain both FW and SWCNTs.

The CNT alignment along the long axis of the yarns was estimated by WAXD. The WAXD spectra showed a broad peak between q = 11.41 and 25.27 nm^{-1}, which corresponds to the planes perpendicular to the CNT axis. The CNT alignments were estimated based on the full width at half maximum (FWHM) of the azimuthal scan of the peaks (Figure 4). The Hamamatsu and Meijo yarns have Herman orientation factors of 0.78–0.86 (Table 1). On the other hand, DexMat and Nanocomp yarns display higher alignments (=0.94). Note that the broad small peak at an azimuth angle = 150–200° is X-ray scattering from the Kapton tape, which is used as a window material for the 2D detector and x-ray guideline.

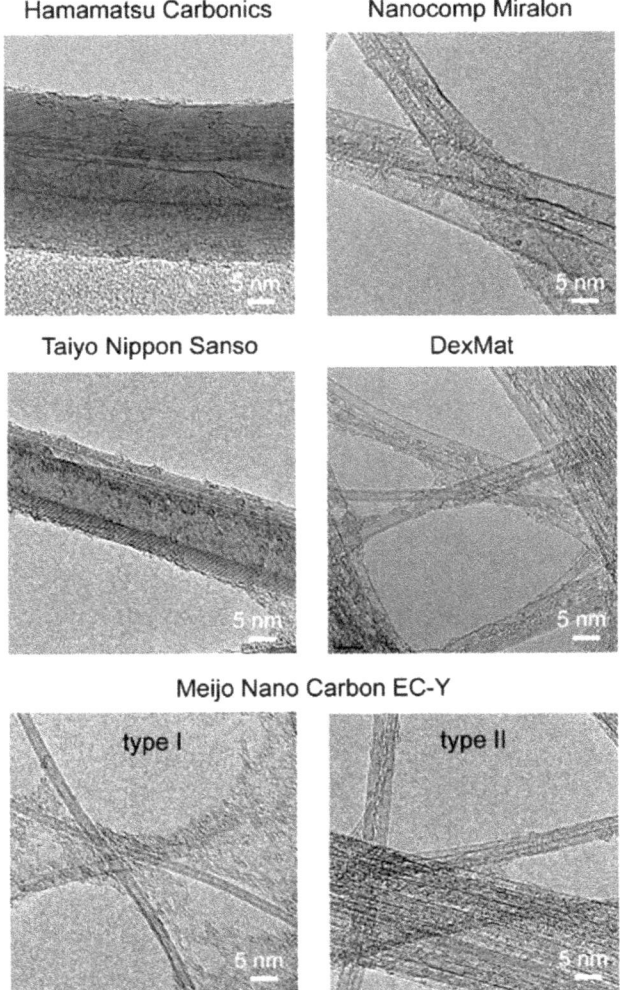

Figure 3. TEM images of constituent CNTs of each yarn.

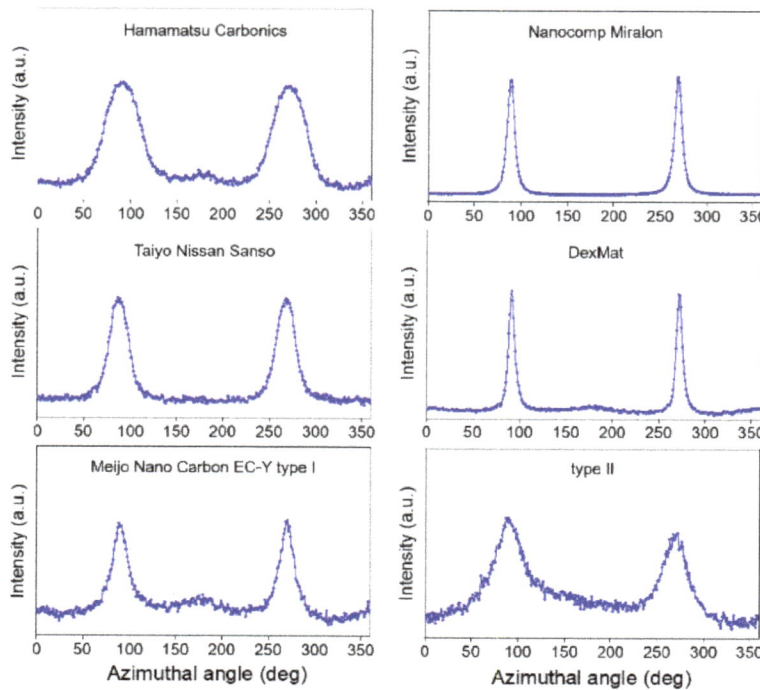

Figure 4. WAXD azimuthal scan of CNT yarns.

3.2. Spectroscopic Characterizations

We performed two spectroscopic characterizations (Figure 5). Resonance Raman spectroscopy is a powerful tool for evaluating CNT-based materials (Figure 5a). The integrated G/D intensity ratio of each yarn ranges from 1.2 to 98. Meijo and DexMat tubes show relatively larger G/D values, whereas the Taiyo Nippon Sanso, Hamamatsu Carbonics, and Nanocomp Miralon samples show smaller values around two (Table 1).

Another parameter is the effective length (L_{eff}), which is estimated from the optical absorption of CNTs in the FIR region (Figure 5b) [19–22]. Here, the length estimated based on a one-dimensional plasmon resonance model [22] corresponds to the length (or the size) in the high-crystallinity region of the CNT. Therefore, L_{eff} directly correlates with the physical properties of the CNT films and wet-spun yarns such as electrical conductivity [16,19]. SW and FWCNTs have a longer L_{eff}, whereas MWCNTs show a shorter L_{eff} (Table 1).

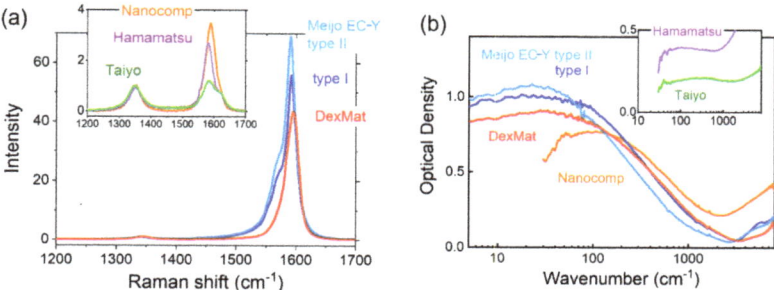

Figure 5. (**a**) Resonance Raman spectra and (**b**) far-infrared spectra of the constituent CNTs in yarns.

3.3. Electrical Conductivity

Figure 6a plots the electrical conductivities of the yarns as functions of L_{eff} of the constituent CNTs. The obtained values are typical for the given spinning method [14,16–18,26,28]. The values range over almost two orders of magnitude. There is no apparent correlation between electrical conductivity and the effective length. However, dividing the CNT yarns into two groups by their density (i.e., high and low densities) reveals a positive dependence on the effective length within the group.

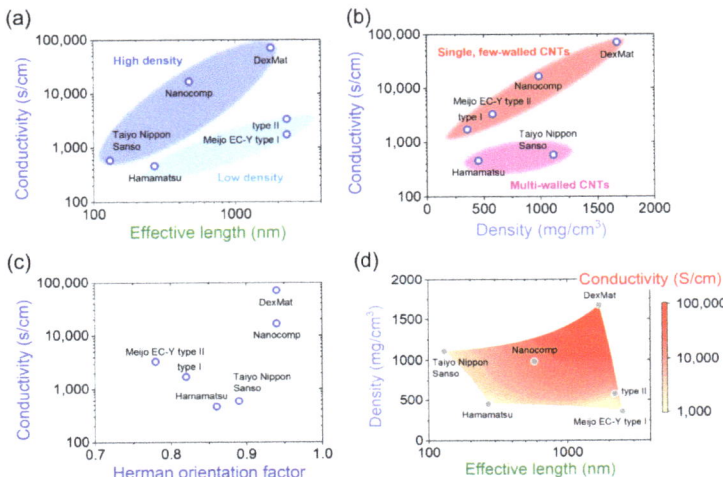

Figure 6. (**a**) Effective length, (**b**) density, and (**c**) Herman orientation factor dependences of the electrical conductivity of yarns. (**d**) 2D contour plot of the yarn conductivity.

The density dependence of the electrical conductivity does not show a clear trend (Figure 6b). Similar to above, classifying the CNT yarns into two groups depending on the number of walls, (i.e., MWCNTs and FW/SWCNTs) highlights an obvious dependence on the yarn density. Since the wall number of CNTs is usually related to the crystallinity [21], the effective lengths of FW and SWCNTs are longer than those of MWCNTs. Consequently, the electrical conductivity depends on both L_{eff} (crystallinity) and the yarn density (Figure 6d). If one of these factors is low, the electrical conductivity will be low. In other words, both the CNT effective length and density must be controlled to improve the conductivity of CNT yarn.

The Herman orientation factor dependence of the electrical conductivity (Figure 6c) is similar to the density dependence (Figure 6b). This is reasonable since highly aligned yarns have larger densities (Figures 1 and 2). Indeed, the calculated correlation coefficient between the Herman orientation factors and the yarn densities is 0.78.

Since L_{eff} represents the length of the high-crystallinity region of CNTs, CNTs with a longer L_{eff} should exhibit a higher conductivity because there are fewer CNT junctions per unit length. On the other hand, a common method to estimate the CNT length is performing counting experiments with atomic force microscopy (AFM) [19,33,34]. Since the length estimated by the AFM observation corresponds to the physically and structurally connected one, we designated it as the 'structural length' (L_{str}) in this paper. As shown in Figure S4a in the Supporting Information, L_{eff} and L_{str} do not show a clear correlation, which is consistent with the previous report [19]. Figure S4b shows the electrical conductivity of the yarns as a function of L_{str} estimated from the AFM observations. The electrical conductivity of Hamamatsu Carbonics yarn is too low for the large L_{str}. This means that although the CNTs are physically connected, defects or kinks on the tube wall cause substantial electrical resistance.

3.4. Tensile Strength

Figure 7a,b show the tensile strengths of CNT yarns as functions of L_{eff} and the yarn density. The obtained tensile strengths are typical values for CNT yarns produced by the corresponding spinning method [14,16–18,26–28,35,36]. Similar to the conductivity, we classified the results into two groups by the yarn density. Although high-density yarns seem to depend on L_{eff}, low-density yarns do not. This means that L_{eff} is a good parameter for the tensile strength only if the yarn is sufficiently dense.

Compared with the electrical conductivity, the tensile strength shows a stronger dependence on the yarn density (Figure 7b) and Herman orientation factor (Figure 7c). Intuitively, CNTs in a dense yarn should have a larger total contact area with neighboring CNTs than those in a low-density yarn. Until the individual CNTs start to slip relative to each other, a greater frictional force is applied to the CNT surface. Thus, dense yarn with a high alignment exhibits a higher tensile strength. Apparently, the increase in friction between CNTs should be higher for longer CNTs. This situation is consistent for CNT yarns with a higher density (Figure 7a).

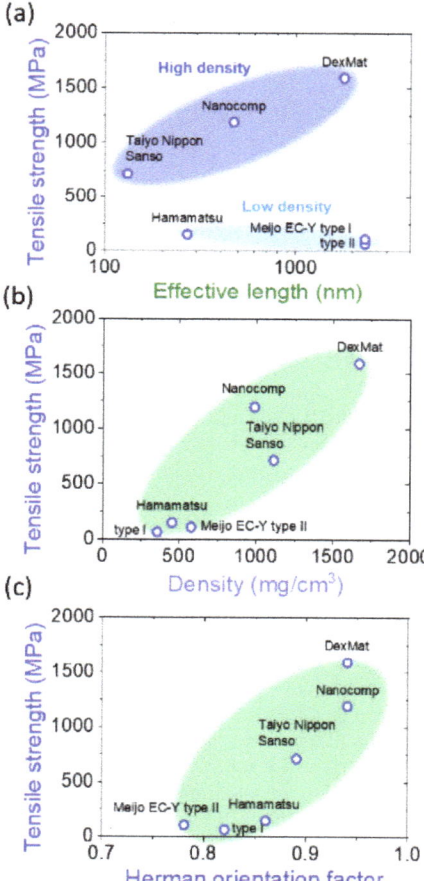

Figure 7. (a) Effective length, (b) density, and (c) Herman orientation factor dependences of the tensile strength of yarns.

3.5. PLS Regression Analysis

The contribution of each physical parameter to the electrical conductivity and the tensile strength was quantitatively analyzed by PLS regression. PLS regression is commonplace in statistics or machine learning, especially for spectroscopic data with high multi-dimensional collinearity [23–25]. PLS regression can efficiently extract information from data even with the multi-dimensional collinearity of variables. Since the absolute values of the regression coefficients of standardized variables ($|\beta|$) are related to the weights/importance, they are good indicators for interpreting or refining the variables. Based on the results (Figure 8), the contributions of the effective length, yarn density, and degree of orientation are higher than the others for the electrical conductivity. This may indicate that it is important to form a current path with the shortest distance without CNT-CNT junctions in CNT yarns.

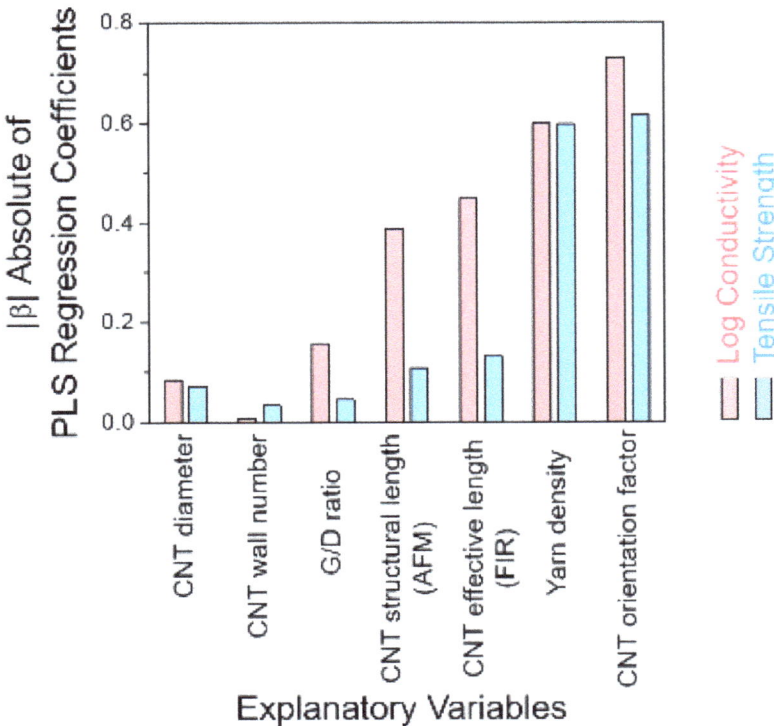

Figure 8. PLS analysis of the electrical conductivity and tensile strength of CNT yarns.

On the other hand, the tensile strength is less dependent on the effective length and the G/D ratio, which are parameters of CNT quality. The density and the Herman orientation factor contribute almost equally. The fracture mechanism of CNT yarn in tension should be dominated by the withdrawal of CNT. Thus, how close a CNT is to other CNTs is more important than the strength of an individual CNT.

4. Conclusions

We here conducted comprehensive characterizations of six CNT yarns produced by the different spinning methods with various CNTs. The structural properties of the yarns were investigated by SEM, laser microscope and WAXD. The properties of the constitute CNTs such as diameter, wall number, and effective length were characterized by TEM, resonance Raman, and FIR spectroscopy. The relationship between the physical properties of the yarns and the obtained structural parameters were then examined. The CNT effective length and the yarn density well characterize the electrical conductivity of all CNT yarns. On the other hand, the tensile strength of the yarns exhibits much stronger dependence on the yarn density. The stronger correlation of the tensile strength to the yarn density indicates the importance of the interfacial interaction with adjacent CNTs (or CNT bundles) in determining the mechanical properties. PLS analysis can explain the above observations quantitatively. The present study offers the scientific perspectives on the neat CNT yarns. One of the important findings is that DexMat yarn is composed of the longest class of CNTs that we have ever measured [16–20], and the packing density is close to the highest limit [6]. Therefore, to improve the properties of the CNT neat yarn further, post-treatment processes such as doping [10] and cross-linking [37] will become crucial. The doping effects on the electronic structures and the electrical conductivities of DexMat CNT yarn have been investigated by the present research group, which will appear in near future.

Supplementary Materials: The following supporting information can be downloaded at: https://www.mdpi.com/article/10.3390/nano12040593/s1, Figure S1: Side-view SEM images of CNT yarn at high magnification; Figure S2: (a) TEM images of CNTs used in each yarn. (b) Histograms of CNT diameter and wall number; Figure S3: (a) AFM images of each CNT. (b) Histograms of CNT length; Figure S4: (a) Relationship between structural length and effective length. (b) Structural length dependences of electrical conductivity; Figure S5: (a) Photograph of custom-made fiber diameter measurement system including high-speed optical micrometer (LS-9006MR series, Keyence). (b) Schematic illustration of how to measure yarn diameter. (c) Cross-sectional SEM image of Nanocomp Miralon yarn. (d) Voids extracted from the SEM image by using ImageJ software; Table S1: Carbonaceous content of CNT yarns estimated by TGA.

Author Contributions: T.W., S.Y. (Satoshi Yamazaki), S.Y. (Satoshi Yamashita), and T.I. performed the experiments and analyzed the data. S.M. performed PLS analysis. T.O. and T.W. co-wrote the original manuscript. T.W., S.Y. (Satoshi Yamazaki), S.Y. (Satoshi Yamashita), T.I., S.M., T.M., K.K. and T.O. discussed the results, commented on the manuscript, and approved its submission. All authors have read and agreed to the published version of the manuscript.

Funding: This work was supported by a project (JPNP16010) commissioned by the New Energy and Industrial Technology Development Organization (NEDO).

Institutional Review Board Statement: Not applicable.

Informed Consent Statement: Not applicable.

Data Availability Statement: Not applicable.

Conflicts of Interest: The authors declare no conflict of interest.

References

1. Li, H.J.; Lu, W.G.; Li, J.J.; Bai, X.D.; Gu, C.Z. Multichannel ballistic transport in multiwall carbon nanotubes. *Phys. Rev. Lett.* **2005**, *95*, 080601. [CrossRef] [PubMed]
2. Bai, Y.; Zhang, R.; Ye, X.; Zhu, Z.; Xie, H.; Shen, B.; Cai, D.; Liu, B.; Zhang, C.; Jia, Z.; et al. Carbon nanotube bundles with tensile strength over 80 GPa. *Nat. Nanotechnol.* **2018**, *13*, 589–595. [CrossRef] [PubMed]
3. Cornwell, C.F.; Welch, C.R. Very-high-strength (60-GPa) carbon nanotube fiber design based on molecular dynamics simulations. *J. Chem. Phys.* **2011**, *134*, 204708. [CrossRef] [PubMed]
4. Zhang, L.; Song, X.; Liu, X.; Yang, L.; Pan, F.; Lv, J. Studies on the removal of tetracycline by multi-walled carbon nanotubes. *Chem. Eng. J.* **2011**, *178*, 26–33. [CrossRef]
5. Tabani, H.; Khodaei, K.; Moghaddam, A.Z.; Alexovič, M.; Movahed, S.K.; Zare, F.D.; Dabiri, M. Introduction of graphene-periodic mesoporous silica as a new sorbent for removal: Experiment and simulation. *Res. Chem. Intermed.* **2019**, *45*, 1795–1813. [CrossRef]
6. Laurent, C.; Flahaut, E.; Peigney, E.A. The weight and density of carbon nanotubes versus the number of walls and diameter. *Carbon* **2010**, *48*, 2994–2996. [CrossRef]
7. Kong, J.; Zhou, C.; Morpurgo, A.; Soh, H.T.; Quate, C.F.; Marcus, C.; Dai, H. Synthesis, integration, and electrical properties of individual single-walled carbon nanotubes. *Appl. Phys. A* **1999**, *69*, 305–308. [CrossRef]
8. Koziol, K.; Vilatela, J.; Moisala, A.; Motta, M.; Cunniff, P.; Sennett, M.; Windle, A. High-performance carbon nanotube fiber. *Science* **2007**, *318*, 1892–1895. [CrossRef]

9. Choo, H.; Jung, Y.; Jeong, Y.; Kim, H.C.; Ku, B.-C. Fabrication and Applications of Carbon Nanotube Fibers. *Carbon Lett.* **2012**, *13*, 191–204. [CrossRef]
10. Behabtu, N.; Young, C.C.; Tsentalovich, D.E.; Kleinerman, O.; Wang, X.; Ma, A.W.K.; Bengio, E.A.; ter Waarbeek, R.F.; de Jong, J.J.; Hoogerwerf, R.E.; et al. Strong, light, multifunctional fibers of carbon nanotubes with ultrahigh conductivity. *Science* **2013**, *339*, 182–186. [CrossRef]
11. Janas, D.; Koziol, K.K. Carbon nanotube fibers and films: Synthesis, applications and perspectives of the direct-spinning method. *Nanoscale* **2016**, *8*, 19475–19490. [CrossRef]
12. Yadav, M.D.; Dasgupta, K.; Patwardhan, A.W.; Joshi, J.B. High Performance Fibers from Carbon Nanotubes: Synthesis, Characterization, and Applications in Composites—A Review. *Ind. Eng. Chem. Res.* **2017**, *56*, 12407–12437. [CrossRef]
13. Xu, W.; Chen, Y.; Zhan, H.; Wang, J.N. High-strength carbon nanotube film from improving alignment and densification. *Nano Lett.* **2016**, *16*, 946–952. [CrossRef] [PubMed]
14. Tsentalovich, D.E.; Headrick, R.J.; Mirri, F.; Hao, J.; Behabtu, N.; Young, C.C.; Pasquali, M. Influence of Carbon Nanotube Characteristics on Macroscopic Fiber Properties. *ACS Appl. Mater. Interfaces* **2017**, *9*, 36189–36198. [CrossRef] [PubMed]
15. Taylor, L.W.; Dewey, O.S.; Headrick, R.J.; Komatsu, N.; Peraca, N.M.; Wehmeyer, G.; Kono, J.; Pasquali, M. Improved properties, increased production, and the path to broad adoption of carbon nanotube fibers. *Carbon* **2021**, *171*, 689–694. [CrossRef]
16. Wu, X.; Morimoto, T.; Mukai, K.; Asaka, K.; Okazaki, T. Relationship between mechanical and electrical properties of continuous polymer-free carbon nanotube fibers by wet-spinning method and nanotube-length estimated by far-infrared spectroscopy. *J. Phys. Chem. C* **2016**, *120*, 20419–20427. [CrossRef]
17. Wu, X.; Mukai, K.; Asaka, K.; Morimoto, T.; Okazaki, T. Effect of surfactants and dispersion methods on properties of single-walled carbon nanotube fibers formed by wet-spinning. *Appl. Phys. Express* **2017**, *10*, 055101. [CrossRef]
18. Tajima, N.; Watanabe, T.; Morimoto, T.; Kobashi, K.; Mukai, K.; Asaka, K.; Okazaki, T. Nanotube Length and Density Dependences of Electrical and Mechanical Properties of Carbon Nanotube Fibres Made by Wet Spinning. *Carbon* **2019**, *152*, 1–6. [CrossRef]
19. Morimoto, T.; Joung, S.-K.; Saito, T.; Futaba, D.N.; Hata, K.; Okazaki, T. Length-Dependent Plasmon Resonance in Single-Walled Carbon Nanotubes. *ACS Nano* **2014**, *8*, 9897–9904. [CrossRef]
20. Morimoto, T.; Ichida, M.; Ikemoto, Y.; Okazaki, T. Temperature dependence of plasmon resonance in single-walled carbon nanotubes. *Phys. Rev. B* **2016**, *93*, 195409. [CrossRef]
21. Morimoto, T.; Okazaki, T. Optical resonance in far-infrared spectra of multiwalled carbon nanotubes. *Appl. Phys. Express* **2015**, *8*, 055101. [CrossRef]
22. Nakanishi, T.; Ando, T. Optical Response of Finite-Length Carbon Nanotubes. *J. Phys. Soc. Jpn.* **2009**, *78*, 114708. [CrossRef]
23. Wold, H. *Multivariate Analysis*; Academic Press: New York, NY, USA, 1966; p. 391.
24. Muroga, S.; Hikima, Y.; Ohsima, M. Visualization of hydrolysis in polylactide using near-infrared hyperspectral imaging and chemometrics. *J. Appl. Polym. Sci.* **2018**, *135*, 45898. [CrossRef]
25. Muroga, S.; Hikima, Y.; Ohsima, M. Near-Infrared Spectroscopic Evaluation of the Water Content of Molded Polylactide under the Effect of Crystallization. *Appl. Spectrosc.* **2017**, *71*, 1300–1309. [CrossRef] [PubMed]
26. Headrick, R.J.; Tsentalovich, D.E.; Berdegué, J.; Bengio, E.A.; Liberman, L.; Kleinerman, O.; Lucas, M.S.; Talmon, Y.; Pasquali, M. Structure–Property Relations in Carbon Nanotube Fibers by Downscaling Solution Processing. *Adv. Mater.* **2018**, *30*, 1704482. [CrossRef]
27. Ghemes, A.; Minami, Y.; Muramatsu, J.; Okada, M.; Mimura, H.; Inoue, Y. Fabrication and mechanical properties of carbon nanotube yarns spun from ultra-long multi-walled carbon nanotube arrays. *Carbon* **2012**, *50*, 4579–4587. [CrossRef]
28. Chaffee, J.; Lashmore, D.; Lewis, D.; Mann, J.; Schauer, M.; White, B. Direct Synthesis of CNT Yarns and Sheets. *Nsti Nanotech* **2008**, *3*, 118–121.
29. Paukner, C.; Koziol, K.K.K. Ultra-pure single wall carbon nanotube fibres continuously spun without promoter. *Sci. Rep.* **2014**, *4*, 3903. [CrossRef]
30. Wang, J.N.; Luo, X.G.; Wu, T.; Chen, Y. High-strength carbon nanotube fibre-like ribbon with high ductility and high electrical conductivity. *Nat. Commun.* **2014**, *5*, 3848. [CrossRef]
31. Zhong, X.-H.; Li, Y.-L.; Liu, Y.-K.; Qiao, X.-H.; Feng, Y.; Liang, J.; Jin, J.; Zhu, L.; Hou, F.; Li, J.-Y. Continuous multilayered carbon nanotube yarns. *Adv. Mater.* **2010**, *22*, 692–696. [CrossRef]
32. Mukai, K.; Asaka, K.; Wu, X.; Morimoto, T.; Okazaki, T.; Saito, T.; Yumura, M. Wet spinning of continuous polymer-free carbon-nanotube fibers with high electrical conductivity and strength. *Appl. Phys. Express* **2016**, *9*, 055101. [CrossRef]
33. Moore, K.E.; Flavel, B.S.; Ellis, A.V.; Shapter, J.G. Comparison of double-walled with single-walled carbon nanotube electrodes by electrochemistry. *Carbon* **2011**, *49*, 2639–2647. [CrossRef]
34. Chapkin, W.A.; Wenderott, J.K.; Green, P.F.; Taub, A.I. Length dependence of electrostatically induced carbon nanotube alignment. *Carbon* **2018**, *131*, 275–282. [CrossRef]
35. Miao, M. Characteristics of carbon nanotube yarn structure unveiled by acoustic wave propagation. *Carbon* **2015**, *91*, 163–170. [CrossRef]
36. Alemán, B.; Reguero, V.; Mas, B.; Vilatela, J.J. Strong Carbon Nanotube Fibers by Drawing Inspiration from Polymer Fiber Spinning. *ACS Nano* **2015**, *9*, 7392–7398. [CrossRef]
37. Park, O.K.; Choi, H.; Jeong, H.; Jung, Y.; Yu, J.; Lee, J.K.; Hwang, J.Y.; Kim, S.M.; Jeong, Y.; Park, C.R.; et al. High-modulus and strength carbon nanotube fibers using molecular cross-linking. *Carbon* **2017**, *118*, 413–421. [CrossRef]

Article

Physicochemical Characterization and Antibacterial Properties of Carbon Dots from Two Mediterranean Olive Solid Waste Cultivars

Giuseppe Nocito [1,†], Emanuele Luigi Sciuto [1,†], Domenico Franco [1], Francesco Nastasi [1], Luca Pulvirenti [2], Salvatore Petralia [3], Corrado Spinella [4], Giovanna Calabrese [1,*], Salvatore Guglielmino [1] and Sabrina Conoci [1,4,5,6,*]

1. Department of Chemical, Biological, Pharmaceutical and Environmental Sciences, University of Messina, Viale Ferdinando Stagno d'Alcontres, 31, 98168 Messina, Italy; gnocito@unime.it (G.N.); emanueleluigi.sciuto@unime.it (E.L.S.); dfranco@unime.it (D.F.); fnastasi@unime.it (F.N.); salvatore.guglielmino@unime.it (S.G.)
2. Department of Chemical Science, University of Catania, Viale A. Doria, 6, 95125 Catania, Italy; luca.pulvirenti@phd.unict.it
3. Department of Drug Science and Health, University of Catania, Viale A. Doria, 6, 95125 Catania, Italy; salvatore.petralia@unict.it
4. Istituto per la Microelettronica e Microsistemi, Consiglio Nazionale delle Ricerche (CNR-IMM) Zona Industriale, VIII Strada 5, 95121 Catania, Italy; corrado.spinella@imm.cnr.it
5. Department of Chemistry "Giacomo Ciamician", University of Bologna, Via Selmi 2, 40126 Bologna, Italy
6. LabSense beyond Nano, URT Department of Physic, CNR Viale Ferdinando Stagno d'Alcontres, 31, 98168 Messina, Italy
* Correspondence: gcalabrese@unime.it (G.C.); sabrina.conoci@unime.it (S.C.)
† These authors contributed equally to this work.

Abstract: Carbon nanomaterials have shown great potential in several fields, including biosensing, bioimaging, drug delivery, energy, catalysis, diagnostics, and nanomedicine. Recently, a new class of carbon nanomaterials, carbon dots (CDs), have attracted much attention due to their easy and inexpensive synthesis from a wide range of precursors and fascinating physical, chemical, and biological properties. In this work we have developed CDs derived from olive solid wastes of two Mediterranean regions, Puglia (CDs_P) and Calabria (CDs_C) and evaluated them in terms of their physicochemical properties and antibacterial activity against *Staphylococcus aureus* (*S. aureus*) and *Pseudomonas aeruginosa* (*P. aeruginosa*). Results show the nanosystems have a quasi-spherical shape of 12–18 nm in size for CDs_P and 15–20 nm in size for CDs_C. UV–Vis characterization indicates a broad absorption band with two main peaks at about 270 nm and 300 nm, respectively, attributed to the π-π* and n-π* transitions of the CDs, respectively. Both samples show photoluminescence (PL) spectra excitation-dependent with a maximum at λ_{em} = 420 nm (λ_{exc} = 300 nm) for CDs_P and a red-shifted at λ_{em} = 445 nm (λ_{exc} = 300 nm) for CDs_C. Band gaps values of $\approx$ 1.48 eV for CDs_P and $\approx$ 1.53 eV for CDs_C are in agreement with semiconductor behaviour. ζ potential measures show very negative values for CDs_C compared to CDs_P (three times higher, −38 mV vs. −18 mV at pH = 7). The evaluation of the antibacterial properties highlights that both CDs have higher antibacterial activity towards Gram-positive than to Gram-negative bacteria. In addition, CDs_C exhibit bactericidal behaviour at concentrations of 360, 240, and 120 µg/mL, while lesser activity was found for CDs_P (bacterial cell reduction of only 30% at the highest concentration of 360 µg/mL). This finding was correlated to the higher surface charge of CDs_C compared to CDs_P. Further investigations are in progress to confirm this hypothesis and to gain insight on the antibacterial mechanism of both cultivars.

Keywords: carbon dots; green synthesis; antibacterial properties; *S. aureus*

1. Introduction

Since their serendipitous discovery as byproducts of the arc-discharged synthesis of single-walled carbon nanotubes [1], carbon dots (CDs) represent an emerging, luminescent, carbon-based nanomaterial. A simple synthetic procedure, wide precursors, and fascinating physical and chemical properties have successfully stimulated researchers in the last years. In the spreading field of nanotechnology, CDs have been utilized as promising tools for many applications [2], such as optical sensors for ions and molecular species [3], photocatalysis [4], optoelectronics [5], biomaterials [6,7], bio-imaging [8], cancer diagnosis and therapy [9], drug delivery in tumours [10,11], and so on.

CDs are quasi-spherical, carbonaceous nanoparticles with sizes generally below 10 nm. They are mainly constituted by a crystalline sp^2 core, surrounded by sp^3 imperfections and high oxygen content on their surface [12]. They are chemically stable, low in toxicity, biocompatible, good conductors/semiconductors, and possess bright luminescence, high photostability, and broadband UV absorption [13]. Their surface is mainly rich with hydroxyls (–OH) and carboxyl/carboxylates (–COOH/–COO$^-$). The latter contribute from 5 to 50% (weight) of their oxygen content and impart excellent water colloidal dispersibility and subsequent easy functionalization or passivation with a great variety of chemical species [14]. All these properties can be modulated by synthetic conditions that produce photophysical behaviour, size, and reactivity. CDs present emission bands that are shift-dependent from excitation wavelength modulated by precursor change, surface passivation or heteroatom doping [15]. The reason for this phenomenon is attributed to both (a) the nanometric size that induces the quantum confinement effect and (b) the chemical composition referred to different surface functionalization groups and π-domain extension, inducing many possible states slightly different in energy between the frontier's orbitals [16].

The synthetic strategies for CD production could be broadly divided into top-down and bottom-up approaches. The first uses physical and chemical methods starting from a wide range of natural or chemical precursors that assemble to produce CDs, such as pyrolysis, hydrothermal treatments, microwave irradiation, and ultrasound. The second uses physical methods to nano-fragment larger, inorganic carbon precursors (graphene, graphite, carbon nanotubes) such as laser ablation, arc discharge, and electrochemistry [9,17,18]. Recently great attention has been also paid to new synthetic methods in terms of green chemistry that address the production of CDs from biomass wastes, cheap or abundant, heterogeneous and biodegradable materials obtained from the manufacturing processes of food, forestry, energy, and many other industrial processes [19]. CDs derived from biomass are greener and, in some cases, better than their chemical counterparts [20] and have been produced using top-down approaches from a lot of precursors, such as papaya [21], spent tea [22], watermelon peels [23], peanut shells [24], wool [25], strawberries [26], olive pits [27], and many others. Further, olive waste management is one of the main ecological issues in the Mediterranean basin, due to the concentration of more than 98% of global olive production and a market in huge expansion over the last two decades [28]. Consequently, their use as precursors for the synthesis of value-added nanomaterials could represent an eco-friendly, economical, and highly available strategy for several applications.

More specifically to the biomedical field, CDs gained a growing interest due to their excellent photoluminescence properties, diverse surface functions, good water solubility, low cytotoxicity, cellular uptake, biocompatibility, microbial adhesion, and theranostic properties [29,30]. Among these properties, antibacterial activity is one of the most appealing features in the design of new biomaterials in which nanotechnology is making fundamental contributions [31–34]. In this context, the specific physicochemical properties of CDs (e.g., size and surface charge) make them promising tools for addressing antibacterial processes, such as drug resistance, biofilms, and intracellular active/latent bacteria [35–39].

In this context, we have developed CDs derived from the olive solid wastes of two Mediterranean regions, Puglia (CDs_P) and Calabria (CDs_C) and evaluated them in terms of their physicochemical properties and antibacterial activity against *Staphylococcus aureus* (*S. aureus*) and *Pseudomonas aeruginosa* (*P. aeruginosa*).

2. Materials and Methods

2.1. Carbon Dots Preparation

CDs were prepared from olive solid wastes collected from two regions of Southern Italy, Puglia (CDs_P) and Calabria (CDs_C), according to the method reported in [40]. Briefly, olive solid wastes were washed several times with boiling water, dried overnight in an oven, and pyrolyzed in a muffle furnace in the absence of air at 600 °C for 1 h. The resulting carbon-based material was finely ground in a mortar and suspended in deionized water (10 mg/mL). The mixture was sonicated in an ultrasonic bath (Bandelin Sonorex RK 100 H, Bandelin electronic GmbH & Co. KG, Berlin, Deutschland) for 10 min; 1 mL of hydrogen peroxide (H_2O_2 sol. 30% (w/w)—Sigma-Aldrich, Milan, Italy) was added, and then it was refluxed for 90 min under stirring. The reaction mixture was purified by centrifugation at 8000 rpm for 20 min (Eppendorf Centrifuge 5430, Eppendorf SE, Hamburg, Deutschland), and the supernatant was syringe-filtered (Sartorius Minisart RC 0.2 µm, Sartorius AG, Göttingen, Deutschland). Lastly, the concentration of CDs' colloidal dispersion was estimated by weighing (Sartorius Quintix balance, Sartorius AG, Göttingen, Deutschland) after evaporation of the solvent under reduced pressure and subsequent drying in the oven (CDs_P $\approx$ 0.7 mg/mL; CDs_C $\approx$ 1.4 mg/mL). The mixture was simply purified by centrifugation and filtration, obtaining the final CD colloidal dispersion. A production yield of about 10% was obtained from each olive solid waste cultivar.

2.2. Chemical and Physical Characterization

UV–Vis absorption spectra were recorded with a Jasco V-560 spectrophotometer, (JASCO Corporation, Tokyo, Japan), steady-state photoluminescence spectra with a Spex Fluorolog-2 (mod. F-111) spectrofluorometer (Horiba Ltd., Kyoto, Japan) in air-equilibrated 1 cm quartz cells.

The isoelectric points of the CDs' colloidal dispersions were estimated by ζ (Zeta) potential pH titration using the dynamic light scattering (DLS) technique with a Malvern Zetasizer Nano ZS90 instrument (Malvern Panalytical Ltd., Malvern, United Kingdom). pH was moved to 8, 10, and 2 by adding, respectively, NaOH 0.1 M (sodium hydroxide—Sigma-Aldrich, Merck KGaA, Darmstadt, Deutschland) and HCl 0.1 M solutions (hydrochloric acid—Sigma-Aldrich, Milan, Italy). The isoelectric point was found by plotting the pH vs. Zeta potential and intercepting the pH value when the ζ potential was zero.

X-ray photoelectron spectroscopy (XPS) was performed on carbon dots deposited on silicon slides using a PHI 5600 multi-technique ESCA-Auger spectrometer (Physical electronics Inc., Chanhassen, MN, USA) equipped with a monochromatic Al-Kα X-ray source. The XPS binding energy (BE) scale was calibrated on the C 1s peak of adventitious carbon at 285.0 eV. Transmission FT–IR measurements on the silicon-deposited carbon dots were obtained using a JASCO FTIR 4600LE spectrometer (JASCO Corporation, Tokyo, Japan) in the spectral range of 560–4000 cm^{-1} (resolution 4 cm^{-1}).

Transmission electron microscopy (TEM) analysis was performed using the bright field in conventional TEM parallel beam mode. An ATEM JEOL JEM 2010 equipped with a 30 mm^2 window energy dispersive X-ray (EDX) spectrometer (JEOL Ltd., Musashino, Akishima, Tokyo, Japan), was used.

2.3. Bacterial Assays

S. aureus (ATCC 29213) was purchased from American Type Culture Collection (LGC Promochem, Milan, Italy) and cultured in tryptone soya broth (TSB, Sigma-Aldrich, Milan, Italy). *P. aeruginosa* (ATCC 27853) was purchased from American Type Culture Collection (LGC Promochem, Milan, Italy) and cultured in Luria–Bertani broth (LB, Sigma-Aldrich, Milan, Italy).

Antibacterial tests were performed in Mueller–Hinton broth (MHB, Sigma-Aldrich, Milan, Italy), a culture medium susceptible to antibiotics.

To evaluate the minimal inhibitory concentration (MIC) of both CDs' colloidal dispersions, the microplate inhibition assay was used. Specifically, semi-exponential cultures of bacterial strains at the final concentration of about 10^5 bacteria per mL were inoculated in MHB in the presence of increasing concentrations of CDs (60–360 μg/mL) in 96-well plates and incubated at 37 °C under shaking overnight. After incubation, the concentrations inhibiting at least 90% and 99.9% of bacteria, MIC_{90} and MIC_{99}, were determined compared to the untreated control.

To evaluate the bacterial cell viability, an MTS assay (CellTiter 96® AQueous One Solution Cell Proliferation Assay, Promega, Milan, Italy) was performed. In more detail, bacterial cultures in the presence of different concentrations of CDs were grown overnight at 37 °C in 96-well plates. Then, MTS reagent was added to the bacteria culture media, incubated for 2 h at 37 °C in static condition, the plate was shaken briefly, and absorbance was measured at 490 nm by using a microtiter plate reader (Multiskan GO, Thermo Scientific, Waltham, MA, USA). The reduction of bacterial viability was evaluated in terms of the percentage of MTS reduction (% MTS_{red}), compared to the untreated control (CTR) using the following equation:

$$MTS_{red}(\%) = \left(\frac{A}{B}\right) \times 100 \quad (1)$$

where A e B are the OD_{490} from the MTS-reduced formazan of condition with CDs and CTR. The samples were analysed in triplicate for each experimental condition.

Figure 1 reports the schematic representation of CD preparation, physico-chemical characterization, and bacterial testing.

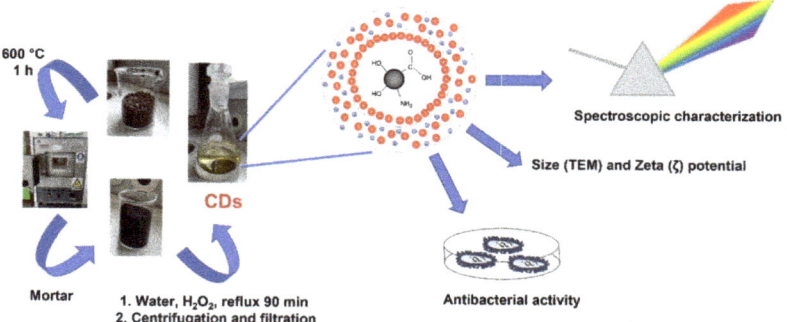

Figure 1. Schematic representation of CD preparation, physico-chemical characterization, and bacterial testing.

3. Results and Discussion

3.1. Physicochemical Characterization of CDs

The optical properties of both synthesized CDs were characterized by the UV–Vis absorbance and photoluminescence emission spectra displayed in Figure 2. The absorption spectrum of CDs_P (Figure 2a) exhibits broadband UV absorption, with a trend compatible with the light scattering operated by a small nanoparticle's colloidal dispersion. The UV–Vis absorbance spectrum of CDs_P shows two detectable peaks as shoulders located around 270 nm and 300 nm, attributed respectively to the π–π* transition of CDs and n–π* transitions of the functional groups present on CDs [41,42]. These two absorption shoulders probably suggest the existence of conjugated structures as well as the presence of functional groups containing oxygen in the CDs [12–14]. Figure 2b shows the PL spectra of CDs_P with the excitation wavelength in the interval between 300 nm and 450 nm. The emission maximum is around λ_{em} = 420 nm for λ_{exc} = 300 nm.

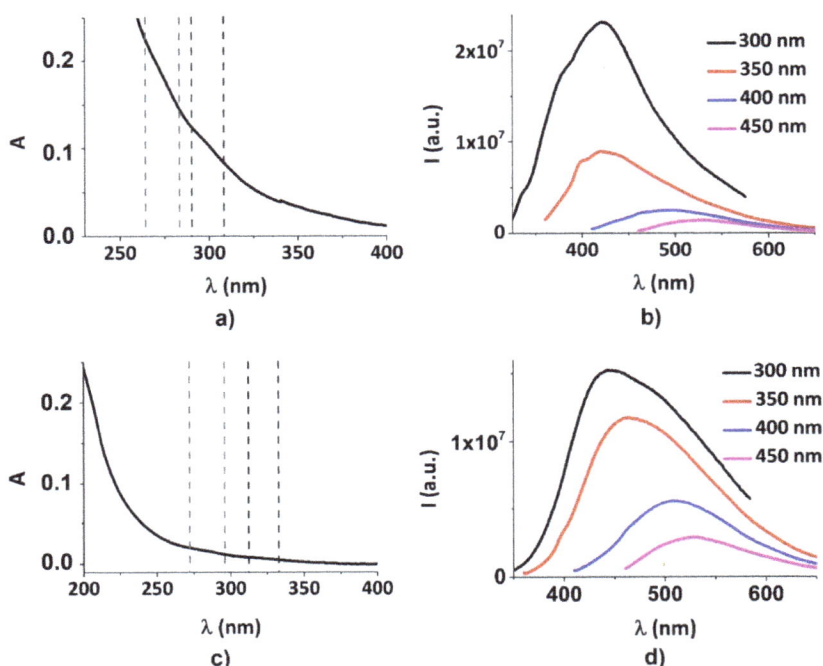

Figure 2. Optical characterization of CDs: (**a**) UV–Vis absorption spectrum of CDs_P; (**b**) Photoluminescence spectra of CDs_P; (**c**) UV–Vis absorption spectrum of CDs_C; (**d**) Photoluminescence spectra of CDs_C.

Although the absorption spectrum of CDs_C showed a trend similar to that exhibited by CDs_P, some differences are present (Figure 2a–c). First, the peaks detected as shoulders are shifted to lower energies, founding them around 280 nm and 320 nm (Figure 2c). Secondly, although CDs_C show PL, the emission spectra registered in the same excitation wavelength interval (300–450 nm) are red-shifted with respect to CDs_P, as for the absorption bands [43]. The emission maximum is around λ_{em} = 445 nm for λ_{exc} = 300 nm (Figure 2d).

To further evaluate the difference between CDs derived from two different Mediterranean olive solid waste cultivars, we calculated the respective band gaps from UV–Vis absorption spectra reported in Figure 2a,c, using a Tauc plot [44] with the formula [45]:

$$(\alpha h\nu)^{\frac{1}{\gamma}} = B(h\nu - E_g) \qquad (2)$$

where α is the absorption coefficient ($\alpha = 2.303$ Acm^{-1}), h is the Plank constant, ν is the frequency of the incident photon. The γ factor depends on the nature of the electronic transition, and in our case for permitted transition, it could be 0.5 for the indirect one and 2 for the direct one. B is a constant assumed to be 1, and E_g is the energy band gap. E_g was calculated using $\gamma = \frac{1}{2}$ for direct electronic transition.

Plots are reported in Figure 3. Band gap values of $\approx$ 1.48 eV for CDs_P and $\approx$ 1.53 eV for CDs_C, respectively, were found. These values are in agreement with semiconductor behaviour, according to similar values found in [46,47].

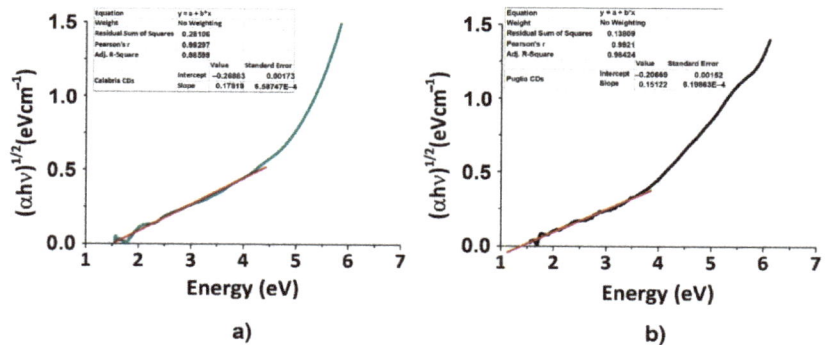

Figure 3. Tauc plots for the optical energy band gap calculation: (**a**) CDs_P; (**b**) CDs_C.

Concerning the chemical characterization of CDs_P, both FTIR and XPS analysis were reported in our previous works [40,48] and showed for FTIR the following peaks: 3424, 3236, 2923/2850, 1656, 1412, 1320, 1116, and 1096 cm^{-1}, attributed to –OH, N–H, C–H, C=O (carbonyl), COO$^-$ (carboxylate), C–OH (hydroxyl), and C–O–C (epoxide) groups, respectively. XPS analysis showed peaks of C1s at 285 eV for C–C and 289 eV for O=C–O respectively. Regarding CD_C, FTIR analysis (see Figure S1) reveals the presence of –OH, C=O (carbonyl), COO$^-$ (carboxylate), C–OH (hydroxyl), and C–O–C (epoxide) groups, while the XPS spectrum (see Figure S2) exhibits the same C1s peaks of CDs_P at a 285 eV for C–C and 289 eV for O=C–O, respectively. The similarity of surface groups for both cultivars also account for similar band gap values (see above).

Figure 4 reports the plot of ζ potential values as a function of pH for both CDs_P and CDs_C. It can be noticed that both cultivars show negative surface charges, but CDs_P exhibits higher values of ζ potential with respect to CDs_P. According to that, the values of the isoelectric point (pH value at 0 charge) correspond to pH $\approx$ 3 for CDs_P and pH $\approx$ 2.4 for CDs_C. More interesting, in the physiological conditions (pH $\approx$ 7), CDs_C feature a charge (ζ potential) about three times more negative than CDs_P (-32 mV vs. -11 mV). This certainly can be a notable point for the antibacterial activity of the nanomaterials (vide infra).

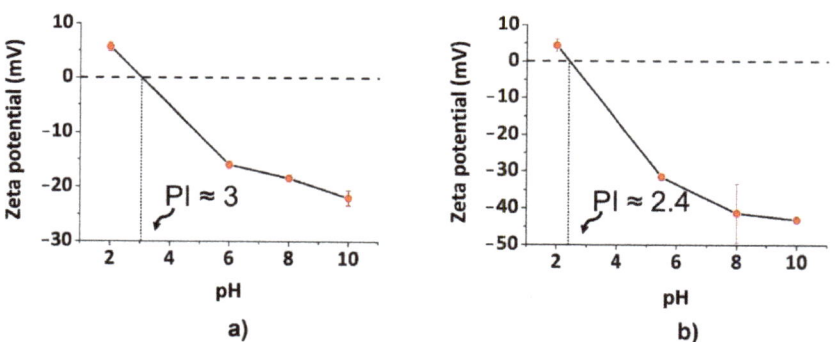

Figure 4. Graphical representation of ζ potential as a function of pH titration: (**a**) CDs_P; (**b**) CDs_C.

Morphological characteristics were also investigated by transmission electron microscopy (TEM). Figure 5 reports TEM images of the two different cultivars. CDs_P displays dispersed quasi-spherical nanoparticles with particle size ranging between 12–18 nm, while CDs_C exhibits similar characteristics, with a mean size of 15–20 nm.

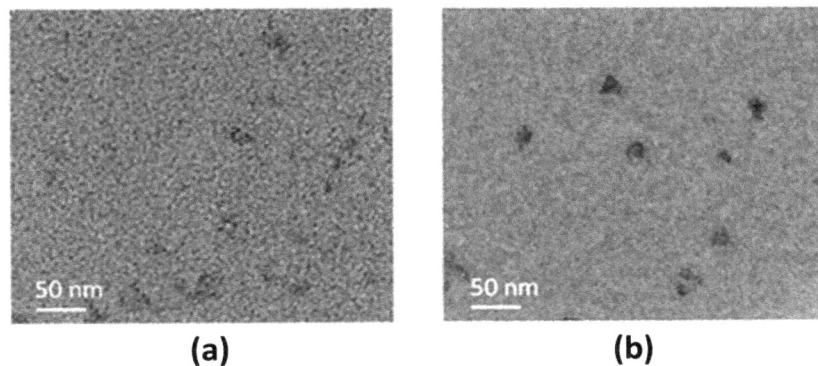

Figure 5. Representative transmission electron microscopy (TEM) images of: (**a**) CDs_P; (**b**) CDs_C.

3.2. Antibacterial properties of Carbon Dots

The bactericidal activity of both CD dispersions was evaluated against both *S. aureus* (Gram-positive) and *P. aeruginosa* (Gram-negative), two of the most common pathogens involved in a wide range of infections. First, values of MIC for both CD types were determined by microplate inhibition assay up to the maximum concentration of 360 μg/mL. In Table 1 are reported the MIC_{90} and MIC_{99} values [μg/mL] for both CDs.

Table 1. MIC values [μg/mL] of *S. aureus* strains in Mueller–Hinton broth (MH).

Bacterial strain	CDs	MIC_{90}	MIC_{99}
S. aureus	CDs_P	-	-
	CDs_C	120 μg/mL	360 μg/mL
P. aeruginosa	CDs_P	-	-
	CDs_C	-	-
Dash (-) = no antibacterial activity up to the concentration of 360 μg/mL.			

Our results showed that CDs_C exhibit an *S. aureus* viability reduction of about 99.9% (MIC_{99}) at 360 μg/mL and 90% bacterial inhibition (MIC_{90}) at about 120 μg/mL, while no bacterial inhibition was observed against *P. aeruginosa*. On the contrary, no bacterial inhibition was observed for CDs_P at all using concentrations (ranging from 60 to 360 μg/mL) against either bacterial strain.

To further evaluate the antibacterial properties of both CD types, we also performed an MTS cell viability assay (Figures 6 and 7).

MTS data showed that CDs_C dispersion reduces almost completely *S. aureus* viability at the higher concentrations (360, 240, and 120 μg/mL), while at the lowest concentration (60 μg/mL), it is reduced by about 50% compared to the control (untreated). On the contrary, MTS data of CDs_P indicated a bacterial cell reduction of only 30% at the highest concentration (360 μg/mL), showing very poor antibacterial activity. In more details, the bacterial viability of CDs_C was found of 47.5 ± 2.37% at 60 μg/mL, 4.75 ± 0.24% at 120 μg/mL, 0.47 ± 0.005% at 240 μg/mL, and 0.05 ± 0.0005% at 360 μg/mL; for CDs_P, it was 94.9 ± 4.74% at 60 μg/mL, 92.6 ± 4.63% at 120 μg/mL, 85.7 ± 4.28% at 240 μg/mL, and 68.4 ± 3.42% at 360 μg/mL. These data were in agreement with MIC results.

On the other hand, MTS data obtained against *P. aeruginosa* showed that only CDs_C at the higher concentration (360 μg/mL) exhibited a bacterial reduction of about 20%. In more detail, the bacterial viability of CDs_C was found to be 98.4 ± 1.17% at 60 μg/mL, 95.75 ± 1.49% at 120 μg/mL, 93.43 ± 1.12% at 240 μg/mL, and 83.81 ± 3.5% at 360 μg/mL.

In addition to the above considerations, we also observed a different level of antibacterial activity between CDs_P and CDs_C, more evident in *S. aureus* than in *P. aeruginosa* probably due to the different cell wall compositions. Several action mechanisms have been suggested to explain the antibacterial activity of CDs closely related to their physicochemical properties, including their dimensionalities, lateral size, shape, number of layers, surface charges, the presence and nature of surface functional groups, and doping [35–38].

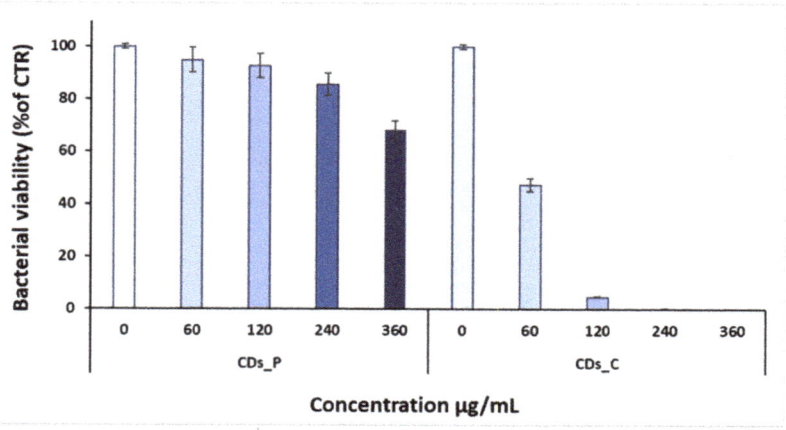

Figure 6. MTS assay of CDs_C and CDs_P against *S. aureus* strain. Data are presented as the mean ± SD from three independent experiments.

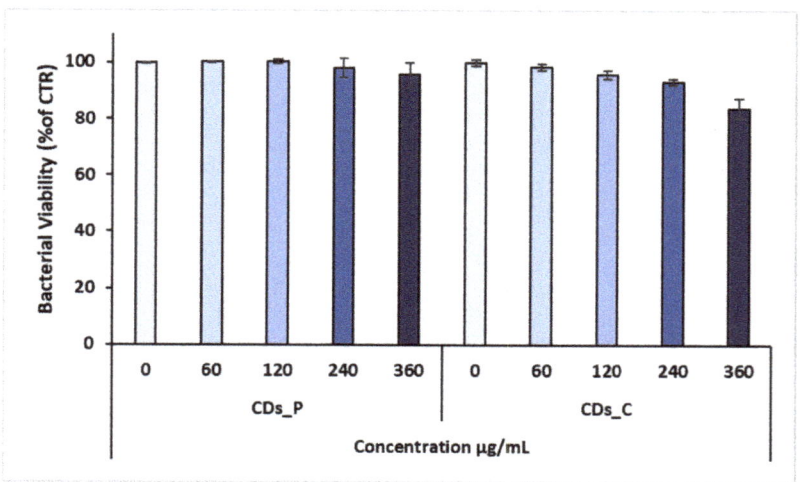

Figure 7. MTS assay of CDs_C and CDs_P against *Pseudomonas aeruginosa* strain. Data are presented as the mean ± SD from three independent experiments.

A hypothesis to explain the dissimilar antibacterial behaviour of the two CD types could be the charge distribution and functional groups of the CDs. Recently, some studies suggested that factors due to particle size and surface functionalization and charges can affect antibacterial effects [35,49]. With specific focus on surface charge, Bing et al. studied the antibacterial capability of CDs with three different surface charges (uncharged, positive, and negative) finding only positive- (spermine derived positive CDs, SC-dots) and negative-charged CDs (candle-soot derived negative CDs, CC-dots) exhibited signs of cell death on *E. coli* (Gram negative), such as DNA fragmentation, extracellular exposure of phosphatidylserine, condensation of the chromosome, and loss of structural integrity, but the same effects did not occur with uncharged glucose carbon dots (GC-dots). Based on these considerations, our results could be compared with negatively charged candle-soot C-dots (CC-dots). Although similar results were obtained against *P. aeruginosa*, we found that *S. aureus* viability was almost totally and slightly reduced when treated with CDs_C and CDs_P, respectively. Since our systems, CDs_C and CDs_P, exhibit similar sizes of nanoparticles, and no surface functionalization was carried out, neither factor is relevant to the findings, indicating that CDs_C show higher antibacterial activity compared with to CDs_P. These results suggest, rather, different antibacterial mechanisms for the two cultivars, probably attributable to the different surface charges and ζ potential. Although both CDs from the two cultivars exhibited negative surface charges, CDs_C have much more of a negative charge than do CDs_P (about -38 mV for CDs_C and -18 mV for CDs_P at pH = 7). Actually, during the bacterial growth, the medium pH decreased over time (0–8 h), reaching more acidic values (from 7.2 to about 5.5) (See Figure S3). By considering this aspect, we can observe that the surface charge of both cultivars reaches about -32 mV (CDs_C) and -11 mV (CDs_P), respectively, at pH = 5.5, highlighting that CDs_C feature a negative surface charge 3 times higher than that of CDs_P. This is probably the reason for the increased antibacterial activity of CDs_C in contrast with CDs_P. Further studies are in progress to validate our hypothesis and gain insights on the active antibacterial mechanism.

4. Conclusions

In this paper, we report on the development of CDs derived from olive solid wastes of two Mediterranean regions, Puglia (CDs_P) and Calabria (CDs_C) and their evaluation in terms of their physicochemical properties and antibacterial activity against *S. aureus*. The UV–Vis characterization shows a typical broad absorption band with two main peaks at about 270 nm and 300 nm, attributed to the π–π^* and n—π^* transitions of CDs, respectively. The PL spectra are excitation-dependent, and CDs_P shows emission maximum at λ_{em} = 420 nm (λ_{exc} = 300 nm), while CDs_C has an emission maximum red-shifted at λ_{em} = 445 nm (λ_{exc} = 300 nm). Band gaps values of ≈ 1.48 eV for CDs_P and ≈ 1.53 eV for CDs_C, respectively, were found, in agreement with semi-conductor behaviour. ζ potential values are negative for both cultivars. However, CDs_C feature a more negative charge, in fact three times more negative than that of CDs_P (-38 mV vs. -18 mV at pH = 7). The TEM morphological inspection shows quasi-spherical nanoparticles 12–18 nm in size for CDs_P and 15–20 nm in size for CDs_C. The evaluation of antibacterial properties highlights that both CDs have higher antibacterial activity towards Gram-positive bacteria than to Gram-negative bacteria. In addition, the evaluation of the antibacterial properties towards *S. aureus* of both CD types highlights that CDs_C exhibit antibacterial properties at concentrations of 360, 240, and 120 µg/mL, while CDs_P shows a bacterial cell reduction of 30% at the highest concentration of 360 µg/mL. This finding was correlated to the highest surface charge of CDs_C compared to CDs_P, which is still very negative during bacterial growth, reaching an acidic pH of 5.5. Further investigations are in progress to confirm this hypothesis and gain insights on the antibacterial mechanism of both cultivars.

Supplementary Materials: The following supporting information can be downloaded at: https://www.mdpi.com/article/10.3390/nano12050885/s1, Figure S1. FTIR Spectra CDs_C; Figure S2. XPS Spectra CDs_C; Figure S3. pH changes during *S. aureus* bacterial growth (OD_{540}) (0–8hrs).

Author Contributions: Conceptualization, S.C., G.C. and S.G.; methodology, G.N., E.L.S., D.F., L.P. and S.P.; investigation, F.N.; resources, S.C. and S.G.; data curation, S.C. and G.C.; writing—original draft preparation, G.N., E.L.S. and D.F.; writing—review and editing, S.G., G.C., C.S. and S.C.; supervision, G.C. and S.C. All authors have read and agreed to the published version of the manuscript.

Funding: This research received no external funding.

Conflicts of Interest: The authors declare no conflict of interest.

References

1. Xu, X.; Ray, R.; Gu, Y.; Ploehn, H.J.; Gearheart, L.; Raker, K.; Scrivens, W.A. Electrophoretic Analysis and Purification of Fluorescent Single-Walled Carbon Nanotube Fragments. *J. Am. Chem. Soc.* **2004**, *126*, 12736–12737. [CrossRef] [PubMed]
2. Liu, J.; Li, R.; Yang, B. Carbon Dots: A New Type of Carbon-Based Nanomaterial with Wide Applications. *ACS Cent. Sci.* **2020**, *6*, 2179–2195. [CrossRef] [PubMed]
3. Xu, D.; Lin, Q.; Chang, H.T. Recent Advances and Sensing Applications of Carbon Dots. *Small Methods* **2020**, *4*, 1900387. [CrossRef]
4. Han, M.; Zhu, S.; Lu, S.; Song, Y.; Feng, T.; Tao, S.; Liu, J.; Yang, B. Recent progress on the photocatalysis of carbon dots: Classification, mechanism and applications. *Nano Today* **2018**, *19*, 201–218. [CrossRef]
5. Stepanidenko, E.A.; Ushakova, E.V.; Fedorov, A.V.; Rogach, A.L. Applications of Carbon Dots in Optoelectronics. *Nanomaterials* **2021**, *11*, 364. [CrossRef]
6. Peng, Z.; Miyanji, E.H.; Zhou, Y.; Pardo, J.; Hettiarachchi, S.D.; Li, S.; Blackwelder, P.L.; Skromne, I.; Leblanc, R.M. Carbon dots: Promising biomaterials for bone-specific imaging and drug delivery. *Nanoscale* **2017**, *9*, 17533–17543. [CrossRef]
7. Su, W.; Wu, H.; Xu, H.; Zhang, Y.; Li, Y.; Li, X.; Fan, L. Carbon dots: A booming material for biomedical applications. *Mater. Chem. Front.* **2020**, *4*, 821–836. [CrossRef]
8. Kasouni, A.; Chatzimitakos, T.; Stalikas, C. Bioimaging applications of carbon nanodots: A review. *J. Carbon Res.* **2019**, *5*, 19. [CrossRef]
9. Nocito, G.; Calabrese, G.; Forte, S.; Petralia, S.; Puglisi, C.; Campolo, M.; Esposito, E.; Conoci, S. Carbon dots as promising tools for cancer diagnosis and therapy. *Cancers* **2021**, *13*, 1991. [CrossRef]
10. Calabrese, G.; De Luca, G.; Nocito, G.; Rizzo, M.G.; Lombardo, S.P.; Chisari, G.; Forte, S.; Sciuto, E.L.; Conoci, S. Carbon Dots: An Innovative Tool for Drug Delivery in Brain Tumors. *Int. J. Mol. Sci.* **2021**, *22*, 11783. [CrossRef]
11. Zeng, Q.; Shao, D.; He, X.; Ren, Z.; Ji, W.; Shan, C.; Qu, S.; Li, J.; Chen, L.; Li, Q. Carbon dots as a trackable drug delivery carrier for localized cancer therapy: In vivo. *J. Mater. Chem. B* **2016**, *4*, 5119–5126. [CrossRef] [PubMed]
12. Baker, S.N.; Baker, G.A. Luminescent carbon nanodots: Emergent nanolights. *Angew. Chem. Int. Ed.* **2010**, *49*, 6726–6744. [CrossRef] [PubMed]
13. Namdari, P.; Negahdari, B.; Eatemadi, A. Synthesis, properties and biomedical applications of carbon-based quantum dots: An updated review. *Biomed. Pharmacother.* **2017**, *87*, 209–222. [CrossRef] [PubMed]
14. Lim, S.Y.; Shen, W.; Gao, Z. Carbon quantum dots and their applications. *Chem. Soc. Rev.* **2015**, *44*, 362–381. [CrossRef] [PubMed]
15. Macarain, J.R.; de Medeiros, T.V.; Gazzetto, M.; Yarur Villanueva, F.; Cannizzo, A.; Naccache, R. Elucidating the mechanism of dual-fluorescence in carbon dots. *J. Colloids Interfaces Sci.* **2022**, *601*, 67–76. [CrossRef]
16. Ding, H.; Li, X.-H.; Chen, X.-B.; Wei, J.-S.; Li, X.-B.; Xiong, H.-M. Surface states of carbon dots and their influences on luminescence. *J. Appl. Phys.* **2020**, *127*, 231101. [CrossRef]
17. Meng, W.; Bai, X.; Wang, B.; Liu, Z.; Lu, S.; Yang, B. Biomass-derived carbon dots and their applications. *Energy Environ. Mater.* **2019**, *2*, 172–192. [CrossRef]
18. Mintz, K.J.; Zhou, Y.; Leblanc, R.M. Recent development of carbon quantum dots regarding their optical properties, photoluminescence mechanism, and core structure. *Nanoscale* **2019**, *11*, 4634–4652. [CrossRef]
19. Kang, C.; Huang, Y.; Yang, H.; Yan, X.F.; Chen, Z.P. A Review of Carbon Dots Produced from Biomass Wastes. *Nanomaterials* **2020**, *10*, 2316. [CrossRef]
20. Wareing, T.C.; Gentile, P.; Phan, A.N. Biomass-based carbon dots: Current development and future perspectives. *ACS Nano* **2021**, *15*, 15471–15501. [CrossRef]
21. Doula, M.K.; Moreno-Ortego, J.L.; Tinivella, F.; Inglezakis, V.J.; Sarris, A.; Komnitsas, K. Olive mill waste: Recent advances for the sustainable development of olive oil industry. In *Olive Mill Waste: Recent Advances for Sustainable Management*; Galanakis, C.M., Ed.; Academic Press: London, UK; Elsevier: London, UK, 2017; Chapter 2; pp. 29–56.
22. Wang, N.; Wang, Y.; Guo, T.; Yang, T.; Chen, M.-L.; Wang, J.-H. Green preparation of carbon dots with papaya as carbon source for effective fluorescent sensing of Iron (III) and Escherichia coli. *Biosens. Bioelectron.* **2016**, *85*, 68–75. [CrossRef] [PubMed]
23. Abbas, A.; Tabish, T.A.; Bull, S.J.; Lim, T.M.; Phan, A.N. High yield synthesis of graphene quantum dots from biomass waste as a highly selective probe for Fe^{3+} sensing. *Sci. Rep.* **2020**, *10*, 21262. [CrossRef] [PubMed]
24. Xue, M.; Zou, M.; Zhao, J.; Zhan, Z.; Zhao, S. Green preparation of fluorescent carbon dots from lychee seeds and their application for the selective detection of methylene blue and imaging in living cells. *J. Mater. Chem. B* **2015**, *3*, 6783–6789. [CrossRef] [PubMed]
25. Xue, M.Y.; Zhan, Z.H.; Zou, M.B.; Zhang, L.L.; Zhao, S.L. Green synthesis of stable and biocompatible fluorescent carbon dots from peanut shells for multicolor living cell imaging. *New J. Chem.* **2016**, *40*, 1698–1703. [CrossRef]
26. Shi, L.; Zhao, B.; Li, X.; Zhang, G.; Zhang, Y.; Dong, C.; Shuang, S. Eco-friendly synthesis of nitrogen-doped carbon nanodots from wool for multicolor cell imaging, patterning, and biosensing. *Sens. Actuators B Chem.* **2016**, *235*, 316–324. [CrossRef]
27. Zhao, Y.; Duan, J.; He, B.; Jiao, Z.; Tang, Q. Improved charge extraction with N-doped carbon quantum dots in dye-sensitized solar cells. *Electrochim. Acta* **2018**, *282*, 255–262. [CrossRef]
28. Algarra, M.; Orfãos, L.D.; Alves, C.S.; Moreno-Tost, R.; Pino-González, M.S.; Jiménez-Jiménez, J.; Rodríguez-Castellón, E.; Eliche-Quesada, D.; Castro, E.; Luque, R. Sustainable Production of Carbon Nanoparticles from Olive Pit Biomass: Understanding Proton Transfer in the Excited State on Carbon Dots. *ACS Sustain. Chem. Eng.* **2019**, *7*, 10493–10500. [CrossRef]
29. Nekoueian, K.; Amiri, M.; Sillanpää, M.; Marken, F.; Boukherroub, R.; Szunerits, S. Carbon-based quantum particles: An electroanalytical and biomedical perspective. *Chem. Soc. Rev.* **2019**, *48*, 4281–4316. [CrossRef]

30. Dong, X.; Liang, W.; Meziani, M.J.; Sun, Y.P.; Yang, L. Carbon dots as potent antimicrobial agents. *Theranostics* **2020**, *10*, 671–686. [CrossRef]
31. Calabrese, G.; Petralia, S.; Franco, D.; Nocito, G.; Fabbi, C.; Forte, L.; Guglielmino, S.; Squarzoni, S.; Traina, F.; Conoci, S. A new Ag-nanostructured hydroxyapatite porous scaffold: Antibacterial effect and cytotoxicity study. *Mater. Sci. Eng. C* **2021**, *118*, 111394. [CrossRef]
32. Franco, D.; Calabrese, G.; Petralia, S.; Neri, G.; Corsaro, C.; Forte, L.; Squarzoni, S.; Guglielmino, S.; Traina, F.; Fazio, E.; et al. Antimicrobial effect and cytotoxic evaluation of Mg-doped hydroxyapatite functionalized with Au-nano rods. *Molecules* **2021**, *26*, 1099. [CrossRef] [PubMed]
33. Hajipour, M.J.; Fromm, K.M.; Ashkarran, A.A.; de Aberasturi, D.J.; de Larramendi, I.R.; Rojo, T.; Serpooshan, V.; Parak, W.J.; Mahmoudi, M. Antibacterial properties of nanoparticles. *Trends Biotechnol.* **2012**, *30*, 499–511. [CrossRef] [PubMed]
34. Leonardi, A.A.; Lo Faro, M.J.; Irrera, A. Biosensing platforms based on silicon nanostructures: A critical review, Analytica. *Chim. Acta* **2021**, *1160*, 338393.
35. Sun, B.; Wu, F.; Zhang, Q.; Chu, X.; Wang, Z.; Huang, X.; Li, J.; Yao, C.; Zhou, N.; Shen, J. Insight into the effect of particle size distribution differences on the antibacterial activity of carbon dots. *J. Colloid Interface Sci.* **2021**, *584*, 505–519. [CrossRef]
36. Xin, Q.; Shah, H.; Nawaz, A.; Xie, W.; Akram, M.Z.; Batool, A.; Tian, L.; Jan, S.U.; Boddula, R.; Guo, B.; et al. Antibacterial carbon-based nanomaterials. *Adv. Mater.* **2019**, *31*, 1804838. [CrossRef]
37. Al-Jumaili, A.; Alancherry, S.; Bazaka, K.; Jacob, M.V. Review on the Antimicrobial Properties of Carbon Nanostructures. *Materials* **2017**, *10*, 1066. [CrossRef]
38. Karahan, H.E.; Wiraja, C.; Xu, C.; Wei, J.; Wang, Y.; Wang, L.; Liu, F.; Chen, Y. Graphene Materials in Antimicrobial Nanomedicine: Current Status and Future Perspectives. *Adv. Healthc. Mater.* **2018**, *7*, 1701406. [CrossRef]
39. Zou, X.; Zhang, L.; Wang, Z.; Luo, Y. Mechanisms of the Antimicrobial Activities of Graphene Materials. *J. Am. Chem. Soc.* **2016**, *138*, 2064–2077. [CrossRef]
40. Sawalha, S.; Moulaee, K.; Nocito, G.; Silvestri, A.; Petralia, S.; Prato, M.; Bettini, S.; Valli, L.; Conoci, S.; Neri, G. Carbon-dots conductometric sensor for high performance gas sensing. *Carbon Trends* **2021**, *5*, 100105. [CrossRef]
41. Zhang, X.; Wang, J.; Liu, J.; Ji, W.; Chen, H.; Bi, H. Design and preparation of a ternary composite of graphene oxide/carbon dots/polypyrrole for supercapacitor application: Importance and unique role of carbon dots. *Carbon* **2017**, *115*, 134–146. [CrossRef]
42. Stachowska, J.D.; Murphy, A.; Mellor, C.; Fernandes, D.; Gibbons, E.N.; Krysmann, M.J.; Kelarakis, A.; Burgaz, E.; Moore, J.; Yeates, S.G. A rich gallery of carbon dots based photoluminescent suspensions and powders derived by citric acid/urea. *Sci. Rep.* **2021**, *11*, 10554. [CrossRef] [PubMed]
43. Holub, J.; Santoro, A.; Lehn, J.-M. Electronic absorption and emission properties of bishydrazone [2 × 2] metallosupramolecular grid-type architectures. *Inorg. Chim. Acta* **2019**, *494*, 223–231. [CrossRef]
44. Tauc, J.; Grigorovici, R.; Vancu, A. Optical properties and electronic structure of amorphous germanium. *Phys. Status Solidi* **1966**, *15*, 627–637. [CrossRef]
45. Makula, P.; Pacia, M.; Macyk, W. How to correctly determine the band gap Energy of modified semiconductor photocatalysts based on UV-Vis Spectra. *J. Phys. Chem. Lett.* **2018**, *9*, 6814–6817. [CrossRef] [PubMed]
46. Wei, S.; Yin, X.; Li, H.; Du, X.; Zhang, L.; Yang, Q.; Yang, R. Multi-Color Fluorescent Carbon Dots: Graphitized sp2 Conjugated Domains and Surface State Energy Level Co-Modulate Band Gap Rather Than Size Effects. *Chem. Eur. J.* **2020**, *26*, 8129–8136. [CrossRef] [PubMed]
47. Choi, J.; Kim, N.; Oh, J.-W.; Kim, F.S. Bandgap engineering of nanosized carbon dots through electron-accepting functionalization. *J. Ind. Eng. Chem.* **2018**, *65*, 104–111. [CrossRef]
48. Sawalha, S.; Silvestri, A.; Criado, A.; Bettini, S.; Prato, M.; Valli, L. Tailoring the sensing abilities of carbon nanodots obtained from olive solid wastes. *Carbon* **2020**, *167*, 696–708. [CrossRef]
49. Bing, W.; Sun, H.; Yan, Z.; Ren, J.; Qu, X. Programmed bacteria death induced by carbon dots with different surface charge. *Small* **2016**, *12*, 4713–4718. [CrossRef]

Article

Structural Study of Sulfur-Added Carbon Nanohorns

Ysmael Verde-Gómez [1,*], Elizabeth Montiel-Macías [1], Ana María Valenzuela-Muñiz [1], Ivonne Alonso-Lemus [2], Mario Miki-Yoshida [3], Karim Zaghib [4], Nicolas Brodusch [5] and Raynald Gauvin [5]

[1] Tecnológico Nacional de México/I.T. de Cancún, Av. Kabah km. 3, Cancún 77500, Q.Roo., Mexico; elizabethmontielmacias@hotmail.com (E.M.-M.); ana.vm@cancun.tecnm.mx (A.M.V.-M.)
[2] CONACyT-CINVESTAV Unidad Saltillo, Sustentabilidad de los Recursos Naturales y Energía, Av. Industria Metalúrgica, Parque Industrial Saltillo-Ramos Arizpe, Ramos Arizpe 25900, Coah., Mexico; ivonne.alonso@cinvestav.edu.mx
[3] Centro de Investigación en Materiales Avanzados S.C., Av. Miguel de Cervantes 120, Chihuahua 31136, Chih., Mexico; mario.miki@cimav.edu.mx
[4] Department of Chemical and Materials Engineering, Concordia University, 1515 Rue Sainte-Catherine O, Montréal, QC H3G 2W1, Canada; karim.zaghib@concordia.ca
[5] Department of Mining and Materials Engineering, McGill University, 3610 University Street, Montréal, QC H3A 0C5, Canada; nicolas.brodusch@mcgill.ca (N.B.); raynald.gauvin@mcgill.ca (R.G.)
* Correspondence: jose.vg@cancun.tecnm.mx; Tel.: +52-998-880-7432

Abstract: In the past few decades, nanostructured carbons (NCs) have been investigated for their interesting properties, which are attractive for a wide range of applications in electronic devices, energy systems, sensors, and support materials. One approach to improving the properties of NCs is to dope them with various heteroatoms. This work describes the synthesis and study of sulfur-added carbon nanohorns (S-CNH). Synthesis of S-CNH was carried out by modified chemical vapor deposition (m-CVD) using toluene and thiophene as carbon and sulfur sources, respectively. Some parameters such as the temperature of synthesis and carrier gas flow rates were modified to determine their effect on the properties of S-CNH. High-resolution scanning and transmission electron microscopy analysis showed the presence of hollow horn-type carbon nanostructures with lengths between 1 to 3 μm and, diameters that are in the range of 50 to 200 nm. Two types of carbon layers were observed, with rough outer layers and smooth inner layers. The surface textural properties are attributed to the defects induced by the sulfur intercalated into the lattice or bonded with the carbon. The XRD patterns and X-ray microanalysis studies show that iron serves as the seed for carbon nanohorn growth and iron sulfide is formed during synthesis.

Keywords: carbon nanohorns; sulfurated nanostructures; iron sulfide nanoparticles; chemical vapor deposition

1. Introduction

Nanostructured carbon (NC) materials have the potential to be used in many fields due to their unique and tunable properties. Synthesis methods and applications of some NCs such as carbon nanotubes, fullerenes, and recently graphene have been widely reported. However, horn-type carbon nanomaterials are less studied and therefore less reported. Initially, carbon nanohorns (CNH) were observed as aggregates of other NCs, and in 1999, Iijima et al. reported their synthesis [1]. CNHs differ from other carbon allotropes by their cone-shaped tips and sp[2]-hybridized carbon structure [2]. In this regard, the single-walled carbon nanohorn (SWCNH) with dahlia-like shape is one of the most studied of this family [3]. SWCNHs are considered to be a promising material in fields such as nanomedicine [4], energy [5], absorbents, and catalysis [6]. However, the development and study of CNHs has been slow, mainly due to spherical agglomerates which form during their synthesis process, hindering their functionalization and use of their entire surface [7]. Notwithstanding the foregoing, the production of CNHs can be cheap and easily scalable, and sometimes a metal precursor is not required [8]. Some companies have CNH production facilities, such as the NEC Corporation using laser ablation [9], Carbonium SRL using Joule heating method [10], and EEnano Tech Ltd. using arc discharge method [11].

The most outstanding properties of pristine CNHs over other carbon allotropes are their unique geometry, their ability to be produced at room temperature, high thermal and chemical stability, high porosity, and roughness, which makes them more reactive than nanostructures such as carbon nanotubes [12–15]. On the other hand, theoretical studies show that the electrostatic dipole moment increases with the number of carbon atoms in the conical tip, which also improves the reactivity of these nanostructures [16].

On the other hand, it is well known that NCs doped with heteroatoms (e.g., S, N, B, and P) enhance their properties [17]. In particular, the doped carbons have proven to be versatile functional materials with a wide range of potential applications, including heterogeneous catalysts for oxygen reduction reactions (ORR) [18–20], anodes for Li-ion batteries [21–23], cathodes for lithium-oxygen batteries [24], supercapacitors [25], adsorbents for hydrogen storage and CO_2 capture [26], adsorption of heavy metals and toxic gases [27], and desulfurization of diesel and crude oil [28] among others.

However, the synthesis of heteroatom-doped CNHs has not been explored despite the potential they could have in multiple applications. For example, sulfur-doped CNHs have shown to be a promising bifunctional catalyst for water splitting [29], while N-O-doped CNHs have high stability and good production rates for the electrosynthesis of hydrogen peroxide [30,31]. Furthermore, it has been reported that S-doped CNH and N-doped CNH have reliable performance with regard to oxygen evolution and reduction reactions, respectively [32,33]. In addition, N-doped SWCNH is a low-cost and high-performance electrode material for Lithium-sulfur battery applications [34].

Thus, the synthesis of heteroatom-doped CNHs by easy and scalable methods is attractive. This work presents the synthesis of novel sulfur-added multiwalled carbon nanohorns (S-CNH) by a modified one-step chemical vapor deposition method, evaluating the effect of the gas flow rate and the synthesis temperature on the physicochemical properties of the S-CNH.

2. Materials and Methods

Synthesis of the sulfur-added carbon nanohorns (S-CNH) was conducted by a Modified Chemical Vapor Deposition method (m-CVD) in a tubular furnace. A Vycor® tube (96% silica glass, Corning, Corning, NY, USA) was used as a substrate for deposition of the S-CNH. A precursor solution of toluene (99.8% Sigma-Aldrich, St. Louis, MO, USA) and thiophene (99% Alfa Aesar, Haverhill, MA, USA) (4:1 %vol.) served as the carbon and sulfur source, respectively. Ferrocene (98% Sigma-Aldrich, St. Louis, MO, USA) was added as an organometallic precursor (19 g/L) that acts as a source of Fe nanoparticles on which the CNH grows. The precursor solution was preheated until vaporization before its introduction into the furnace. The furnace temperature was 800 and 900 °C, while the carrier gas (argon) flow rate was 0.5 and 1 L/min. After the precursor solution was completely vaporized (50 mL), the tube was kept under an inert atmosphere until it cooled to room temperature.

After the synthesis, the samples were treated to remove residual iron and amorphous carbon using a reflux system with concentrated nitric acid for 12 h; the S-CNH-acid solution was continuously stirred. Following this procedure, the S-CNHs were recovered by filtration and washed with distilled water to remove all residual acid. Finally, the samples were dried overnight at 80 °C. The samples were labeled by the following code: SCNA8, SCNA9, SCNB8, and SCNB9, where A is carrier gas flow of 1.0 L/min, and B is carrier gas flow of 0.5 L/min; 8 and 9 correspond to the synthesis temperature of 800 °C and 900 °C, respectively.

The structural properties were analyzed by X-ray diffraction (XRD) and Raman spectroscopy. XRD analysis was performed with a Bruker diffractometer, model D8 Advance using Bragg–Brentano scan mode, LIXEYE detector, a step size of 0.01°, and step time of

5 s in a 2θ range from 10 to 85°. Raman analysis was performed using a Micro Raman Horiba spectrometer, Labram model, with a He-Ne laser (632.8 nm). The morphology and nanostructure of the samples were characterized by transmission electron microscopy (TEM) with a JEOL JEM 2200FS + CS at an accelerating voltage (E0) of 200 kV. A HITACHI SU-8230 (Hitachi, Tokyo, Japan) was used for scanning electron microscopy (SEM) equipped with a FlatQuad 5060F silicon drift detector (SDD) as an X-ray energy dispersive spectroscopy (EDS) detector from Bruker Nano (Billerica, MA, USA) with 1.2 steradians of solid angle, making it possible to achieve high X-ray count rates even at low primary energy. A TESCAN model VEGA3 SBU EasyProbe (Brno, The Czech Republic) with a Bruker EDS detector (Billerica, MA, USA) was used at 20 kV to determine the bulk chemical composition of the S-CNHs.

3. Results and Discussion

Figure 1 compares the XRD patterns of the synthesized S-CNH samples. In all samples, the diffraction patterns showed evidence of graphitic structure (002) at 2θ = 26.3° with an interplanar distance d = 0.338 nm [35–37]. Also, the peaks corresponding to Fe$_3$C were identified with planes (121), (002), (201), (211), (102), (112), (131), (221), (122), and (230). All samples showed a high-intensity peak at 2θ = 44.7° and a second characteristic peak at 2θ = 65.05° (marked by X), which corresponds to metallic Fe that remained encapsulated in the core of the carbon nanostructures due to their growth mechanisms [38]. Finally, samples SCNA9 and SCNB9 display peaks for FeS at 2θ = 29.9°, 33.7°, 43.1°, 53.1°, corresponding to the crystalline planes (110), (112), (114), and (300), respectively. The presence of FeS in these samples is attributed to the high temperature (900 °C), which favors its formation [39].

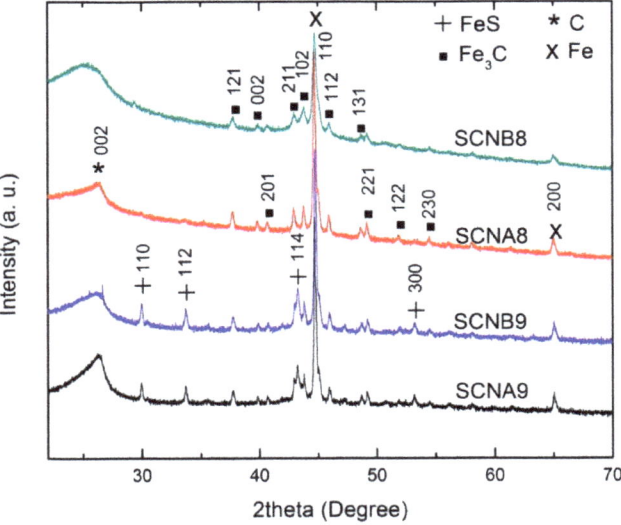

Figure 1. X-ray diffraction patterns of S added carbon nanohorns (S-CNHs).

The elemental composition obtained by EDS of bulk nanohorn samples shows that sulfur content was around 0.3 wt% in SCNA8 and SCNB8, and close to 1.0 wt% for SCNA9 and SCNB9 (Table S1 in the Supplementary Materials). An increase of the sulfur content is observed as the temperature of synthesis increased which is in agreement with the XRD results due to the FeS formation. The four samples presented the same morphology as revealed from a high-resolution field emission scanning electron microscope (FE-SEM) analysis. Examples of this morphology are given in Figure 2 (more HRSEM representative images of each sample are shown in the Supplementary Materials). By using a landing voltage as low as EL = E0 − Edecc = 2.0 − 1.5 = 0.5 kV (where EL is the landing voltage and Edecc is the deceleration voltage) and the high-resolution secondary electron (SE) signal collected by in-lens detectors, the surface morphology of the S-CNH was revealed. From Figure 2a,b, the horn-like structures are observed (also see Figures S1e, S2e, S3c and S4f in the Supplementary Materials). The structures have a length in the range of 1 to 3 µm and diameters in the range of from 50 to 200 nm. Along with the structures, the diameter decreases, hence showing the formation of a horn-like nanostructure. Micrographs obtained at low voltages revealed the surface textural properties of the carbon structures which present two types of layers. The outer layer exhibits a rough surface, indicating the possible presence of defects probably due to the intercalation of the sulfur atoms into the graphite lattice. On the other hand, the inside layer presents a smooth surface, a well-known characteristic of carbon nanotubes. Figure 2c is an enlarged view of the interface between the inner layer and the outer layer. In addition, Figures S1a–d, S3a–c and S4e in the Supplementary Materials show the two contrasting layers. It can be noticed that round particles of nanometer-scale (<5 nm) are unevenly dispersed at the nanohorn (Figures S1d and S3b in the Supplementary Materials) as well as at the outer layer surface (Figures S1b, S2f, S3d and S4d in the Supplementary Materials). The large scale or bulk structural patterns of the carbon nanohorn after the mechanical removal from the Vycor tube are homogenous as is observed in Figure 2a and the low magnification SEM images of the Supplementary Materials (Figures S2a,b, S3a and S4a,b).

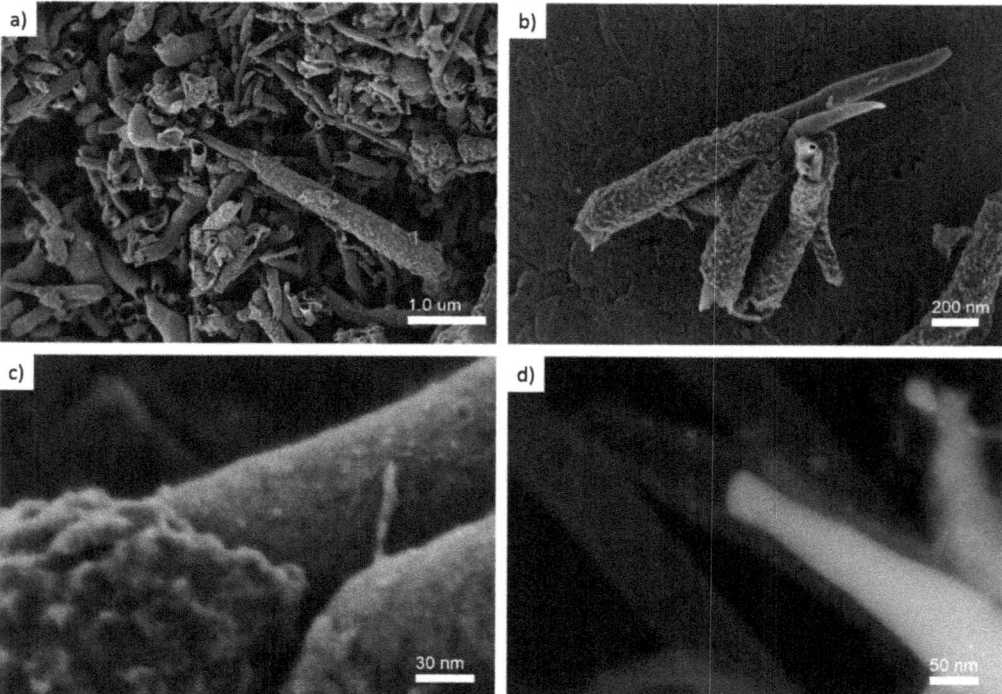

Figure 2. High resolution scanning electron microscopy micrographs of the SCNA9 (**a**), SCNA8 (**b**–**d**) samples. Using in-lens secondary electron detector with EL = E0 − Edecc = 2.0 − 1.5 = 0.5 kV (**a**–**c**) and dark-field transmitted electron detector (STEM-DF) at E0 = 30 kV (**d**).

Hollow tubular structures were observed from the dark-field scanning transmission electron microscopy (STEM-DF) obtained with E0 = 30 kV (Figure 2d, Figures S1f and S3f in the Supplementary Materials). Rod-shaped metallic nanoparticles with high atomic number contrast were also observed inside the tubular structure in Figure 2d which might be iron-based seeds from the growth process. The nanoparticles observed in Figure 2c were also noticed in the STEM-DF micrograph (Figure 2d) with the same size but their distribution looked more uniform in the outer layer.

An X-ray map acquired with an annular SDD detector is presented in Figure 3. It was obtained in STEM-DF mode with E0 = 20 kV on an area with both hollow and filled nanohorns. As expected, the rod-like particles inside the tubular structure were composed of a high atomic number material. From the Fe, C, and S maps, it can be deduced that these structures are either Fe_3C or FeS which is in agreement with the XRD results. Sulfur can also be observed in the carbon area. This was confirmed by generating the net intensities Fe/S ratio maps (Figure 4), from the data used in Figure 3 in addition to that from the SCNA9 sample (Figure S5 in the ESM shows the complete X-ray map for this sample). This demonstrated that the metal particles contained inside the nanohorns were a mixture of iron carbide and iron sulfide for all samples. The samples SCNA8 and SCNB8 presented, however, a smaller amount of iron sulfide compared to iron as deduced from Figure 1. In addition, the small nanoparticles observed in Figure 2d were also observed in the outer layer of the nanohorn analyzed in Figure 4 (bottom) and were identified as a mixture of Fe and S, possibly FeS. Note that, as a result of the thick Mylar window protecting the SDD crystal from beam damage due to the backscattered electrons (BSE), the carbon signal is increased when the BSE signal increases, typically when the Fe and S signals are strong. The carbon map may thus not be fully representative of the real carbon signal.

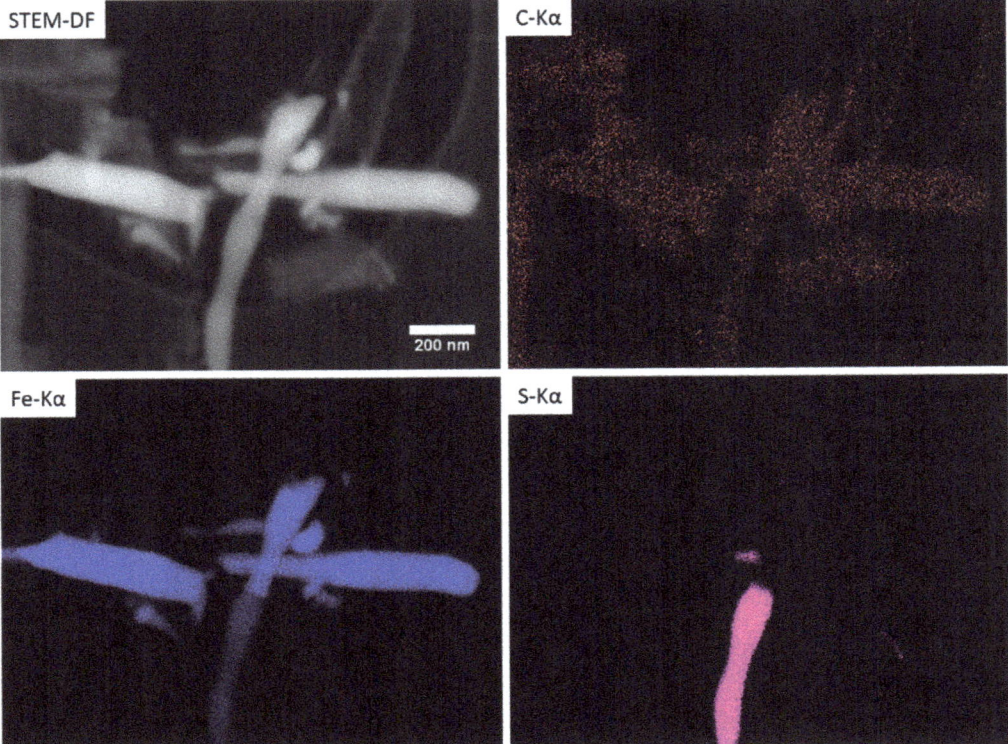

Figure 3. Elemental X-ray map of SCNA8 sample in STEM mode at E0 = 20 kV.

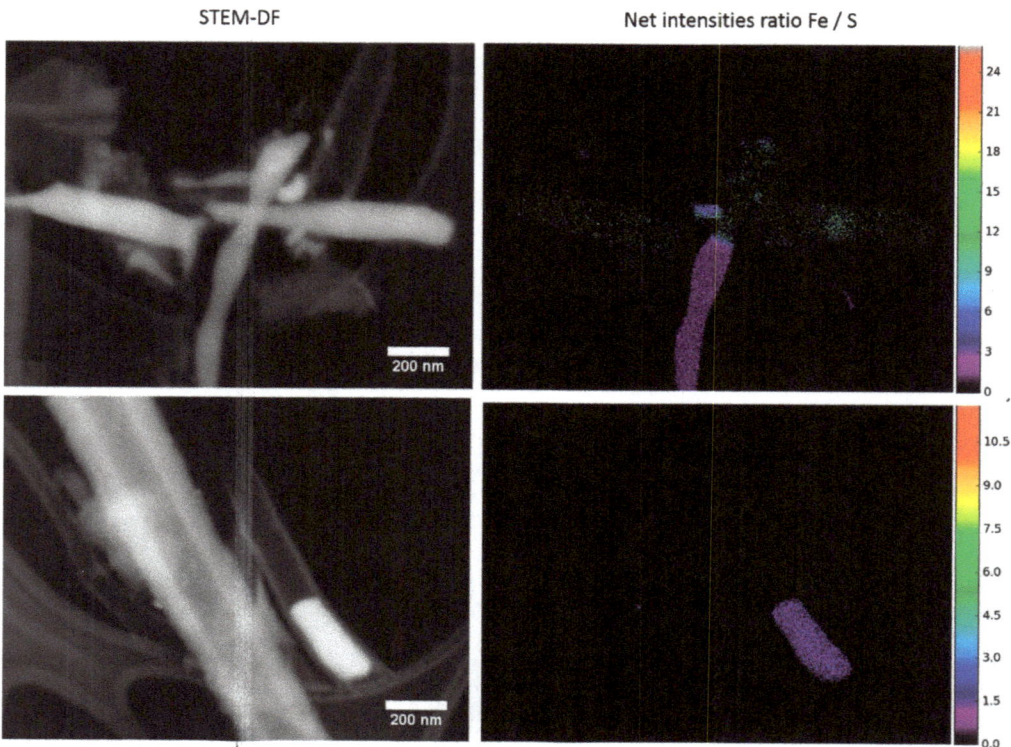

Figure 4. STEM-DF images (**left** side) and net intensities Fe/S ratio maps (**right** side) of the SCNA8 (**top**) and SCNA9 (**bottom**) samples at E0 = 20 kV.

Transmission electron microscopy images of SCNA8, SCNA8, SCNB8, and SCNB9 are shown in Figure 5. The microscopy images (Figure 5a,c,e,g) clearly show the presence of long tubular carbon nanostructures with diameters between 50 and 200 nm. It can be noted that, along with the structures the diameters tending to decrease, forming horn-type shapes, this behavior was observed for all the experimental conditions.

In addition, layers formed on the S-CNH external walls (ordered and amorphous) are also noticed. This behavior could be attributed to the insertion of sulfur in the structure. In the images obtained at higher magnification (Figure 5b,d,f,h), the atomic layers (lattice fringes) are evident. The lattice distances were measured in different areas of the samples. However, it was not possible to identify any clear tendency; the ordered layers (close to the core of the S-CNH), as well as the disordered layers (on top of the ordered ones), showed a mixture of lattice distances mostly of two groups, 0.355 ± 0.001 nm, and 0.296 ± 0.002 nm. Different authors have reported that the interplanar distance for graphite is in the range of 0.335 nm to 0.340 nm [40,41]. According to the XRD analysis, the first group of lattice distances is associated with modified graphite planes, possibly due to the incorporation of S. While the second group is attributed to the formation of an iron sulfide phase [42], which is in agreement with the FeS nanoparticles observed by microscopy.

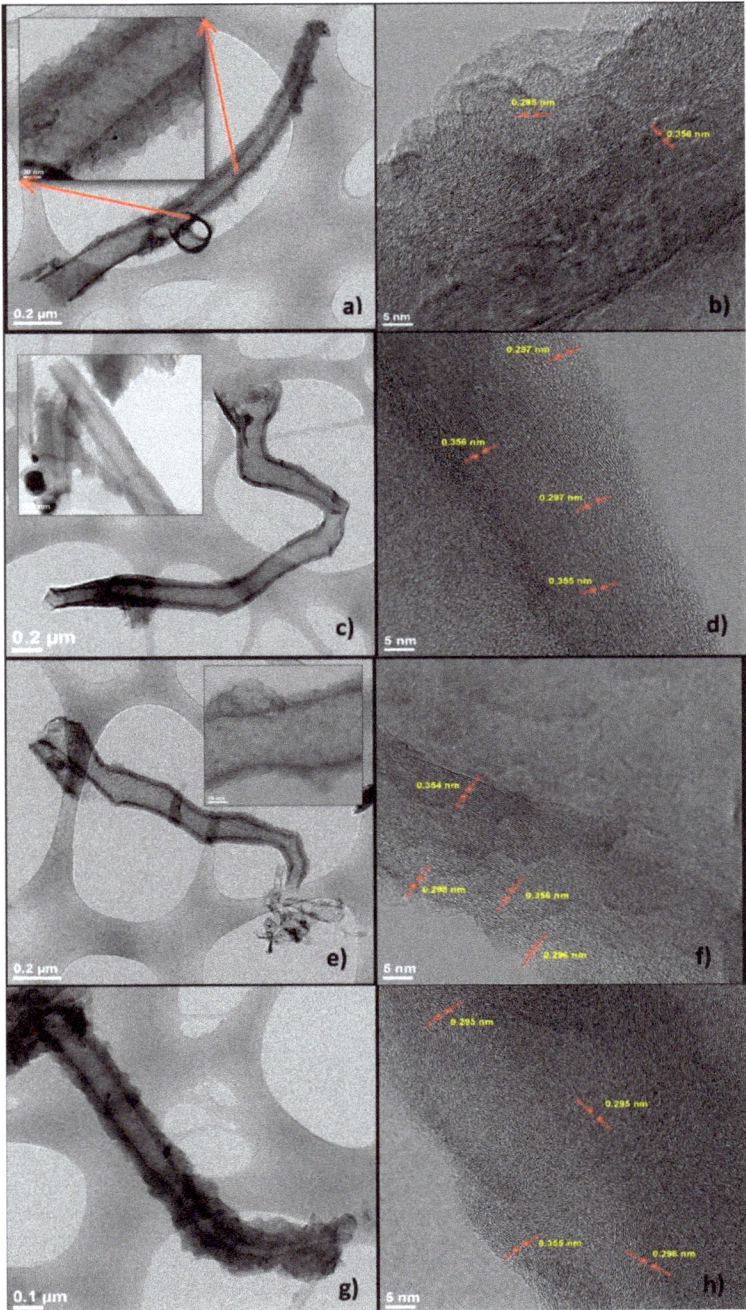

Figure 5. TEM images of SCNA8 (**a**,**b**), SCNA9 (**c**,**d**), SCNB8 (**e**,**f**), and SCNB9 (**g**,**h**).

The Raman spectra presented in Figure 6, show intensities that are associated with the D band (1300 cm^{-1}) and G band (1600 cm^{-1}). The width of the bands, present in all samples, is significantly greater than the width that is found in polycrystalline graphite [43]. Hence, the increment in the width of the peak can be attributed to the curvature of the multilayers. Moreover, the 2D band is extremely weak; a low intensity of this band was observed in MWCNT [44]. The D band is indicative of defects in the carbon layers and the intensity ratio (I_D/I_G) between the D and G bands provides information about the number of defects. The calculated ratio I_D/I_G is 1.37 >1.16 >1.15 >1.05 for samples SCNB8, SCNA9, SCNA8, and SCNB9, respectively. A large number of defects in all of the samples may be due to the presence of sulfur in the carbon atomic lattice; also, the sulfur incorporation in the carbon network leads to a structural modification that can induce an additional band near the G band. However, there may be a slight influence of the synthesis temperature and flow rate of carrier gas on the graphitization degree. Lower synthesis temperature and low gas flow rate promote the formation of lattice defects. Finally, the Raman spectra show a group of very intense bands between 100 and 200 cm^{-1}, they are mainly attributed to the radial breathing mode (RBM) associated with the inner tube's small diameter when a suitable resonance condition is established [45]. Hence, the Raman results obtained in this band range could be attributed to the narrower part of the horns where there are a few carbon layers and small diameters, observed in the HRTEM images.

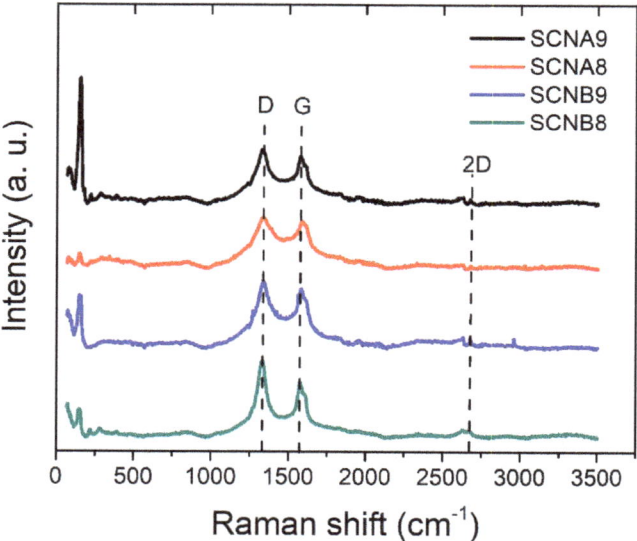

Figure 6. Raman spectra of sulfur-added carbon nanostructures synthesized at different temperatures and carrier gas flow rates.

4. Conclusions

In summary, multiwalled carbon nanohorns with sulfur were synthesized by a single-step reaction using a modified CVD technique. The hollow carbon nanostructures have diameters that tend to decrease (in the range from 50 to 200 nm), causing horn-type shapes. During synthesis, sulfur is intercalated into the carbon lattice. Also, the S reacts with Fe forming iron sulfide nanoparticles that serve as seeds for the carbon nanostructures' growth. High-resolution microscopy analyses show that FeS nanoparticles are detached and intercalated in the carbon lattices. The inner layers of the carbon are very well-ordered graphite layers; however, the outer layers of nanohorn surfaces have defects due to the influence of sulfur. The physical and chemical properties of this novel material are expected to be attractive for biological applications, drug carriers, adsorbents, catalysis, sensors, and energy application in electrochemical devices such as fuel cells, supercapacitors, and batteries.

Supplementary Materials: The following supporting information can be downloaded at: https://www.mdpi.com/article/10.3390/ma15103412/s1, Figure S1: High resolution scanning electron microscopy micrographs of the SCNA8 sample using the upper (secondary electrons, a, c, and e) and top (energy-filtered backscattered electrons, b and d) in-lens detectors at E0 = 0.5 kV (a–d) and E0 = 0.2 kV (e), and a dark-field transmitted electron detector (STEM-DF) at E0 = 30 kV (f); Figure S2: High-resolution scanning electron microscopy micrographs of the SCNB8 sample using the upper (secondary electrons, a, c, and e) and top (energy-filtered backscattered electrons, b, d, and f) in-lens detectors at E0 = 0.5 kV; Figure S3: High-resolution scanning electron microscopy micrographs of the SCNA9 sample using the upper (secondary electrons, a, c, and e) and top (energy-filtered backscattered electrons, b, and d) in-lens detectors at E0 = 0.5 kV, and a dark-field transmitted electron detector (STEM-DF) at E0 = 30 kV (f); Figure S4: High-resolution scanning electron microscopy micrographs of the SCNB9 sample using the upper (secondary electrons, a, c, e, and f) and top (energy-filtered backscattered electrons, b and d) in-lens detectors at E0 = 0.5 kV.; Figure S5: Elemental X-Ray map of SCNA9 sample in STEM mode at E0 = 20 kV; Table S1: Elemental Analysis (EDS) from the Bulk of the Carbon Nanohorn.

Author Contributions: The credit authorship contribution is: Y.V.-G.: Conceptualization, funding acquisition, investigation, methodology, supervision, writing—original draft, writing—review and editing; E.M.-M.: Conceptualization, investigation, writing—original draft; A.M.V.-M.: Validation, writing—review and editing; I.A.-L.: Methodology, writing—review and editing; M.M.-Y.: Investigation, writing—review and editing, resources, formal analysis; K.Z.: Writing—review and editing, resources; N.B.: Data curation, formal analysis, writing—review and editing; R.G.: Writing—review and editing, resources. All authors have read and agreed to the published version of the manuscript.

Funding: This research was funded by Tecnológico Nacional de México under grant No. 5528.15-P.

Institutional Review Board Statement: Not applicable.

Informed Consent Statement: Not applicable.

Data Availability Statement: The data are available from the corresponding author upon reasonable request.

Acknowledgments: The authors thank Pedro Pisa and Carlos Ornelas from CIMAV for technical assistance in Raman Spectroscopy and HRTEM, respectively.

Conflicts of Interest: The authors declare no conflict of interest.

References

1. Iijima, S.; Yudasaka, M.; Yamada, R.; Bandow, S.; Suenaga, K.; Kokai, F.; Takahashi, K. Nano-aggregates of single-walled graphitic carbon nano-horns. *Chem. Phys. Lett.* **1999**, *309*, 165–170. [CrossRef]
2. Pagona, G.; Mountrichas, G.; Rotas, G.; Karousis, N.; Pispas, S.; Tagmatarchis, N. Properties, applications and functionalization of carbon nanohorns. *Int. J. Nanotechnol.* **2009**, *6*, 176–195. [CrossRef]
3. Parasuraman, P.S.; Parasuraman, V.R.; Anbazhagan, R.; Tsai, H.C.; Lai, J.Y. Synthesis of "Dahlia-Like" Hydrophilic Fluorescent Carbon Nanohorn as a Bio-Imaging PROBE. *Int. J. Mol. Sci.* **2019**, *20*, 2977. [CrossRef] [PubMed]
4. Yamashita, T.; Yamashita, K.; Nabeshi, H.; Yoshikawa, T.; Yoshioka, Y.; Tsunoda, S.; Tsutsumi, Y. Carbon Nanomaterials: Efficacy and Safety for Nanomedicine. *Materials* **2012**, *5*, 350–363. [CrossRef]
5. Zhang, Z.C.; Han, S.; Wang, C.; Li, J.P.; Xu, G.B. Single-Walled Carbon Nanohorns for Energy Applications. *Nanomaterials* **2015**, *5*, 1732–1755. [CrossRef]
6. Szymański, G.S.; Kaczmarek-Kędziera, A.; Zięba, M.; Kowalczyk, P.; Terzyk, A.P. Insight into the Mechanisms of Low Coverage Adsorption of N-Alcohols on Single Walled Carbon Nanohorn. *Materials* **2021**, *14*, 4001. [CrossRef]
7. Karousis, N.; Suarez-Martinez, I.; Ewels, C.P.; Tagmatarchis, N. Structure, Properties, Functionalization, and Applications of Carbon Nanohorns. *Chem. Rev.* **2016**, *116*, 4850. [CrossRef]
8. Amazi, T.; Kasuya, D.; Yuge, R.; Yudasaka, M.; Iijima, S.; Yoshitake, T.; Kubo, Y. Large-Scale production of Single-Wall Carbon Nanohorns with high purity. *J. Phys. Chem. C* **2008**, *112*, 1330–1334.
9. NEC. Available online: http://www.nec.com/en/global/prod/cnh/ (accessed on 1 April 2022).
10. Carbonium SRL. Available online: http://www.carbonium.it (accessed on 1 April 2022).
11. TIE GmbH. Available online: http://www.t-i-e.eu (accessed on 1 April 2022).
12. Murata, K.; Kaneko, K.; Steele, W.A.; Kokai, F.; Takahashi, K.; Kasuya, D.; Hirahara, K.; Yudasaka, M.; Iijima, S. Molecular Potential Structures of Heat-Treated Single-Wall Carbon Nanohorn Assemblies. *J. Phys. Chem. B* **2001**, *105*, 10210. [CrossRef]

13. Bekyarova, E.; Kaneko, K.; Yudasaka, M.; Kasuya, D.; Iijima, S.; Huidobro, A.; Rodriguez-Reinoso, F. Controlled Opening of Single Wall Carbon Nanohorns by Heat Treatment in Carbon Dioxide. *J. Phys. Chem. B* **2003**, *107*, 4479. [CrossRef]
14. Li, T.; He, C.; Zhang, W. Rational design of porous carbon allotropes as anchoring materials for lithium sulfur batteries. *J. Energy Chem.* **2021**, *52*, 121. [CrossRef]
15. Bakoglidis, K.D.; Palisaitis, J.; Santos, R.B.d.; Rivelino, R.; Persson, P.O.Å.; Gueorguiev, G.K.; Hultman, L. Self-Healing in Carbon Nitride Evidenced As Material Inflation and Superlubric Behavior. *ACS Appl. Mater. Interfaces* **2018**, *10*, 16238–16243. [CrossRef] [PubMed]
16. Heiberg-Andersen, H.; Skjeltorp, A.T.; Sattler, K. Carbon Nanocones: A Variety of Non-Crystalline Graphite. *J. Non-Cryst. Solids* **2008**, *354*, 5247. [CrossRef]
17. Paraknowitsch, J.P.; Thomas, A. Doping carbons beyond nitrogen: An overview of advanced heteroatom doped carbons with boron, sulphur and phosphorus for energy applications. *Energy Environ. Sci.* **2013**, *6*, 2839–2855. [CrossRef]
18. Yang, Z.; Nie, H.G.; Chen, X.; Chen, X.H.; Huang, S.M. Recent progress in doped carbon nanomaterials as effective cathode catalysts for fuel cell oxygen reduction reaction. *J. Power Sources* **2013**, *236*, 238–249. [CrossRef]
19. Yu, D.S.; Nagelli, E.; Du, F.; Dai, L.M. Metal-Free Carbon Nanomaterials Become More Active than Metal Catalysts and Last Longer. *J. Phys. Chem. Lett.* **2010**, *1*, 2165–2173. [CrossRef]
20. Yang, L.J.; Jiang, S.J.; Zhao, Y.; Zhu, L.; Chen, S.; Wang, X.Z.; Wu, Q.; Ma, J.; Ma, Y.W.; Hu, Z. Boron-Doped Carbon Nanotubes as Metal-Free Electrocatalysts for the Oxygen Reduction Reaction. *Angew. Chem. Int. Ed.* **2011**, *50*, 7132–7135. [CrossRef]
21. Yue, L.; Zhong, H.X.; Tang, D.P.; Zhang, L.Z. Porous Si coated with S-doped carbon as anode material for lithium ion batteries. *J. Solid State Electrochem.* **2013**, *17*, 961–968. [CrossRef]
22. Yue, L.; Wang, S.Q.; Zhao, X.Y.; Zhang, L.Z. Nano-silicon composites using poly(3,4-ethylenedioxythiophene):poly(styrenesulfonate) as elastic polymer matrix and carbon source for lithium-ion battery anode. *J. Mater. Chem.* **2012**, *22*, 1094–1099. [CrossRef]
23. Pappas, G.S.; Ferrari, S.; Huang, X.; Bhagat, R.; Haddleton, D.M.; Wan, C. Heteroatom doped-carbon nanospheres as anodes in lithium ion batteries. *Materials* **2016**, *9*, 35. [CrossRef]
24. Li, Y.L.; Wang, J.J.; Li, X.F.; Geng, D.S.; Banis, M.N.; Tang, Y.J.; Wang, D.N.; Li, R.Y.; Sham, T.K.; Sun, X.L. Discharge product morphology and increased charge performance of lithium-oxygen batteries with graphene nanosheet electrodes: The effect of sulphur doping. *J. Mater. Chem.* **2012**, *22*, 20170–20174. [CrossRef]
25. Zhao, X.C.; Zhang, Q.; Chen, C.M.; Zhang, B.S.; Reiche, S.; Wang, A.Q.; Zhang, T.; Schlogl, R.; Su, D.S. Aromatic sulfide, sulfoxide, and sulfone mediated mesoporous carbon monolith for use in supercapacitor. *Nano Energy* **2012**, *1*, 624–630. [CrossRef]
26. Seema, H.; Kemp, K.C.; Le, N.H.; Park, S.-W.; Chandra, V.; Lee, J.W.; Kim, K.S. Highly selective CO_2 capture by S-doped microporous carbon materials. *Carbon* **2014**, *66*, 320–326. [CrossRef]
27. Morris, E.A.; Kirk, D.W.; Jia, C.Q.; Morita, K. Roles of Sulfuric Acid in Elemental Mercury Removal by Activated Carbon and Sulfur-Impregnated Activated Carbon. *Environ. Sci. Technol.* **2012**, *46*, 7905–7912. [CrossRef] [PubMed]
28. Seredych, M.; Bandosz, T.J. Removal of dibenzothiophenes from model diesel fuel on sulfur rich activated carbons. *Appl. Catal. B Environ.* **2011**, *106*, 133–141. [CrossRef]
29. Kagkoura, A.; Arenal, R.; Tagmatarchis, N. Sulfur-Doped Carbon Nanohorn Bifunctional Electrocatalyst for Water Splitting. *Nanomaterials* **2020**, *10*, 2416. [CrossRef] [PubMed]
30. Chakthranont, P.; Nitrathorn, S.; Thongratkaew, S.; Khemthong, P.; Nakajima, H.; Supruangnet, R.; Butburee, T.; Sano, N.; Faungnawakij, K. Rational Design of Metal-free Doped Carbon Nanohorn Catalysts for Efficient Electrosynthesis of H_2O_2 from O_2 Reduction. *ACS Appl. Energy Mater.* **2021**, *4*, 12436–12447. [CrossRef]
31. Iglesias, D.; Giuliani, A.; Melchionna, M.; Marchesan, S.; Criado, A.; Nasi, L.; Bevilacqua, M.; Tavagnacco, C.; Vizza, F.; Prato, M.; et al. N-Doped Graphitized Carbon Nanohorns as a Forefront Electrocatalyst in Highly Selective O_2 Reduction to H_2O_2. *Chem* **2018**, *4*, 106–123. [CrossRef]
32. Montiel-Macias, E.; Valenzuela-Muñiz, A.M.; Alonso-Núñez, G.; Farías-Sánchez, M.H.; Gauvin, R.; Verde-Gómez, Y. Sulfur doped carbon nanohorns towards oxygen reduction reaction. *Diamod Relat. Mater.* **2020**, *103*, 107671. [CrossRef]
33. Nan, Y.L.; He, Y.Y.; Zhang, Z.H.; Jian, W.; Zhang, Y.B. Controllable synthesis of N-doped carbon nanohorns: Tip from closed to half-closed, used as efficient electrocatalysts for oxygen evolution reaction. *RCS Adv.* **2021**, *11*, 35463–35471. [CrossRef]
34. Gulzar, U.; Li, T.; Bai, X.; Colombo, M.; Ansaldo, A.; Marras, S.; Prato, M.; Goriparti, S.; Capiglia, C.; Zaccaria, R.P. Nitrogen-Doped Single-Walled Carbon Nanohorns as a Cost-Effective Carbon Host toward High-Performance Lithium–Sulfur Batteries. *ACS Appl. Mater. Interfaces* **2018**, *10*, 5551–5559. [CrossRef] [PubMed]
35. Khai, T.V.; Kwak, D.S.; Kwon, Y.J.; Cho, H.Y.; Huan, T.N.; Chung, H.; Ham, H.; Lee, C.; Dan, N.V.; Tung, N.T.; et al. Direct production of highly conductive graphene with a low oxygen content by a microwave-assisted solvothermal method. *Chem. Eng. J.* **2013**, *232*, 346–355. [CrossRef]
36. Peng, H.; Hou, S.; Dang, D.; Zhang, B.; Liu, F.; Zheng, R.; Luo, F.; Song, H.; Huang, P.; Liao, S. Ultra-high-performance doped carbon catalyst derived from o-phenylenediamine and the probable roles of Fe and melamine. *Appl. Catal. B Environ.* **2014**, *158–159*, 60–69. [CrossRef]
37. Zhang, L.W.; Gao, A.; Liu, Y.; Wang, Y.; Ma, J.T. PtRu nanoparticles dispersed on nitrogen-doped carbon nanohorns as an efficient electrocatalyst for methanol oxidation reaction. *Electrochim. Acta* **2014**, *132*, 416–422. [CrossRef]
38. Dresselhaus, M.S.; Dresselhaus, G.; Avouris, P. *Carbon Nanotubes: Synthesis, Structure, Properties, and Applications*, 1st ed.; Springer: Berlin/Heidelberg, Germany, 2001.

39. Kiciński, W.; Szala, M.; Bystrzejewski, M. Sulfur-doped porous carbons: Synthesis and applications. *Carbon* **2014**, *68*, 1–32. [CrossRef]
40. Wissler, M. Graphite and carbon powders for electrochemical applications. *J. Power Sources* **2006**, *156*, 142–150. [CrossRef]
41. Popov, V.N. Carbon nanotubes: Properties and application. *Mater. Sci. Eng. R Rep.* **2004**, *43*, 61–102. [CrossRef]
42. Ohfuji, H.; Rickard, D. High resolution transmission electron microscopic study of synthetic nanocrystalline mackinawite. *Earth Planet. Sci. Lett.* **2006**, *241*, 227–233. [CrossRef]
43. Zhang, H.-B.; Lin, G.-D.; Zhou, Z.-H.; Dong, X.; Chen, T. Raman spectra of MWCNTs and MWCNT-based H_2-adsorbing system. *Carbon* **2002**, *40*, 2429–2436. [CrossRef]
44. Gomez, J.; Verde, Y.; Lara-Romero, J.; Alonso-Nuñez, G. In-Situ Deposition of Nickel Nanoparticles on Carbon Nanotubes by Spray Pyrolysis. *Fuller. Nanotub. Carbon Nanostruct.* **2009**, *17*, 507–518. [CrossRef]
45. Puech, P.; Bassil, A.; Gonzalez, J.; Power, C.; Flahaut, E.; Barrau, S.; Demont, P.; Lacabanne, C.; Perez, E.; Bacsa, W.S. Similarities in the Raman RBM and D bands in double-wall carbon nanotubes. *Phys. Rev. B* **2005**, *72*, 155436. [CrossRef]

Article

The Effect of Carbon Nanotubes on the Strength of Sand Seeped by Colloidal Silica in Triaxial Testing

Weifeng Jin [1,*], Ying Tao [1], Xin Wang [1] and Zheng Gao [2]

[1] School of Civil Engineering and Architecture, Zhejiang University of Science and Technology, Hangzhou 310023, China; 212002814015@zust.edu.cn (Y.T.); 211602814002@zust.edu.cn (X.W.)
[2] Hydro China Huadong Engineering Corporation, Hangzhou 310014, China; gao_z3@ecidi.com
* Correspondence: 112014@zust.edu.cn

Abstract: Colloidal silica can quickly seep through sand and then form silica gels to cement sand particles. To improve the strength of sand seeped by colloidal silica, carbon nanotubes were dispersed in the colloidal silica to form carbon-nanotube-reinforced sand-gel composites. Then triaxial tests were performed to explore how carbon nanotube content affects shear strength. The test results showed that: (1) with the increase of colloidal silica concentration, the shear strength significantly increased with the same carbon nanotube content (especially the low concentration of 10 wt. % colloidal silica, which showed almost no reinforcing effect with carbon nanotubes) while 40 wt. % colloidal silica plus 0.01 wt. % carbon nanotube caused the maximum increase of shear strength by up to 93.65%; (2) there was a concentration threshold of colloidal silica, above which the shear strength first increased to the peak value and then decreased with increasing carbon nanotube content (and we also established a formula to predict such phenomenon); and (3) SEM images showed that carbon nanotubes were connected as several ropes in the micro-cracks of the silica gel, resulting in greater macroscopic shear strength. Our new method of mixing carbon nanotubes and colloidal silica to seep through sand can contribute to sandy ground improvement.

Keywords: colloidal-silica-stabilized sand; deviatoric stress at failure; carbon-nanotube-reinforced; triaxial test; shear strength prediction

1. Introduction

A fast and non-toxic way to stabilize loose sand is the seepage of colloidal silica through sand. The colloidal silica has a low viscosity, which allows its rapid seepage through sand [1,2]. In an alkaline environment, the silica nano-particles in the colloidal silica are stably suspended against gelation due to the repulsive force between the particles. When the pH of colloidal silica is changed from an alkaline to an acid, the repulsive force between the particles decreases, and the silica nano-particles agglomerate to form silica gels which can cement sand particles [3]. Additionally, by adjusting the pH of the colloidal silica, the colloidal silica can maintain its fluidity for a certain period and then become a solid gel [4]. The above characteristics make colloidal silica ideal for stabilizing loose sand by rapid seepage.

Currently, for the effect of adding carbon nanotubes on the sand seeped by colloidal silica, no references are available. For sand seeped by colloidal silica, the existing literature has not provided answers the following two questions: (1) as the carbon nanotube content increases, what are the trends of shear strength at a certain concentration of colloidal silica; and (2) what is the optimal ratio of carbon nanotube to colloidal silica resulting in the maximum shear strength? We describe current knowledge related to our research from three aspects. The first aspect focused on the sand stabilized by colloid silica, especially the seepage and mechanical characteristics of colloidal-silica-stabilized sand. For instance, Gallagher and Lin [1] studied colloidal silica's ability to transport through sand in an adequate concentration; Fujita and Kobayashi [3] focused on the fundamental transport behaviors of silica particles influenced by pH conditions; Agapoulaki and Papadimitriou [5] studied travel distance and the effect of temperature on viscosity-versus-time curves; Saiers et al. [6] performed tests on transport through heterogeneous porous media; Hamderi and Gallagher [7] studied the simulation of optimum coverage; Hamderi et al. [8] studied the numerical model for simulating colloidal silica transport through sand columns; Hamderi and Gallagher [9] studied the effect of injection rate on the degree of grout penetration; Gallagher et al. [2] used field tests under explosion to study liquefaction mitigation of colloidal-silica-stabilized sand; Gallagher et al. [4] used centrifuge tests to study colloidal-silica-stabilized sand against liquefaction; Pamuk et al. [10] performed centrifuge tests on colloidal-silica-stabilized site; Conlee et al. [11] performed centrifuge modeling for liquefaction mitigation; Conlee [12] performed centrifuge model tests and full-scale field tests on colloidal-silica-stabilized soil; Kodaka et al. [13] modeled strength and cyclic deformation characteristics of colloidal-silica-stabilized sand; Antonio-Izarraras et al. [14] and Díaz-Rodríguez et al. [14] used shear tests to investigate the cyclic strength of sand stabilized with colloidal silica; Kakavand and Dabiri [15] and Wong et al. [16] performed shear tests on sandy soil improved by colloidal silica; Gallagher et al. [17] used triaxial tests to investigate the influence of colloidal silica on cyclic undrained behavior of sand; Mollamahmutoglu and Yilmaz [18] used triaxial tests to studypre- and post-cyclic strength of colloidal-silica-stabilized sand; Persoff et al. [19] investigated the influence of dilution and contaminants on the strength of colloidal-silica-stabilized sand; Pavlopoulou et al. [20] used monotonic and cyclic loading tests to compare the differences between colloidal-silica-stabilized sand and untreated sand in the stress-strain relationships and cumulative strain curves; Triantafyllos et al. [21] used triaxial compression tests to find important changes in the colloidal-silica-stabilized-sand's mechanical behavior such as increase in stress ratio and relocation of sand's critical state line in the e-p' plane; Krishnan et al. [22] obtained the gel time by adding salt and varying pHs and used direct shear tests and triaxial tests to analyze the mechanical enhancement of colloidal-silica-stabilized sand; Vranna et al. [23] used undrained monotonic and cyclic triaxial tests to obtain monotonic and cyclic strength of colloidal-silica-stabilized sand; and Ghadr et al. [24] used undrained triaxial tests to determine the shear strength and the critical-state-line changes of colloidal-silica-stabilized sand. However, the above researchers did not extended their research to the application of carbon nanotubes in colloidal-silica-stabilized sand, so this literature cannot answer how carbon nanotubes affect the strength of colloidal-silica-stabilized sand. The second aspect of the state of the art related to our research is the silica gel containing carbon nanotubes. Gavalas et al. [25] described a class of composite materials designed by combining multiwall carbon nanotubes and silica gel. However, this gel has not been used to stabilize sand. The third aspect of the state of the art related to our research is the application of carbon nanotubes in the Portland cement matrix. In previous literatures, carbon nanotubes have been dispersed in Portland cement and used to bridge micro-cracks in the matrix. For instance, Wang et al. [26] studied the durability of cement doped with carbon nanotubes; Liu et al. [27] studied the shrinkage and crack resistance of carbon-nanotube-reinforced cement; Cwirzen et al. [28] investigated the mechanical properties of cement reinforced by surface-decorated carbon nanotubes; Li et al. [29] studied the pressure-sensitive properties and microstructures of carbon-nanotube-reinforced cement; Nochaiya and Chaipanich [30] investigated the effect of carbon nanotubes on the porosity and microstructure of cement; Rana and Fangueiro [31] review the dispersion and mechanical properties of carbon-nanotube-reinforced cement; and Li et al. [32] studied the coupling effect of nano-silica sol-gel and carbon nanotubes on the cement based materials. However, the previous literature mixed carbon-nanotube-dispersed Portland cement and sand by stirring, and such cement is not suitable to seep through sand due to its high viscosity. So, the main difference from these previous literatures is that we transport carbon nanotubes to the pores between sand particles by seepage instead of stirring. In conclusion, the research gap between the previous literature and our work is that no references elaborated the influence of carbon nanotube on the strength of sand seeped by colloidal silica, since previous studies did not extend the application of carbon nanotubes to the colloidal-silica-stabilized sand. Besides, the case of carbon-nanotube-dispersed Portland cement cannot be a direct reference in our case, since the mixing method of Portland cement and sand was stirring, and not the seepage as in our case.

The contribution of this paper is that, for sand seeped by colloidal silica, we obtained laws concerning how carbon nanotube content affects the strength of colloidal-silica-stabilized sand. That is, we obtained the trends of strength with increasing carbon nanotube content at different concentrations of colloidal silica. We also obtained the optimal ratio of carbon nanotube to colloidal silica resulting in the maximum shear strength. Additionally, we established a formula to predict strength varying with carbon nanotube content at 40 wt. % colloidal silica. The motivation of this paper is that, based on the images of the scanning electron microscope (SEM), there are a large number of micro-cracks in the silica gel which bonds sand particles. These micro-cracks reduce the strength of the gel-sand composite. So, we attempted to add carbon nanotubes to colloidal silica in the hope that carbon nanotubes in the micro cracks could result in the improvement of the macro shear strength of the gel-sand composite. The reason for doing this research is that the effects of carbon nanotube content on the shear strength can be clarified through tests. In addition, what this article would add to the current available literature are the laws of how carbon nanotube content affects colloidal-silica-stabilized sand, as well as the prediction formula at high concentration of colloidal silica (thisis the first time that the characteristics of the trends of shear strength with increasing carbon nanotubes are clarified). So, this paper is suitable for the application in the field of ground improvement by using colloidal silica, since we present the optimal ratio of colloidal silica to carbon nanotubes resulting in the maximum shear strength as well as a prediction formula for the effect of carbon nanotubes on the strength of colloidal-silica-stabilized sand.

In this paper, based on the triaxial tests, we first explore the influence of carbon nanotube content on the shear strength of sand seeped with different concentration levels of colloidal silica; second, we introduce physically meaningful variables to propose a formula for predicting the maximum shear strength; finally, based on the SEM photos, from the micro scale perspective, we explore the morphology of carbon nanotubes in the micro-cracks of silica gel, which results in greater macro shear strength compared to sand stabilized without carbon nanotubes.

2. Experimental Details

Gallagher et al. [2] described various factors influencing the gel time of colloidal silica such as concentration, nano-silica particle size, ionic strength and pH. Gel time decreases with increasing concentration, increasing the size of the nano-silica particle, and increasing ionic strength, and gel time reaches its minimum when pH ranges from 5 to 7. In our tests, we adjusted the pH value to control the gel time, since the particle size was not changed and we did not add salt to control ionic strength. For the four concentrations of colloidal silica used in our tests, when the pH is adjusted to 5~6, colloidal silica forms solids and cement sand particles within one day. So, we adjusted the pH value to 5~6 before using colloidal silica to seep through sand specimens. After the sand specimens were seeped by colloidal silica, the sand specimens were cured for three days to ensure that the specimens were stabilized. Then we performed triaxial tests on these specimens.

2.1. Experimental Materials

The sand used was Pingtan standard sand from Fujian, China. Index properties of the sand, including relative density, are listed in Table 1.

The multi-walled carbon nanotubes (MWCNTs), produced by Suzhou Hengqiu Graphene Technology Co., Ltd., (Suzhou, China) were used as reinforcing fibers. The physical and structural properties of MWCNTs are listed in Table 2. The MWCNTs have inner diameters of 3–5 nm, outer diameters of 8–15 num, and lengths of 3–12 ìm. The specific surface area is >230 m^2/g according to the manufacturer.

Table 1. Properties of sand sample.

Specific Gravity, G_s	Max. Dry Density (kg/m^3)	Min. Dry Density (kg/m^3)	Relative Density	Coefficient of Uniformity, C_u	Coefficient of Curvature, C_c
2.65	1750	1510	0.43	4	0.51

Table 2. Properties of multi-walled carbon nanotubes.

Inner Diameters (nm)	Outer Diameter (nm)	Length (μm)	Specific Surface Area (m^2/g)	Density (g/cm^3)
3–5	8–15	3–12	>230	2.1

The colloidal silica used was provided by Qingdao Maike Silica Gel Dessicant Co., Ltd., and four concentrations of 10, 20, 30, and 40% by weight were used. The physical properties of the colloidal silica are shown in Table 3. The silicanano-particles, which are suspended under an alkaline environment, have diameters in the range of 10 to 20 nm.

Table 3. Physical properties of the colloidal silica.

SiO$_2$ (%)	pH	Density (g/cm^3)	Viscosity (Pa·s)	Average Particle Size (nm)
10%	8.5~9.5	1.08~1.10	3.0×10^{-3}	10~20
20%	8.5~10.0	1.12~1.14	5.0×10^{-3}	10~20
30%	8.5~10.0	1.19~1.21	7.0×10^{-3}	10~20
40%	9.0	1.28~1.3	25.0×10^{-3}	10~20

2.2. Specimen Preparation and Testing Apparatus

For the method of specimen preparation, we used the carbon-nanotube-dispersed colloidal-silica to seep through and then stabilize the sand. The key point, which was to disperse carbon nanotubes in colloidal-silica, was fulfilled within two dispersion steps: first, we put carbon nanotubes in the container filled with colloidal-silica, and then used a rotating bar to mechanically stir the colloidal-silica and carbon nanotubes. Secondly, ultrasonic vibration conducted across the container and make carbon nanotubes more uniformly dispersed in colloidal-silica. The duration time of each of the above dispersion step was calibrated by trial. We used the above two steps to make sure that carbon nanotubes were well dispersed, which was why we used the above two steps to disperse carbon nanotubes.

For the method of testing shear strength, since triaxial testing apparatuses can maintain a user-defined stress in the horizontal direction and apply load in the vertical direction, we utilized triaxial apparatuses to obtain the shear strength with different horizontal stresses.

The sand specimens were treated using the following steps. First, the pH of colloidal silica was adjusted to 5.0–5.5 by adding acetic acid, then the colloidal silica was magnetically stirred with carbon nanotubes for 30 min (see Figure 1a). Secondly, carbon nanotubes were further dispersed by ultrasonic dispersion for 60–120min (see Figure 1b). Finally, by a peristaltic pump, carbon-nanotube-dispersed colloidal-silica was slowly injected into the sand from the bottom of a cylindrical mold (see Figure 1c). The stabilized specimens are shown in Figure 2. The reason for moisture proof membranes is to prevent the moisture in the specimen from evaporating. Although it is not necessary to immerse the specimen in water, we still ensure that the specimen is in a wet state and moisture will not leak from the membrane joints, so we still placed the specimen wrapped with the membrane in water.

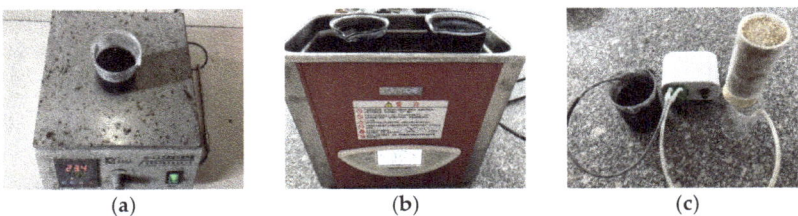

Figure 1. Specimen preparation: (**a**) Magnetically stirred mixture of carbon nanotubes and colloidal silica, (**b**) ultrasonic dispersion of carbonnanotubes in colloidalsilica, (**c**) thecarbon-nanotube-dispersed colloidal-silica seeped through sand.

Figure 2. Stabilized specimens.

AGDS advanced triaxial system (see Figure 3) was used to carry out undrained triaxial compression tests on specimens. The diameter and height of the test specimen are 38 mm and 76 mm, respectively.

Figure 3. Specimen for triaxial compression test.

2.3. Experimental Plan

The experimental plan, as summarized in Table 4, is mainly aimedat investigating the effect of the content of carbon nanotube on the strength of the stabilized specimens. Sands were seeped and stabilizedby the mixture of colloidal silica and carbon nanotubes. The concentrations of colloidal silica are 10, 20, 30 and 40 wt. %, respectively. The contents of carbon nanotubes are 0, 0.01, 0.02, 0.03, 0.04 and 0.05% by weight of colloidal silica, respectively. The initial mean effective stresses in undrained triaxial tests for these specimens were 50, 80 and 110 kPa, respectively. During each triaxial test, the confining pressure (horizontal stress) is set equal to the initial mean effective stress. The molds containing the stabilized specimens were wrapped in moisture proof membranes and immersed in water to cure for three days before testing. Thus, there are a total of 72 specimens tested.

Table 4. Experimental plan.

Colloidal Silica Concentration (wt. %)	Carbon Nanotube Content (wt. %)	Initial Mean Effective Stress (kPa)	Curing Period (Days)
10, 20, 30 and 40	0, 0.01, 0.02, 0.03, 0.04 and 0.05	50, 80, and 110	3

3. Testing Results and Analysis

Figure 4 shows the failure modes of the specimens after the triaxial tests. The specimens in Figure 4a–c were reinforced with 10, 20 and 30 wt. % colloidal silica respectively, while the sample in Figure 4d was reinforced with 40 wt. % colloidal silica mixed with 0.02% carbon nanotubes. It can be seen from Figure 4a–c that as the concentration of colloidal silica increases, the slip surface becomes more obvious. In Figure 4d, the stabilized specimen with carbon nanotubes shows the most obvious slip surface.

Figure 4. Failure modes of specimens after triaxial tests, and (**a**) 10 wt. % colloidal silica, (**b**) 20 wt. % colloidal silica, (**c**) 30 wt. % colloidal silica, and (**d**) 40 wt. % colloidal silica +0.02 wt. % carbon nanotubes.

3.1. Effect of Carbon Nanotubes on the Shear Strength

Since we use deviatoric stress at failure as shear strength, Figure 5a–c show the curves of shear strength versus carbon nanotube content under different initial mean effective stresses of 50, 80 and 110 kPa. The corresponding data are shown in Table 5. As shown in Figure 5, when the concentration of colloidal silica is 10 wt. %, carbon nanotubes have no reinforcement effect on the specimens. For 20 wt. % colloidal silica, carbon nanotubes cause a slight increase in the deviatoric stress at failure, which is less than the increase in 30 wt. % colloidal silica. In the case of 40 wt. % colloidal silica, 0.01–0.02 wt. % carbon nanotubes cause the greatest reinforcement effect on the specimens (i.e., under the initial mean effective stresses of 50, 80 and 110 kPa, the deviatoric stresses at failure increase by 83.6, 93.7 and 78.2%, respectively). For 0.01 wt. % carbon nanotubes dispersed in 10, 20, 30 and 40 wt. % colloidal silica, according to Table 5 or Figure 5, the shear strength increased by up to 9.93, 13.1, 21.37 and 93.65%, respectively. That is, as the concentration of colloidal silica increases from 10 to 40 wt. %, after adding 0.01 wt. % carbon nanotubes, and the maximum percentage of shear strength increment increases from 9.93 to 93.65%.

Therefore, the reinforcement effect of carbon nanotubes increases with increasing colloidal silica concentration. Additionally, for 40% colloidal silica, based on different horizontal stress levels, there are optimal contents of carbon nanotubes between 0.01–0.02 wt. % which lead to the peak shear strengths.

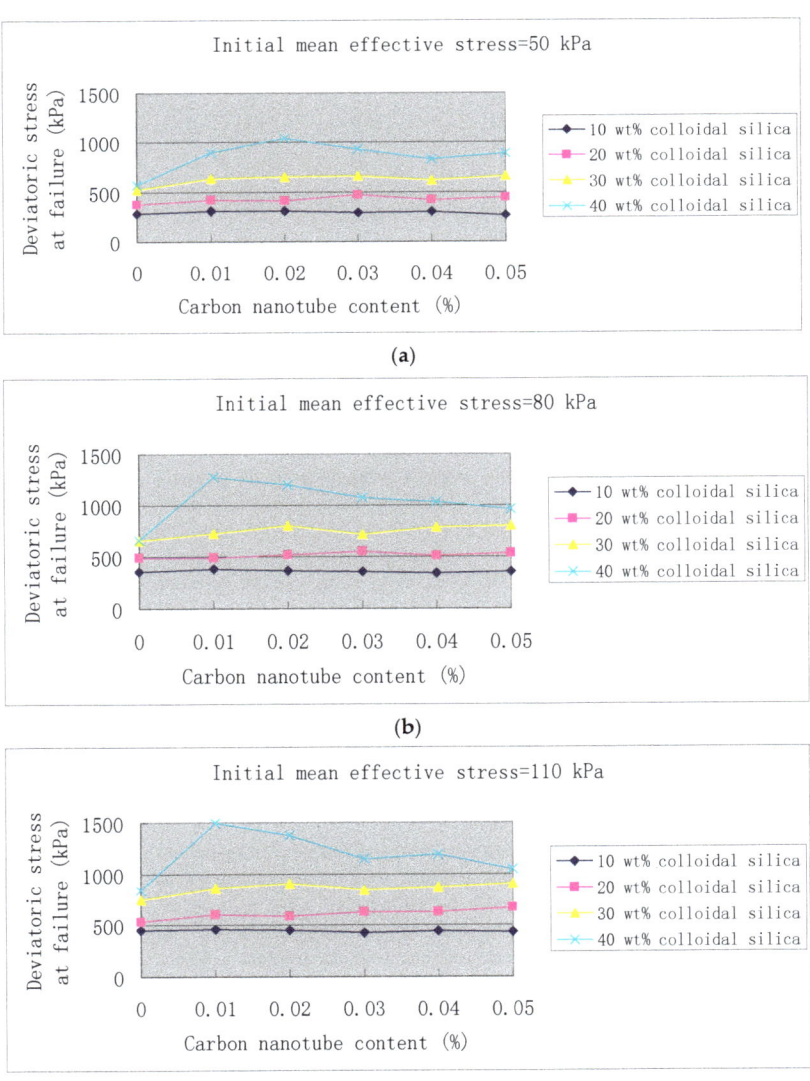

Figure 5. Deviatoric stress at failure vs. carbon nanotube content, and (a) initial mean effective stress= 50 kPa, (b) initial mean effective stress = 80 kPa, and (c) initial mean effective stress = 110 kPa.

Since the above-mentioned optimal carbon nanotube contents only appear in 40 wt. % colloidal silica, and does not appear in 10, 20 and 30 wt. % colloidal silica, we can conclude that there is a threshold for the concentration of colloidal silica, above which there exists the phenomenon of the optimal carbon nanotube content leading to the peak of shear strength. Obviously, 40 wt. % is above this threshold of colloidal silica concentration.

Table 5. Deviatoric stress at failure.

Colloidal Silica Concentration	Curing Period (Days)	Initial Mean Effective Stress (kPa)	Deviatoric Stress at Failure (kPa)					
			Carbon Nanotube 0 wt. %	Carbon Nanotube 0.01 wt. %	Carbon Nanotube 0.02 wt. %	Carbon Nanotube 0.03 wt. %	Carbon Nanotube 0.04 wt. %	Carbon Nanotube 0.05 wt. %
10 wt. %	3	50	282	310	305	292	302	262
10 wt. %	3	80	360	385	368	355	340	358
10 wt. %	3	110	452	461	454	425	440	434
20 wt. %	3	50	374	423	408	472	417	448
20 wt. %	3	80	497	492	523	555	510	535
20 wt. %	3	110	535	601	591	630	630	670
30 wt. %	3	50	524	636	650	656	620	660
30 wt. %	3	80	660	725	800	716	780	803
30 wt. %	3	110	752	860	905	840	870	902
40 wt. %	3	50	572	900	1050	935	830	890
40 wt. %	3	80	661	1280	1200	1070	1030	960
40 wt. %	3	110	842	1500	1382	1150	1190	1050

Previous literature has shown that, for Portland-cement-based matrix, with the increase of carbon nanotube content there are three possible trends in the compressive strength: (1) the compressive strength remains unchanged or even decreases [33,34]; (2) the compressive strength increases [35–42]; or (3) the compressive strength increases first and then decreases [43–45]. Our study shows similar trends in carbon-nanotube-reinforced sand-gel composites, and such trends depend on the concentrations of colloidal silica, as shown in Figure 5: (1) at 10 wt. % colloidal silica concentration, the shear strength even decreases with increasing carbon nanotube content; (2) when the colloidal silica concentration is 20 wt. % or 30 wt. %, the increase of carbon nanotube content causes the shear strength to increase; and (3) at 40 wt. % colloidal silica concentration, with the increase of carbon nanotube content, the shear strength first increases and then decreases (i.e., for a given horizontal stress, there is an optimal carbon nanotube content that leads to the peak shear strength during vertical compression).

Figure 6 shows the curves of the internal friction angle versus carbon nanotube content, and the corresponding data are shown in Table 6. Figure 7 shows the curves of cohesion versus carbon nanotube content, and the corresponding data are shown in Table 7. If the data point of 0.05% carbon nanotubes in 40 wt. % colloidal silica is not considered, then it can be concluded that as the carbon nanotube content increases, the internal friction angle first decreases and then increases, which reaches its minimum at 0.03% carbon nanotube content (see Figure 6). On the contrary, at 0.03% carbon nanotube content, cohesion reaches its maximum (see Figure 7). That is, as the content of carbon nanotubes increases, and the internal friction angle and cohesion have opposite development trends.

In the next section, for the case where there is an optimal carbon nanotube content in 40 wt. % colloidal silica, a formula reflecting the peak of shear strengthwith the optimal carbon nanotube content is established.

3.2. Shear Strength Model Reflecting Optimal Carbon Nanotube Content
3.2.1. Model Establishment

Here we present a formula for the effect of the optimal content of carbon nanotubes on the strength of sands stabilized with the mixture of colloidal silica and carbon nanotubes. Since the optimal carbon nanotube content which leads tothe peak shear strength is only observed in the case of 40 wt. % colloidal silica, while this effect of the optimal carbon nanotube content is not observed at lower colloidal silica concentrations (see Figure 5), we establish the formula only for sands stabilized by colloidal silica at the concentration of 40 wt. %.

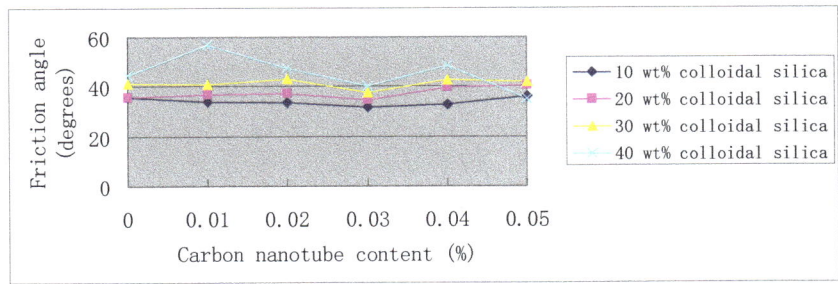

Figure 6. Friction angle vs. carbon nanotube content.

Table 6. Internal friction angle.

Colloidal Silica Concentration	Curing Period (Days)	Internal Friction Angle (Degrees)					
		Carbon Nanotube 0 wt. %	Carbon Nanotube 0.01 wt. %	Carbon Nanotube 0.02 wt. %	Carbon Nanotube 0.03 wt. %	Carbon Nanotube 0.04 wt. %	Carbon Nanotube 0.05 wt. %
10 wt. %	3	35.9	33.9	33.7	31.8	32.9	36.1
20 wt. %	3	35.8	36.8	37.4	34.8	39.8	40.6
30 wt. %	3	41.1	40.8	43	37.7	42.8	42.1
40 wt. %	3	44.3	56.7	47.3	40.2	48.6	34.9

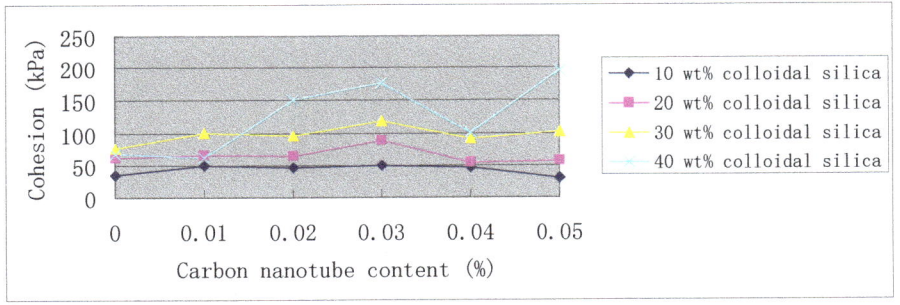

Figure 7. Cohesion vs. carbon nanotube content.

Table 7. Cohesion.

Colloidal Silica Concentration	Curing Period (Days)	Cohesion (kPa)					
		Carbon Nanotube 0 wt. %	Carbon Nanotube 0.01 wt. %	Carbon Nanotube 0.02 wt. %	Carbon Nanotube 0.03 wt. %	Carbon Nanotube 0.04 wt. %	Carbon Nanotube 0.05 wt. %
10 wt. %	3	35.1	49.1	47.1	50.1	46.2	30.8
20 wt. %	3	62.2	66.4	64.3	88.9	54.8	57.9
30 wt. %	3	77.1	100.5	96.3	119.4	91.2	103
40 wt. %	3	67.9	61.4	149.9	176.8	101	196.3

Suppose that the shear strength of cemented sand is composed of two parts, the basic part is caused by the combination of colloidal silica and sand particles, and the remaining part is caused by carbon nanotubes.

$$q_f = q_{f/CS} + q_{f/CNTs} \tag{1}$$

where q_f is the deviatoric stress at failure of cemented sand, $q_{f/CS}$ is the part caused by the combination of colloidal silica and sand, and $q_{f/CNTs}$ is the part caused by carbon nanotubes.

Next, we give the expressions of $q_{f/CS}$ and $q_{f/CNTs}$.

First, we express $q_{f/CS}$ as a function of friction angle and colloidal silica concentration, which is inspired by the expression of cemented sand [46].

$$q_{f/CS} = \frac{2\sin\varphi}{1-\sin\varphi} p'_{initial} + k_1 C_{CS} \tag{2}$$

where φ is the internal frictional angle of sands treated with 40 wt. % colloidal silica plus 0% carbon nanotubes, p'_{inital} is the initial mean effective stress, k_1 is the parameter relating the colloidal silica concentration to the deviatoric stress at failure, and C_{CS} is the colloidal silica concentration.

Then we provide a function of $q_{f/CNTs}$ to express the effect of carbon nanotubes on the shear strength by introducing two key parameters. The first key parameter is the optimal carbon nanotube content, which is denoted as C^o_{CNTs} since, as shown in Figure 5, for a given initial mean effective stress, there is an optimal carbon nanotube content to maximize the shear strength (deviatoric stress at failure) at 40 wt. % colloidal silica concentration. The second key parameter is the increment of deviatoric stress at failure with the optimal carbon nanotube content, which is denoted by q_Δ and expressed as follows

$$q_\Delta = q_{C_{CNTs}=C^o_{CNTs}} - q_{C_{CNTs}=0\%} \tag{3}$$

$$q_{f/CNTs} = \frac{2 q_\Delta \cdot C^o_{CNTs} \cdot C_{CNTs}}{(C_{CNTs})^2 + (C^o_{CNTs})^2} \tag{4}$$

where $q_{C_{CNTs}=C^o_{CNTs}}$ is the deviatoric stress at failure with the optimal carbon nanotube content, while $q_{C_{CNTs}=0\%}$ is the deviatoric stress at failure without carbon nanotubes. By introducing the above two key parameters, $q_{f/CNTs}$, which reflects the effect of carbon nanotubes on the shear strength, is expressed as follows

This is the first time that Equation (4) is designed in order to fulfill the requirement of describing the phenomenon that the shear strength first increases and then decreased with increasing carbon nanotube content in out tests.

Here is how we arrived at the idea to present Equation (4). As shown in Figure 8, with the increase of carbon nanotube content, the shear strength increases first and then decreases according to the test data. Therefore, in order to show the above characteristics, when we tried to express $q_{f/CNTs}$ (part of shear strength caused by carbon nanotubes) as a function of C_{CNTs} (carbon nanotube content), we found that the reciprocal of hyperbolic function can describe the characteristics of the peak shear strength. So we designed Equation (4) as the form of the reciprocal of hyperbolic function. Then Equation (4) can simultaneously describe the optimal carbon nanotube content and its corresponding peak shear strength.

3.2.2. Determination of Parameters

φ is the internal frictional angle of sand treated with 40 wt. % colloidal silica plus 0% carbon nanotubes, and the value of φ can be obtained from the triaxial compression tests.

In order to obtain the value of k_1, Equation (2) is rewritten as Equation (5) to calculate k_1. The variables on the right side of Equation (5) are all obtained from tests on specimens stabilized only with colloidal silica. Utilizing Equation (5), k_1 is obtained as the average value of tests with different initial mean effective stresses.

$$k_1 = \frac{1}{C_{CS}} \left(q_{f/CS} - \frac{2\sin\varphi}{1-\sin\varphi} p'_{initial} \right) \tag{5}$$

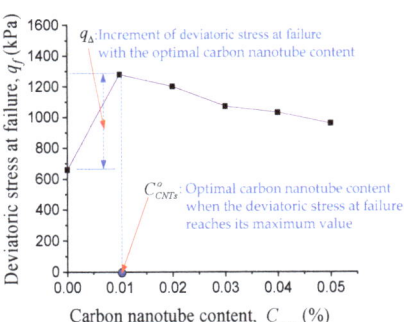

Figure 8. Schematic view of defining two key parameters in $q_{f/CNTs}$ (Equation (4)).

C^o_{CNTs} is the optimal carbon nanotube content when the deviatoric stress at failure reaches its maximum value. As shown in Figure 5, in the case of 40 wt. % colloidal silica, for the initial mean effective stresses of 50, 80 and 110 kPa, the deviatoric stresses at failure reaches their maximum values at the carbon nanotube contents of 0.02, 0.01 and 0.01 wt. %, respectively. We use Equation (6) to obtain C^o_{CNTs} in the prediction.

$$C^o_{CNTs} = \frac{m_1}{(p'_0)^{m_3}} + m_2 \quad (6)$$

where m_1, m_2, and m_3 are parameters. By curve fitting as shown in Figure 9, m_1, m_2 and m_3 are determined as 1.95×10^{11}, 9.99×10^{-5} and 9, respectively.

q_Δ is the increment of deviatoric stress at failure with the optimal carbon nanotube content. In specimens stabilized with 40 wt. % colloidal silica, for the same initial mean effective stress, substituting the values of $q_{C_{CNTs}=C^o_{CNTs}}$ and $q_{C_{CNTs}=0\%}$ from Figure 5 into Equation (3) yields the value of q_Δ. There are three values of q_Δ for the three different initial mean effective stresses $p'_{initial}$. We express q_Δ as a polynomial function of $p'_{initial}$, as shown in Equation (7), and Figure 10 shows that this polynomial form fits the test data well.

$$q_\Delta = a \times (p'_{initial})^2 + b \times p'_{initial} + c \quad (7)$$

where a, b, and c are parameters. By curve fitting, a, b, and c are determined as -0.057, 12.07, and 16.77, respectively.

Parameters used for the prediction of variation of shear strength with the content of carbon nanotubes are listed in Table 8.

3.2.3. Prediction of Shear Strength

The prediction of variations of deviatoric stresses at failure with the content of carbon nanotubes is in good agreement with the tests, as shown in Figure 11, reflecting the main characteristic of the effect of carbon nanotube content on the shear strength: there exists an optimal carbon nanotube content which leads to the peak shear strength. That is, by introducing two physically meaningful parameters (namely the optimal carbon nanotube content and the increment of deviatoric stress at failure with the optimal carbon nanotube content), our model depicts the phenomenon well: for the concentration of 40% colloidal silica, as the carbon nanotube content increases, the shear strength first increases and then decreases.

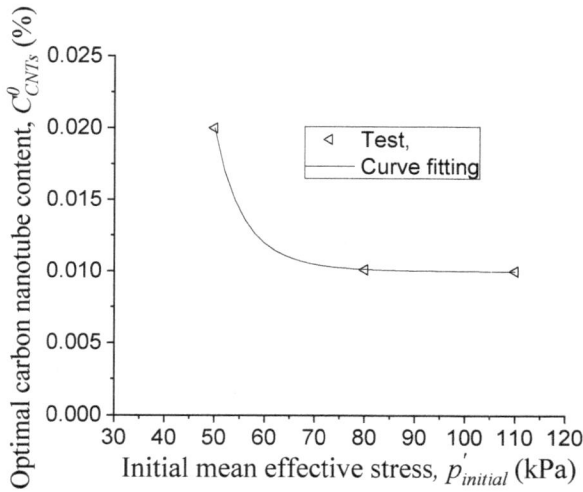

Figure 9. Optimal carbon nanotube content vs. initial mean effective stress.

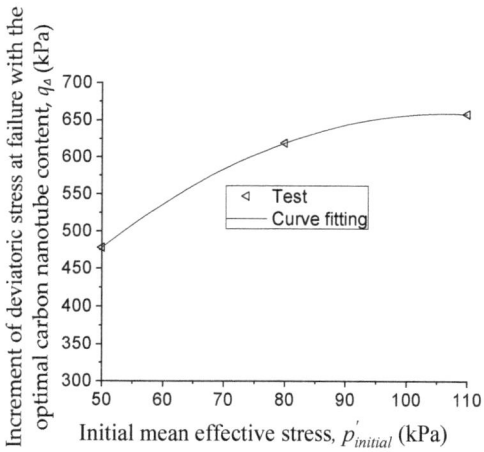

Figure 10. Increment of shear strength with the optimal carbon nanotube content vs. initial mean effective stress.

Table 8. Parameters for predicting the shear strength.

Internal Friction Angle, φ (Degrees)	k_1 (kPa)	Optimal Carbon Nanotube Content, C^o_{CNTs} (wt. %)			Increment of Deviatoric Stress at Failure, q_Δ (kPa)		
		m_1	m_2	m_3	a	b	c
44.3	802.8	1.95×10^{11}	9.99×10^{-5}	9	−0.057	12.07	16.77

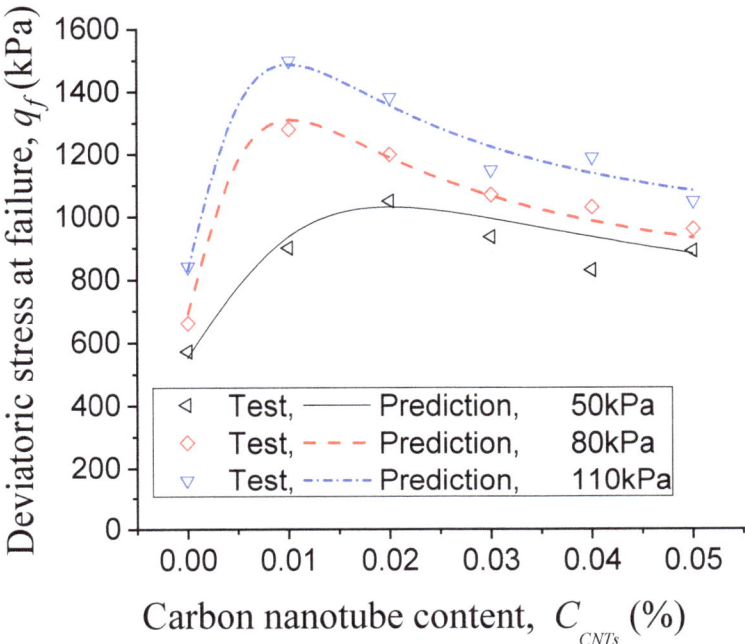

Figure 11. Effects of carbon nanotube content on the shear strength (sands seeped with 40 wt. % colloidal silica).

4. Microscale Analysis

Figure 12a,b shows the scanning electron microscope images of the stabilized sand with and without carbon nanotubes. Obviously, there are cracks in the silica gel around the sand particles. However, the cracks in the specimen stabilized by the mixture of colloidal silica and carbon nanotubes are not obviously less than that in the specimen without carbon nanotubes, which cannot definitely indicate that the addition of carbon nanotubes resulted in fewer cracks in the silica gel. Due to the uneven height caused by sand particles, it is difficult to directly focus on the cracks after continuing to zoom in and take pictures. So, it is difficult to observe the carbon nanotubes in the cracks at this time.

As a comparison, silica gel was formed by 40% colloidal silica without carbon nanotubes, whose cracks are shown in Figure 13. In order to further observe the carbon nanotubes in the silica gel cracks, silica gel was formed by dispersing 0.02% carbon nanotubes into 40% silica colloidal without sand, whose cracks are shown in Figure 14.

The phenomenon that carbon nanotubes act as bridges across cracks and inhibit cracking can be seen in Portland-cement-based matrix [29,31]. Similarly, Figure 14 shows that in the gel matrix, the carbon nanotubes are entangled into ropestobridge the crack. From the perspective of the micro-scale, the addition of carbon nanotubes results in ropes entangled by carbon nanotubes in the micro-cracks, and the corresponding macro phenomenon is that shear strength increases with the addition of carbon nanotubes.

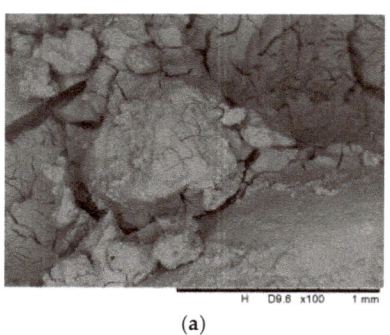 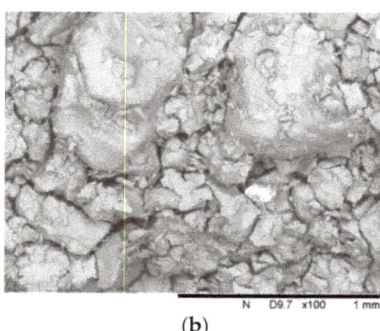

(a)   (b)

Figure 12. SEM photographs (100× enlargement) of stabilized sand with and without carbon nanotubes: (**a**) sand stabilized by 40 wt. % colloidal silica and (**b**) sand stabilized by 40 wt. % colloidal silica + 0.02% carbon nanotubes.

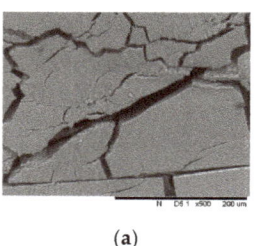 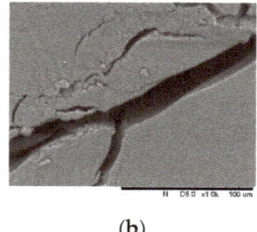 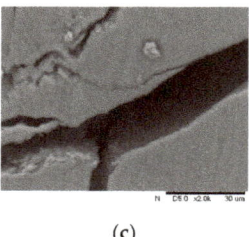

(a)   (b)   (c)

Figure 13. Cracks in silica gel (formed by 40 wt. % colloidal silica): (**a**) 500× enlargement, (**b**) 1000× enlargement, and (**c**) 2000× enlargement.

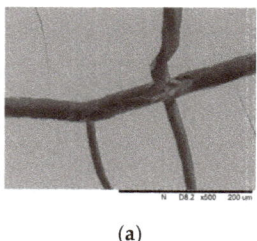 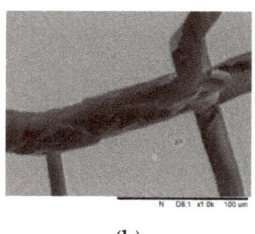 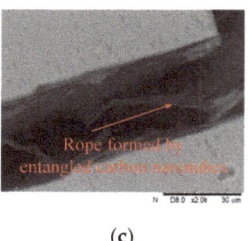

(a)   (b)   (c)

Figure 14. Cracks in silica gel (formed by 40 wt. % colloidal silica + 0.02% carbon nanotubes): (**a**) 500× enlargement, (**b**) 1000× enlargement, and (**c**) 2000× enlargement.

5. Conclusions

The study of sands seeped and stabilized with the mixture of colloidal silica and carbon nanotubes has allowed the authors to establish the following conclusions:

1. The degree of reinforcement of carbon nanotubes depends on the concentration of colloidal silica; when the same content of carbon nanotubes is added, shear strength increases with increasing concentration of colloidal silica. The low concentration (10 wt. %) of colloidal silica especially shows almost no reinforcement effect after adding carbon nanotubes. However, when the concentration of colloidal silica is increased to 40 wt. %, the shear strength, which can be represented by the deviator stress at failure, can be increased by up to 93.65% after adding 0.01 wt. % carbon nanotubes.
2. There is an optimal carbon nanotube content leading to the peak shear strength, which only occurs when the colloidal silica concentration is 40 wt. %. That is, there is a

concentration threshold of colloidal silica, above which the shear strength increases first and then decreases with increasing content of carbon nanotubes.
3. A formula is established to predict the peak effect of shear strength wellin 40 wt. % colloidal silica by introducing two physically meaningful parameters, which are the optimal carbon nanotube content and the increment of deviatoric stress at failure with the optimal carbon nanotube content.
4. SEM photos show that ropes formed by entangled carbon nanotubes result in enhancing macro strength.
5. The present work is helpful to the design and engineering application of sandy ground stabilization. That is, for the project to improve the strength of the sandy ground, based on the stress level, we present the optimal combination of 40 wt. % colloidal silica and 0.01–0.02 wt. % carbon nanotubes. Besides, the formula of shear strength that varies with carbon nanotube content can be applied to site reinforcement design.

In this paper, there are at least two limitations to the trends of the effects of carbon nanotube content on colloidal-silica-stabilized sand: (1) the carbon nanotubes used here are multi-walled carbon nanotubes, while we do not know how single-walled carbon nanotubes would affect the shear strength of colloidal-silica-stabilized sand; and (2) here we only use magnetically stirring and ultrasonic vibration to achieve the dispersion of carbon nanotubes in colloidal silica. However, different dispersion methods, such as treating carbon nanotubes with acid or using non-covalent surface modification for carbon nanotube surfaces [41], may lead to different trends of shear strength varying with increasing carbon nanotubes.

The guiding research question is how carbon nanotubes affect the shear strength of colloidal-silica-stabilized sand. For this research question, we performed triaxial tests on these carbon-nanotube-reinforced sand-gel composites and obtained the shear strength varying with different concentrations of colloidal silica combined with different carbon nanotube contents. In addition, the optimal ratio of colloidal silica to carbon nanotubes was obtained and a corresponding prediction formula was presented.

For future research, we can try single-walled carbon nanotubes and different methods of dispersing carbon nanotubes in the colloidal silica to obtain how dispersion methods affect the trends of the shear strength of the carbon-nanotube-reinforced colloidal-silica-stabilized sand.

Author Contributions: W.J., conceptualization, formal analysis, funding acquisition, methodology, supervision, and writing—original draft; Y.T., methodology, writing—review and editing; X.W., data curation, formal analysis, and writing—original draft; Z.G., funding acquisition, methodology. All authors have read and agreed to the published version of the manuscript.

Funding: This research was funded by the Natural Science Foundation of Zhejiang Province (Grant No. LHY19E090001) and the National Natural Science Foundation of China (Grant No. 51408547).

Institutional Review Board Statement: Not applicable.

Informed Consent Statement: Not applicable.

Data Availability Statement: All the test data for strength curves in Figures 5–7 are shown in Tables 5–7.

Conflicts of Interest: The authors declare no conflict of interest.

References

1. Gallagher, P.M.; Lin, Y. Colloidal silica transport through liquefiable porousmedia. *J. Geotech. Geoenviron. Eng.* **2009**, *135*, 1702–1712. [CrossRef]
2. Gallagher, P.M.; Conlee, C.T.; Rollins, K.M. Full-scale field testing of colloidal silica grouting for mitigation of liquefaction risk. *J. Geotech. Geoenviron. Eng.* **2007**, *133*, 186–196. [CrossRef]
3. Fujita, Y.; Kobayashi, M. Transport of colloidal silica in unsaturated sand: Effect of charging properties of sand and silica particles. *Chemosphere* **2016**, *154*, 179–186. [CrossRef] [PubMed]
4. Gallagher, P.M.; Pamuk, A.; Abdoun, T. Stabilization of liquefiable soils using colloidal silica grout. *J. Mater. Civ. Eng.* **2007**, *19*, 33–40. [CrossRef]

5. Agapoulaki, G.I.; Papadimitriou, A.G. Rheological Properties of Colloidal Silica Grout for Passive Stabilization Against Liquefaction. *J. Mater. Civ. Eng.* **2018**, *30*, 04018251. [CrossRef]
6. Saiers, J.E.; Hornberger, G.M.; Harvey, C. Colloidal silica transport through structured, heterogeneous porous media. *J. Hydrol.* **1994**, *163*, 271–288. [CrossRef]
7. Hamderi, M.; Gallagher, P.M. An optimization study on the delivery distance of colloidal silica. *Sci. Res. Essays* **2013**, *8*, 1314–1323.
8. Hamderi, M.; Gallagher, P.M.; Lin, Y. Numerical model for colloidal silica injected column tests. *Vadose Zone J.* **2014**, *13*, vzj2013.07.0138. [CrossRef]
9. Hamderi, M.; Gallagher, P.M. Pilot-scale modeling of colloidal silica delivery to liquefiable sands. *Soils Found.* **2015**, *55*, 143–153. [CrossRef]
10. Pamuk, A.; Gallagher, P.M.; Zimmie, T.F. Remediation of piled foundations against lateral spreading by passive sites tabilization technique. *Soil Dyn. Earthq. Eng.* **2007**, *27*, 864–874. [CrossRef]
11. Conlee, C.T.; Gallagher, P.M.; Boulanger, R.W.; Kamai, R. Centrifuge Modeling for Liquefaction Mitigation Using Colloidal Silica Stabilizer. *J. Geotech. Geoenviron. Eng.* **2012**, *138*, 1334–1345. [CrossRef]
12. Conlee, C.T. Dynamic Properties of Colloidal Silica Soils Using Centrifuge Model Testsanda Full-Scale Field Test. Ph.D. Thesis, Drexel University, Philadelphia, PA, USA, 2010.
13. Kodaka, T.; Oka, F.; Ohno, Y.; Takyu, T.; Yamasaki, N. Modeling of Cyclic Deformation and Strength Characteristics of Silica Treated Sand. In *Geomechanics: Testing, Modeling, and Simulation: Proceedings of the first Japan-U.S. Workshop on Testing, Modeling, and Simulation*; Yamamuro, J.A., Koseki, J., Eds.; Geotechnical Special Publication: Boston, MA, USA, 2005; No. 143; pp. 205–216.
14. Antonio-Izarrraras, V.M.; Díaz-Rodríguez, J.A.; Bandini, P.; López-Molina, J.A. Cyclic strength of a natural liquefiable sand stabilized with colloidal silica grout. *Can. Geotech. J.* **2008**, *45*, 1345–1355.
15. Kakavand, A.; Dabiri, R. Experimental study of applying colloidal nano Silica in improving sand-silt mixtures. *Int. J. Nano Dimens.* **2018**, *9*, 357–373.
16. Wong, C.; Pedrotti, M.; ElMountassir, G.; Lunn, R.J. A study on the mechanical interaction between soil and colloidal silica gel for ground improvement. *Eng. Geol.* **2018**, *243*, 84–100. [CrossRef]
17. Gallagher, P.M.; Mitchell, J.K. Influence of colloidal silica grout on liquefaction potential and cyclic undrained behavior of loose sand. *Soil Dyn. Earthq. Eng.* **2002**, *22*, 1017–1026. [CrossRef]
18. Mollamahmutoglu, M.; Yilmaz, Y. Pre-and post-cyclic loading strength of silica-grouted sand. *Geotech. Eng.* **2010**, *163*, 343–348. [CrossRef]
19. Persoff, P.; Apps, J.; Moridis, G.; Whang, J.M. Effect of dilution and contaminants on sand grouted with colloidal silica. *J. Geotech. Geoenviron. Eng.* **1999**, *125*, 461–469. [CrossRef]
20. Gavalas, V.G.; Andrews, R.; Bhattacharyya, D.; Bachas, L.G. Carbon nanotube sol-gel composite materials. *Nano Lett.* **2001**, *1*, 719–721. [CrossRef]
21. Wang, B.M.; Liu, S.; Han, Y.; Leng, P. Preparation and Durability of Cement-Based Composites Doped with Multi-Walled Carbon Nanotubes. *Nanosc. Nanotech. Lett.* **2015**, *7*, 411–416. [CrossRef]
22. Liu, Y.; Shi, T.; Zhao, Y.; Gu, Y.; Zhao, Z.; Chen, J.; Zheng, B. Autogenous shrinkage and crack resistance of carbon nanotubes reinforced cement based composites. *Int. J. Concr. Struct. Mater.* **2020**, *14*, 43. [CrossRef]
23. Cwirzen, A.; Habermehl-Cwirzen, K.; Penttala, V. Surface decoration of carbon nanotubes and mechanical properties of cement/carbon nanotube composites. *Adv. Cem. Res.* **2008**, *20*, 65–74. [CrossRef]
24. Li, G.Y.; Wang, P.M.; Zhao, X. Pressure-sensitive properties and microstructure of carbon nanotube reinforced cement composites. *Cem. Concr. Compos.* **2007**, *29*, 377–382. [CrossRef]
25. Nochaiya, T.; Chaipanich, A. Behavior of multi-walled carbon nanotubes on the porosity and microstructure of cement-based materials. *Appl. Surf. Sci.* **2011**, *257*, 1941–1945. [CrossRef]
26. Rana, S.; Fangueiro, R. A Review on Nanomaterial Dispersion, Microstructure, and Mechanical Properties of Carbon Nanotube and Nanofiber Reinforced Cementitious Composites. *J. Nanomater.* **2013**, *2013*, 71075.
27. Li, W.; Weiming, J.; Forood, T.I.; Yaocheng, W.; Gengying, L.; Yi, L.; Feng, X. Nano-Silica Sol-Geland Carbon Nanotube Coupling Effect on the Performance of Cement-Based Materials. *Nanomaterials* **2017**, *7*, 185. [CrossRef] [PubMed]
28. María, D.C.C.; Galao, O.; Baeza, F.; Zornoza, E.; Garcés, P. Mechanical properties and durability of CNT cement composites. *Materials* **2014**, *7*, 1640–1651.
29. Collins, F.; John, L.; Duan, W.H. The influences of admixtures on the dispersion, workability, and strength of carbon nanotube-OPC paste mixtures. *Cem. Concr. Compos.* **2012**, *34*, 201–207. [CrossRef]
30. Han, B.; Yu, X.; Ou, J. *Multifunctional and Smart Carbon Nanotube Reinforced Cement-Based Materials*; Springer: Berlin/Heidelberg, Germany, 2011; pp. 1–47.
31. Xu, S.; Liu, J.; Li, Q. Mechanical properties and microstructure of multi-walled carbon nanotube-reinforced cement paste. *Constr. Build. Mater.* **2015**, *76*, 16–23. [CrossRef]
32. Musso, S.; Tulliani, J.M.; Ferro, G. Influence of carbon nanotubes structure on the mechanical behavior of cement composites. *Compos. Sci. Tech.* **2009**, *69*, 1985–1990. [CrossRef]
33. Pavlopoulou, E.E.; Georgiannou, V.N. Effect of Colloidal Silica Aqueous Gel on the Monotonic and Cyclic Response of Sands. *J. Geotech. Geoenviron. Eng.* **2021**, *147*, 04021122. [CrossRef]

34. Triantafyllos, P.K.; Georgiannou, V.N.; Pavlopoulou, E.; Dafalias, Y.F. Strength and Dilatancy of Sand before and after Stabilisation with Colloidal-Silica Gel. *Géotechnique* **2021**, 1–15, Published Online: 18 February. [CrossRef]
35. Krishnan, J.; Sharma, P.; Shukla, S. Experimental Investigations on the Mechanical Properties of Sand Stabilized with Colloidal Silica. *Iran. Sci. Technol. Trans. Civ. Eng.* **2021**, *45*, 1737–1758. [CrossRef]
36. Vranna, A.; Tika, T.; Papadimitriou, A. Laboratory investigation into the monotonic and cyclic behaviour of a clean sand stabilised with colloidal silica. *Géotechnique* **2020**, 1–14, Published online: 14 December. [CrossRef]
37. Ghadr, S.; Assadi-Langroudi, A.; Hung, C.; Bahadori, H. Stabilization of Sand with Colloidal Nano-Silica Hydrosols. *Appl. Sci.* **2020**, *10*, 5192. [CrossRef]
38. Li, G.Y.; Wang, P.M.; Zhao, X. Mechanical behavior and microstructure of cement composites incorporating surface-treated multi-walled carbon nanotubes. *Carbon* **2005**, *43*, 1239–1245. [CrossRef]
39. Nochaiya, T.; Tolkidtikul, P.; Singjai, P. Microstructure and characterizations of Portland—carbon nanotubes pastes. *Adv. Mater. Res.* **2008**, *55*, 549–552. [CrossRef]
40. Yu, X.; Kwon, E. A Carbon nanotube/cement composite with piezoresistive properties. *Smart Mater. Struct.* **2009**, *18*, 055010. [CrossRef]
41. Torkittikul, P.; Chaipanich, A. Bioactivity properties of white Portland cement paste with carbon nanotubes. In Proceedings of the INEC 2010 3rd International Nanoelectronics Conference, HongKong, China, 3–8 January 2010; pp. 838–839.
42. Xu, S.L.; Gao, L.L.; Jin, W.J. Production and mechanical properties of aligned multiwalled carbon nanotubes-M140 composites. *Sci. Chin. Ser. E Tech. Sci.* **2009**, *52*, 2119–2127. [CrossRef]
43. Hu, S.; Xu, Y.; Wang, J.; Zhang, P.; Guo, J. Modification effects of carbon Nanotube dispersion on the mechanical properties, pore structure, and microstructure of cement mortar. *Materials* **2020**, *13*, 1101. [CrossRef]
44. Morsy, M.S.; Alsayed, S.H.; Aqel, M. Hybrid effect of carbon nanotube and nano-clay on physico-mechanical properties of cement mortar. *Constr. Build. Mater.* **2011**, *25*, 145–149. [CrossRef]
45. Li, W.W.; Wei-Ming, J.; Yao-Cheng, W.; Yi, L.; Ruo-Xu, S.; Feng, X. Investigation on the Mechanical Properties of a Cement-Based Material Containing Carbon Nanotube under Drying and Freeze-Thaw Conditions. *Materials* **2015**, *8*, 8780–8792. [CrossRef] [PubMed]
46. Schnaid, F.; Prietto, P.D.M.; Consoli, N.C. Characterization of cemented sand in triaxial compression. *J. Geotech. Geoenviron. Eng.* **2001**, *127*, 857–868. [CrossRef]

Article

Development of Multi-Scale Carbon Nanofiber and Nanotube-Based Cementitious Composites for Reliable Sensing of Tensile Stresses

Shama Parveen [1,*], Bruno Vilela [2], Olinda Lagido [2], Sohel Rana [1,*] and Raul Fangueiro [3,*]

1. Department of Fashion and Textiles, School of Arts and Humanities, University of Huddersfield, Huddersfield HD1 3DH, UK
2. Department of Civil Engineering, University of Minho, Campus de Azurém, 4800-058 Guimarães, Portugal; bmnvilela@gmail.com (B.V.); olinda.lagido@live.com.pt (O.L.)
3. Department of Mechanical Engineering, University of Minho, Campus de Azurém, 4800-058 Guimarães, Portugal
* Correspondence: parveenshama2011@gmail.com (S.P.); S.Rana@hud.ac.uk (S.R.); rfangueiro@dem.uminho.pt (R.F.)

Abstract: In this work, multi-scale cementitious composites containing short carbon fibers (CFs) and carbon nanofibers (CNFs)/multi-walled carbon nanotubes (MWCNTs) were studied for their tensile stress sensing properties. CF-based composites were prepared by mixing 0.25, 0.5 and 0.75 wt.% CFs (of cement) with water using magnetic stirring and Pluronic F-127 surfactant and adding the mixture to the cement paste. In multi-scale composites, CNFs/MWCNTs (0.1 and 0.15 wt.% of cement) were dispersed in water using Pluronic F-127 and ultrasonication and CFs were then added before mixing with the cement paste. All composites showed a reversible change in the electrical resistivity with tensile loading; the electrical resistivity increased and decreased with the increase and decrease in the tensile load/stress, respectively. Although CF-based composites showed the highest stress sensitivity among all specimens at 0.25% CF content, the fractional change in resistivity (FCR) did not show a linear correlation with the tensile load/stress. On the contrary, multi-scale composites containing CNFs (0.15% CNFs with 0.75% CFs) and MWCNTs (0.1% MWCNTs with 0.5% CFs) showed good stress sensitivity, along with a linear correlation between FCR and tensile load/stress. Stress sensitivities of 6.36 and 11.82%/MPa were obtained for the best CNF and MWCNT-based multi-scale composite sensors, respectively.

Keywords: multi-scale composites; carbon fibers; cement; carbon nanotubes; stress sensing

1. Introduction

Cementitious composites are extensively used in civil infrastructures and are susceptible to deterioration of their properties over time. Therefore, health monitoring of cement-based buildings and infrastructures at periodic intervals is an important requirement to ensure the safety of the occupants, as well as to extend the lifespan of the infrastructures. The monitoring of real-time conditions and performance of structures, which is known as structural health monitoring (SHM), is performed mainly in the critical zones of the structures using various sensors [1,2]. The collected data are used to evaluate the health conditions of structures in order to take timely maintenance actions. SHM is frequently performed using various sensors such as optical fiber sensors, electrical resistance strain gauges, piezoelectric (PZT) ceramics, etc., each one of which has their own limitations [1,2] and, consequently, a great deal of research is currently underway to find an affordable, reliable and easy-to-use technique for SHM of civil infrastructures.

From past few years, investigations on the piezoresistive cementitious composites (i.e., composites which show change in their electrical resistivity with mechanical stress/strain) for SHM applications have accelerated considerably [1–3]. Piezoresistive cementitious sensors have better compatibility with civil structures and are durable [1–3]. These sensors were initially developed using short carbon fibers (SCFs) [4,5]. However, researchers are currently utilizing various electrically conductive nanofillers to introduce piezoresistivity into cementitious composites [6–8]. Nanomaterials are preferred over carbon fibers (CFs), as they are required at much lower concentrations and provide a positive influence on other properties of cementitious composites (e.g., mechanical properties, microstructure, thermal properties, etc.) due to their high surface area and aspect (i.e., length/diameter) ratio [9–11]. Extensive studies have been carried out to date on developing piezoresistive cementitious composites using different nanomaterials such as multi-wall carbon nanotubes (MWCNTs), graphene, nano graphite platelets, spiky spherical nickel powders containing nano tip, carbon nano fibers (CNF), nano carbon black (NCB), etc. [6,12,13].

Formation of a percolating electrical network is required to achieve piezoresistive properties in cementitious composites. Percolation threshold is defined as the critical concentration of conductive fillers to enable non-conductive cementitious matrices to show conductivity [7]. The percolation threshold of CFs (5 mm long and 10 μm diameter) within a cementitious matrix was found to be between 0.5 (~0.3 wt.%) and 1 vol.% (~0.58 wt.%) [4,5]. The percolation threshold of MWCNT in cementitious composites was also found in the similar range (between 0.3 and 0.6 wt.% of cement) [14]. However, according to Yoo et al., the optimum concentration of MCNTs to introduce piezoresistivity into cementitious matrices was found to be 1 wt.% (with respect to cement) [7,8]. For graphene-based cementitious composites, the percolation threshold was found to be between 1 and 5 wt.% of graphene and the resulting composites showed good piezoresistive properties [12]. However, MWCNTs were considered as superior and more effective nanofillers for the fabrication of piezoresistive cementitious composites when compared to graphite nanofibers and graphene when used in similar concentrations [8]. More recently, the use of hybrid conductive fillers, i.e., the combination of two different fillers proved more effective in achieving superior conductivity and sensing properties in cementitious materials (these composites are known as multi-scale composites, as they are developed using hybrid reinforcements with micro- and nano-scale diameters [9]). For example, the use of CFs (15 mm in length and 5–7 μm in diameter) in combination with MWCNTs improved the stability of electrical resistivity of cementitious composites [14]. Reliable sensing of compressive loads and strains of cementitious composites was also achieved with hybrid conductive fillers composed of CFs and MWCNTs [15]. Cementitious composites containing these hybrid fillers demonstrated superior repeatability of sensing results when compared to CF-based cementitious composites [14]. Zhang et al. recently reported that a hybrid filler system containing CNT/NCB (40:60) showed a percolation threshold of 0.39 to 1.49 vol.% (of mortar) and the resulting cementitious composites demonstrated a stable and sensitive piezoresistive property [16]. The observed percolation threshold and piezoresistivity of CNTs (and other nanomaterials), CFs and hybrid fillers in cementitious composites were different in different studies and this was attributed to the use of different types of nanomaterials and CFs (possessing different diameters and aspect ratios), their dispersion states, as well as different cement/water ratios and compositions used for the development of cementitious composites. Besides piezoresistivity, the hybrid filles were also found to be effective in improving the physical and mechanical properties of cementitious composites. For example, the use of 0.25 wt.% SCFs (of cement) with 0.75 wt.% of MWCNTs (of cement) improved the flexural strength by ~243%, flexural modulus by 200% and toughness by 672% of plain cement-based composites [17]. Also, cementitious composites with 2.25 wt.% SCF and 0.5 wt.% MWCNTs improved the tensile strength of plain cement composites by ~53%, tensile modulus by 60% and failure strain by 44% [18].

Due to the growing interest and prospect of piezoresistive cementitious composites in the civil engineering sector, a cost-effective and an easy fabrication method to develop these composites is highly desirable. Although cementitious composites containing hybrid fillers showed superior results, only a few studies have been carried out to date. Also, to the best of the authors' knowledge, the existing studies investigated the sensing properties of hybrid cementitious composites under compression loading and no study has been conducted to date under tensile loading mode. The present study, therefore, investigated and compared the piezoresistive properties of CF-reinforced, CF-MWCNT and CF-CNF hybrid filler-reinforced cementitious composites under tensile loading. CNFs and MWCNTs were selected as the conductive filles for developing these hybrid cementitious composites due to their high electrical conductivity (possessing electrical resistivity as low as 1×10^{-4} and $2 \times 10^{-3} - 1 \times 10^{-4}$ Ωcm, respectively), relatively low cost when compared to other conductive nanomaterials such as single-walled CNTs, graphene, etc., high aspect ratio (250–2000 and 100–10,000, respectively), as well as their high mechanical properties (tensile strength of 2.92 and 10–60 GPa, respectively and tensile modulus of 240 and 1000 GPa, respectively] [19]. The comparison of the piezoresistive behavior of CF-MWCNT and CF-CNF hybrid filler-reinforced cementitious composites has also not been addressed in the existing literature. Moreover, a non-ionic surfactant, Pluronic F-127, was used for the first time to ensure proper dispersion of the fillers in the developed sensing cementitious composites. The piezoresistive properties of the composites were studied under cyclic tensile loading at different loading conditions. The fractional change in resistivity (FCR), its correlation with the applied load and stress sensitivity were determined and discussed in detail.

2. Materials and Methods

2.1. Raw Materials

Cementitious composite specimens were fabricated using the Portland cement CEM I 42.5 R (purchased by Lisbon, Secil, Portugal). The properties of this cement are summarized in Table 1. Short CFs (Tenax®, diameter: 7.0 μm, Length: 5 mm) were supplied by Teijin Carbon Europe GmbH (Wuppertal, Germany) and MWCNTs and CNFs were supplied by Nanostructured & Amorphous Materials, Inc. (Houston, TX, USA). Their physical and mechanical properties are summarized in Table 2. CNFs had an electrical conductivity of more than 100 S/cm. CNFs may contain significant amount of amorphous carbon, as well as residual catalysts and other inorganic impurities such as Fe, Co, S, etc. and a trace amount of Mg, Cl, Ca, Cr, etc. [20] and these residual catalyst particles and other impurities can significantly influence the electrical conductivity of CNFs. Pluronic F-127 (the chemical structure is provided in Figure 1a), a non-ionic surfactant, was used to disperse MWCNTs/CNFs and CFs in water and was purchased from Sigma Aldrich (Algés, Portugal). A defoamer, tri-butyl phosphate (the chemical structure is provided in Figure 1b), was supplied by Acros Organics (Thermo Fischer Scientific, Porto Salvo, Portugal).

Table 1. Composition and properties of cement used in the present study.

	Composition	95–100% Clinker + 0–5% Minor Additional Components
Ordinary Portland Cement (CEM I 42.5 R) [1]	Loss on ignition	≤5%
	Insoluble residue	≤5%
	Sulphur trioxide (SO_3)	≤4.0%
	Chloride (Cl^-)	≤0.1%
	Initial setting time	≥60 min
	Soundness	≤10 mm
	2 days compressive strength	≥20.0 MPa
	28 days compressive strength	≥42.5 MPa ≤ 62.5 MPa

[1] Source:www.secil.pt, accessed on 12 January 2013, Lisbon, Portugal

Table 2. Properties of carbon fibers, CNFs, MWCNTs, Pluronic F-127 and defoamer.

Materials	Physical and Mechanical Properties				
Tenax-e HTA40 E13 6k 400 tex	Tensile Strength (MPa)	Tensile Modulus (GPa)	Elongation at Break (%)	Filament Diameter(μm)	Density(g/cm^3)
	4100	240	1.7	7	1.77
	Physical Properties				
	Diameter (nm)		Length (μm)	Surface Area (m^2/g)	Purity (%)
	Inside	Outside			
MWCNT [1]	2–5	<8	10–30	350–420	>95%
CNF		200–600 nm	5–50 μm	>18	>70 wt.%, Ash: <5 wt.%
Pluronic F-127	Non-ionic surfactant, molecular weight: 12,500 g/mol, CMC: 950–1000 ppm				

[1] Source: Nanostructured & Amorphous Materials, Inc., Katy, TX, USA.

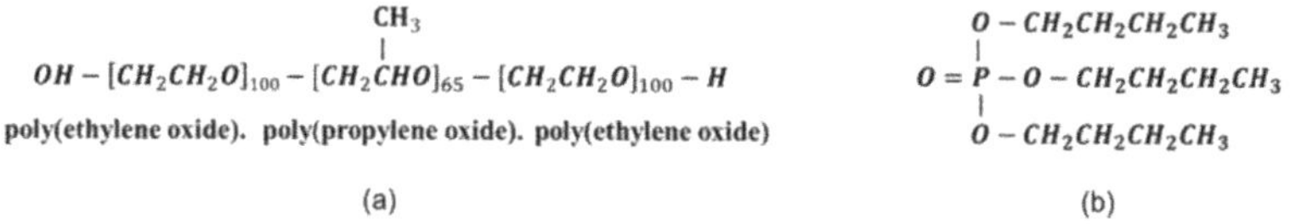

$$\begin{array}{c} CH_3 \\ | \\ OH - [CH_2CH_2O]_{100} - [CH_2CHO]_{65} - [CH_2CH_2O]_{100} - H \end{array}$$

poly(ethylene oxide). poly(propylene oxide). poly(ethylene oxide)

(a)

$$\begin{array}{c} O - CH_2CH_2CH_2CH_3 \\ | \\ O = P - O - CH_2CH_2CH_2CH_3 \\ | \\ O - CH_2CH_2CH_2CH_3 \end{array}$$

(b)

Figure 1. Chemical structure of (**a**) Pluronic F-127 and (**b**) Tri butyl Phosphate.

2.2. Characterization of Morphology of Carbon Nanofibers and Nanotubes

Scanning Electron Microscopy (FEG-SEM, NOVA 200 Nano SEM, FEI, Hillsboro, Oregon, USA) at an acceleration voltage of 10 kV was used to study the morphology of MWCNTs and CNFs. To avoid the charging of samples during SEM, they were coated with a 30 nm film of Au-Pd in a high-resolution sputter coater (208HR Cressington, Watford, UK).

2.3. Preparation of Aqueous Suspensions of Carbon Fibers, Carbon Nanofibers and Nanotubes

The schematic diagram, showing the preparation of various aqueous suspensions, is shown in Figure 2. The aqueous suspensions of CF, using 5 wt.% of Pluronic F-127 (on the weight of water), were prepared by first mixing Pluronic F-127 in water with the help of magnetic stirring for 10 min. CFs were then added in the surfactant solution and mixed with the help of magnetic stirring for another 10 min. In case of aqueous suspensions containing CFs, along with MWCNTs or CNFs, MWCNT or CNF powder was first added to the surfactant solution and then magnetic stirring was carried out for 10 min to ensure that there were no big lumps of MWCNTs/CNFs in the aqueous suspensions. The MWCNT/CNF surfactant suspensions were then kept in a bath ultrasonicator (Sonica Ultrasonicator 3200 S3, Milan, Italy) operated at 40 kHz frequency and 180 W power for 1 h. After removing the MWCNT/CNF suspensions from the ultrasonicator, CFs were added and mixed using magnetic stirring for 10 min. The defoamer (in the weight ratio of 1:0.5 with respect to Pluronic F-127) was then added to the suspensions, which were used later for the fabrication of cementitious composites. For the characterization of CNT/CNF dispersion in aqueous suspensions, the defoamer and CFs were not added to avoid film formation by the defoamer and agglomeration caused by CFs during the characterization of aqueous suspensions. Figure 2 shows the magnetic stirring and ultrasonication processes of aqueous suspensions and Figure 3c shows a suspension containing MWCNTs and CFs.

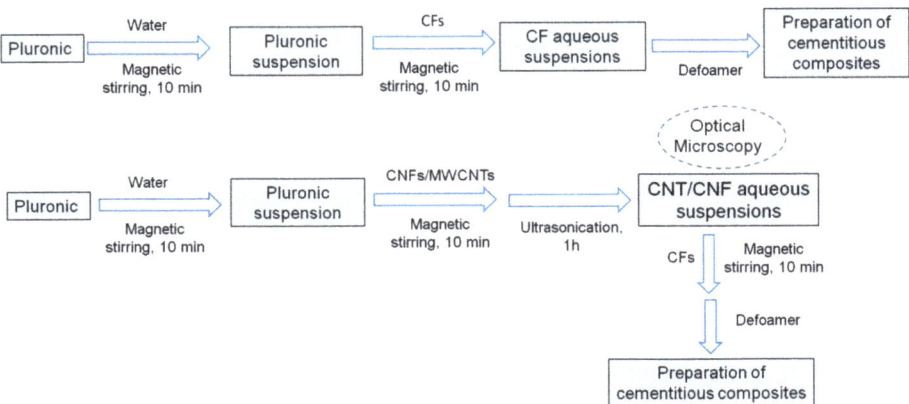

Figure 2. Schematic diagram showing the preparation of aqueous suspensions for fabrication of cementitious composites.

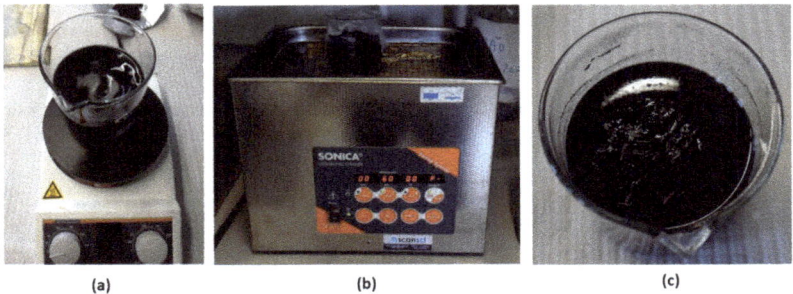

Figure 3. Preparation of aqueous suspensions: (**a**) magnetic stirring, (**b**) ultrasonication and (**c**) the aqueous suspension containing CFs and MWCNTs.

2.4. Characterization of Aqueous Suspensions

Optical microscopy (Olympus BH2 Microscope, Hamburg, Germany) was used to identify the CNT/CNF agglomerates in the aqueous suspensions. This characterization was performed to also study the homogeneity of the prepared suspensions. To carry out the optical microscopic analysis, a drop of MWCNT/CNFs (without CF) suspension was placed on a glass slide and covered with a cover slip. The images were captured in 5 different places of the drop. The analysis was repeated for 3 times and the images were captured in two different magnifications to clearly understand the quality of the prepared suspensions.

2.5. Preparation of Cementitious Composites

Cementitious composites were fabricated using aqueous suspensions of CF or CF with MWCNTs/CNFs (315 mL) and cement (900 g) following EN 196-1:2006 standard. A set of unreinforced samples, i.e., plain mortars, were also prepared using water (315 mL) to compare with the reinforced cementitious composites. The weighed amount of cement was mixed with the aqueous suspensions using a Hobart mixer; the mixer was set for 1.5 min at a slow speed (140 ± 5 rpm) and then 1.5 min at a high speed (285 ± 10 rpm). The mixtures were then poured into the molds (three samples were prepared for each mixture). The specimens were prepared in a dog-bone shaped mold (having a cross section of 30 mm × 20 mm, the distance between the inner grid for voltage measurement was 70 mm and the outer grid for passing current was 80 mm), as shown in Figure 4, in order to perform the piezoresistive measurement. The grids were made of copper foils having 30 mm × 15 mm dimension.

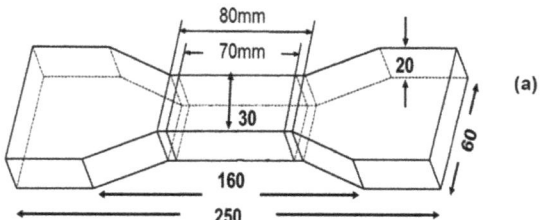

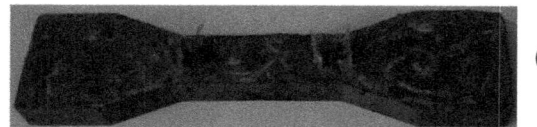

Figure 4. (**a**) Dimension of the dog bone samples. (**b**) Samples used for piezoresistive characterization.

The molds were placed on a jolting machine for 30 s to remove the entrapped air. The molded samples were then covered by cellophane and placed in a chamber with a moist atmosphere for 24 h. The samples were demolded after 24 h and kept submerged in water for 28 days at 25 °C. The samples were taken out 4 h before the test, wiped with a cotton cloth and kept at the room temperature prior to the piezoresistive characterization. The compositions of different samples prepared are listed in Table 3.

Table 3. Composition of different samples prepared for piezoresistive characterization (each sample contains 900 g of cement).

Samples	CF (wt.% of Cement)	CNF (wt.% of Cement)	MWCNT (wt.% of Cement)
Plain Mortar	0	0	0
0.25 CF	0.25	0	0
0.5% CF	0.50	0	0
0.75% CF	0.75	0	0
0.25% CF 0.1% CNF	0.25	0.10	0
0.5% CF 0.1% CNF	0.50	0.10	0
0.75% CF 0.1% CNF	0.75	0.10	0
0.25% CF 0.15% CNF	0.25	0.15	0
0.5% CF 0.15% CNF	0.50	0.15	0
0.75% CF0.15% CNF	0.75	0.15	0
0.25% CF 0.1% CNT	0.25	0	0.10
0.5% CF 0.1% CNT	0.50	0	0.10
0.75% CF 0.1% CNT	0.75	0	0.10
0.25% CF 0.15% CNT	0.25	0	0.15
0.5% CF 0.15% CNT	0.50	0	0.15
0.75% CF 0.15% CNT	0.75	0	0.15

2.6. Characterization of Piezoresistive Properties

The test setups for the characterization of the electrical resistance and stress sensing properties of cementitious composites are shown in Figure 5. Stress sensing properties were characterized by measuring the electrical resistivity of samples using a digital multimeter (Agilent 34460a, Santa Clara, CA, USA) in the elastic regime of tensile loading with three different loading rates: 20, 30 and 40 N.s^{-1}. For each loading rate, the load was increased from 0 N up to 10 s and then decreased to 0 N at the same unloading rate. 5 cycles of loading and unloading (20 s per cycle) were studied at each loading rate to verify the repeatable performance of the developed cementitious composites. The DC electrical resistance was measured simultaneously during the mechanical testing using a four-probe method. Fractional change in the resistivity (FCR) and stress sensitivity of the composites in each cycle were calculated using the following equations:

$$FCR = \frac{final\ resistivity\ (\rho) - initial\ resistivity\ (\rho_0)}{initial\ resistivity (\rho_0)} \quad (1)$$

$$Stress\ Sensitivity = \frac{100\ x\ FCR}{Applied\ tensile\ stress} \quad (2)$$

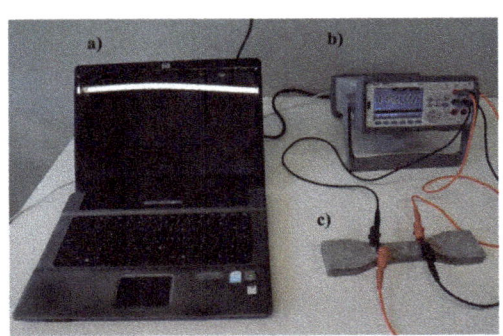

Figure 5. Measurement of (**a**) electrical resistance and (**b**,**c**) piezoresistive property of cementitious composites (**d**) piezoresistive property of cementitious composites in tensile mode.

2.7. Microstructural Characterization of Developed Cementitious Composites

The fractured surfaces of developed specimens were analyzed by using SEM (FEG-SEM, NOVA 200 Nano SEM, FEI) using the secondary electron mode and an acceleration voltage of 10 kV after coating with a thin film (30 nm) of Au-Pd in a high-resolution sputter coater (208HR Cressington, Watford, UK).

3. Results

3.1. Morphology of Carbon Fibers, Nanofibers and Nanotubes

The image of short CFs is shown in Figure 6a. The SEM micrographs of CNF and MWCNTs are shown in Figure 6b,c, respectively. Significant entanglements or agglomeration can be observed in the case of CNFs. However, MWCNTs showed the highest degree of agglomeration and clustering in the powder, as can be seen from Figure 6c. Transmission electron microscope (TEM) images of MWCNT aqueous suspensions also showed clustering and entanglements of nanotubes, as shown in Figure 6d. Therefore, to break these MWCNT/CNF agglomerates and disperse them homogeneously within the cementitious matrix, a combination of magnetic stirring (10 min) and ultrasonication (1 h) was used.

Figure 6. Reinforcement of cementitious composites: (**a**) short CFs, (**b**) SEM micrograph of CNFs and (**c**) SEM micrograph MWCNTs and (**d**) TEM image of MWCNTs, taken from manufacturer website.

3.2. Aqueous Suspensions of MWCNT and CNT

The optical micrographs of aqueous suspensions of MWCNT and CNF are shown in Figure 7. It is clear from Figure 7a that MWCNTs could be homogeneously dispersed in water using a Pluronic F-127-assisted ultrasonication process. MWCNTs were dispersed without any noticeable agglomeration. Homogeneous dispersion of CNTs and CNFs is prerequisite for developing high performance cementitious composites [9] and the use of Pluronic F-127 was proven to be effective in achieving homogenous CNT dispersion in previous studies also [10,11]. It can also be clearly observed from Figure 7c that dispersed MWCNTs formed electrically conductive pathways within the aqueous medium. The aqueous suspension of CNF also showed homogeneous dispersion free from noticeable CNF agglomerates. To the best of authors' knowledge, Pluronic F-127 has been utilized to disperse CNFs for the first time in the present study and a homogeneous dispersion was obtained due to the steric stabilization induced by Pluronic F-127 molecules, as previously reported for CNTs [10,11]. However, in this case, a few CNFs were seen to form bundles with each other and formed a greater number of noticeable electrically conductive pathways when compared to MWCNTs within the aqueous medium, as can be seen from Figure 7d.

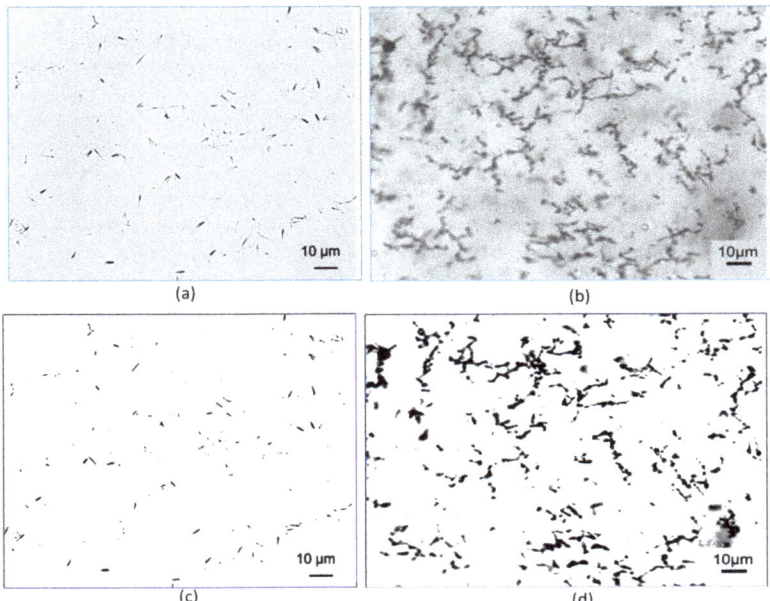

Figure 7. Optical Micrographs of the aqueous suspensions of (**a**) 0.2% MWCNT% with 5 wt.% Pluronic F-127, (**b**) 0.2% CNF with 5 wt.% Pluronic F-127, (**c,d**) are the high contrast micrographs of (**a,b**), respectively, showing the conductive networks formed by CNFs and MWCNTs.

3.3. Electrical Resistance of Cementitious Composites

The electrical resistivity of mortar containing only CF and mortar containing CF and different concentrations of CNF and MWCNT is presented in Figure 8. It can be noticed that mortars containing 0.25 wt.% CF had an electrical resistivity of 3.7 Ω.m. The electrical resistivity decreased with the increase in the CF wt.% and the sample containing 0.75 wt.% CF showed an electrical resistivity of 0.6 Ω.m. The electrical resistivity obtained in this case was lower when compared to previously reported mortar samples containing CF, carbon black and other nanomaterials [21–24].

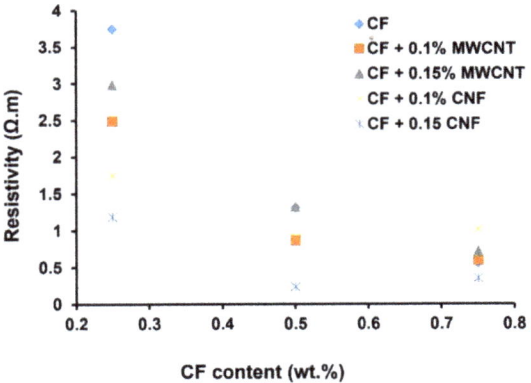

Figure 8. Influence of CF, CNF and MWCNT on the electrical resistance of cementitious composites.

The use of Pluronic F-127 in the present study was believed to improve the dispersion of CFs within the mortar paste, leading to the reduced clustering of CFs and resulting in a lower porosity of the reinforced cementitious composites. This significantly improved the electrical conductivity of cementitious composites. It is clear from Figure 8 that the addition of MWCNT and CNF significantly reduced the electrical resistivity of CF-reinforced cementitious composites. This was attributed to the fact that the highly conducting MWCNTs and CNFs can bridge the conducting paths formed by CFs and, hence, improved the conductivity of composites [15]. It can also be observed that composites with CNFs at 1.5 wt.% showed a lower resistivity when compared to composites prepared with MWCNTs. This could be due to the better formation of conductive paths with CNFs due to their larger dimensions, as also observed in Figure 7d. Composites with MWCNTs, on the other hand, showed a higher resistivity at the higher concentration, i.e., 1.5 wt.%, probably due to formation of CNT agglomerates, which resulted in the increased porosity and higher resistivity of composites. This observation agrees with the previous studies on CNT-reinforced cementitious composites [25].

3.4. Response of Cementitious Composites to Cyclic Tensile Loading

3.4.1. Effect of Cyclic Tensile Loading on Electrical Resistivity

The electrical response of CF-reinforced cementitious composites containing 0.25, 0.5 and 0.75 wt.% CF to cyclic tensile loading is shown in Figure 9. These responses were achieved at five loading-unloading cycles at three different loading rates: 20, 30 and 40 N/s. Each loading and unloading cycle took 10 s and therefore, the loading cycles reached 200, 300 and 400 N, respectively, for these three loading rates and came back to 0 N after each unloading cycle. It can be clearly noticed that the electrical resistivity showed a reversible change with the tensile loading, i.e., the electrical resistivity increased with the increase in loading and decreased when the loading decreased in the unloading cycles. The increase in electrical resistivity of short CF-reinforced cementitious composites with increased tensile loading has been previously reported [4,5]. The extension of composites due to tensile loading caused a reduction in the electrical contact points between short CFs, leading to a reduction in the conductive pathways and an increase in the electrical resistivity of composites [4,5]. The change in resistivity in all three studied load levels (i.e., 200, 300 and 400 N) was reversible, indicating that these load levels were within the elastic regime of the composites and did not introduce a permanent damage within the composite structure.

Composites containing different CF contents showed a similar behavior, except a flattening of electrical resistivity at the maximum load was noticed for the composites containing 0.25 wt.% CF (Figure 9a1–a3). This delayed electrical response from the composites in the region when the loading cycle reversed was probably attributed to a relatively lesser number of electrical contacts between CFs in the case of 0.25 wt.% CF. The electrical response of hybrid cementitious composites containing CF and different concentrations of CNF and MWCNT are shown in Figures 10–13. It can be noticed that these composites also showed similar trends of changing their electrical resistivity with tensile loading, i.e., the electrical resistivity increased and decreased reversibly with the increase and decrease in the tensile loading, respectively. However, the extent of electrical resistivity change in different cycles was dependent on the loading rates/maximum load as well as on the composition of the composites.

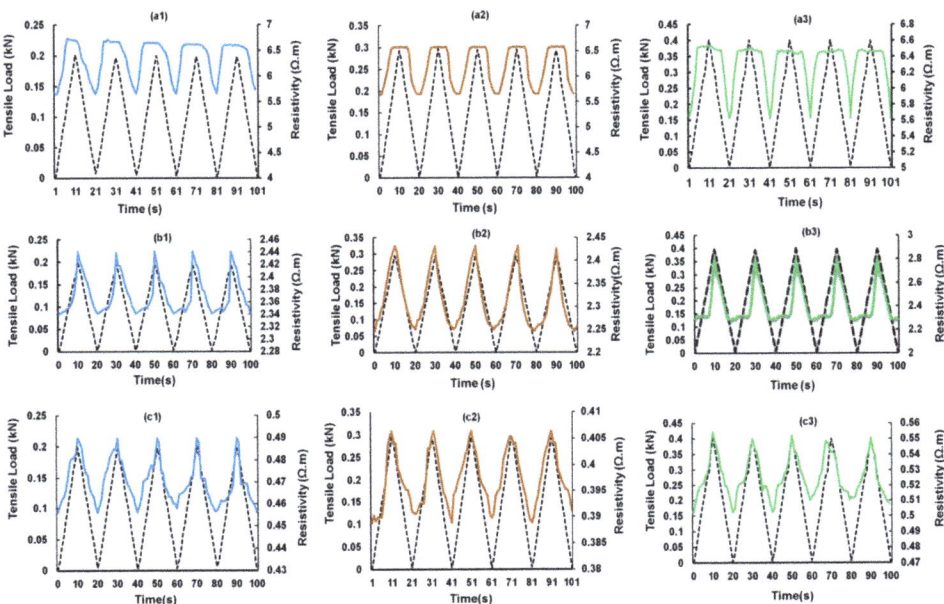

Figure 9. Change of electrical resistance with cyclic tensile load of CF reinforced cementitious composites at different peak loads containing 0.25% CF (**a1–a3**), 0.50% CF (**b1–b3**) and 0.75% CF (**c1–c3**).

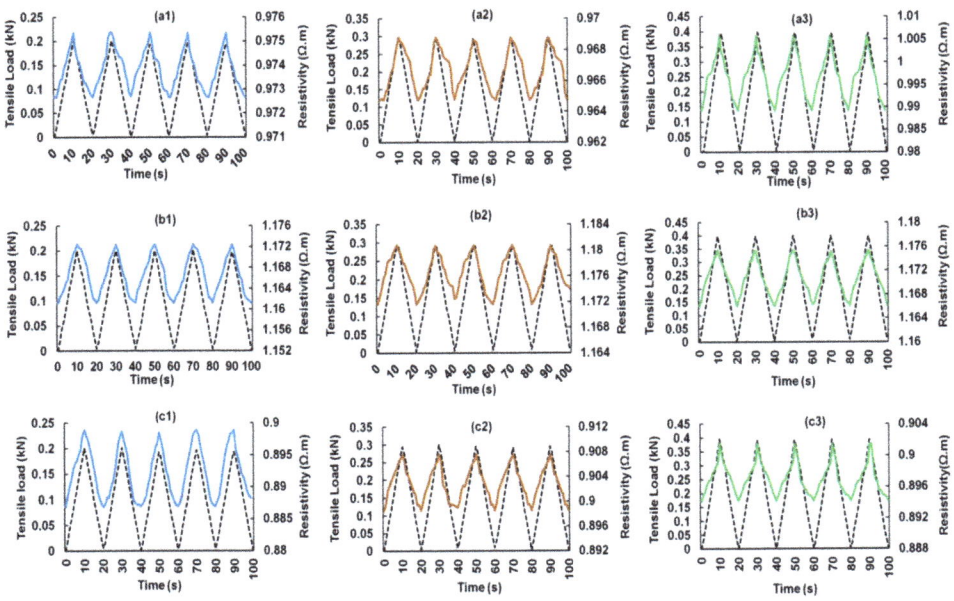

Figure 10. Change of electrical resistance with cyclic tensile load of 0.1% CNF-reinforced cementitious composites at different peak loads containing 0.25% CF (**a1–a3**), 0.50% CF (**b1–b3**) and 0.75% CF (**c1–c3**).

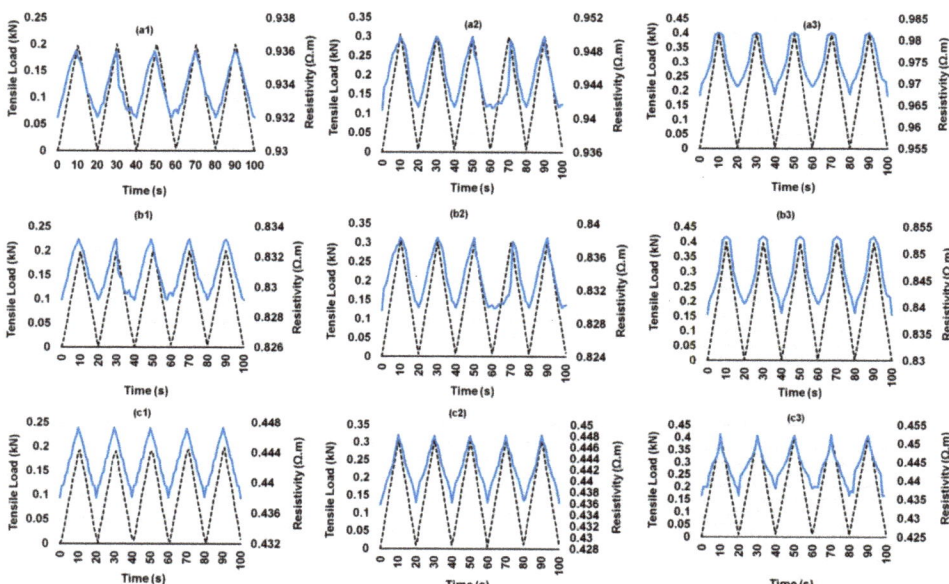

Figure 11. Change of electrical resistance with cyclic tensile load of 0.15% CNF-reinforced cementitious composites at different peak loads containing 0.25% CF (**a1–a3**), 0.50% CF (**b1–b3**) and 0.75% CF (**c1–c3**).

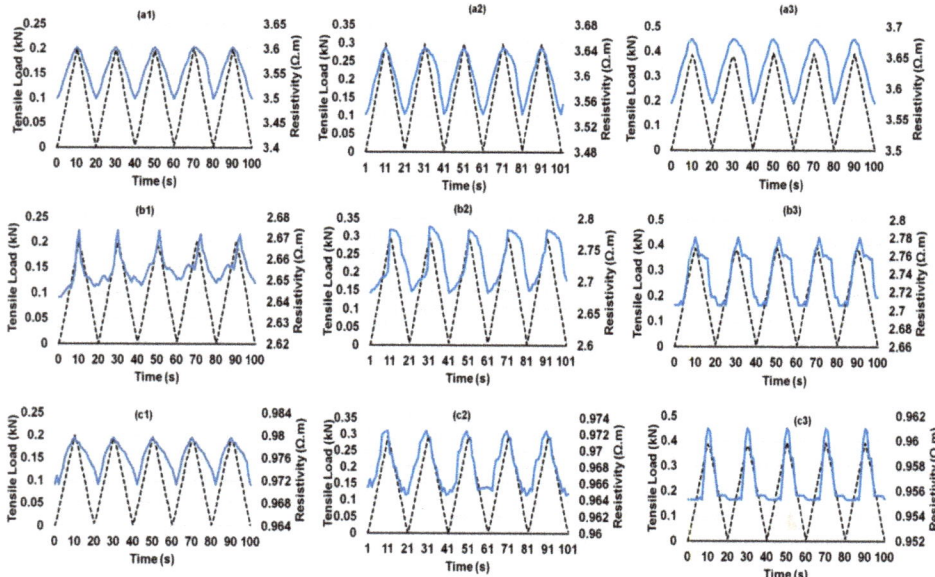

Figure 12. Change of electrical resistance with cyclic tensile load of 0.1% MWCNT-reinforced cementitious composites at different peak loads containing 0.25% CF (**a1–a3**), 0.50%SCF (**b1–b3**) and 0.75% SCF (**c1–c3**).

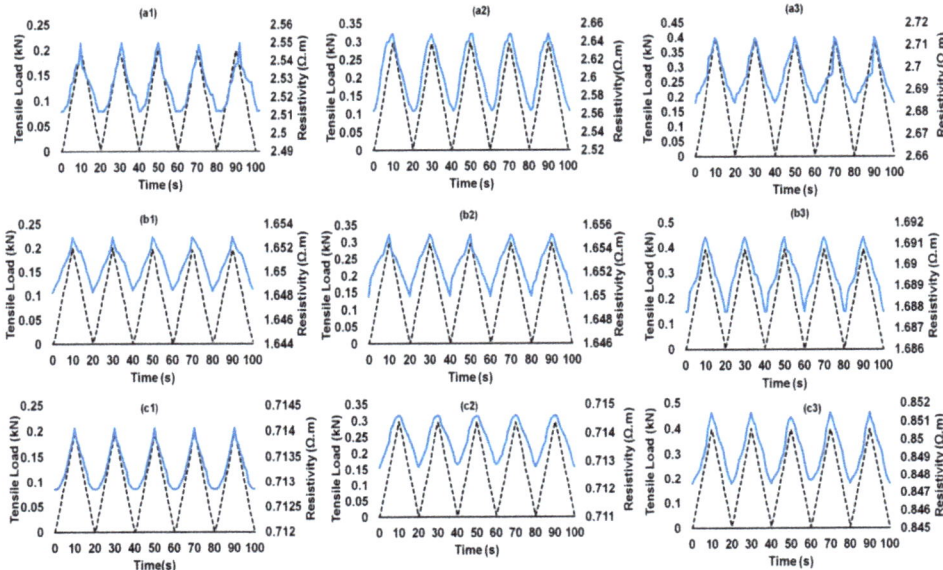

Figure 13. Change of electrical resistance with cyclic tensile load of 0.15% MWCNT-reinforced cementitious composites at different peak loads containing 0.25% CF (**a1–a3**), 0.50%SCF (**b1–b3**) and 0.75% SCF (**c1–c3**).

3.4.2. Fractional Change in Electrical Resistivity and Stress Sensitivity

The average FCR values of five loading cycles for different composites and at different loading conditions are listed in Tables 4 and 5. It is clear from Table 4 that the average FCR values of CF-reinforced cementitious composites were dependent on the loading rate (and maximum load) and on the amount of CF in these composites.

Table 4. Fractional resistance change of composites containing CF.

Samples	Loading Rate (N.s^{-1}), Max Load (N)	FCR
Mortar with 0.25% CF	20, 200	0.177
	30, 300	0.165
	40, 400	0.148
Mortar with 0.5% CF	20, 200	0.042
	30, 300	0.082
	40, 400	0.231
Mortar with 0.75% CF	20, 200	0.074
	30, 300	0.045
	40, 400	0.097

Table 5. Fractional resistance change of composites containing CF along with CNF/MWCNT.

Samples	Loading Rate (N.s^{-1}), Max. Load (N)	FCR CNF	FCR MWCNT
Mortar + 0.25% CF + 0.1% CNF/MWCNT	20, 200 N	0.003	0.030
	30, 300 N	0.004	0.029
	40, 400 N	0.017	0.029
Mortar + 0.25% CF + 0.15% CNF/MWCNT	20, 200 N	0.004	0.015
	30, 300 N	0.009	0.033
	40, 400 N	0.015	0.011
Mortar + 0.5% CF + 0.1% CNF/MWCNT	20, 200 N	0.010	0.039
	30, 300 N	0.008	0.038
	40, 400 N	0.008	0.028
Mortar + 0.5% CF + 0.15% CNF/MWCNT	20, 200 N	0.005	0.003
	30, 300 N	0.010	0.003
	40, 400 N	0.017	0.002
Mortar + 0.75%CF + 0.1% CNF/MWCNT	20, 200 N	0.013	0.008
	30, 300 N	0.010	0.007
	40, 400 N	0.008	0.006
Mortar + 0.75%CF + 0.15% CNF/MWCNT	20, 200 N	0.021	0.002
	30, 300 N	0.028	0.002
	40, 400 N	0.037	0.005

The influence of the loading rate and maximum load on FCR, however, did not show any clear trend. Previous studies on hybrid nano-carbon containing piezoresistivity cementitious composites showed a decrease in FCR with the increasing rate of compressing loading [13]. This was attributed to the reduced compressive strains with the increasing loading rates, resulting in the reduced resistivity change in each cycle. However, the effect of loading rates on the resistivity change of short CF-based cement composites has not been reported to date. In the present study, both the loading rate and maximum load were changed simultaneously, and this made it difficult to understand their individual effects. An increase in FCR with the increase in the maximum tensile strain per cycle in the elastic regime was previously observed in short CF-based piezoresistive composites [4,5]. In the present case, the increase in loading rates (from 20 to 40 N/s) could decrease the tensile strain of composites, as observed previously in cement-based composites [26]. However, the increase in the maximum load (from 200 to 400 N) could also increase the tensile strain at the same time and, therefore, no clear trend was observed due to these two opposing effects. Further studies by changing the loading rates while maintaining the same maximum load could help to properly understand this phenomenon.

It can also be observed from Table 4 that the FCR values were higher in the case of composites containing 0.25% CFs when compared to composites with 0.5 and 0.75% CFs. CFs formed an electrical percolation network at 0.25% as evidenced by the low resistivity values (see Figure 8). Similar resistivity values were reported previously for CF-based and other cementitious composites above the percolation threshold of the conductive fillers [21–24]. Therefore, the composites containing 0.25% CFs provided high values of FCR. Previously reported CF-based composites also showed similar FCR values under tensile loading [4,5]. The decrease in FCR at higher CF contents could be attributed to the fact the higher amount of CFs resulted in more touching of CFs and a high number of conductive pathways. This resulted in lower change in the conducting network under tensile deformation and consequently, a lower change in the electrical resistance. This phenomenon was previously observed in the case of CF-reinforced piezoresistive polymeric composites [27,28].

It is clear from Table 5 that the cementitious composites with hybrid fillers, i.e., CFs along with CNFs/CNTs showed lower FCR values when compared to only CF-based composites. The reason for this is the same as discussed above for 0.5 and 0.75% CF-based cementitious composites. The presence of CNF and MWCNTs significantly increased the number of conductive pathways and more touching between the conductive fillers, making the conductive network more stable and less sensitive to the mechanical deformations. Figure 14 explains this phenomenon schematically. It can be observed that when the CF content is low (Figure 14a), the conductive network becomes extended under tensile deformation, increasing the distance between CFs (indicated by dotted circles) and leading to a significant increase in the electrical resistivity. However, at higher CF contents, e.g., 0.75% (Figure 14b), due to higher number of electrical contacts the extension of the conductive network does not significantly change the electrical contact points and, therefore, the change in electrical resistivity is limited. Similarly, the presence of MWCNTs/CNFs in the case of 0.25% CF (Figure 14c) can maintain the electrical contacts between CFs when the conductive network extends under tensile deformation (as can be seen in the dotted circles in Figure 14c). Therefore, the presence of MWCNTs/CNFs results in a significantly lower change in the electrical resistivity under tensile loading. Observations made by Kim et al. support this hypothesis as it was noticed that the hybrid fillers composed of CFs and CNTs made the electrically resistivity less sensitive to the change in water to cement ratio, temperature and evaporation of electrolytic pore solution, due to the extended conductive pathways [14]. However, the authors did not evaluate the effect of hybrid fillers on the piezoresistive properties of the composites.

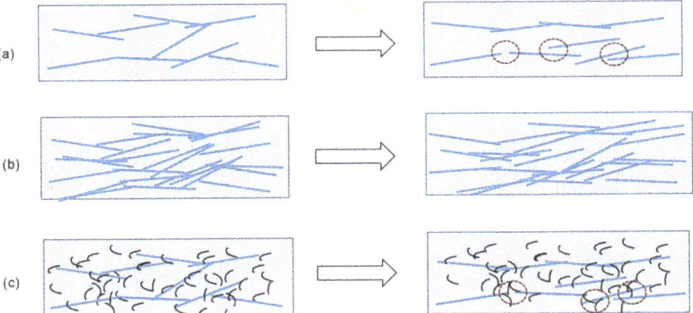

Figure 14. Change in the electorally conductive network under tensile deformation of (**a**) 0.25% CF, (**b**) 0.75% CF and (**c**) 0.25% CF with CNFs/MWCNTs.

FCR values of cementitious composites containing different concentrations of hybrid fillers are compared in Table 5. It can be noticed that the hybrid composites showed different FCR values depending on the CF, CNF and MWCNT concentrations, as well as the loading conditions, i.e., the rate and maximum load. Similar to CF-based composites, the influence of loading conditions on FCR values did not show any clear trend. The FCR values have been further converted into stress sensitivity according to Equation (2) and are presented in Figure 15. Stress sensitivity is a better parameter to compare the load/stress sensing behavior of cementitious composites containing different concentrations of conductive fillers. It can be observed that the CF-based composites showed a considerably higher stress sensitivity when compared to the hybrid composites due to the higher FCR values of the former, as explained earlier.

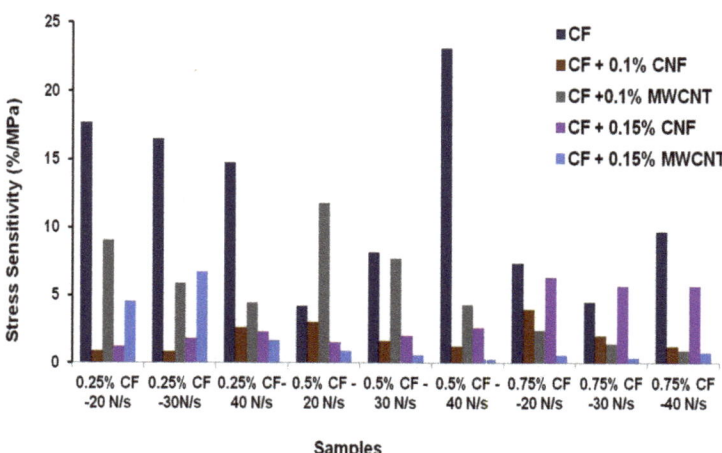

Figure 15. Stress sensitivity of various cement based sensors.

In the cases of hybrid composites with 0.25 and 0.5% CF, the use of MWCNTs led to superior stress sensitivity when compared to CNFs. On the contrary, the stress sensitivity was higher with CNFs when compared to MWCNTs when a higher amount of CF, i.e., 0.75 wt.% CF was used in the composites. The superior sensitivity of MWCNT-based composites at lower CF contents could be attributed to the superior ability of MWCNTs to form a piezoresistive network due to their higher electrical conductivity and smaller dimensions. At higher CF contents, a saturation in the electrically conducting pathways resulted in a lower sensitivity of these composites. Also, the stress sensitivity was superior with 0.1% MWCNTs when compared to 0.15% and this could be due to the increased agglomeration of MWCNTs at 0.15% when dispersed using the same process. On the contrary, composites with 0.15% CNF showed superior stress sensitivity when compared to those with 0.1% CNF. This could be attributed to the fact that CNFs could be more homogeneously dispersed at higher concentrations when compared to MWCNTs using the same dispersion process, due to their lower surface area and agglomeration tendency [29].

The highest stress sensitivities achieved with hybrid composites containing 0.15% CNFs and 0.1% CNTs were 6.36 and 11.82%/MPa at 0.75 and 0.5% CF contents, respectively. The observed stress sensitivity values were much higher when compared to those previously reported for hybrid nano-carbon-reinforced cementitious composites under compression loading [13].

3.5. FCR-Load Correlations of Developed Sensing Composites

The relationships between FCR and load for cementitious composites containing different CF contents at different loading conditions are shown in Figure 16. It can be observed that the change in FCR with the tensile load was dependent on the loading conditions. From the values of linear regression coefficients (R^2), it can be commented that the cementitious composites containing only CFs did not show a good linear correlation between FCR and load (and stress, as stress is proportional to the applied load), making the calibration of these sensors difficult and leading to measurements with high error values. The large scatter of the FCR values indicates an uneven and random change in the electrical resistivity of the samples in different loading and unloading cycles. In CF-based cementitious composites, the loading and unloading cycles caused random and large changes in the electrical contact points between the CFs (schematically shown in Figure 14), making electrical resistivity change and FCR not linearly dependent on the applied load or stress. Further, presence of CF clusters could also lead to unpredictable and random change in the electrical network and, consequently, in the electrical resistivity of the composites. As a result, the scatter in the FCR data was high. As a reliable and accurate measurement with a low scattering of FCR values is required for practical applications, these CF-based sensors are not, therefore, suitable for sensing of tensile stresses in civil engineering structures. A high scatter of FCR values in the case of CF-based cementitious composites was previously observed in the case of compressive loading [15].

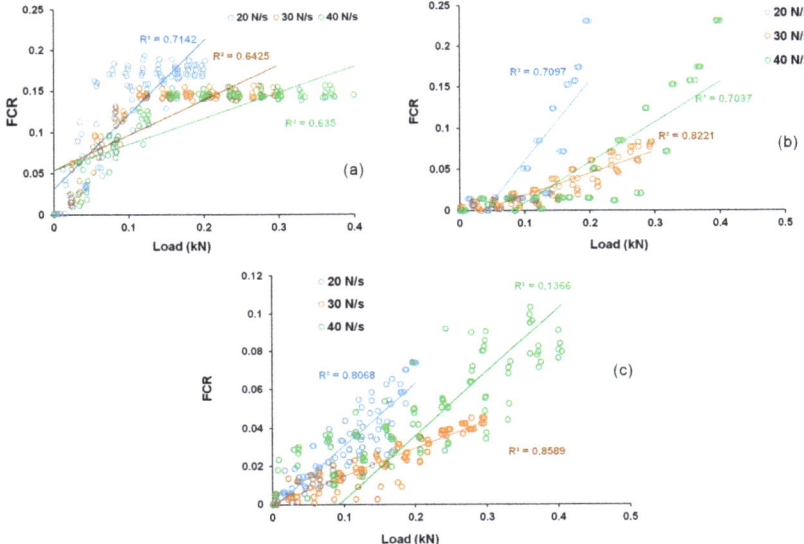

Figure 16. FCR-load correlation of cementitious composites containing different carbon fiber contents: (**a**) 0.25 wt.%, (**b**) 0.5 wt.% and (**c**) 0.75 wt.%.

The correlation between FCR and tensile load for CNF-based hybrid cementitious composites are shown in Figure 17. It can be observed that hybrid composites containing CNFs presented superior linear correlation between FCR and tensile load (and therefore, with tensile stress) and a lower scatter in FCR values when compared to the composites containing only CFs. It can also be noticed that among different samples, the sample containing 0.15% CNF with 0.75% CFs showed a good linear correlation with a low scatter of data. This composite also presented the best stress sensitivity (6.36%/MPa) and therefore, can be considered as the optimized sample for developing CNF-based hybrid stress sensors for construction applications.

Figure 18 shows the FCR-load correlation for hybrid cementitious composites containing MWCNTs. It is clear that the correlation was much better with a lower scatter of data for these composites when compared to only CF-based and CNF-based hybrid cementitious composites. It can also be observed that the composites with 0.1% MWCNT and 0.5% CF showed a good linear correlation along with high stress sensitivity (11.82%/MPa), as observed in Figure 15 and therefore, these composites can be considered as the best CNT-based hybrid cementitious sensors for the sensing of stresses in civil engineering structures.

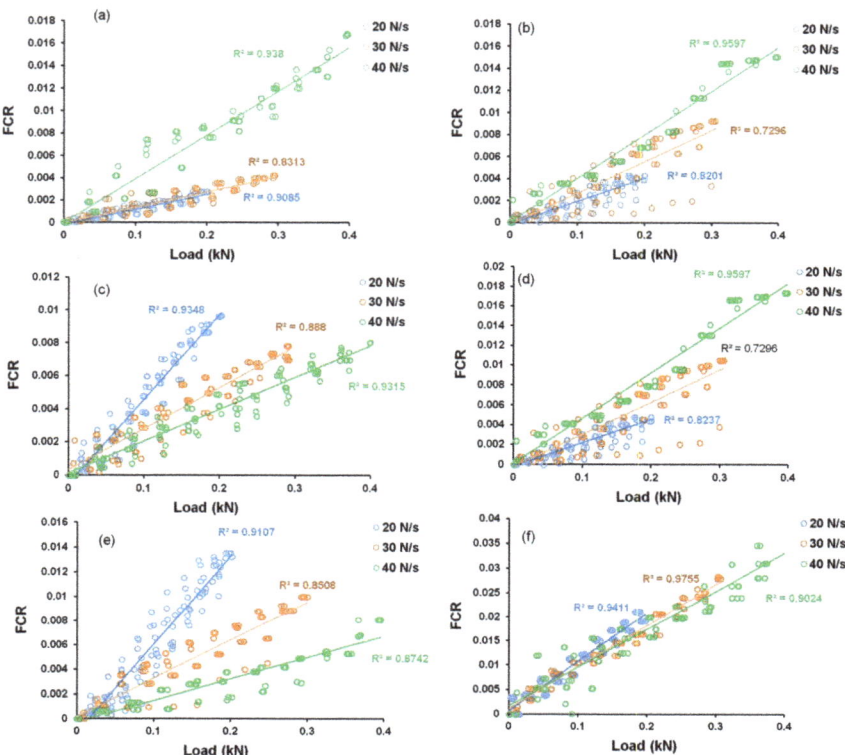

Figure 17. FCR-load correlation of CNF-based hybrid cementitious composites: (**a**) 0.25% CF + 0.1% CNF, (**b**) 0.25% CF + 0.15% CNF, (**c**) 0.5% CF + 0.1% CNF, (**d**) 0.5% CF + 0.15% CNF, (**e**) 0.75% CF + 0.1% CNF and (**f**) 0.75% CF + 0.15% CNF.

As discussed earlier, the presence of CNFs and MWCNTs in CF-based cementitious composites increased the conductive pathways preventing random and abrupt changes in the electrical network under tensile loading. Therefore, a more stable, accurate and reliable piezoresistive behavior was obtained in the presence of MWCNTs/CNFs. The results were superior with MWCNTs at low concentrations when compared to CNFs, due to the higher electrical conductivity of MWCNTs and owing to their superior ability to form percolating and piezoresistive electrical networks when dispersed homogeneously, due to their smaller dimensions. It can be noticed in Figures 17 and 18 that the slopes of the FCR-load curves changed with the loading rates, indicating that the stress sensitivity was dependent on the loading rates. As discussed in Section 3.4.2, the loading rate could influence the tensile strain of the developed cementitious composites and this, in turn, influenced the change in the electrical resistivity. A higher loading rate was expected to reduce the tensile strain, thereby reducing the change in the electrical resistivity and stress sensitivity of the composites. However, in the present study, the applied load was also changed along with the loading rates, and this resulted in an opposite effect on stress sensitivity, i.e., an increase in the load could increase the tensile strain and increase the stress sensitivity. Due to this reason, the effect of the loading rate on the stress sensitivity of the developed composites was not very clear in the present study and needs to be further investigated.

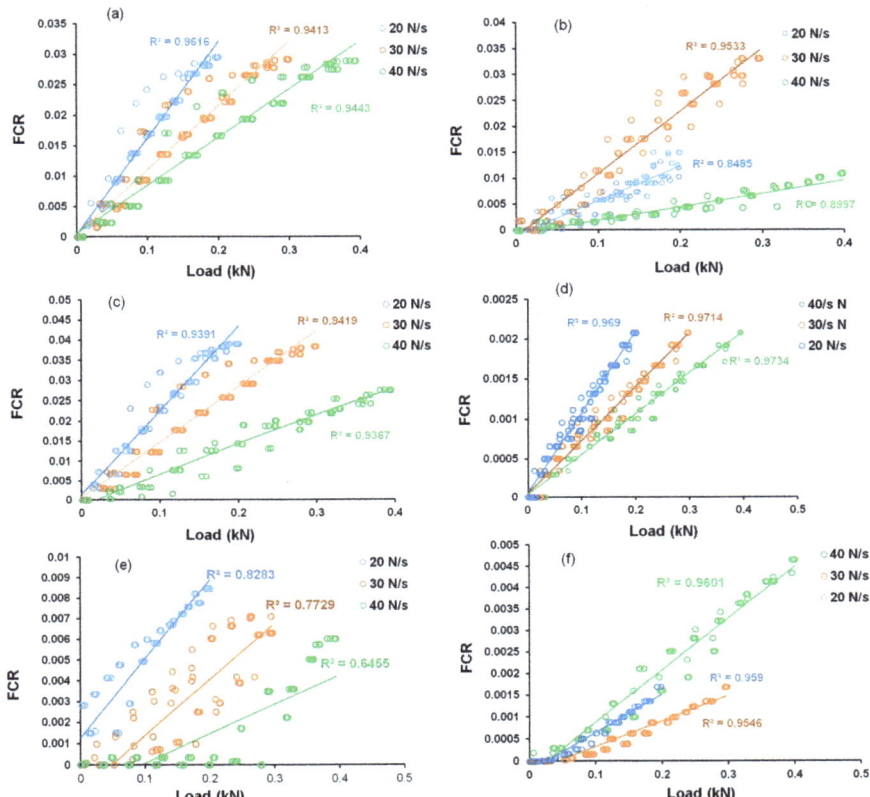

Figure 18. FCR-load correlation of MWCNT-based hybrid cementitious composites: (**a**) 0.25% CF + 0.1% MWCNT, (**b**) 0.25% CF + 0.15% MWCNT, (**c**) 0.5% CF + 0.1% MWCNT, (**d**) 0.5% CF + 0.15% MWCNT, (**e**) 0.75% CF + 0.1% MWCNT and (**f**) 0.75% CF + 0.15% MWCNT.

3.6. Microstructure of Developed Cementitious Composites

The fracture surfaces of broken samples in tensile tests (up to failure) were studied using SEM and are shown in Figure 19. It can be clearly observed from Figure 19a,b that CFs (indicated by arrows) were uniformly dispersed within cementitious composites. This confirmed that the used dispersion process using magnetic stirring along with Pluronic F-127 surfactant was able to ensure the homogeneous dispersion of CFs. CNFs and MWCNTs can also be observed in the fracture surface of hybrid composites, as can be seen in Figure 19c–f. It is interesting to note from Figure 19d–f that both CNFs and MWCNTs formed electrical connections between CFs and cement hydration products and helped to form the percolating and piezoresistive networks, as discussed earlier. The SEM micrographs of fracture surfaces support the mechanism of piezoresistivity, as illustrated in Figure 14.

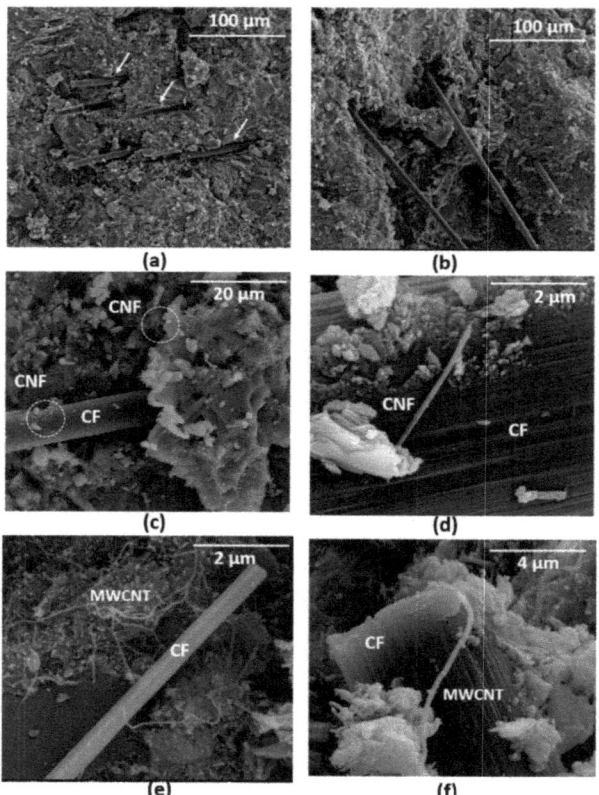

Figure 19. Fracture surface of a plain mortar (**a,b**), cementitious composite with CNFs (**c,d**) and a cementitious composite with MWCNTs (**e,f**) at different magnifications.

4. Conclusions

In this research, multi-scale cementitious composites were developed using CFs along with CNFs or MWCNTs and their stress-sensing behavior was characterized and compared with CF-based cementitious composites. The following conclusions can be made from the present study:

(1) The electrical resistivity of CF-based composites decreased with the increase in CF content from 0.25 to 0.75%. The incorporation of CNF and MWCNT (0.1 and 0.15% of cement weight) in CF-based composites led to a significant decrease in the electrical resistivity of cementitious composites.

(2) CF-based cementitious composites showed a reversible increase in electrical resistivity with the cyclic tensile load. The highest value of FCR was achieved at the lower CF content, i.e., 0.25% and an increase in the CF content resulted in a decrease in FCR due to the saturation in the electrical contact points reducing the stress sensitivity of the composites.

(3) Multi-scale cementitious composites also showed a reversible increase in the electrical resistivity with tensile loads. Overall, the multi-scale composites showed a lower FCR when compared to CF-based composites, due to an increase in the electrically conducting pathways. MWCNTs and CNFs formed connections between well-dispersed CFs and cement hydration products, forming a well-connected percolation network.

(4) Although CF-based composites presented good stress sensitivity, the FCR-load correlation was not good and a high scatter in FCR values was noticed. This makes the CF-based cement sensors not suitable for accurately measuring tensile stresses in practical applications. On the contrary, the multi-scale composite sensors showed a good linear correlation between FCR and tensile loads with a low scatter of data. Superior results were obtained in the case of MWCNT-based multi-scale composites when compared to the CNF-based composites. The best CNF- and MWCNT-based sensors provided stress sensitivity of 6.36 and 11.82%/MPa, respectively.

Author Contributions: S.P.: involved in conceptualization, investigation, result analysis, writing—original draft, project management. B.V.: investigation, result analysis, O.L.: investigation, result analysis, S.R.: involved in conceptualization, results analysis, review and editing, project management, funding acquisition. R.F.: involved in conceptualization, project administration, funding acquisition. All authors have read and agreed to the published version of the manuscript.

Funding: The research is funded by Fibrenamics, Institute of Innovation in Fiber-based Materials and Composites and University of Minho—Guimarães, Portugal, Project UID/CTM/00264/2019 of 2C2T—Centro de Ciência e Tecnologia Têxtil, funded by National Funds through FCT/MCTES. The APC is funded by the School of Arts and Humanities, University of Huddersfield and Fibernamics, Institute of Innovation in Fiber-based Materials and Composites and University of Minho—Guimarães, Portugal, Project UID/CTM/00264/2019 of 2C2T—Centro de Ciência e Tecnologia Têxtil, funded by National Funds through FCT/MCTES.

Data Availability Statement: The raw/processed data required to reproduce these findings cannot be shared at this time as the data also forms part of an ongoing study. The relevant data can be made available on request.

Conflicts of Interest: The authors declare no conflict of interest.

References

1. Brownjohn, J.M.W. Structural health monitoring of civil infrastructure. *Philos. Trans. R. Soc. A Math. Phys. Eng. Sci.* **2007**, *365*, 589–622. [CrossRef]
2. Rana, S.; Subramani, P.; Fangueiro, R.; Correia, A.G. A review on smart self-sensing composite materials for civil engineering applications. *AIMS Mater. Sci.* **2016**, *3*, 357–379. [CrossRef]
3. Dong, W.; Li, W.; Tao, Z.; Wang, K. Piezoresistive properties of cement-based sensors: Review and perspective. *Constr. Build. Mater.* **2019**, *203*, 146–163. [CrossRef]
4. Wen, S.; Chung, D.D.L. Uniaxial tension in carbon fiber reinforced cement, sensed by electrical resistivity measurement in longitudinal and transverse directions. *Cem. Concr. Res.* **2000**, *30*, 1289–1294. [CrossRef]
5. Chung, D.D.L. Piezoresistive cement-based materials for strain sensing. *J. Intell. Mater. Syst. Struct.* **2002**, *13*, 599–609. [CrossRef]
6. Dong, W.; Li, W.; Wang, K.; Guo, Y.; Sheng, D.; Shah, S.P. Piezoresistivity enhancement of functional carbon black filled cement-based sensor using polypropylene fibre. *Powder Technol.* **2020**, *373*, 184–194. [CrossRef]
7. D'Alessandro, A.; Rallini, M.; Ubertini, F.; Materazzi, A.L.; Kenny, J.M. Investigations on scalable fabrication procedures for self-sensing carbon nanotube cement-matrix composites for SHM applications. *Cem. Concr. Compos.* **2016**, *65*, 200–213. [CrossRef]
8. Yoo, D.Y.; You, I.; Lee, S.J. Electrical and piezoresistive sensing capacities of cement paste with multi-walled carbon nanotubes. *Arch. Civ. Mech. Eng.* **2018**, *18*, 371–384. [CrossRef]
9. Parveen, S.; Rana, S.; Fangueiro, R. A Review on Nanomaterial Dispersion, Microstructure, and Mechanical Properties of Carbon Nanotube and Nanofiber Reinforced Cementitious Composites. *J. Nanomater.* **2013**, *2013*, 710175. [CrossRef]
10. Parveen, S.; Rana, S.; Fangueiro, R.; Paiva, M.C. Microstructure and mechanical properties of carbon nanotube reinforced cementitious composites developed using a novel dispersion technique. *Cem. Concr. Res.* **2015**, *73*, 215–227. [CrossRef]
11. Parveen, S.; Rana, S.; Fangueiro, R.; Paiva, M.C. Characterizing dispersion and long term stability of concentrated carbon nanotube aqueous suspensions for fabricating ductile cementitious composites. *Powder Technol.* **2017**, *307*, 15–24. [CrossRef]
12. Zheng, Q.; Han, B.; Cui, X.; Yu, X.; Ou, J. Graphene-engineered cementitious composites: Small makes a big impact. *Nanomater. Nanotechnol.* **2017**, *7*, 1847980417742304. [CrossRef]
13. Han, B.; Wang, Y.; Ding, S.; Yu, X.; Zhang, L.; Li, Z.; Ou, J. Self-sensing cementitious composites incorporated with botryoid hybrid nano-carbon materials for smart infrastructures. *J. Intell. Mater. Syst. Struct.* **2017**, *28*, 699–727. [CrossRef]
14. Kim, G.; Park, S.; Ryu, G.; Lee, H.-K. Electrical characteristics of hierarchical conductive pathways in cementitious composites incorporating CNT and carbon fiber. *Cem. Concr. Compos.* **2017**, *82*, 65–75. [CrossRef]
15. Azhari, F.; Banthia, N. Cement-based sensors with carbon fibers and carbon nanotubes for piezoresistive sensing. *Cem. Concr. Compos.* **2012**, *34*, 866–873. [CrossRef]

16. Zhang, L.; Ding, S.; Li, L.; Dong, S.; Wang, D.; Yu, X.; Han, B. Effect of characteristics of assembly unit of CNT/NCB composite fillers on properties of smart cement-based materials. *Compos. Part A Appl. Sci. Manuf.* **2018**, *109*, 303–320. [CrossRef]
17. Hunashyal, A.M.; Quadri, S.S.; Banapurmath, N.R. Experimental investigation on the study of mechanical and microstructural properties of hybrid composite cement beams reinforced with multi-walled carbon nanotubes and carbon fibres. *Proc. Inst. Mech. Eng. Part N J. Nanoeng. Nanosyst.* **2012**, *226*, 135–142. [CrossRef]
18. Hunashyal, A.M.; Tippa, S.V.; Quadri, S.S.; Banapurmath, N.R. Experimental investigation on effect of carbon nanotubes and carbon fibres on the behavior of plain cement mortar composite round bars under direct tension. *Int. Sch. Res. Not.* **2011**, *2011*, 856849. [CrossRef]
19. Al-Saleh, M.H.; Sundararaj, U. A review of vapor grown carbon nanofiber/polymer conductive composites. *Carbon* **2009**, *47*, 2–22. [CrossRef]
20. Tessonnier, J.P.; Rosenthal, D.; Hansen, T.W.; Hess, C.; Schuster, M.E.; Blume, R.; Girgsdies, F.; Pfänder, N.; Timpe, O.; Su, D.S.; et al. Analysis of the structure and chemical properties of some commercial carbon nanostructures. *Carbon* **2009**, *47*, 1779–1798. [CrossRef]
21. Ozbulut, O.E.; Jiang, Z.; Harris, D.K. Exploring scalable fabrication of self-sensing cementitious composites with graphene nanoplatelets. *Smart Mater. Struct.* **2018**, *27*, 115029. [CrossRef]
22. Monteiro, A.O.; Cachim, P.B.; Costa, P.M.F.J. Self-sensing piezoresistive cement composite loaded with carbon black particles. *Cem. Concr. Compos.* **2017**, *81*, 59–65. [CrossRef]
23. Liu, C.; Liu, G.; Ge, Z.; Guan, Y.; Cui, Z.; Zhou, J. Mechanical and Self-Sensing Properties of Multiwalled Carbon Nanotube-Reinforced ECCs. *Adv. Mater. Sci. Eng.* **2019**, *2019*, 2646012. [CrossRef]
24. Donnini, J.; Bellezze, T.; Corinaldesi, V. Mechanical, electrical and self-sensing properties of cementitious mortars containing short carbon fibers. *J. Build. Eng.* **2018**, *20*, 8–14. [CrossRef]
25. Kim, G.M.; Yang, B.J.; Cho, K.J.; Kim, E.M.; Lee, H.K. Influences of CNT dispersion and pore characteristics on the electrical performance of cementitious composites. *Compos. Struct.* **2017**, *164*, 32–42. [CrossRef]
26. Yang, E.H.; Li, V.C. Strain-rate effects on the tensile behavior of strain-hardening cementitious composites. *Constr. Build. Mater.* **2014**, *52*, 96–104. [CrossRef]
27. Bakis, C.E.; Nanni, A.; Terosky, J.A.; Koehler, S.W. Self-monitoring, pseudo-ductile, hybrid FRP reinforcement rods for concrete applications. *Compos. Sci. Technol.* **2001**, *61*, 815–823. [CrossRef]
28. Rana, S.; Zdraveva, E.; Pereira, C.; Fangueiro, R.; Correia, A.G. Development of hybrid braided composite rods for reinforcement and health monitoring of structures. *Sci. World J.* **2014**, *2014*, 170187. [CrossRef]
29. Rana, S.; Alagirusamy, R.; Joshi, M. A review on carbon epoxy nanocomposites. *J. Reinf. Plast. Compos.* **2009**, *28*, 461–487. [CrossRef]

Article

DNAzyme-Amplified Electrochemical Biosensor Coupled with pH Meter for Ca²⁺ Determination at Variable pH Environments

Hui Wang [1], Fan Zhang [1,2], Yue Wang [1,3], Fangquan Shi [4], Qingyao Luo [1], Shanshan Zheng [1], Junhong Chen [5], Dingzhen Dai [5], Liang Yang [1,*], Xiangfang Tang [1,*] and Benhai Xiong [1,*]

1. State Key Laboratory of Animal Nutrition, Institute of Animal Science, Chinese Academy of Agricultural Sciences, Beijing 100193, China; wanghui_lunwen@163.com (H.W.); zhangfan19@139.com (F.Z.); wangyue9313@163.com (Y.W.); luoqingyao@caas.cn (Q.L.); zhengshanshan@caas.cn (S.Z.)
2. College of Animal Science and Technology, China Agricultural University, Beijing 100193, China
3. College of Animal Science and Technology, Northwest A&F University, Yangling 712100, China
4. Animal Husbandry and Veterinary Station of Xihe County, Longnan 742100, China; sfquan@163.com
5. Department of Animal Science and Technology, Jinling Institute of Technology, Nanjing 211169, China; chenjunhong@jit.edu.cn (J.C.); dzdai@163.com (D.D.)
* Correspondence: yangliang@caas.cn (L.Y.); tangxiangfang@caas.cn (X.T.); xiongbenhai@caas.cn (B.X.); Tel.: +86-010-62811680 (B.X.)

Abstract: For more than 50% of multiparous cows, it is difficult to adapt to the sudden increase in calcium demand for milk production, which is highly likely to cause hypocalcemia. An electrochemical biosensor is a portable and efficient method to sense Ca^{2+} concentrations, but biomaterial is easily affected by the pH of the analyte solution. Here, an electrochemical biosensor was fabricated using a glassy carbon electrode (GCE) and single-walled carbon nanotube (SWNT), which amplified the impedance signal by changing the structure and length of the DNAzyme. Aiming at the interference of the pH, the electrochemical biosensor (GCE/SWNT/DNAzyme) was coupled with a pH meter to form an electrochemical device. It was used to collect data at different Ca^{2+} concentrations and pH values, and then was processed using different mathematical models, of which GPR showed higher detecting accuracy. After optimizing the detecting parameters, the electrochemical device could determine the Ca^{2+} concentration ranging from 5 µM to 25 mM, with a detection limit of 4.2 µM at pH values ranging from 4.0 to 7.5. Finally, the electrochemical device was used to determine the Ca^{2+} concentrations in different blood and milk samples, which can overcome the influence of the pH.

Keywords: electrochemical sensor; single-walled carbon nanotube; DNAzyme; dairy cow; hypocalcemia

1. Introduction

Calcium is an essential macronutrient in living organisms that plays an irreplaceable role in physiological and biochemical functions [1,2]. Calcium ions (Ca^{2+}) can maintain the biological potential on both sides of the cell membrane and normal nerve conduction function [3,4]. For dairy cows, a tremendous amount of calcium is required every day for daily milk production. To maintain the balance of calcium in the blood, dairy cows must be supplemented with a certain amount of calcium from food or absorb it from the intestines and the kidneys [5]. Inappropriate Ca concentrations affect the function and motility of the rumen, abomasum, intestines, and uterus, with severe consequences on energy metabolism [6]. At the onset of lactation, more than 50% of multiparous cows have difficulty adapting to the sudden increase in calcium demand by the mammary gland for milk production [7,8], which might cause subclinical hypocalcemia without any symptoms [9].

In general, the blood calcium concentration of healthy dairy cows is 2.1 ~ 2.5 mM, of which the ionic calcium occupies about 50%. When the blood Ca^{2+} is in the range of 0.7 ~ 1.05 mM, it is called subclinical hypocalcemia. If the blood Ca^{2+} concentration of dairy cows is lower than 0.7 mM without the supplement, it is called clinical hypocalcemia, which can result in reduced feed intake, poor rumen and intestine motility, an increased risk of a displaced abomasum, reduced milk yield, an increased susceptibility to infectious diseases, and an increased risk of early lactation removal from the herd [10,11]. The loss of a single cow can reach thousands of dollars, mainly in feeding and milk production. Thus, it is of great importance to detect the blood Ca^{2+} content to evaluate hypocalcemia.

Currently, there is no commercial device for Ca^{2+} detection with a low cost, portability, and high precision, and is easy to use. If a veterinarian wants to know the Ca^{2+} level of a dairy cow, he has to send the samples to a professional analysis organization, which is time-consuming, high-cost, and not real-time. It is difficult to meet the requirements of low-cost and fast detection. According to the preliminary research, the Ca^{2+} content accounts for about half of the blood Ca in dairy cow [12]. The electrochemical biosensor [13] has gained the most attention in the analysis field, because its outstanding advantages include a rapid and effective response, being simple to use, low cost, and miniaturization [14], which can measure the blood Ca^{2+} with simple treatment.

A DNAzyme is a DNA molecule with catalytic activity [15], mainly consisting of two parts: an enzyme strand composed of one catalytic core and two arms, and a substrate strand with a single RNA linkage (rA). When the target substance binds to the enzyme, the substrate strand can be cleaved at the rA site [16]. In the past two decades, many DNAzymes have been developed that show excellent selectivity and sensitivity to the metal ions, such as Na^+ [17], K^+ [18], Ag^+ [19], Hg^{2+} [20], Mg^{2+} [21], Cu^{2+} [22], Pb^{2+} [23], and Cr^{3+} [24]. Owing to their excellent stability, cost-effectiveness, high catalytic efficiency, and programmability [25], many DNAzyme-based biosensors have been explored for metal ion determination [26,27]. Zhou [28] previously studied a DNAzyme named EtNa that can be cleaved by Na^+ and Ca^{2+}. To improve the specificity, they further optimized the structure of EtNa to produce EtNa-C5T, which exhibited higher activity and better Ca^{2+} selectivity than the original EtNa. Considering the metal concentration difference (500-fold), the cleavage rate of EtNa was 11,200-fold higher with Ca^{2+} than with Na^+, whereas that of EtNa-C5T was 257,000-fold higher with Ca^{2+} than with Na^+ [29]. However, the cleavage efficiency of EtNa-C5T was easily affected by the pH that showed a positive correlation. For real samples, the pH values are different, which might affect the detecting accuracy of electrochemical biosensors.

In this work, an electrochemical biosensor was fabricated using a glassy carbon electrode, single-walled carbon nanotubes, and EtNa-C5T, but the electrochemical response was weak. To improve the sensitivity and amplify the impedance signal, the structure and length of EtNa-C5T were designed and optimized. In addition, the electrochemical biosensor was combined with a pH meter to form an electrochemical device, and the data at various pH values were used to build a mathematical model. The proposed method can simplify the pH adjustment during sample pretreatment and measure Ca^{2+} concentrations accurately.

2. Materials and Methods

2.1. Materials and Reagents

Single-walled carbon nanotubes were purchased from Jiangsu Xianfeng Nanomaterial Technology Co., Ltd. (Nanjing, China). Tween 20 (98%), 3-Aminopropyltriethoxysilane (APTES, 99%), and ethanolamine (EA, 99%) were provided by Thermo Fisher Technology Co., Ltd. (Beijing, China). $NaNO_3$, KNO_3, $Mg(NO_3)_2$, $Ca(NO_3)_2$, $Cu(NO_3)_2$, $Fe(NO_3)_2$, $Zn(NO_3)_2$, $Fe(NO_3)_3$, and $Sn(NO_3)_2$ were purchased from Sigma Aldrich Company (Beijing, China). N, N-Dimethylformamide (DMF), and ethanol were bought from Macklin Inc. (Shanghai, China). N-Hydroxysuccinimide 1-pyrene butyric acid (PBASE) was obtained from Invitrogen Company (Tianjing, China). Ethanol, hydrochloric acid (HCl), and acetic acid (CH_3COOH) were obtained from Beijing Chemical Industry Co., Ltd. (Beijing, China). Tris(hydroxymethyl)aminomethane (Tris) was purchased from Titan Scientific Co., Ltd. (Shanghai, China). The standard stock solution of Ca^{2+} was offered by the China National

Research Center for Reference Materials (Beijing, China) and diluted to different Ca^{2+} concentrations. All of the chemicals and reagents in this study were of analytical grade, which were used without further purification. Ultrapure water was produced by a Millipore Milli-Q system (18.2 MΩ cm^{-3}). Both Tris-HCl buffer and potassium acetate were chosen as the buffer solution. A dry powder of oligonucleotides was synthesized by Sangon Biotech Co., Ltd. (Shanghai, China), which was dispersed in sterilized ultrapure water to 50 nM. The oligonucleotide sequences are listed in Table 1.

Table 1. Sequences of the modified DNA.

	DNA Name	Sequences and Modifications (Starting from 5 Terminal ′)
Ca_sub	NO-substrate	NH_2-($CH_2$6)GCGGTAGAAGGATATCACTGAGCACTG
	NS-substrate	NH_2-(C6)GCGGTAGAAGG/rA/TATCACTGAGCACTG
	DS-substrate	NH_2-(C6)GCGGTAGAAGG/rA/TATCACTGAGCACTGGG/rA/TAAGCGG TAGAACTCACAATGTATAATGCGCGCATTATACATTGTGAGT
CAZyme	EtNa-C5T	CAGTGCTCAGTGATTGTTGGAATGGCTCATGCCACACTCTTTTCTACCGC
	sEtNa-C5T	TCTACCGCTTATCCCAGTGCTCAGTGATTGTTGGAATGGCTC ATGCCACACTCTTTTCTACCGC
	dEtNa-C5T	TCTACCGCTTTGTTGGAATGGCTCATGCCACACTCTTCAGTGCT CAGTGATTGTTGGAATGGCTCATGCCACACTCTTTTCTACCGC

2.2. Apparatus and Measurements

A field emission scanning electron microscope (FE-SEM SU8040) was used to obtain SEM images at an accelerating voltage of 10 kV. The UV-visible absorption spectra of GCEs modified with different materials were recorded using a Beckman Kurt DU800 UV-visible spectrophotometer (Kraemer Boulevard Brea, CA, USA) at wavelengths ranging from 200 nm to 800 nm. Raman spectra were measured through Renishaw inVia with an imaging microscope (532 nm diode and Ar ion lasers). A portable pH meter (Testo 206PH) was bought from Testo AG (Beijing, China). The electrochemical measurements included cyclic voltammetry (CV) and electrochemical impedance spectroscopy (EIS), which were performed using an electrochemical cell on a CHI 760E electrochemical workstation (Shanghai Chenhua Instrument, Shanghai, China). A three-electrode system consisted of a modified glassy carbon electrode (GCE/SWNT/DNAzyme, Φ = 3 mm), a platinum wire, and Ag/AgCl (saturated with KCl), which served as the working electrode, counter electrode, and reference electrode, respectively. The CV measurement was swept from -0.2 V to $+0.6$ V with a scan rate of 50 mV s^{-1}, and the frequency range of electrochemical impedance spectroscopy (EIS) ranged from 10^5 to 0.1 Hz with the potential of 0.20 V (versus Ag/AgCl) and the amplitude of 5 mV, which were carried out in a mixture solution of 5.0 mM of $[Fe(CN)_6]^{3-/4-}$ and 0.1 M KCl. A centrifugal machine (SN-LSC-40) was purchased from Shanghai Shangyi Instrument Equipment Co., Ltd. (Shanghai, China).

2.3. Fabrication of the Biosensor

Single-walled carbon nanotubes (SWNTs) were carboxylated according to the following protocol. Briefly, the SWNTs were refluxed with a mixed solution of concentrated HNO_3 and H_2SO_4 (v/v = 1:3) for 6 h. After that, the SWNTs were rinsed with enough Milli-Q purified water to reach a neutral pH. The treated SWNTs were placed in a vacuum oven and heated at 80 °C for 60 min. The SWNTs (1 mg) were dissolved in 2 mL of DMF to form a homogeneous solution.

Figure 1 illustrates the modification process as follows. First, a bare GCE was sequentially polished using 0.3 μm and 0.05 μm of Al_2O_3 powder, and then sonicated in DI water several times. After that, the cleaned GCE was immersed in 5% APTES in ethanol for 30 min, and then washed with sufficient ultrapure water. The GCE surface was covered with a 3 μL SWNTs solution, annealed in air at 60 °C for 30 min, and then rinsed to remove residual SWNTs. After that, the GCE/SWNT was immersed in 6 mM of PBASE in dimethylformamide for 60 min under a normal temperature. In a high-humidity environment, the PBASE-modified GCE/SWNT was covered with a 5 µL Ca substrate solution overnight at 4 °C to immobilize the Ca substrate on the SWNT surface through the amide bonds between the amine at the 5′ end and the ester groups of PBASE. The modified electrode was immersed in 0.1 mM of EA and 0.1% Tween 20 to block the excess ester groups on the PBASE and SWNT surfaces. Finally, the Ca substrate on the SWNT surface was hybridized with CAZyme to form a GCE/SWNT/DNAzyme.

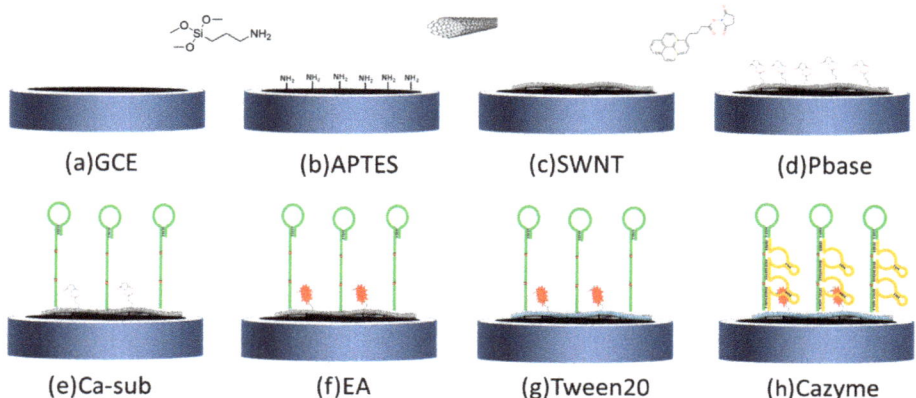

Figure 1. GCE functionalized with APTES, SWNT, Pbase, Ca_sub, EA, Tween20, and Cazyme.

2.4. Electrochemical Measurements

To determine the Ca^{2+} concentration, the R_{ct} value of the electrochemical biosensor must be calculated before and after measuring the real sample. The biosensor was covered with the real sample for 9 min, and then rinsed with sufficient water to remove the residue. After that, the biosensor was immersed in 5 mM $[Fe(CN)_6]^{4-/3-}$ and 0.1 M KCl, and EIS was performed. The pH of the serum or milk samples was directly measured using a Testo 206PH.

The relative resistance was calculated using the following equation:

$$\text{Relative resistance} = \frac{R_{ct0} - R_{ct}}{R_{ct0}} \times 100\% \tag{1}$$

where R_{ct0} is the initial resistance value, and R_{ct} is the resistance after exposure to the real sample.

2.5. Sensing of Real Samples

Blood and milk samples were collected from dairy cows at the China–Israel demonstration dairy farm located in southeastern Beijing. The blood samples were collected from the vein of the cow's tail using a centrifugal tube with heparin lithium, and the supernatant was obtained after centrifugation. The supernatant of centrifugation was directly used for Ca^{2+} detection. In addition, the milk samples were collected through milking equipment and treated with HNO_3. Briefly, 15 g of a milk sample was evaporated to dryness by using an electric furnace, and then ashed in a muffle furnace at 550 °C for 5 h. After that, the sample was mixed with 5 mL of HCl (20%). The treated solution was placed into a 50 mL volumetric flask, and the volume was fixed with Tris-HCl buffer.

3. Results

3.1. Choice of Materials

EtNa-C5T (CAZyme) as a molecular recognition probe specifically hybridizes the complementary substrate (Ca_sub) to form the DNAzyme. After exposure to Ca^{2+}, CAZyme can bind to Ca^{2+}, and then cleave the complementary substrate (Ca_sub) at the rA site. Due to the poor conductivity of the DNAzyme, the structure change affects the impedance response of GCE/SWNTs, which can be used for Ca^{2+} detection with high selectivity. In addition, the cleavage efficiency of CAZyme is easily affected by the pH that shows a positive correlation. Therefore, a GCE/SWNTs/DNAzyme combined with a pH meter can detect Ca^{2+} concentration at variable pH environments.

3.2. Characterization

Morphological analysis was performed by field emission scanning electron microscopy. The GCE/SWNT before and after being functionalized with a DNAzyme is shown in Figure S1. A dense network structure was evenly distributed on the GCE surface in Figure S1A, which indicated that SWNTs were immobilized on the GCE surface. Compared to Figure S1B, there was no obvious change, indicating that it is difficult to distinguish DNAzyme modification with SEM.

The blank quartz was selected to replace the GCE because of its excellent light transmittance. Figure 2A shows the UV-visible absorption spectrum of the blank quartz functionalized with SWNTs, PBASE, and DNAzymes in the range from 190 nm to 800 nm. To reduce the interference, the absorption spectrum of blank quartz was chosen as the background, since the curve was almost a straight line. For the GCE/SWNT, a strong absorption peak was located at 280 nm, which was ascribed to SWNTs. When PBASE was modified on the GCE/SWNT, there were three absorption peaks located at 240 nm, 275 nm, and 335 nm that illustrated that PBASE had attached on the SWNT surface through CC interaction between the carbon atom of the pyrene ring and the carbon atom of SWNTs. After the DNAzyme was further functionalized on the GCE/SWNT/PBASE, there were two weak absorption peaks observed on the spectrum, contributing that the absorption peak of the DNAzyme and GCE/SWNT/PBASE was superimposed because of the UV-visible absorption peak of the DNAzyme at 260 nm. As shown in Figure 2B, Raman spectroscopy was used to investigate the change of different GCE-modified materials using 532 nm laser excitation from 100 to 3500 cm^{-1}. For a bare GCE, three weak peaks were found at 1350, 1595, and 2687 cm^{-1}. After the GCE was treated with SWNTs, then RBM, D, G, and 2D peaks were located at 180, 1350, 1595, and 2675 cm^{-1}, which was consistent with the characteristic peaks of SWNT [30]. While the GCE/SWNT was treated with PBASE and a DNAzyme, the ratios of D to G markedly decreased, and G peak of the GCE/SWNT/DNAzyme shifted in the negative direction, indicating that PBASE and the DNAzyme were immobilized on the SWNT surface [31].

3.3. CV and EIS Characterization

Electrochemical impedance spectroscopy (EIS) was employed as a highly sensitive technique to monitor the interfacial properties of each modified step, which can characterize the fabrication and assembly process of a biosensor. Each modification of the GCE/SWNT/DNAzyme was monitored by EIS analysis in 5 mM $[Fe(CN)_6]^{4-/3-}$ and 0.1 M KCl at ambient temperature. Figure S2 shows the Randle equivalent circuit over a range of frequencies from 100 kHz to 1 Hz, including charge transfer resistance (R_{ct}), the capacitance (C_{dl}), electrolyte resistance (R_s), and the Warburg element (Z_w). In the Nyquist plot, the R_{ct} directly controls the transfer kinetics of the redox-probe electron at the electrode surface, corresponding to the diameter of the semicircle. As shown in Figure 3, there was a small semicircle in the impedance spectra on the bare GCE, where the R_{ct} was about 574 Ω. While SWNT was deposited on the GCE surface, the semicircle almost decreased into a linear plot in the impedance spectra, and the R_{ct} was zero, ascribing to the deposition that SWNT possesses excellent electrical conductivity. After PBASE was assembled on the surface of the GCE/SWNT, there was still no obvious semicircle, but the slope of the linear plot in the impedance spectra greatly decreased. The semicircle reappeared when Ca_sub was functionalized on the SWNT surface through a peptide bond, where the R_{ct} reached 437 Ω. The main reason is the accumulation of negatively charged probe Ca_sub that introduces much stronger steric hindrance and electrostatic repulsion to the diffusion process of redox probe $[Fe(CN)_6]^{3-/4-}$ toward the electrode surface. When the specific sites of PBASE were blocked through EA, the semicircle diameter increased to 504 Ω. The semicircle rose again to where the R_{ct} was 856 Ω when Tween20 was attached to the naked SWNT surface. Finally, when CAZyme was incubated on the modified electrode from the hybridization of DNAzyme, a significantly increased semicircle could be observed, where the R_{ct} was about 1843 Ω.

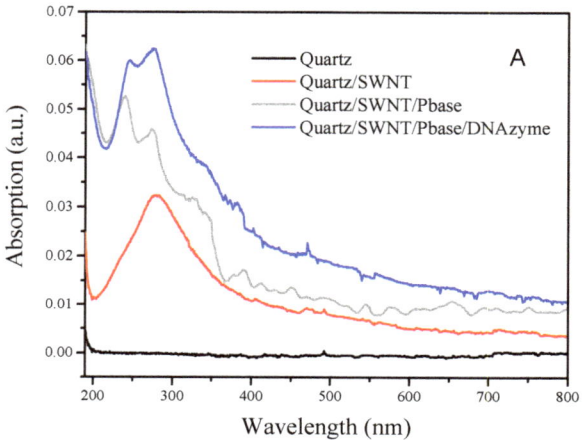

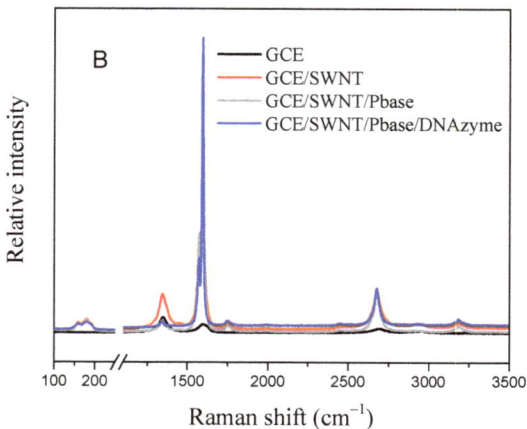

Figure 2. (**A**) UV-visible spectrum of blank quartz modified with SWNT, PBASE, and DNAzyme; (**B**) Raman spectrum of GCE modified with SWNT, PBASE, and DNAzyme.

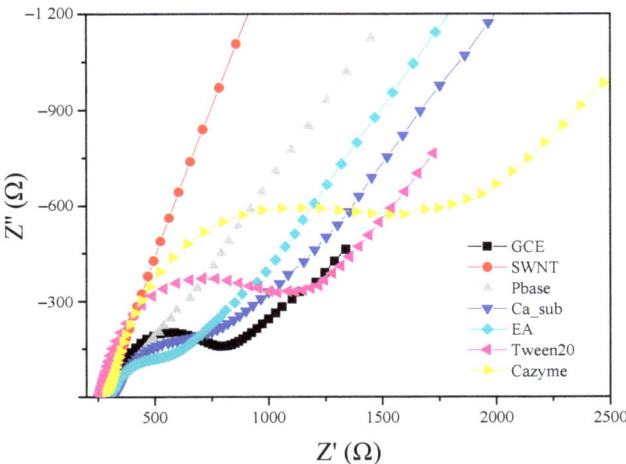

Figure 3. Electrochemical impedance spectra of GCE modified successively with SWNTs, PBASE, Ca_sub, EA, Tween 20, and CAZyme obtained at frequencies ranging from 1 to 10^5 Hz (potential = 0.2 V) in 5 mmol/L $[Fe(CN)_6]^{3-/4-}$ and 0.1 mol/L KCl.

Figure 4 shows the cyclic voltammograms of the GCE functionalized successively with SWNTs, PBASE, Ca_sub, EA, Tween 20, and CAZyme in a mixed solution of 5 mmol/L $[Fe(CN)_6]^{3-/4-}$ and 0.1 mol/L KCl. All the CV curves exhibited a pair of well-defined redox peaks, but differences in the peak currents and the peak-to-peak separation were observed. After the SWNTs were immobilized on the GCE surface, the peak current increased significantly to 73 µA. After incubation with PBASE, the peak current decreased to 57 µA. When the negatively charged Ca_sub was assembled on the SWNT surface, the peak current was continuously reduced. After the EA and non-electroactive Tween 20 were employed to block the non-specific sites of PBASE and the bare SWNT, the peak current and peak-to-peak separation did not vary significantly. Finally, the electrode was incubated in 50 mM to form the DNAzyme, and the peak current decreased obviously. These results were consistent with the EIS results. Thus, we could preliminarily conclude that the proposed biosensor based on a DNA hydrogel was successfully prepared.

3.4. Optimization

To improve the sensitivity and amplify the impedance signal, several detection parameters of the GCE/SWNT/DNAzyme were optimized at different Ca^{2+} concentrations.

The incubation time can significantly affect the sensitivity of the biosensor. Figure 5 shows that the relative resistances changed with the incubation time when the GCE/SWNT/DNAzyme was immersed in different Ca^{2+} concentrations. It was clear that the relative resistance increased as the incubation time ranged from 1 to 13 min, which was attributed to the continuous cleavage of the DNAzyme by Ca^{2+}. However, the growth rate of the relative resistance greatly decreased when the incubation time exceeded 7 min, especially in 10 mM of Ca^{2+}. This result was mainly due to the limited DNAzyme on the GCE surface. To balance the sensitivity and detection time, 7 min was selected as the incubation time for further measurements.

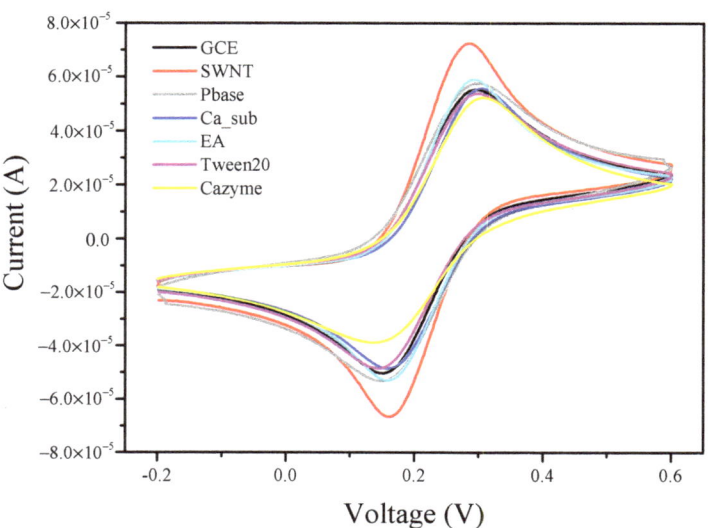

Figure 4. Cyclic voltammograms of the GCE before and after modification with SWNTs, PBASE, Ca_sub, EA, Tween 20, and CAZyme obtained in 5.0 mmol/L [Fe(CN)$_6$]$^{3-/4-}$ and 0.1 mol/L KCl solutions at a scan rate of 50 mV/s.

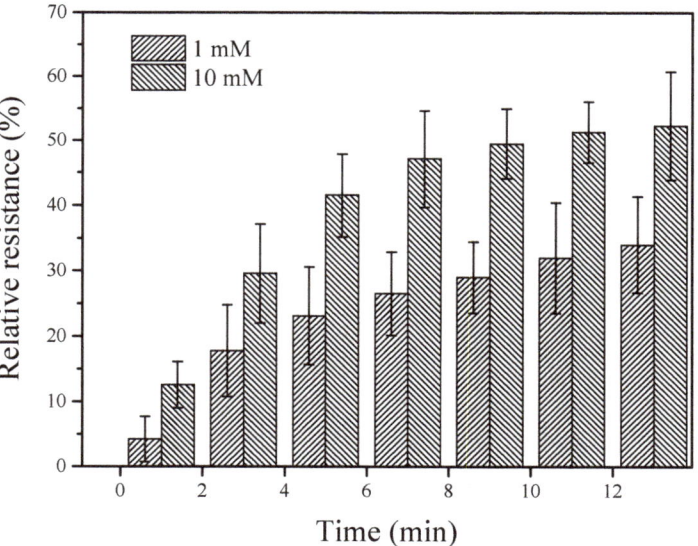

Figure 5. Relative resistance of the GCE/SWNT/DNAzyme after incubation in different Ca^{2+} concentrations for different times.

The relative resistance of the GCE/SWNT/DNAzyme was strongly affected by the pH of the electrolytic solution. Figure 6 shows the results for the GCE/SWNT/DNAzyme incubated in different Ca^{2+} concentrations at various pHs. The relative resistance increased gradually as the pH ranged from 4.0 to 7.0. One reason for this result was that the cleavage efficiency of the DNAzyme increased with the pH. In addition, hydrogen ions were absorbed on the surface of the GCE/SWNT/DNAzyme, affecting the conductivity. Therefore, it was necessary to develop a method for detecting the Ca^{2+} concentration at different pHs to improve practicality.

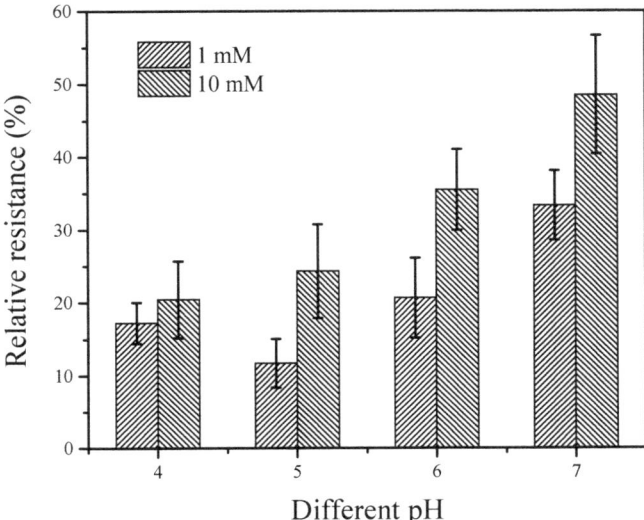

Figure 6. Relative resistance of the GCE/SWNT/DNAzyme after incubation in different Ca^{2+} concentrations at various pH values.

The structure and length of the DNAzyme are important factors influencing the sensitivity of the GCE/SWNT/DNAzyme. As shown in Figure 7, four different DNAzymes were designed for functionalizing the GCE/SWNT, and the functionalized GCE/SWNTs were used to detect different Ca^{2+} concentrations in an electrolytic solution at pH 7.5. For the DNAzyme without rA, the relative resistance was almost zero, indicating that Ca_sub was not cleaved by the Ca/CAZyme. The relative resistances of the GCE/SWNT/DNAzyme were approximately 3.5 and 9.32 at Ca^{2+} concentrations of 1 mM and 10 mM, respectively. When the length of the DNAzyme was designed to be longer, the relative resistances reached 18 and 30 at Ca^{2+} concentrations of 1 mM and 10 mM, respectively, due to the poor conductivity of the DNA molecule. To improve the sensitivity, the two enzymes were used for the DNAzyme, and the detection response increased. Thus, the DNAzyme with two enzymes was selected as the biomaterial for this work.

3.5. Selectivity

Anti-interference is an important performance factor for electrochemical biosensors. Many metal ions, such as Na^+, K^+, Mg^{2+}, Zn^{2+}, Cu^{2+}, Fe^{2+}, Fe^{3+}, and Cr^{3+}, are present in the dairy cows' blood and milk samples in Table S1, which might interfere with the relative resistance of a GCE/SWNT/DNAzyme. Different GCE/SWNT/DNAzymes were applied to measure 10 mM of other metal ions in an electrolytic solution at pH 7.5, and the detection results are shown in Figure 8. The relative resistance of the GCE/SWNT/DNAzymes was approximately 48 for 10 mM Ca^{2+}. In comparison, the relative resistances for Na^+ and Mg^{2+} were much higher than those for the rest of the metal ions. The four metal ion solutions had slightly higher relative resistances than the blank buffer solution, indicating that most metal ions did not interfere with the GCE/SWNT/DNAzymes.

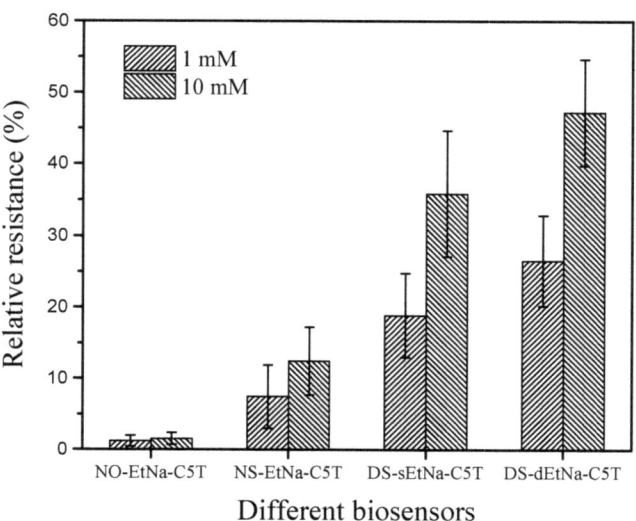

Figure 7. Relative resistances of GCE/SWNT/DNAzymes after incubation in different Ca^{2+} concentrations at various pH values.

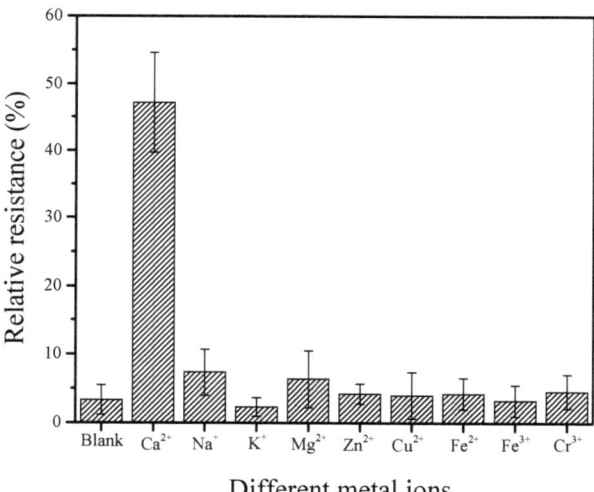

Figure 8. Relative resistances of the GCE/SWNT/DNAzymes before and after incubation in 10 mM solutions of different metal ions including, Na^+, K^+, Mg^{2+}, Zn^{2+}, Cu^{2+}, Fe^{2+}, Fe^{3+}, and Cr^{3+}.

3.6. Ca^{2+} Sensing

The linear relationship between the Ca^{2+} concentration and relative resistance was determined in the presence of a 10 mM Tris-HCl buffer solution via electrochemical impedance spectroscopy.

The GCE/SWNT/DNAzyme measured different Ca^{2+} concentrations ranging from 0 μM to 25 mM under the optimized parameters, as shown in Figure 9A, which reveals that the semicircle diameter decreased with increasing Ca^{2+} concentrations. The calculated relative resistances are listed in Figure 9B. When the Ca^{2+} concentration was less than 500 μM, the relative resistance increased gradually, indicating that the cleavage efficiency of the DNAzyme was slow at low Ca^{2+} concentrations. The relative resistance exhibited a linear relationship with the logarithm of the Ca^{2+} concentration in the range of 5 μM to 500 μM. The linear equation was Relative resistance $= 1.8898 + 6.4947 \log_{10}\left[Ca^{2+}\right]$ (μM) with a linear regression coefficient of 0.991, and the detection limit was calculated to be 4.2 μM (S/N = 3). At higher Ca^{2+} concentrations, the relative resistance increased rapidly as the Ca^{2+} concentrations increased from 500 μM to 25 mM, and the linear regression equation was Relative resistance $= -43.02 + 22.68 \log_{10}\left[Ca^{2+}\right]$ (μM) with a linear regression coefficient of 0.991.

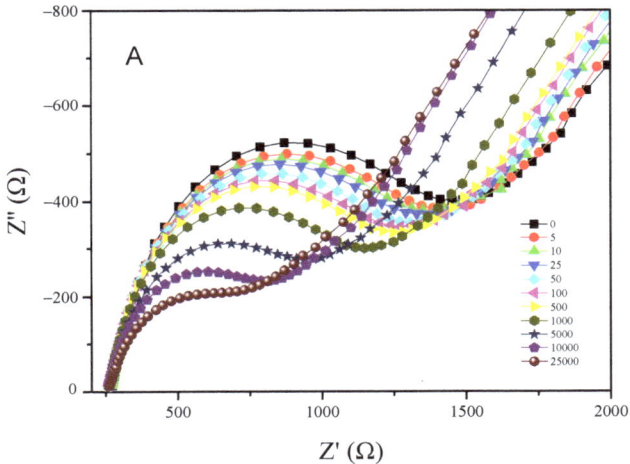

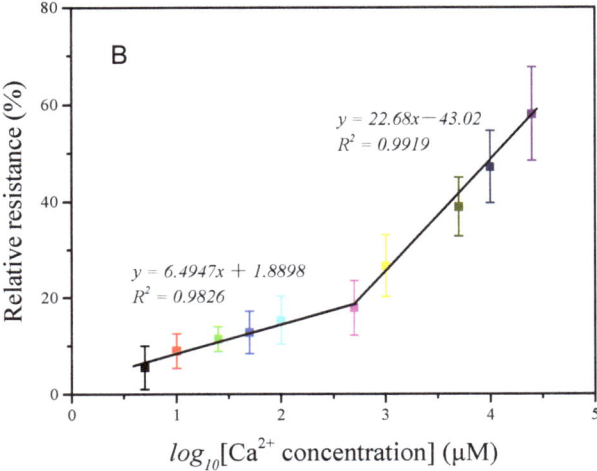

Figure 9. (**A**) Electrochemical impedance spectra of the GCE/SWNT/DNAzyme for different Ca^{2+} concentrations in the range of 1 μM to 25 mM; (**B**) the linear relationship between the relative resistance and the logarithm of the Ca^{2+} concentration.

3.7. Mathematical Model

The pH of the electrolyte solution can affect the relative resistance of a GCE/SWNT/DNAzyme and thus interfere with the detection accuracy. To overcome this shortcoming, the GCE/SWNT/DNAzyme was coupled with a pH meter to fabricate an electrochemical sensor array that could process detection data using different mathematical models, as shown in Figure 10.

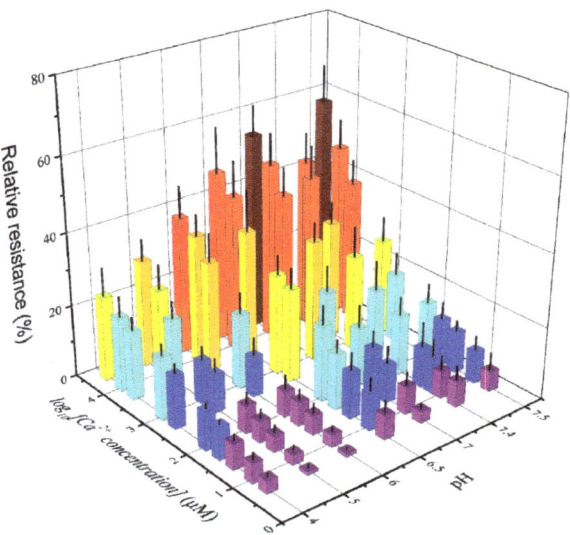

Figure 10. The relative resistances of the GCE/SWNTs/DNAzyme after exposure to different Ca^{2+} concentrations with various pH value. (The ratio of pH vs. Ca^{2+} concentration: 4.0–5, 4.0–10, 4.0–25, 4.0–50, 4.0–100, 4.0–500, 4.0–1000, 4.0–5000, 4.0–10000, 4.0–25000; 5.0–5, 5.0–10, 5.0–25, 5.0–50, 5.0–100, 5.0–500, 5.0–1000, 5.0–5000, 5.0–10000, 5.0–25000; 6.0–5, 6.0–10, 6.0–25, 6.0–50, 6.0–100, 6.0–500, 6.0–1000, 6.0–5000, 6.0–10000, 6.0–25000; 6.5–5, 6.5–10, 6.5–25, 6.5–50, 6.5–100, 6.5–500, 6.5–1000, 6.5–5000, 6.5–10000, 6.5–25000; 7.0–5, 7.0–10, 7.0–25, 7.0–50, 7.0–100, 7.0–500, 7.0–1000, 7.0–5000, 7.0–10000, 7.0–25000; 7.4–5, 7.4–10, 7.4–15, 7.4–25, 7.4–100, 7.4–250, 7.4–750, 7.4–2500, 7.4–7500; 7.5–5, 7.5–10, 7.5–25, 7.5–50, 7.5–100, 7.5–500, 7.5–1000, 7.5–5000, 7.5–10000, 7.5–25000.) The error bars indicate the standard deviations of five biosensors.

The X and Y matrices were pretreated using the following equations:

$$X = [1, x_1, x_2, x_1^2, x_2^2, x_1 \times x_2] \\ Y = \log_{10}[C] \quad (2)$$

where x_1 and x_2 represent the relative resistance of the GCE/SWNT/DNAzyme and the pH value, respectively, and C is the Ca^{2+} concentration.

3.7.1. Stepwise Linear Regression

Stepwise linear regression (SLR) [32], one of several linear regression methods, is a statistical technique that selects independent variables to predict a response variable's outcome. It can be applied to the training dataset for the Ca^{2+} concentration using statistical methods to remove the redundant variables. The basic idea is to add variables or remove them from a model step by step while performing an F-test and t-test for the selected variables individually. Variables that do not significantly change the SLR are removed to ensure that each significant variable is included. This is an iterative process that repeats until significant variables are no longer added to the SLR, and independent variables with negligible contributions are no longer removed.

First, each of the independent variables is related to the dependent variable, the value of which is calculated by the F-test.

$$Y = \beta_0 + \beta_i X_i + \in, \ i = 1, \cdots, p \qquad (3)$$

$$F_{i_1}^{(1)} = max\left\{ F_1^{(1)}, \cdots, F_1^{(1)} \right\} \qquad (4)$$

where Y is the dependent variable, X_i are the independent variable, p is the number of the independent variables, β_i are regression coefficients, and $F_{i_1}^{(1)}$ is the value of F-test or T-test. If $F_{i_1}^{(1)}$ was greater than a certain significant level, X_{i_1} was selected for the SLR.

Then, different sets of independent variables were chosen, such as $\{X_{i_1}, X_1\}, \cdots, \{X_{i_1}, X_{i_1-1}\}, \{X_{i_1}, X_{i_1+1}\}, \cdots, \{X_{i_1}, X_p\}$, and a binary regression with the dependent variable was performed.

$$F_{i_2}^{(2)} = max\left\{ F_1^{(2)}, \cdots, F_{i_1}^{(2)}, F_{i_1+1}^{(2)}, \cdots, F_p^{(2)} \right\} \qquad (5)$$

where $F_{i_2}^{(2)}$ is the value of the F-test or T-test. If $F_{i_2}^{(2)}$ was greater than a certain significance level, X_{i_2} was selected for the SLR. The rest of the independent variables were determined in turn.

The X matrix was chosen as the independent variable, and the Y matrix was selected as the dependent variable. The matrices were processed using SLR. The actual response vs. the predicted response is shown in Figure S3 and Figure 11. Based on the computation of the model of statistics, the coefficients of determination (R-squared) for the actual and predicted Ca^{2+} concentrations was approximately 0.95. The root means square error (RMSE) of the true and predicted Ca^{2+} concentrations was 0.26127. The mean square error (MSE) and mean absolute error (MAE) were 0.068261 and 0.19161, respectively.

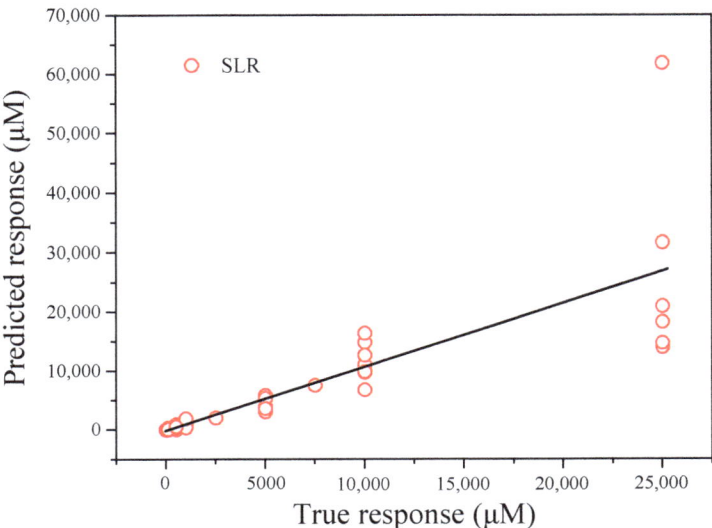

Figure 11. Predicted Ca^{2+} concentrations obtained by stepwise linear regression vs. the actual Ca^{2+} concentrations.

3.7.2. Neural Network Fitting

Neural network fitting (NNF) [33] is a non-parametric non-linear method that combines the advantages of neural networks and regression. The most important feature is that the mapping equation between the input and output does not need to be determined during learning. The NNF procedure can be divided into three layers, namely input, hidden, and output layers.

The X matrix was chosen as the input layer, the corresponding Y matrix was selected as the output layer, and the mapping relationship between the input and output layers was established through the hidden layer. The whole dataset included 70 samples, which were divided randomly into three groups: the proportions of the training, validation, and test sets were 70%, 15%, and 15%, respectively. As shown in Figure S4, the number of hidden layer nodes was set to 10, based on the optimization of the hidden nodes with the Levenberg–Marquardt (LM) algorithm. The predicted Ca^{2+} concentrations versus the actual Ca^{2+} concentrations are shown in Figure S5 and Figure 12. An RMSE of 0.386, MSE of 0.1481, and R-squared value of 0.9926 were obtained for the training dataset.

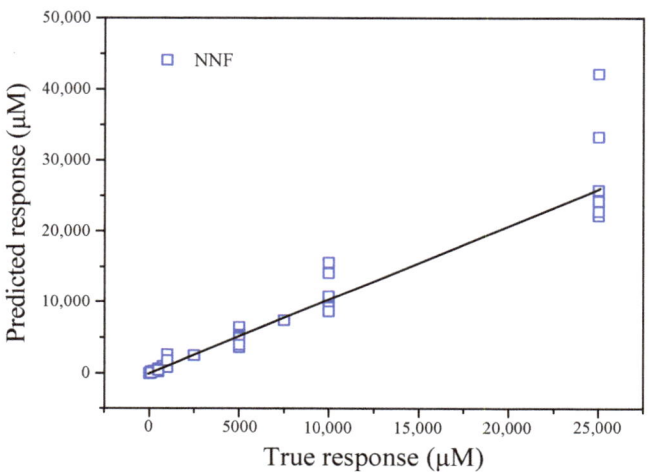

Figure 12. Predicted Ca^{2+} concentrations obtained by the established NNF model vs. the actual Ca^{2+} concentrations.

3.7.3. Gaussian Process Regression

Gaussian process regression [34,35] is a non-parametric statistical method for functional regression analysis that focuses on problems involving functional response variables, as well as mixed functional and scalar covariates. GPR has several advantages, for example, it works well with small datasets and can provide uncertainty measurements for predictions.

With the dataset $D = \{X, Y\} = \{x_i, y_i\}_{i=1}^n$, the regression model can be formulated as

$$y = f(x) + \varepsilon \tag{6}$$

where $f(\cdot)$ denotes an unknown regression function, and ε is the Gaussian noise with a zero mean and variance σ_n^2. From the funtion space view, the Gaussian process is completely specified by its mean function $m(x)$ and covariance function $C(x, x')$ from the function space view, which are defined as follows:

$$m(x) = E[f(x)] \tag{7}$$

$$C(x, x') = E[(f(x) - m(x))(f(x') - m(x'))] \tag{8}$$

Then, the Gaussian process can be described by:

$$f(x) \sim GP(m(x), C(x, x')) \tag{9}$$

Usually, the data are normalized for notation simplicity, and then the output observations follow a Gaussian distribution as

$$y \sim GP(0, C(x, x')) \qquad (10)$$

When a query input x_* is received, the joint distribution of the training outputs y and test output y^* based on the previous test is

$$\begin{bmatrix} y \\ y_x \end{bmatrix} \sim N\left(0, \begin{bmatrix} C & k^* \\ k^T & C(x_*, x_*) \end{bmatrix}\right) \qquad (11)$$

where $k_* = [C(x_*, x_1), \cdots, C(x_*, x_n)]^T$. Then, the prediction output (y_*) and variance (σ_*^2) can be given as:

$$\begin{cases} y^* = k_*^T \\ \sigma_*^2 = C(x_*, x_*) - k_*^T C^{-1} k_* \end{cases} \qquad (12)$$

The X matrix was chosen as the workspace variable, and the Y matrix was selected as the response. Cross-validation was set as 10 folds. The predicted concentrations obtained after modelling the training dataset vs. the actual concentrations of Ca^{2+} are shown in Figure S6 and Figure 13. The RMSE, MSE, MAE, and R-squared values were calculated as 0.23753, 0.056421, 0.18097, and 0.96, respectively.

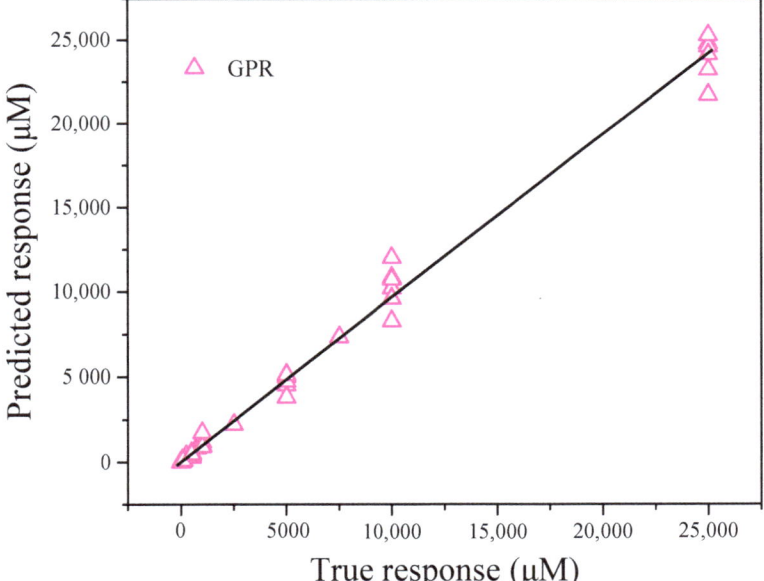

Figure 13. Predicted Ca^{2+} concentrations obtained by the established GPR model vs. the actual Ca^{2+} concentrations.

3.8. Comparison of Detection Performance

The parameters of different sensors for Ca^{2+} determination are listed in Table 2. Clearly, sensors based on different principles have various advantages during usage. Most of the ion-selective electrodes based on membrane potentials had a fast detection time and wide linear range, but the selectivity was lower than the GCE/SWNTs/CAZyme. For spectroscopic and electrochemical methods, the detection time and pH range had no obvious advantages, but the linear range and detection limit were similar to those of the ion-selective electrode. For this work, the electrochemical device (GCE/SWNTs/CAZyme and pH meter) overcome the shortcoming that biosensors can only detect Ca^{2+} in a specific pH environment, making it suitable for different biological samples without the need for complicated pretreatment.

Table 2. Comparison of the performances of electrochemical device and other reported sensors for Ca^{2+} determination.

Sensor	Method	pH	Time (min)	Linear Range (μM)	LOD (μM)	Ref.
CD-EGTA	Fluorescent	7	240	15 ~ 300 μM	0.38 μM	[8]
DNAzyme/SWNT/FET	Current	6.9	9	10 ~ 1000	7.2	[27]
DLLME	UV-visible	12.8	30	1.5 ~ 37.5	0.425	[36]
SD-ISO	Fluorescence	6.0–8.0	5	10 ~ 10^6	9.3	[37]
Ca-ISE	Potential	-	1.67	$10^{-2.5}$ ~ $10^{-1.5}$ M	-	[38]
Ca^{2+}-SCISE	Potential	-	1	10^{-1} ~ 10^4	5	[39]
Fe_2O_3-ZnO NRs/FET	Current	7.6	-	0.01 ~ 3.0 mM	0.05	[40]
Electrochemical device	EIS	4.0–7.5	7	5 ~ 2.5×10^4	4.2	This work

CD-EGTA: carbon dot-ethylenebis(oxyethylenenitrilo)tetraacetic acid; ISO: ion-selective optode; SD: solvatochromic dye; ISE: potentiometry based on a calcium-selective electrode; SCISE: solid-contact calcium-selective electrode.

3.9. Detection of Ca^{2+} in Real Sample

A comparison of the three methods revealed significant differences in the model parameters. The RMSE and MSE values for GPR were lower than those for SLR and NNR. Based on the formula calculations, the R-squared value for NNR was much higher than those for SLR and GPR. In addition to the inverse transformation of the Ca^{2+} concentrations for each method, these results are shown in Figures 11–13. Therefore, GPR was chosen as the optimal analysis model.

The electrochemical device was used to investigate three different blood and milk samples collected from dairy cows. As previously described, the blood and milk samples were pretreated following the protocol in Section 2.5. The extracted samples were studied by the proposed method and atomic absorption spectrometry (AAS), and the detection results are shown in Table 3. We found that the detection accuracy of the proposed method was lower than that of the AAS. The average recoveries ranged from 93.42% to 113.28%, showing that the electrochemical device can be used to determine Ca^{2+} concentrations in the blood and milk samples. A t-test was used to analyze the results of the two methods, and $p = 0.956 > 0.05$. In addition, the proposed method only requires a portable instrument that is inexpensive and easy to use. These results confirmed that the proposed method could be successfully used to determine Ca^{2+} concentrations in blood and milk samples of dairy cows.

Table 3. Ca^{2+} concentrations in the real samples determined by the electrochemical device and AAS.

Sample		Electrochemical Device (mM)	AAS (mM)	Recovery (%)
Blood	1	0.78	0.73	106.85
	2	1.15	1.12	102.68
	3	0.81	0.88	92.05
	4	1.09	1.17	93.16
Milk	1	27.32	29.20	93.56
	2	31.34	28.45	105.34
	3	21.87	22.78	96.01
	4	34.34	37.87	90.68

4. Conclusions

In this study, a novel method was proposed to establish an electrochemical device using an electrochemical biosensor and a pH meter, which overcome the electrochemical biosensor interfering with the pH in an analyte. The electrochemical biosensor was fabricated by modifying a glassy carbon electrode with SWNTs and a Ca^{2+}-cleaved DNAzyme. The sensitivity of the GCE/SWNT/DNAzyme was improved by changing the structure and length of the DNAzyme. The linear range was from 5 µM to 25 mM, with a detection limit of 4.2 µM. The GCE/SWNT/DNAzyme was combined with a pH meter to form an electrochemical device, and the data of different Ca^{2+} concentrations at different pHs were processed using three different mathematical models, namely MLR, NNF, and GPR. Among them, GPR showed a comparatively excellent performance in predicting the Ca^{2+} concentration at pH values ranging from 4.0 to 7.5. The prepared electrochemical device was applied to detect Ca^{2+} in blood and milk samples with good results.

Supplementary Materials: The following are available online at https://www.mdpi.com/article/10.3390/nano12010004/s1, Figure S1: SEM image of GCE/SWNT/DNAzyme, Figure S2: Randle equivalent circuit, Figure S3: predicted Ca^{2+} concentration against true Ca^{2+} concentration using the stepwise linear regression, Figure S4: overall structure of the NNF model with the LM algorithm, Figure S5: predicted Ca^{2+} concentration against true Ca^{2+} concentration using the established NNF model, Figure S6: predicted Ca^{2+} concentration against true Ca^{2+} concentration using the established GPR model, Table S1: Related indexes in dairy blood.

Author Contributions: Conceptualization, H.W., D.D. and J.C.; methodology, Q.L. and F.Z.; writing—original draft preparation, H.W.; writing—review and editing, S.Z., F.Z. and Y.W.; supervision, F.S.; project administration, L.Y.; funding acquisition, X.T. and B.X. All authors have read and agreed to the published version of the manuscript.

Funding: This research was funded by Key Realm R&D Program of Guangdong Province (No. 2019B020215004 and No. 2019B020215002), State Key Laboratory of Animal Nutrition (No. 2004DA12 5184G2104), Science and Technology Innovation Project of Institute of Animal Sciences (No. 2021-YWF-ZYSQ-05), and Beijing Dairy Industry Innovation Team (No. bjcystx-ny-1).

Institutional Review Board Statement: Not applicable.

Informed Consent Statement: Not applicable.

Data Availability Statement: Not applicable.

Conflicts of Interest: The authors declare no conflict of interest.

References

1. Balk, E.M.; for the International Osteoporosis Foundation Calcium Steering Committee; Adam, G.P.; Langberg, V.N.; Earley, A.; Clark, P.; Ebeling, P.R.; Mithal, A.; Rizzoli, R.; Zerbini, C.A.F.; et al. Global dietary calcium intake among adults: A systematic review. *Osteoporos. Int.* **2017**, *28*, 3315–3324. [CrossRef]
2. Giorgi, C.; Marchi, S.; Pinton, P. The machineries, regulation and cellular functions of mitochondrial calcium. *Nat. Rev. Mol. Cell Biol.* **2018**, *19*, 713–730. [CrossRef]
3. Maltsev, V.; Lakatta, E.G. Dynamic interactions of an intracellular Ca^{2+} clock and membrane ion channel clock underlie robust initiation and regulation of cardiac pacemaker function. *Cardiovasc. Res.* **2008**, *77*, 274–284. [CrossRef]
4. Xu, Y.; Ye, J.; Zhou, D.; Su, L. Research progress on applications of calcium derived from marine organisms. *Sci. Rep.* **2020**, *10*, 1–8. [CrossRef] [PubMed]
5. Daresjö, S. Determinants for Milk Fever: An Epidemiological Study of Swedish Dairy Cows. 2020. Available online: https://stud.epsilon.slu.se/15725/1/daresjo_s_200120.pdf (accessed on 13 December 2021).
6. Tesfaye, A. Calcium requirement in relation to milk fever for high yielding dairy cows: A review. *J. Food Agric. Environ.* **2019**, *18*, 1–12.
7. Kaufman, E.; Asselstine, V.; LeBlanc, S.; Duffield, T.; Devries, T. Association of rumination time and health status with milk yield and composition in early-lactation dairy cows. *J. Dairy Sci.* **2018**, *101*, 462–471. [CrossRef]
8. Yue, J.; Li, L.; Cao, L.; Zan, M.; Yang, D.; Wang, Z.; Chang, Z.; Mei, Q.; Miao, P.; Dong, W.-F. Two-step hydrothermal preparation of carbon dots for calcium ion detection. *ACS Appl. Mater. Interfaces* **2019**, *11*, 44566–44572. [CrossRef]
9. Hernández-Castellano, L.E.; Hernandez, L.L.; Weaver, S.; Bruckmaier, R.M. Increased serum serotonin improves parturient calcium homeostasis in dairy cows. *J. Dairy Sci.* **2017**, *100*, 1580–1587. [CrossRef]

10. Goff, J.P.; Hohman, A.; Timms, L.L. Effect of subclinical and clinical hypocalcemia and dietary cation-anion difference on rumination activity in periparturient dairy cows. *J. Dairy Sci.* **2020**, *103*, 2591–2601. [CrossRef] [PubMed]
11. Nedić, S.; Palamarević, M.; Arsić, S.; Jovanović, L.; Prodanović, R.; Kirovski, D.; Vujanac, I. Parathyroid hormone response in treatment of subclinical hypocalcemia in postpartum dairy cows. *Res. Vet. Sci.* **2020**, *132*, 351–356. [CrossRef]
12. Ibrahim, N.; Kirmani, M.A. Milk Fever in Dairy Cows: A Systematic Review. Available online: https://www.researchgate.net/profile/Nuraddis-Ibrahim/publication/350942379_ (accessed on 13 December 2021).
13. Uniyal, S.; Sharma, R.K. Technological advancement in electrochemical biosensor based detection of organophosphate pesticide chlorpyrifos in the environment: A review of status and prospects. *Biosens. Bioelectron.* **2018**, *116*, 37–50. [CrossRef]
14. Beitollahi, H.; Mohammadi, S.Z.; Safaei, M.; Tajik, S. Applications of electrochemical sensors and biosensors based on modified screen-printed electrodes: A review. *Anal. Methods* **2020**, *12*, 1547–1560. [CrossRef]
15. Liang, G.; Man, Y.; Li, A.; Jin, X.; Liu, X.; Pan, L. DNAzyme-based biosensor for detection of lead ion: A review. *Microchem. J.* **2017**, *131*, 145–153. [CrossRef]
16. Zhang, J. RNA-cleaving DNAzymes: Old catalysts with new tricks for intracellular and in vivo applications. *Catalysts* **2018**, *8*, 550. [CrossRef]
17. Zhou, W.; Saran, R.; Ding, J.; Liu, J. Two completely different mechanisms for highly specific Na^+ recognition by DNAzymes. *ChemBioChem* **2017**, *18*, 1828–1835. [CrossRef]
18. He, Y.; Chen, D.; Huang, P.-J.J.; Zhou, Y.; Ma, L.; Xu, K.; Yang, R.; Liu, J. Misfolding of a DNAzyme for ultrahigh sodium selectivity over potassium. *Nucleic Acids Res.* **2018**, *46*, 10262–10271. [CrossRef]
19. Wang, H.; Liu, Y.; Liu, G. Label-free biosensor using a silver specific RNA-cleaving DNAzyme functionalized single-walled carbon nanotube for silver ion determination. *Nanomaterials* **2018**, *8*, 258. [CrossRef]
20. Wang, H.; Liu, Y.; Liu, G. Electrochemical biosensor using DNA embedded phosphorothioate modified RNA for mercury ion determination. *ACS Sens.* **2018**, *3*, 624–631. [CrossRef] [PubMed]
21. Yin, H.-S.; Li, B.-C.; Zhou, Y.-L.; Wang, H.-Y.; Wang, M.-H.; Ai, S.-Y. Signal-on fluorescence biosensor for microRNA-21 detection based on DNA strand displacement reaction and Mg^{2+}-dependent DNAzyme cleavage. *Biosens. Bioelectron.* **2017**, *96*, 106–112. [CrossRef] [PubMed]
22. Wang, H.; Liu, Y.; Wang, J.; Xiong, B.; Hou, X. Electrochemical impedance biosensor array based on DNAzyme-functionalized single-walled carbon nanotubes using Gaussian process regression for Cu (II) and Hg (II) determination. *Microchim. Acta* **2020**, *187*, 1–9. [CrossRef] [PubMed]
23. Ren, W.; Huang, P.J.; He, M.; Lyu, M.; Wang, C.; Wang, S.; Liu, J. Sensitivity of a classic DNAzyme for Pb^{2+} modulated by cations, anions and buffers. *Analyst* **2020**, *145*, 1384–1388. [CrossRef]
24. Zhu, P.; Shang, Y.; Tian, W.; Huang, K.; Luo, Y.; Xu, W. Ultra-sensitive and absolute quantitative detection of Cu^{2+} based on DNAzyme and digital PCR in water and drink samples. *Food Chem.* **2017**, *221*, 1770–1777. [CrossRef]
25. Gong, L.; Zhao, Z.; Lv, Y.-F.; Huan, S.-Y.; Fu, T.; Zhang, X.-B.; Shen, G.-L.; Yu, R.-Q. DNAzyme-based biosensors and nanodevices. *Chem. Commun.* **2015**, *51*, 979–995. [CrossRef] [PubMed]
26. Ren, W.; Huang, P.-J.J.; de Rochambeau, D.; Moon, W.J.; Zhang, J.; Lyu, M.; Wang, S.; Sleiman, H.; Liu, J. Selection of a metal ligand modified DNAzyme for detecting Ni^{2+}. *Biosens. Bioelectron.* **2020**, *165*, 112285. [CrossRef] [PubMed]
27. Wang, H.; Luo, Q.; Zhao, Y.; Nan, X.; Zhang, F.; Wang, Y.; Wang, Y.; Hua, D.; Zheng, S.; Jiang, L.; et al. Electrochemical device based on nonspecific DNAzyme for the high-accuracy determination of Ca^{2+} with Pb^{2+} interference. *Bioelectrochemistry* **2021**, *140*, 107732. [CrossRef] [PubMed]
28. Zhou, W.; Saran, R.; Huang, P.-J.J.; Ding, J.; Liu, J. An exceptionally selective DNA cooperatively binding two Ca^{2+} ions. *ChemBioChem* **2017**, *18*, 518–522. [CrossRef] [PubMed]
29. Yu, T.; Zhou, W.; Liu, J. Screening of DNAzyme mutants for highly sensitive and selective detection of calcium in milk. *Anal. Methods* **2018**, *10*, 1740–1746. [CrossRef]
30. Hu, Y.; Chen, S.; Cong, X.; Sun, S.; Wu, J.-B.; Zhang, D.; Yang, F.; Yang, J.; Tan, P.-H.; Li, Y. Electronic Raman Scattering in Suspended Semiconducting Carbon Nanotube. *J. Phys. Chem. Lett.* **2020**, *11*, 10497–10503. [CrossRef]
31. Dugasani, S.R.; Gnapareddy, B.; Kesama, M.R.; Ha, T.H.; Park, S.H. DNA and DNA–CTMA composite thin films embedded with carboxyl group-modified multi-walled carbon nanotubes. *J. Ind. Eng. Chem.* **2018**, *68*, 79–86. [CrossRef]
32. Zhou, N.; Pierre, J.W.; Trudnowski, D. A stepwise regression method for estimating dominant electromechanical modes. *IEEE Trans. Power Syst.* **2011**, *27*, 1051–1059. [CrossRef]
33. Wang, H.; Ramnani, P.; Pham, T.; Villarreal, C.C.; Yu, X.; Liu, G.; Mulchandani, A. Gas biosensor arrays based on single-stranded DNA-functionalized single-walled carbon nanotubes for the detection of volatile organic compound biomarkers released by huanglongbing disease-infected citrus trees. *Sensors* **2019**, *19*, 4795. [CrossRef]
34. Yang, K.; Jin, H.; Chen, X.; Dai, J.; Wang, L.; Zhang, D. Soft sensor development for online quality prediction of industrial batch rubber mixing process using ensemble just-in-time Gaussian process regression models. *Chemom. Intell. Lab.* **2016**, *155*, 170–182. [CrossRef]
35. Deringer, V.L.; Bartók, A.P.; Bernstein, N.; Wilkins, D.M.; Ceriotti, M.; Csányi, G. Gaussian Process Regression for Materials and Molecules. *Chem. Rev.* **2021**, *121*, 10073–10141. [CrossRef] [PubMed]
36. Peng, B.; Zhou, J.; Xu, J.; Fan, M.; Ma, Y.; Zhou, M.; Li, T.; Zhao, S. A smartphone-based colorimetry after dispersive liquid–liquid microextraction for rapid quantification of calcium in water and food samples. *Microchem. J.* **2019**, *149*, 104072. [CrossRef]

37. Shibata, H.; Ikeda, Y.; Hiruta, Y.; Citterio, D. Inkjet-printed pH-independent paper-based calcium sensor with fluorescence signal readout relying on a solvatochromic dye. *Anal. Bioanal. Chem.* **2020**, *412*, 3489–3497. [CrossRef]
38. Liu, S.; Ding, J.; Qin, W. Current pulse based ion-selective electrodes for chronopotentiometric determination of calcium in seawater. *Anal. Chim. Acta* **2018**, *1031*, 67–74. [CrossRef]
39. Ocaña, C.; Abramova, N.; Bratov, A.; Lindfors, T.; Bobacka, J. Calcium-selective electrodes based on photo-cured polyurethane-acrylate membranes covalently attached to methacrylate functionalized poly (3, 4-ethylenedioxythiophene) as solid-contact. *Talanta* **2018**, *186*, 279–285. [CrossRef]
40. Ahmad, R.; Tripathy, N.; Ahn, M.-S.; Yoo, J.-Y.; Hahn, Y.-B. Preparation of a highly conductive seed layer for calcium sensor fabrication with enhanced sensing performance. *ACS Sens.* **2018**, *3*, 772–778. [CrossRef]

Article

Experimental Study on the Salt Freezing Durability of Multi-Walled Carbon Nanotube Ultra-High-Performance Concrete

Guifeng Liu [1], Huadi Zhang [1], Jianpeng Liu [1], Shuqi Xu [2] and Zhengfa Chen [1,*]

[1] Department of Civil Engineering, Changzhou University, Changzhou 213164, China; lgfcczu@163.com (G.L.); zhd690660085@163.com (H.Z.); ljp@cczu.edu.cn (J.L.)
[2] Anhui Jiaxiong Construction Engineering Co., Ltd., Hefei 230000, China; xsqahjx@163.com
* Correspondence: chenzhengfa@cczu.edu.cn

Abstract: Ultra-high-performance concrete (UHPC) is a new type of high-performance cement-based composite. It is widely used in important buildings, bridges, national defense construction, etc. because of its excellent mechanical properties and durability. Freeze thaw and salt erosion damage are one of the main causes of concrete structure failure. The use of UHPC prepared with multi-walled carbon nanotubes (MWCNTs) is an effective method to enhance the durability of concrete structures in complex environments. In this work, the optimal mix proportion based on mechanical properties was obtained by changing the content of MWCNTs and water binder ratio to prepare MWCNTs UHPC. Then, based on the changes in the compressive strength, mass loss rate, and relative dynamic modulus of elasticity (RDME), the damage degree of concrete under different salt erosion during 1500 freeze-thaw (FT) cycles was analyzed. The changes in the micro pore structure were characterized by scanning electron microscope (SEM) and nuclear magnetic resonance (NMR). The test results showed that the optimum mix proportion at the water binder ratio was 0.19 and 0.1% MWCNTs. At this time, the compressive strength was 34.1% higher and the flexural strength was 13.6% higher than when the MWCNTs content was 0. After 1500 salt freezing cycles, the appearance and mass loss of MWCNTs-UHPC prepared according to the best ratio changed little, and the maximum mass loss was 3.18%. The higher the mass fraction of the erosion solution is, the lower the compressive strength and RDME of concrete after FT cycles. The SEM test showed that cracks appeared in the internal structure and gradually increased due to salt freezing damage. However, the microstructure of the concrete was still relatively dense after 1500 salt freezing cycles. The NMR test showed that the salt freezing cycle has a significant influence on the change in the small pores, and the larger the mass fraction of the erosion solution, the smaller the change in the proportion of pores. After 1500 salt freezing cycles, the samples did not fail, which shows that MWCNTs UHPC with a design service life of 150 years has good salt freezing resistance under the coupling effect of salt corrosion and the FT cycle.

Keywords: multi-walled carbon nanotubes; ultra-high-performance concrete; mechanical properties; salt erosion and freeze-thaw coupling; durability; microstructure

1. Introduction

In areas with a cold climate and salinization, concrete materials are inevitably tested by salt erosion and freeze-thaw environments in various bridges, tunnels, hydraulic structures, and other projects. Freeze thaw and salt erosion damage are the main causes of concrete structure deterioration [1].

Scholars have investigated the durability of concrete under the action of a single factor or a combination of multiple adverse factors. Paul Brown et al. [2] studied concrete soaked in a sodium sulfate solution, and revealed change in the microstructure of concrete and the formation mechanism of erosion products. Gastaldini et al. [3] found that with the increase in freeze-thaw cycles, the chloride penetration resistance of concrete decreased in a freeze-thaw and chloride coupling environment. With an increase in the corrosion time, the chloride ion permeability of concrete also decreased gradually. Jan et al. [4] analyzed the mechanical properties of slag concrete in a salt freezing environment and found that the compressive strength of slag concrete gradually decreased under salt freezing. Sahmaran et al. [5] found that with an increase in the freeze-thaw cycles and corrosion time, the water cement ratio of concrete was small, and its durability was excellent. Ji et al. [6] studied the performance changes in carbon fiber-reinforced polymer (CFRP) in a chloride salt and freezing coupling environment. The chloride ion diffusion test showed that the chloride ion concentration of the ordinary specimen was at least 200 times higher than that of the fully strengthened specimen.

However, research on the frost resistance durability of concrete under the coupling environment of freeze-thaw and erosion has mostly focused on ordinary concrete. The water binder ratio of high-performance cement-based materials (e.g., UHPC) is low compared with that of ordinary concrete, which is mainly attributed to the compact arrangement of particles [7]. Therefore, UHPC not only has good mechanical properties (compressive strength, flexural strength) but also has excellent durability [8]. However, UHPC is usually produced without coarse aggregate, so the price of UHPC is more expensive than ordinary concrete. Dong et al. [9] found that the initial construction cost of bridges constructed with UHPC was increased by about 80%. The life cycle cost of the two bridges was also studied. When the service life of these 2 bridges was less than about 120 years, the life cycle cost of UHPC was higher than that of conventional bridges. However, considering the equivalent annual cost, the cost of the UHPC bridge is less than that of an ordinary bridge because the UHPC bridge can prolong the service life of the structure. Therefore, UHPC is mostly used in aggressive environmental structures or important buildings that need to be highly durable [10].

In order to prepare cement-based materials with high strength and toughness, many researchers have focused on carbon nanomaterials. Carbon nanotubes (CNTs) are a carbon material discovered by the Japanese scientist Iijima in the 1990s [11]. According to the number of graphene layers, they are divided into single-walled carbon nanotubes (SWCNTs) and multi-walled carbon nanotubes (MWCNTs). Carbon nanotubes have excellent mechanical properties, with an elastic modulus of about 1tpa and a yield strain of 10%–20% [12–14]. Therefore, carbon nanotubes are an ideal composite-reinforcing material [15].

The research on carbon nanotube cement-based composites is still in the exploratory stage [16,17]. Scholars found that the mechanical properties of carbon nanotube cement-based composites are closely related to the content of carbon nanotubes. The flexural and compressive strength of the composite can be significantly improved when the content is about 0.08–0.2%, but the optimal content and reinforcement effect are not unified [18,19]. Morteza et al. [20] found that the optimum content of carbon nanotubes in cement matrix is about 0.2wt%, and the compressive strength is 23.7% higher than that of ordinary mortar samples. Wang et al. [21] found that the addition of treated carbon nanotubes significantly improved the fracture energy and flexural toughness index of cement paste. Guan et al. [22] found that when the content of carbon nanotubes was 0.05%, the early flexural and compressive strength increased by 70.7% and 25.6%, respectively. The addition of carbon nanotubes also improved the later flexural and compressive strength, with a maximum increase of 16.9% and 11.6%, respectively. The addition of carbon nanotubes also results in significant changes in the microstructure of concrete, which improves the mechanical properties and durability and prolongs the service life. Zhang et al. [23] found that the fiber bridging effect of carbon nanotubes makes the cement paste structure denser, reducing holes and cracks, and improving the toughness and strength of cement. Using the by mercury intrusion porosity method (MIP), Guan et al. [22] found that MWCNTs can reduce the total porosity and improve the pore size distribution, reducing the number of mesopores. In addition, crack bridging, filling, and bonding between MWCNTs and cement matrix were also found in an SEM test, which had a positive impact on the mechanical properties. Fakhim et al. [24] observed that MWCNTs functioned to bridge cracks in cement-based materials with broken fibers. When the content of MWCNTs is 0.1–0.3%, cement-based materials show the best impermeability. An excessive MWCNTs content will produce agglomeration and lead to large holes. Therefore, the use of carbon nanotubes to prepare UHPC has important research significance and social value to enhance the durability of concrete structures in complex environments.

In this paper, MWCNTs were used as fiber reinforcement to prepare UHPC, and the effects of its content and water binder ratio on the mechanical properties were studied. In addition, the time-varying deterioration law of the freeze-thaw durability of MWCNTs UHPC under the coupling effect of salt erosion and freeze-thaw is discussed. The microstructure damage mechanism was analyzed by field SEM and NMR.

2. Materials and Methods

2.1. Sample Preparation

The carbon nanotube material was provided by a nano company in Shandong, and its physical parameters are shown in Table 1. Ordinary Portland cement P.O 52.5, complying with the Chinese standard GB175-2007 [25], was used in the experiment. Polyvinyl pyrrolidone (PVP) produced by Guangdong Co., Ltd. (Guangdong, China) was used as dispersant. A light-yellow viscous liquid polycarboxylic acid was used as a high-efficiency water-reducing agent, which was obtained from Qingdao Co., Ltd. (Qingdao, China). Local river sand 0.16–0.6 mm in size was used as fine aggregate. The granulometry curve is shown in Figure 1.

Table 1. Physical parameters of MWCNTs [26].

Pipe Diameter /nm	Tube Length /μm	Purity /%	Ash /%	Specific Surface Area /$m^2 \cdot g^{-1}$	Packing Density /$g \cdot cm^{-3}$
10–20	5–50	>85	<2.0	200–300	0.006–0.09

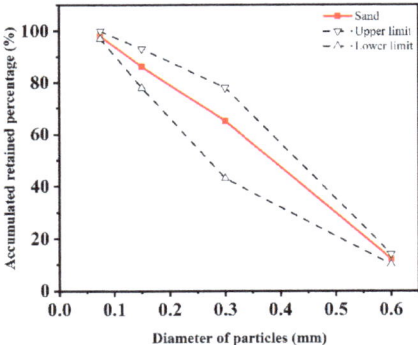

Figure 1. Granulometry curve of sand.

Based on the results of previous studies [27], carbon nanotubes were initially mixed with water, and then magnetically stirred with PVP of 30 °C for 15 min to promote surfactant dissolution, followed by ultrasonic treatment for 40 min. Finally, a carbon nanotube dispersion with different dosages was obtained.

In total, 5 MWCNT UHPC with different water binder ratios were prepared, and MWCNTs was added at 0%, 0.02%, 0.05%, 0.10%, 0.15%, 0.20%, and 0.30% by cement weight. The mix proportions of all specimens are listed in Table 2. The materials were poured into 100 mm × 100 mm× 100 mm and 100 mm× 100 mm × 400 mm molds. All specimens were demolded after 24 h and then subjected to 28 days of standard curing (20 ± 2°C and 95% relative humidity).

Table 2. Mix proportions of the specimens (kg/m³).

Sample	Cement	Silica Fume	Fly Ash	Slag Powder	Sand	W/B	Water Reducer	MWCNTs Content (%)
M1	540	100	180	100	1080	0.16	2.5%	0/0.02/0.05/0.10/0.15/0.20/0.30
M2	540	100	180	100	1080	0.17	2.5%	0/0.02/0.05/0.10/0.15/0.20/0.30
M3	540	100	180	100	1080	0.18	2.5%	0/0.02/0.05/0.10/0.15/0.20/0.30
M4	540	100	180	100	1080	0.19	2.5%	0/0.02/0.05/0.10/0.15/0.20/0.30
M5	540	100	180	100	1080	0.20	2.5%	0/0.02/0.05/0.10/0.15/0.20/0.30

Referring to the actual working conditions in the typical permafrost area of northeast China, a composite salt erosion solution containing sodium sulphate, sodium chloride, and sodium bicarbonate with a mass fraction of 3.4% [28] was prepared. In order to analyze the erosion damage degree and mechanism of various single salts of concrete, sodium bicarbonate, sodium chloride, and sodium sulphate solutions were prepared accordingly. Clear water solution was used as contrast. The composition and mass fraction of the five groups of erosion solution are shown in Table 3.

Table 3. Composition and mass fraction of erosion solution.

Sample	Erosion Solution Type	Type and Dosage of Salt /g·L⁻¹			Solution Mass Fraction/%
		$NaHCO_3$	NaCl	Na_2SO_4	
F1	Composite salt	14.38	7.46	13.36	3.4
F2	Bicarbonate	14.38	0	0	1.42
F3	Chloride salt	0	7.46	0	0.74
F4	Sulphate	0	0	13.36	1.32
F5	Clean water	0	0	0	0

2.2. Test Apparatus

The main apparatus used in the test is shown in Table 4.

Table 4. Main test apparatus.

Name	Model Parameters
Collector type constant temperature heating magnetic stirrer (Shanghai Yuhua Instrument Co., Ltd, Shanghai, China)	DF-101S
Ultrasonic cleaner (Shenzhen yuanpin Instrument Co., Ltd, Shenzhen, China)	KQ-250B
Single horizontal shaft forced concrete mixer (Wuxi Jianyi Instrument Machinery Co., Ltd, Wuxi, China)	HJW-60
Microcomputer controlled electro-hydraulic pressure testing machine (Shanghai Linjia science and Education Instrument Co., Ltd, Shanghai, China)	TYW-2000
Constant loading pressure testing machine (Wuxi xinluda Instrument Equipment Co., Ltd, Wuxi, China)	EHDC
Concrete freeze-thaw testing machine (Shanghai Sanhao refrigeration equipment factory, Shanghai, China)	CDR-5
Dynamic elastic modulus tester (Tianjin Yaxing Automation Experimental Instrument Co., Ltd, Tianjin, China)	DT-20W
Electron scanning microscope (Shanghai Baihe Instrument Technology Co., Ltd, Shanghai, China)	JSM-IT100(L)
Nuclear magnetic resonance imaging analyzer (Suzhou niumag Analytical Instrument Co., Ltd, Suzhou, China)	MesoMR12-060H-I

2.3. Experiment Program

After 28 days, the compressive strength and flexural strength of the specimen were tested with different water binder ratios and different contents according to the Chinese standard GB/T 50081-2019 [29]. The loading speed was 0.05–0.08 MPa/s.

Specimens with a size of 100 mm × 100 mm × 400 mm were prepared based on the optimal water binder ratio and the optimal content of MWCNTs. The samples were subjected to the rapid FT cycle test according to the Chinese standard GB/T 50082-2009 [30]. The temperature range of the fast FT test was $-18 \pm 2 \sim 5 \pm 2$ °C. The ratio of one FT cycle test in the laboratory to the number of FT cycles in the natural environment is generally

10~15, and typically 12 [31]. The annual freezing and thawing times in northeast China is 120 combined with statistical data [32,33]. Therefore, the number of laboratory FT cycles is 1500 for structures with a design service life of 150 years. After every 500 FT cycles, the micro morphology was observed, and the compressive strength was tested. After every 50 FT cycles, the mass loss rate and RDME were tested until the FT cycle reached 1500 times or the test piece failed (mass loss $\geq$ 5% or RDEM $\leq$ 60%).

A diamond saw was used to prepare a specimen slice with a size of 8 mm × 8 mm × 5 mm. The sample slices were cleaned with an ultrasonic wave to prevent dust pollution affecting the test results. The slices were dried for 24 h and then sprayed gold. The micro morphology of MWCNTs UHPC in high vacuum mode was photographed by SEM.

After the FT cycle of the specimen, a 40 mm × 40 mm × 40 mm cube specimen was prepared using the cutting mechanism. A nuclear magnetic resonance imaging analyzer was used to scan the concrete samples. The resonance frequency was 12.00 MHz, the magnet temperature was 32.00 ± 0.01 °C, and the diameter of the probe coil was 25 mm. Vacuum pressure saturation was used to pressurize the concrete sample at 10 MPa and saturate it for 24 h. The pore size distribution of MWCNTs UHPC before and after the freeze-thaw was characterized by the NMR relaxation method. In the NMR test, the relaxation time T_2 value is positively correlated with the pore size. The larger the peak value of the T_2 spectrum, the larger the pore size in the sample. The greater the amplitude, the greater the number of pores in the material [34,35].

The test flow chart is shown in Figure 2.

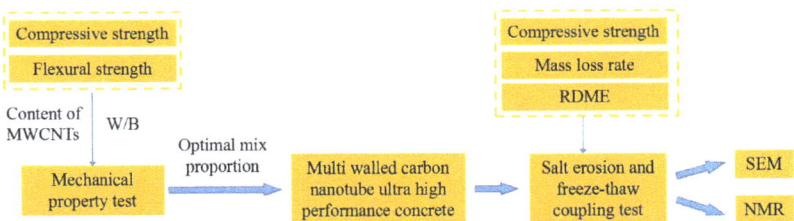

Figure 2. Test flow chart.

3. Results and Discussion

3.1. Optimal Mix Proportion Based on Mechanical Properties in a Non-Salt Freezing Environment

The compressive strength and flexural strength of the specimens with different water binder ratios and different contents without a salt freezing cycle were tested to investigate the effects of the water binder ratio and carbon nanotube content on the mechanical properties. Figure 3 shows that the addition of MWCNTs significantly improved the compressive strength and flexural strength of concrete. After increasing the carbon nanotubes content, the compressive strength and flexural strength increased at first and then decreased, which is consistent results from the literature [36–38]. The maximum compressive strength and flexural strength were 123.2 and 13.9 MPa, respectively, which was shown by the specimen with a water binder ratio of 0.19 and MWCNTs content of 0.1%. The strength was increased by 34.6% and 11.2%, respectively, compared with the control group with an MWCNTs content of 0. When the water binder ratio was 0.19 and the MWCNTs content was 0.1%, the compressive strength of the specimen reached the second highest value of 122.7 MPa and the flexural strength reached the highest value of 14.2 MPa. The strength was increased by 34.1% and 13.6%, respectively, compared with the control group.

This is because MWCNTs are evenly dispersed in concrete when the MWCNTs content is small (0–0.1%). Moreover, the uniformly dispersed MWCNTs fill the concrete interface transition zone, pores, and cracks, enhancing the mechanical properties. When the MWCNTs content is too high (0.1–0.3%), it is difficult for the MWCNTs suspension system to maintain stability and the MWCNTs can easily entangle and agglomerate with each other. As a result, there are a large number of non-dense and honeycomb pores and holes in cement hydration products. This decreases the interface compatibility between MWCNTs and concrete matrix, which is unfavorable regarding the mechanical properties.

At the same time, it can be seen from the figure that the compressive strength and flexural strength of the specimen first increased and then decreased with the increase in the water binder ratio, and reached a peak value when the water binder ratio was 0.18 or 0.19. Generally, the water binder ratio of UHPC should be less than 0.2. However, the fluidity of the concrete mixture decreases, and the interface is incompatible as there is less water in the concrete mixture when the water binder ratio is very low (such as when the water binder ratio was 0.16 and 0.17). A large number of pores will form in the interior due to the non-compactness of the mixture during the preparation of concrete specimens, resulting in a reduction in the UHPC strength. However, the fluidity of UHPC increases when the water binder ratio increases. Cementitious materials and aggregates are closely connected, which greatly improves the homogeneity of UHPC. Therefore, the strength of concrete is improved. However, the amount of mixing water increases and the strength of concrete decreases when the water binder ratio increases further.

Similar conclusions have been made in the studies of other scholars. Ju et al. [39] stated that an increase in W/B in a certain range results in the hydration of cementitious materials being more complete, thereby increasing the content of C-S-H gel and AFt in cement hydration products. To increase the strength of the test piece, Lu et al. [40] found that the early compressive strength of UHPC is largest when the water binder ratio is 0.18 while the flexural strength is highest when the water binder ratio is 0.2. The is because when the water binder ratio is small, there is not enough movable free water in the system. With the increase of water binder ratio, the movable free water in the system increases.

Little difference was observed in the mechanical properties of the concrete specimens when the water binder ratio was 0.19 and the MWCNTs content was 0.1% and 0.15%. Therefore, considering the cost factor (when the strength difference is small, the MWCNTs content should be as small as possible), a water binder ratio of 0.19 and MWCNTs content of 0.1% were selected as the optimal mix proportion. Then, based on the optimal mix proportion of MWCNTs UHPC, the freeze-thaw durability of concrete in a salt erosion and freeze-thaw coupling environment was investigated.

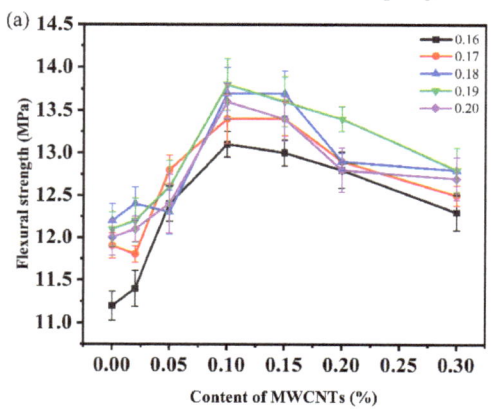 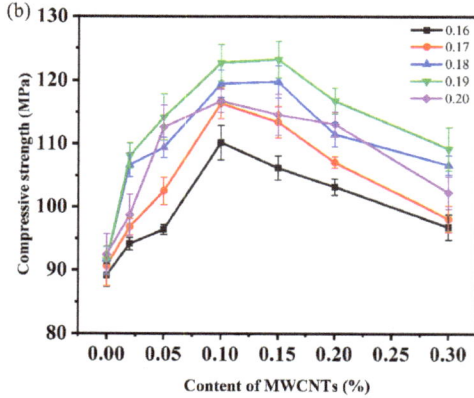

Figure 3. Mechanical properties of MWCNTs UHPC with different water binder ratios (0.16–0.20) and different MWCNTs contents (0–0.30%): (**a**) Flexural strength; (**b**) Compressive strength.

3.2. Rapid FT Test

3.2.1. Morphological and Compressive Strength

Concrete specimens with different solutions and different FT cycles were photographed and recorded. The micro morphologies of specimens after 0, 500, 1000, and 1500 freeze-thaw cycles in a salt freezing-thawing environment were compared and analyzed. The morphological changes are shown in Figure 4. The dotted box shows the morphological changes of some obvious pores after different salt freezing cycles. It can be seen that the appearance of the concrete specimens did not change significantly during the FT cycle. After zero salt freezing cycles, the surface of the specimen was smooth, and only small pores were generated in the process of preparation and curing. With an increase in the salt freezing cycles, small pores on the surface of the specimen gradually developed into medium pores and large pores. Moreover, the number of small pores increased gradually. The mortar was gradually exposed, and the surface became uneven due to salt freezing damage. However, no large cracks appeared on the surface of the specimen and the integrity of the specimen was not damaged after 1500 salt freezing cycles, indicating that MWCNTs-UHPC has good salt freezing resistance.

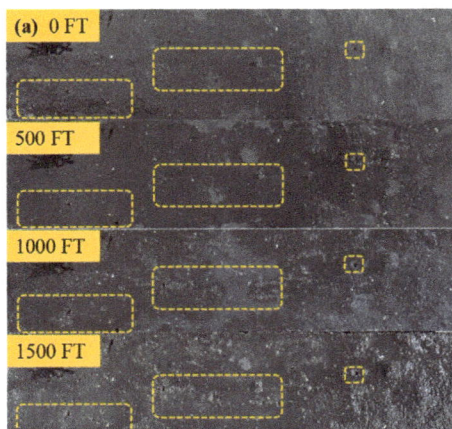
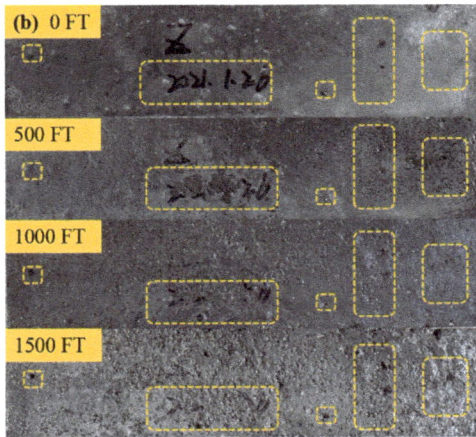
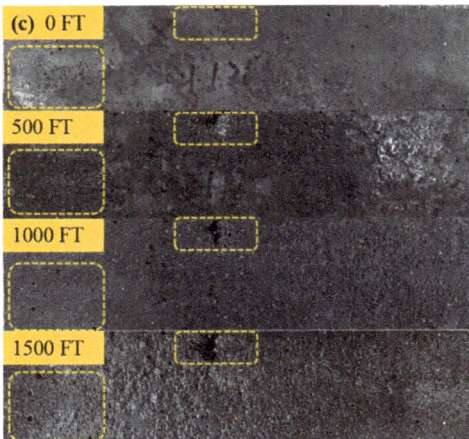

Figure 4. Specimen morphology of MWCNTs UHPC under different salt freezing cycles: (**a**) Bicarbonate and FT coupling; (**b**) Chloride salt and FT coupling; (**c**) Clean water and FT coupling.

Compressive strength is an important index that measures the bearing capacity of concrete. The change in the compressive strength with the number of salt freezing cycles is shown in Figure 5. It can be seen from the figure that the compressive strength of concrete decreased with the increase in the salt freezing cycles. The loss rates of compressive strength were 47%, 40%, 13%, 24%, and 14%, respectively. The variation range of the compressive strength from large to small is: composite salt and FT coupling > bicarbonate and FT coupling > sulphate and FT coupling > chloride and FT coupling > clean water and FT coupling. This shows that the greater the mass fraction of the erosion solution, the greater the loss of compressive strength.

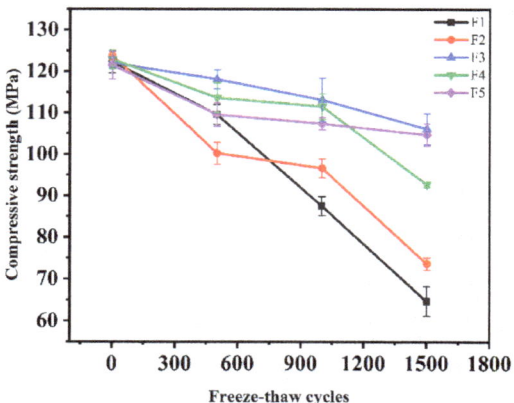

Figure 5. Variation in the compressive strength during salt freezing cycles. (F1: Composite salt and FT coupling; F2: Bicarbonate and FT coupling F3: Chloride salt and FT coupling; F4: Sulphate and FT coupling; F5: Clean water and FT coupling).

3.2.2. Mass Loss Rate

Quality change is an important indicator of concrete deterioration in salt frozen areas [41]. The peeling of mortar on the concrete surface caused by the FT effect is the main reason for a change in the quality [42]. The mass change trend of the test pieces under different solution FT cycles is shown in Figure 6. With the salt freezing cycles, the mass loss rate of all specimens decreased first and then increased. After 1500 FT cycles, the mass loss rates of the specimens were −1.09%, −0.85%, 0.73%, 1.73%, and 3.18%, respectively. The curve decreased rapidly in the early stage of erosion and the range of change was small in the later stage. This is because in the early stage of the FT cycle, cracks appeared and expanded on the interior and surface of the specimen, so water continued to penetrate the interior of the specimen. Salt solution crystallization attached to the surface of the specimen. In addition, the mortar on the surface of the specimen peeled less, and the peeling quality was much lower than that of the water penetrating inside the specimen. Therefore, the mass of the specimen increased, indicating that the mass loss rate was negative, and continuously decreased. As shown in Figure 5, the mortar of the F5 specimen peeled off, resulting in its quality degradation after 150 salt freezing cycles. However, under the subsequent FT cycle, the change in the mass loss rate was insignificant. In the later stage of the FT cycle, the mortar continued to peel off due to the effect of salt freezing damage. The mass of the specimen decreased gradually, which shows that the mass loss rate increased. Nevertheless, the mass loss rate of all specimens did not reach 5% after 1500 salt freezing cycles. This indicates that all specimens were unbroken according to GB/T 50082-2009 [30], and no specimen failed after 1500 salt freezing cycles. The results demonstrated that MWCNTs-UHPC has good resistance to salt freezing erosion under long-term salt freezing.

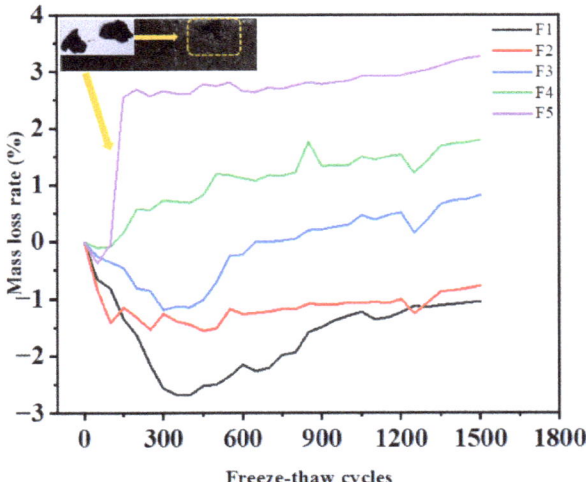

Figure 6. The mass loss rate of MWCNTs UHPC during salt freezing cycles. (F1: Composite salt and FT coupling; F2: Bicarbonate and FT coupling F3: Chloride salt and FT coupling; F4: Sulphate and FT coupling; F5: Clean water and FT coupling).

Compared with the compressive strength loss rate and mass loss rate of UHPC after FT cycles reported by Lu et al. [43–46], as shown in Table 5, it was found that the mass loss rate of MWCNTs-UHPC in the composite salt freezing-thawing environment in this experiment was lower. In addition, a greater loss in compressive strength was found in this experiment. This is because the coupling effect of salt corrosion and freeze-thaw causes more serious damage to the concrete compared with the single freeze-thaw environment and reduces the strength.

Table 5. The referred compressive strength loss rate and mass loss rate of UHPC after FT cycles.

Number	Author	Number of FT Cycles	Compressive Strength Loss Rate (%)	Mass Loss Rate (%)
1	Lu et al. [43]	300	27.5	−0.65
2	Lee et al. [44]	300	−3	-
		600	1	-
		1000	6	-
3	Ji et al. [45]	500	4.9–17.8	0.57–0.95
4	Li et al. [46]	800	0.869–1.501	-

3.2.3. Relative Dynamic Modulus of Elasticity

When the size and quality are certain (no serious peeling occurs), RDME is only affected by the compactness of concrete. In order to better reflect the internal compactness and damage of MWCNTs UHPC specimens after salt freezing cycles, RDME under different solution freezing and thawing cycles was tested. The test results are shown in Figure 7.

RDEM of the specimen decreased obviously in the early stage of the salt freezing cycles, and then increased slightly. After 1500 salt freezing cycles, RDME of the composite salt invasion and single salt invasion specimens was 91.3%, 94.3%, 99.36%, 99.03%, and 99.55%, respectively. The larger the mass fraction of the salt solution, the smaller the slope value of the downward trend line and the more obvious the downward trend. The minimum slope of the F1 compound salt solution is −0.2144 while the maximum slope of the F5 clear water solution is −0.0123. In the rising stage, the larger the mass fraction of the salt solution, the smaller the slope of the rising trend line. The minimum slope of the F1 compound salt solution is 7.4865×10^{-4} while the slope value of the F5 clear water solution is 0.0035. This is because the salt freezing damage of the concrete is related to the type of erosion salt, the mass fraction of the erosion solution, and other factors. The more complex the salt composition, the greater the mass fraction of the erosion solution, and the more serious the frost heave damage to the concrete. This rule is consistent with the change in the compressive strength.

The rising process of RDEM may be due to secondary hydration [47]. A large number of cementitious materials were not fully hydrated during the test because of the low water cement ratio of UHPC [48]. With the increase in the salt freezing cycles, the frost heaving force of water increases. The internal porosity of concrete increases and absorbs more water during salt freezing cycles, which promotes the hydration reaction of cementitious materials. The continuous hydration of the specimen matrix gradually increases the compactness of the specimen, which plays a role in improving the relative dynamic elastic modulus. It shows that MWCNTs UHPC has good frost resistance and salt erosion resistance under the combined action of composite salt and low temperatures.

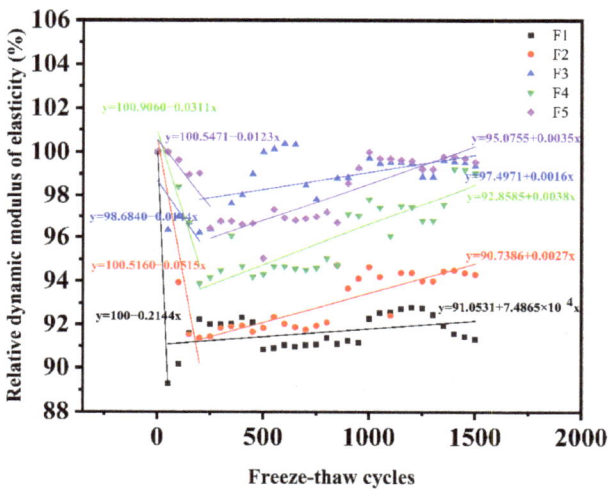

Figure 7. RDME of MWCNTs UHPC during salt freezing cycles (F1: Composite salt and FT coupling; F2: Bicarbonate and FT coupling F3: Chloride salt and FT coupling; F4: Sulphate and FT coupling; F5: Clean water and FT coupling).

3.3. Microscopic Analysis

3.3.1. Scanning Electron Microscope Observation

Figure 8 shows the microstructure of concrete under different salt freezing cycles. It can be seen from the figure that MWCNTs mostly overlap and form bridges with hydration products in the form of a single root to form multiphase composites and improve the mechanical properties of concrete [49]. Before the salt freezing cycles, the microstructure of concrete is dense without obvious pores and cracks. With the increase in the salt freezing cycles, cracks appear in the internal structure and further expand, and the connection between hydration products becomes loose and honeycomb. However, after 1500 salt freezing cycles, the hydration products are still relatively dense, and there is no looseness or porosity. This shows that for MWCNTs UHPC, due to the compactness of its structure, the diffusion rate of the erosion medium is reduced, and then the deterioration process of concrete FT damage is prolonged.

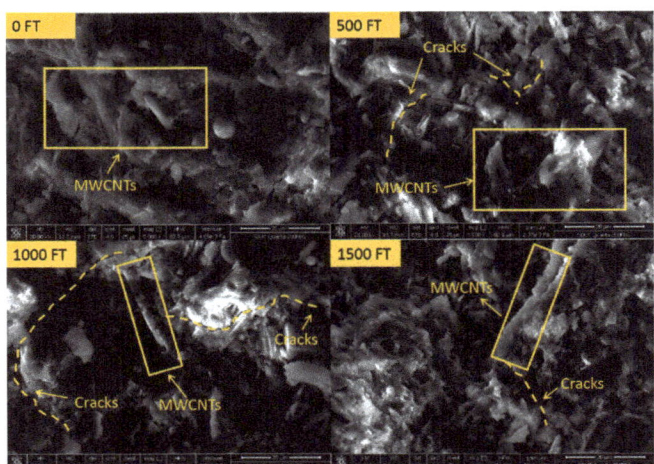

Figure 8. SEM of MWCNTs UHPC under different salt freezing cycles.

3.3.2. Nuclear Magnetic Resonance Test

Figure 9 shows the T_2 distribution curve of the MWCNTs UHPC samples. As can be seen from the figure, there are three peaks in the T_2 distribution diagram, in which the left peak represents the distribution of small holes, the middle peak represents the medium pores, and the right peak represents the large pores. As can be seen from the figure, the middle peak and right peak change little. This shows that salt freezing has little effect on these two peaks, so it is ignored. After 1500 salt freezing cycles, the number of small pores in the concrete samples in the single salt freezing environment increased while the number of small pores in the composite salt freezing environment decreased. This may be because in the composite salt erosion environment, corrosion crystals are more likely to be generated inside the sample and fill in small pores, thus reducing the number of small pores.

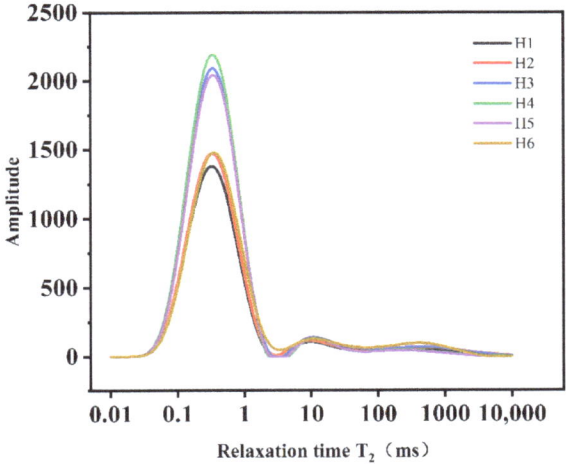

Figure 9. T_2 distribution curve of the NMR test (H1-H5 is the sample after 1500 salt freezing cycles under 5 erosion solutions, and H6 is the control sample without salt freezing cycles).

In order to reflect the proportion change in the pores with different sizes, the peak data were analyzed. The analysis results are shown in Figure 10. As can be seen from Figure 10, the proportion of medium pores and macro pores in the sample after the FT cycle is slightly reduced while the proportion of small pores is increased compared with the control sample without the FT cycle. This is because the concrete will be continuously damaged with the development of the FT cycles, and more micro pores will form. Therefore, the proportion of small holes will increase.

In the composite salt, bicarbonate erosion, and FT coupling environment, the area of the T_2 distribution spectrum decreases and the range of change in the pore ratio is small. In the chloride, sulfate, and clean water erosion and FT coupling environment, the area of the T_2 distribution spectrum increases and the pore ratio changes significantly. This is because the mass fraction of the salt solution is large, and the salt composition is complex in the composite salt, bicarbonate erosion, and freeze-thaw coupling environment. With the salt freezing cycle, the ions in the salt solution enter the concrete and react chemically with the internal materials. The internal pores are constantly filled by corrosion crystals, which causes frost heave damage. At the same time, the macro pores gradually transform into small pores due to the accumulation of crystals. Therefore, the range of change in the pore ratio is small. The mass fraction of the salt solution is relatively small in the chloride, sulfate, clean water erosion, and freeze-thaw coupling environment. Although some crystals fill in the pores, the frost heaving damage to the concrete is more serious, and small holes are constantly generated with the progression of the FT cycle. Therefore, the proportion of small holes changes significantly.

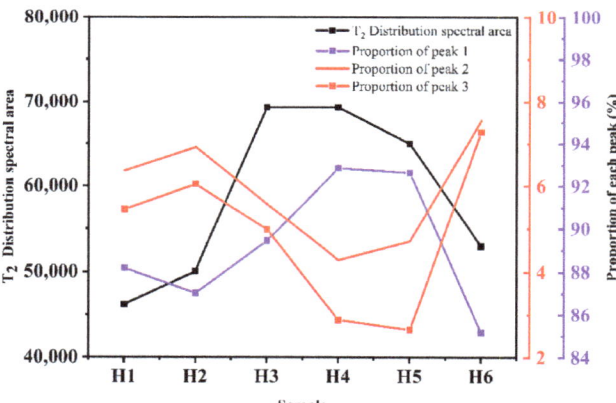

Figure 10. T_2 spectrum area and proportion of the peak value of the T_2 distribution.

4. Conclusions

In this study, the mechanical properties and frost resistance durability of UHPC containing MWCNTs were studied, and the micro analysis was carried out combined with SEM and NMR. The conclusions are as follows:

1. The addition of carbon nanotubes significantly improved the compressive strength and flexural strength of concrete. With the increase in the MWCNTs content, the variation curves of the compressive strength and flexural strength of the specimens increased at first and then decreased. Considering the cost factor, the optimum mix proportion was found to be a 0.19 water binder ratio and 0.1% carbon nanotube content. At this time, the compressive strength of the specimen was 122.7 MPa and the flexural strength was 9.2 MPa. The strength was increased by 34.1% and 13.6%, respectively, compared with the control group with an MWCNTs content of 0.
2. The MWCNTs UHPC prepared based on the optimal mix proportion showed good frost resistance and salt erosion resistance under the combined action of salt erosion

and low temperatures. After 1500 salt freezing cycles, the appearance and mass loss of concrete did not change, and the maximum quality loss was 3.18%. The more complex the salt composition and the greater the mass fraction of the erosion solution, the higher the loss rate of compressive strength, up to 40%. This reduces RDME to 91.3%.
3. After 1500 salt freezing cycles, the microstructure of concrete was still dense. The salt freezing cycle has a significant influence on the change in the small pores but has little influence on the change in the medium pores and large pores. The larger the mass fraction of the erosion solution, the smaller the change in the pore proportion.

Author Contributions: Conceptualization, G.L. and Z.C.; data curation, H.Z. and J.L.; resources, S.X. and Z.C.; writing—original draft preparation, G.L. and H.Z.; writing—review and editing, J.L., Z.C. and S.X. All authors have read and agreed to the published version of the manuscript.

Funding: This work was funded by the National Natural Science Foundation of China (No. 51908075).

Institutional Review Board Statement: Not applicable.

Informed Consent Statement: Not applicable.

Data Availability Statement: The data used to support the findings of this study are available from the corresponding author upon request.

Acknowledgments: Authors greatly acknowledge Shandong Dazhan carbon nano technology Co., Ltd. (Shandong, China) for providing the MWCNTs.

Conflicts of Interest: The authors declare no conflict of interest.

References

1. Wang, B.X.; Pan, J.J.; Fang, R.C.; Wang, Q. Damage model of concrete subjected to coupling chemical attacks and freeze-thaw cycles in saline soil area. *Constr. Build. Mater.* **2020**, *242*, 118205. [CrossRef]
2. Paual, B.; Hooton, R.D.; Boyd, C. Microstructual changes in concretes with sulfate exposure. *Cem. Concr. Compos.* **2004**, *26*, 993–999.
3. Gastaldini, A.L.G.; Isaia, G.C.; Saciloto, A.P.; Missau, F.; Hoppe, T.F. Influence of curing time on the chloride penetration resistance of concrete containing rice husk ash: A technical and economic feasibility study. *Cem. Concr. Compos.* **2010**, *32*, 783–793. [CrossRef]
4. Jan, D. Freezing and deicing salt resistance of blast furnace slag concretes. *Cem. Concr. Compos.* **2003**, *25*, 357–361.
5. Sahmaran, M.; Erdem, T.K.; Yarnan, L.O. Sulfate resistance of plain and blended cements exposed to wetting-drying and heating-cooling environments. *Constr. Build. Mater.* **2007**, *21*, 1771–1778. [CrossRef]
6. Ji, Y.; Liu, W.; Jia, Y.; Li, W. Durability Investigation of Carbon Fiber Reinforced Concrete under Salt-Freeze Coupling Effect. *Materials* **2021**, *14*, 6856. [CrossRef]
7. Shi, C.J.; Wu, Z.M.; Xiao, J.F.; Wang, D.H.; Huang, Z.Y.; Fang, Z. A review on ultra-high performance concrete: Part I. Raw materials and mixture design. *Constr. Build. Mater.* **2015**, *101*, 741–751. [CrossRef]
8. Zhou, M.; Lu, W.; Song, J.W.; Lee, G.C. Application of Ultra-High Performance Concrete in bridge engineering. *Constr. Build. Mater.* **2018**, *186*, 1256–1267. [CrossRef]
9. Dong, Y. Performance Assessment and Design of Ultra-High Performance Concrete (UHPC) Structures Incorporating Life-Cycle Cost and Environmental Impacts. *Constr. Build. Mater.* **2018**, *167*, 414–425. [CrossRef]
10. Marvila, M.T.; Azevedo, A.R.G.; Matos, P.R.; Monteiro, S.N.; Vieira, C.M.F. Materials for Production of High and Ultra-High Performance Concrete: Review and Perspective of Possible Novel Materials. *Materials* **2021**, *14*, 4304. [CrossRef]
11. Iijima, S. Helical microtubules of graphitic carbon. *Nature* **1991**, *354*, 56–58. [CrossRef]
12. Yu, M.F.; Lourie, O.; Dyer, M.J.; Moloni, K.; Kelly, T.F.; Ruoff, R.S. Strength and breaking mechanism of multiwalled carbon nanotubes under tensile load. *Science* **2000**, *287*, 637–640. [CrossRef] [PubMed]
13. Huang, J.Y.; Chen, S.; Wang, Z.Q.; Kempa, K.; Wang, Y.M.; Jo, S.H.; Chen, G.; Dresselhaus, M.S.; Ren, Z.F. Superplastic carbon nanotubes. *Nature* **2006**, *439*, 281. [CrossRef] [PubMed]
14. Han, B.; Sun, S.; Ding, S.; Zhang, L.; Yu, X.; Ou, J. Review of nanocarbon-engineered multifunctional cementitious composites. *Compos. Part A Appl. Sci. Manuf.* **2015**, *70*, 69–81. [CrossRef]
15. Liew, K.M.; Kai, M.F.; Zhang, L.W. Carbon nanotube reinforced cementitious composites: An overview. *Compos. Part A Appl. Sci. Manuf.* **2016**, *91*, 301–323. [CrossRef]
16. Kim, H. Chloride penetration monitoring in reinforced concrete structure using carbon nanotube/cement composite. *Constr. Build. Mater.* **2015**, *96*, 29–36. [CrossRef]
17. Wang, J.; Bai, X.S.; Zhao, J.Y.; Gao, Z.Y. Carbon nanotubes enhance RPC bending fatigue performance. *J. Build. Mater.* **2020**, *23*, 1345–1349.

18. Niu, X.J.; Peng, G.F.; He, J.; Lei, Z.H. Effect of multi-scale steel fiber combination and carbon nanotubes on mechanical properties of RPC. *J. Build. Mater.* **2020**, *23*, 216–223.
19. Konsta-Gdoutos, M.S.; Metaxa, Z.S.; Shah, S.P. Highly dispersed carbon nanotube reinforced cement based materials. *Cem. Concr. Res.* **2016**, *40*, 1052–1059. [CrossRef]
20. Morteza, M.S.; Mahdi, M.; Hassan, A. Effect of functionalized multi-walled carbon nanotubes on mechanical properties and durability of cement mortars. *J. Build. Mater.* **2021**, *41*, 102407.
21. Wang, B.M.; Han, Y.; Liu, S. Effect of highly dispersed carbon nanotubes on the flexural toughness of cement-based composites. *Constr. Build. Mater.* **2013**, *46*, 8–12. [CrossRef]
22. Guan, X.C.; Bai, S.; Li, H.; Ou, J.P. Mechanical properties and microstructure of multi-walled carbon nanotube-reinforced cementitious composites under the early-age freezing conditions. *Constr. Build. Mater.* **2020**, *223*, 117–317. [CrossRef]
23. Zhang, D.; Lu, F.L.; Liang, Y.J. Effect of carbon nanotubes on mechanical properties and durability of cement. *Concrete* **2019**, *11*, 10–14.
24. Fakhim, B.; Hassani, A.; Rashidi, A. Preparation and microstructural properties study on cement composites reinforced with multi-walled carbon nanotubes. *J. Compos. Mater.* **2015**, *49*, 85–98. [CrossRef]
25. *GB175-2007*; General Portland Cement. Chinese National Standards: Beijing, China, 2007.
26. Liu, G.F.; Zhu, D.; Chen, Z.F.; Zhang, H.D. Experimental study on mechanical properties of carbon nanotube activated powder concrete. *China Concr. Cem. Prod.* **2021**, *8*, 5–9.
27. Zhu, D. Experimental Study on Mechanical Properties and Sulfate Resistance of Carbon Nanotube Ultra-High Performance Concrete. Master's Thesis, Changzhou University, Changzhou, China, 2021.
28. Sun, H.Y.; Su, X.P.; Wang, X.P. Study on salt frost resistance durability of ordinary concrete under different salt environments. *J. Changchun Insti. Eng.* **2016**, *17*, 10–15.
29. *GB/T 50081-2019*; Standard for Test Methods of Physical and Mechanical Properties of Concrete. Chinese National Standards: Beijing, China, 2019.
30. *GB/T 50082-2009*; Standard for Test Methods of Long-Term Performance and Durability of Ordinary Concrete. Chinese National Standards: Beijing, China, 2009.
31. Chen, F.L.; Qiao, P.Z. Probabilistic damage modeling and service-life prediction of concrete under freeze-thaw action. *Mater. Struct.* **2015**, *48*, 2697–2711. [CrossRef]
32. Zhou, S.L.; Wang, H.; Wang, Q.S.; Wang, W.S.; Zhang, W. Mechanical properties and freeze-thaw durability of low-sand Replacement rate pervious concrete. *Highway Eng.* **2022**, *47*, 135–141.
33. Zhang, J. Study on the effect of fiber types on freeze-thaw splitting strength of asphalt concrete. *Highway Eng.* **2019**, *44*, 193–196.
34. Liu, J.P.; Yang, P.; Yang, Z.H. Experimental study on deformation characteristics of chloride silty clay during freeze-thaw in an open system. *Cold Reg. Sci. Technol.* **2022**, *197*, 103518. [CrossRef]
35. Liu, J.P.; Yang, P.; Yang, Z.H. Water and salt migration mechanisms of saturated chloride clay during freeze-thaw in an open system. *Cold Reg. Sci. Technol.* **2021**, *186*, 103277.
36. Lu, L.; Ouyang, D.; Xu, W. Mechanical Properties and Durability of Ultra High Strength Concrete Incorporating Multi-Walled Carbon Nanotubes. *Materials* **2016**, *9*, 419. [CrossRef] [PubMed]
37. Li, G.Y.; Wang, P.M.; Zhao, X. Mechanical behavior and microstructure of cement composites incorporating surface-treated multi-walled carbon nanotubes. *Carbon* **2005**, *43*, 1239–1245. [CrossRef]
38. Cwirzen, A.; Habermehl-Cwirzen, K.; Penttala, V. Surface decoration of carbon nanotubes and mechanical properties of cement/carbon nanotube composites. *Adv. Cem. Res.* **2008**, *20*, 65–73. [CrossRef]
39. Ju, Y.Z.; Shen, T.; Wang, D.H. Bonding behavior between reactive powder concrete and normal strength concrete. *Constr. Build. Mater.* **2020**, *242*, 118024. [CrossRef]
40. Lu, Z.; Feng, Z.G.; Yao, D.D.; Ji, H.R.; Qin, W.J.; Yu, L.M. Analysis on Influencing Factors of workability and strength of ultra-high performance concrete. *Mater. Rep.* **2020**, *34*, 203–208.
41. Olanike, A.O. Experimental investigation into the freeze-thaw resistance of concrete using recycled concrete aggregates and admixtures. *Civ. Eng. Archit.* **2014**, *2*, 176–180. [CrossRef]
42. Jiang, W.; Shen, X.; Xia, J.; Mao, L.; Yang, J.; Liu, Q. A numerical study on chloride diffusion in freeze-thaw affected concrete. *Constr. Build. Mater.* **2018**, *179*, 553–565. [CrossRef]
43. Lu, Z.; Feng, Z.G.; Yao, D.D.; Li, X.J.; Ji, H. Freeze-thaw resistance of Ultra-High performance concrete: Dependence on concrete composition. *Constr. Build. Mater.* **2021**, *293*, 123523. [CrossRef]
44. Lee, M.G.; Wang, Y.C.; Chiu, C.T. A preliminary study of reactive powder concrete as a new repair material. *Constr. Build. Mater.* **2007**, *21*, 182–189. [CrossRef]
45. Ji, Y.Y. *Experiment on Durability of Reactive Powder Concrete in Marine Environment*; Harbin Institute of Technology: Harbin, China, 2011.
46. Li, Y.F.; Wang, X.B.; Wei, F.T.; Sun, J.; Ma, J.D. Experimental study on freeze-thaw resistance of ultra-high strength concrete with different mineral admixtures. *China Sci.* **2017**, *12*, 2632–2636.
47. Zhen, Q.M.; He, B.; Li, C.; Jiang, Z.W. Research Progress on crack self-healing of ultra-high performance concrete. *J. Chin. Ceram. Soc.* **2021**, *49*, 2450–2461.

48. Wang, C.L.; Zhong, S.R.; Gao, R.Q.; Qu, Y.Y.; Huang, Z.Q. Analysis on Influencing Factors of ultra-high performance concrete preparation. *Sichuan Build. Mater.* **2020**, *46*, 1–3.
49. Metaxa, Z.S.; Konsta-Gdoutos, M.S.; Shah, S.P. Carbon nanotubes reinforced concrete. *ACI Spec. Publ.* **2009**, *267*, 11–20.

MDPI
St. Alban-Anlage 66
4052 Basel
Switzerland
www.mdpi.com

MDPI Books Editorial Office
E-mail: books@mdpi.com
www.mdpi.com/books

Disclaimer/Publisher's Note: The statements, opinions and data contained in all publications are solely those of the individual author(s) and contributor(s) and not of MDPI and/or the editor(s). MDPI and/or the editor(s) disclaim responsibility for any injury to people or property resulting from any ideas, methods, instructions or products referred to in the content.

www.ingramcontent.com/pod-product-compliance
Lightning Source LLC
LaVergne TN
LVHW070241100526
838202LV00015B/2163